The Maillard Reaction in Foods and Nutrition

George R. Waller, EDITOR
Oklahoma State University

Milton S. Feather, EDITOR
University of Missouri

Based on a symposium
jointly sponsored by the Divisions of
Agricultural and Food Chemistry
and Carbohydrate Chemistry
at the 183rd Meeting of the
American Chemical Society,
Las Vegas, Nevada,
March 28–April 2, 1982

ACS SYMPOSIUM SERIES 215

AMERICAN CHEMICAL SOCIETY
WASHINGTON, D.C. 1983

Library of Congress Cataloging in Publication Data

Maillard reaction in foods and nutrition.
(ACS symposium series, ISSN 0097-6157; 215)

"Based on a symposium jointly sponsored by the Divisions of Agricultural and Food Chemistry, and Carbohydrate Chemistry at the 183rd meeting of the American Chemical Society, Las Vegas, Nevada, March 28–April 2, 1982."
Includes bibliographies and index.

1. Maillard reaction—Congresses. 2. Food industry and trade—Congresses.
I. Waller, George R. II. Feather, Milton S., 1936– . III. American Chemical Society. Division of Agricultural and Food Chemistry. IV. American Chemical Society. Division of Biological Chemistry. V. Series.

TP372.55.M35M34 1983 664'.028 83-3852
ISBN 0-8412-0769-0

PRINTED IN THE UNITED STATES OF AMERICA

ACS Symposium Series

M. Joan Comstock, *Series Editor*

FOREWORD

The ACS SYMPOSIUM SERIES was founded in 1974 to provide a medium for publishing symposia quickly in book form. The format of the Series parallels that of the continuing ADVANCES IN CHEMISTRY SERIES except that in order to save time the papers are not typeset but are reproduced as they are submitted by the authors in camera-ready form. Papers are reviewed under the supervision of the Editors with the assistance of the Series Advisory Board and are selected to maintain the integrity of the symposia; however, verbatim reproductions of previously published papers are not accepted. Both reviews and reports of research are acceptable since symposia may embrace both types of presentation.

CONTENTS

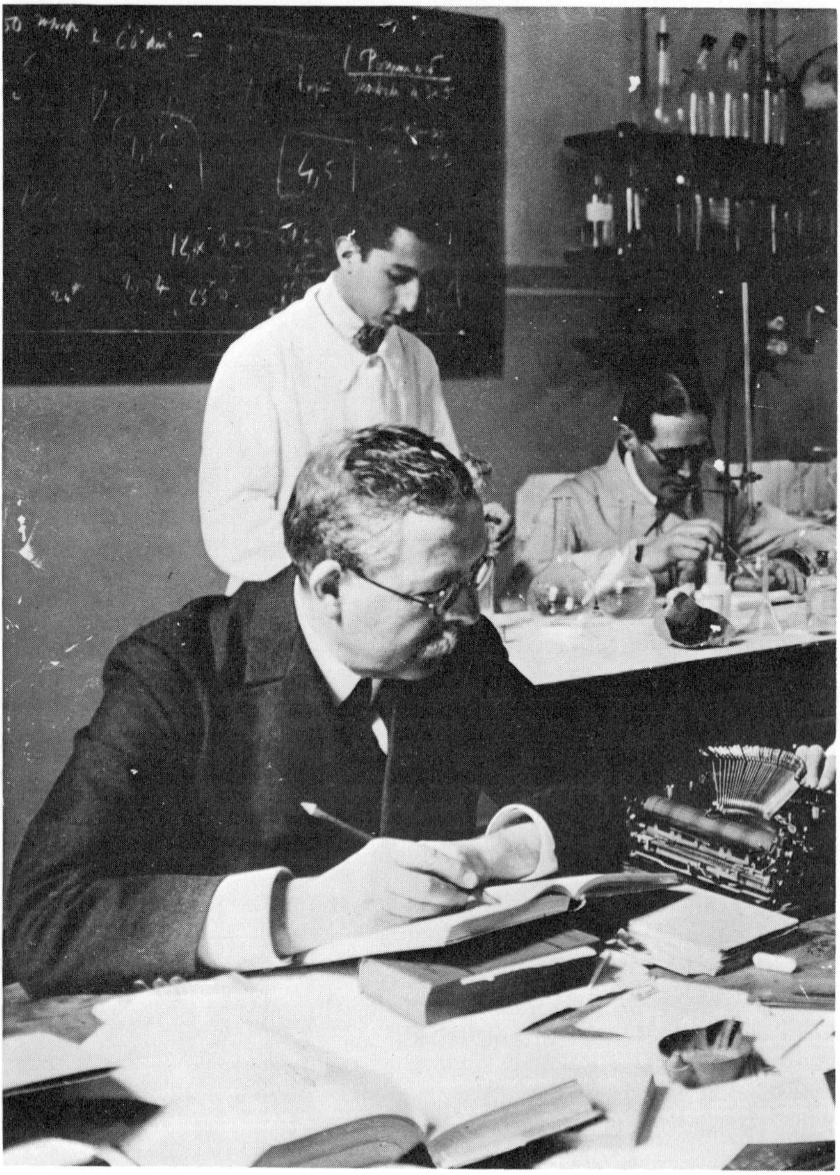

Louis-Camille Maillard (1878–1936) in a photograph taken around 1915.

PREFACE

LOUIS-CAMILLE MAILLARD, a French scientist, first observed the reaction that now bears his name seventy years ago. This reaction occurs between sugars and amino acids, polypeptides, or proteins; and between polysaccharides and polypeptides or proteins. Examples of the reaction are nonenzymatic browning reactions and change of taste when a steak is cooked, and the coloring of a slice of apple exposed to the air. During heat treatment such as frying, roasting, and baking, the Maillard reaction improves food in taste, flavor, and color. When foods are stored however, the reaction often gives unfavorable effects, such as decreased nutritional value and color deterioration. Thus, better and safer preservation of food and its nutritional value is the goal of scientists studying the Maillard reaction. In the last few years there has been an increased interest in the Maillard reaction from scientists in the fields of agriculture, foods, nutrition, and carbohydrate chemistry.

This book is a result of the first symposium on the Maillard reaction held in this country (the second in the world[1]). Panel discussions with audience participation were held on the subjects of food and nutritional benefits of Maillard reaction products and the toxicology of Maillard reaction products but are not reported here. Interaction between participants during this meeting helped cement relations for continued help and perhaps will promote some new areas for cooperative research between American and foreign scientists. This symposium was also an important element in the training of graduate students; it provided them with increased awareness of the breadth of the scientific field. We hope that it is true of each of you!

Although this book cannot give a complete account of the meeting, we hope that it will serve as a starting point to help guide the research of others. We are pleased that the Maillard reaction and its products will now be the subject of on-going international meetings.

[1] The first symposium was held in Uddevalla, Sweden in 1979 and was published as "Maillard Reactions in Food: Chemical, Physiological, and Technological Aspects," Eriksson, C. E. Ed., (Progress in Food and Nutrition Science, Vol. 5, Pergamon Press, Oxford, 1981). The next meeting will be held in 1985, organized by Professor Masao Fujimaki of Tokyo, Japan.

We acknowledge with sincere appreciation the help of the following companies for financial support for this symposium: Bayer AG/Cutter/Miles, Campbell Institute for Research and Technology, Firmenich, Inc., Frito-Lay, Inc., General Mills, Inc., Kellogg Company, Hershey Foods Corporation, M&M/Mars, Miller Brewing Company, Monsanto Company, National Starch & Chemical Corporation, Pfizer Research, Standard Brands, Inc., Strohs Brewery, The Coca-Cola Company, The Procter and Gamble Company, Thomas J. Lipton, Inc., and Wm. Wrigley Jr. Company.

GEORGE R. WALLER
Oklahoma State University

MILTON S. FEATHER
University of Missouri

January 4, 1983

HISTORICAL

1

Seventy Years of the Maillard Reaction

SIN'ITIRO KAWAMURA
Kagawa-ken Meizen Junior College, Kameoka 1-10, Takamatu, 760 Japan

A historical review with 107 references. Life and work of Louis-Camille Maillard (Feb. 4, 1878 - May 12, 1936) are described. The first use of the index term Maillard reaction in Chemical Abstracts was in 1950. German scientists with early interest in this reaction were Lintner (1912) and Ruckdeschel (1914). Several aspects of this reaction are reviewed with emphasis on the work of Japanese scientists.

About the Year 1912

Ten years ago Kawamura (1) published a brief historical review on this reaction in memory of the sixtieth anniversay of its first report by Louis-Camille Maillard (2). The first Maillard paper was presented on January 8, 1912, by Armand Emile Justin Gautier (1837-1920) in a session of the Academy of Sciences in Paris. Six weeks earlier (November 27, 1911) a remarkable study was reported by Maillard (3) on the condensation of amino acids by use of glycerol. The method of peptide synthesis by Emil Hermann Fischer (1835-1919) was known to him. However, Maillard searched for milder conditions. Thus he wished to condense amino acids by use of glycerol as a condensing agent. He thus obtained cycloglycylglycine and pentaglycylglycine.

Maillard (2) then used sugars instead of glycerol to investigate the formation of polypeptides by the reaction of amino acids with alcohols. It was found that the aldehyde group (of an aldose) had more intense effect on amino acids than did the hydroxyl groups. This led to the discovery of this reaction.

In a later report Maillard (4) cited a paper (5) by Arthur Robert Ling (1861-1937). Ling, lecturer on brewing and malting at the Sir John Cass Institute, presented a paper on

0097-6156/83/0215-0003$06.00/0

malting at the meeting held at the Criterion Restaurant, Piccadilly, London, on May 11, 1908. He noted a remarkable effect of kilning or heat drying. "At the second stage of kilning when the range of temperature is from 120 to 150°C, this mellowing by 'autodigestion' is continued.... Flavouring and colouring matters are produced....When these amino-compounds produced from proteins are heated at 120-140°C with sugars such as ordinary glucose or maltose, which are produced at this stage of process, combination occurs. The precise nature of the compounds produced is unknown to me, but they are probably glucosamine-like bodies." He further described the reaction of heating glucose with asparagine, which produced darkening in color.

Carl Joseph Ludwig Lintner (1855-1926), leader of the Scientific Station for Brewing (Wissenschaftliche Station fuer Brauerei) in Munich was studying the formation of malt aroma. He gave a lecture (6) at the 36th Meeting of the Station in November, 1912. As soon as he knew of the report of Maillard (2), he made some experiments by the process of Maillard (nach dem Vorgange von Maillard). He obtained dark reaction products which were responsible also for flavor and aroma (cf. 7).

Ame Pictet (1857-1937) of the University of Geneva reported the formation of pyridine and isoquinoline bases from acid hydrolyzate of casein in the presence of formaldehyde in 1916 (8). Maillard (9) claimed priority over him in discovering the condensation of amino acids with aldehydes or sugars to yield pyridine bases by citing his own paper (10). We should note that World War I went on from July 1914 to November 1918; Swiss scientists could continue research, whereas French ones could not do so easily.

Life and Work of Maillard (1878-1936)

Louis-Camille Maillard was born on February 4, 1878 in Pont-a-Mousson (Meuthe et Moselle) (48.55°N , 6.03°E). He went to Nancy, where he obtained the degrees of M. Sc. in 1897 and Dr. Med. in 1903. Thereafter he worked in the Chemical Division of the School of Medicine, University of Nancy. In 1914 he moved to Paris and the young doctor worked as head of a biological group in the Chemical Laboratory, University of Paris (7).

In 1911 his first report (3) on peptide synthesis was presented and in 1912 his first report (2) on the sugar-amino acid reaction was published. He completed a book (11) in 1913.

After World War I in 1919 he was appointed professor of biological and medical chemistry at the University of Algiers. In the same year he became a corresponding member of the Division of Pharmacy, Academy of Medicine.

In May 1936 he was invited to Paris for a lecture. On

May 12, 1936, he died suddenly during a meeting, where he was a judge of the competition for fellowship (12).

There are at least 6 papers of Maillard (3, 13-17) on peptide synthesis and at least 8 (2, 4, 9, 10, 18-21) on the sugar-amino acid reaction.

This reaction, reported briefly in 1912 (2), was described in detail later (20). The evolution of carbon dioxide during the reaction was attributed to the carboxyl group of the amino acid after very careful quantitative experiments.

The melanoidin prepared from glucose and glycine was soluble in the early stage and then became insoluble in the later stage of heating. The insoluble melanoidin he obtained contained C 58.85, H 4.92, N 4.35, and O 31.88%; thus the empirical formula was $C_{16}H_{15}NO_6$.

He proceeded to carry out the reactions of various amino acids with glucose at 100°C, and reactions of glycine with various sugars (arabinose, xylose, glucose, mannose, galactose, fructose, maltose, lactose, and sucrose) at 100°C. In the first experiments the activity in this reaction was in the order (highest first): alanine, valine, glycine, glutamic acid, leucine, sarcosine, and tyrosine. This is nearly the same as the order given more recently by Kato (22). In later experiments Maillard (20) showed that sucrose, a nonreducing sugar, gave no browning upon short heating with glycine, but the mixture showed the beginning of a darkening reaction after 3 hr of heating, which was attributed to hydrolysis of sucrose. Pentoses, especially xylose, gave higher velocity in this reaction than hexoses.

Being a biochemist, Maillard (20) studied the reaction of glycine with xylose or glucose at 40° and then at 34°C, in order to know the possibility of the change *in vivo*.

The last chapter of the same report (20) deals with reaction of glycylglycine with xylose at 75, 40, and 34°C, and three commercial peptone preparations with xylose at 110°C. The latter combinations darkened after 45 to 90 min.

The last report of Maillard (21) is rather a review in nature, with more than 50 references. The chemical natures not only of humus in soil but also of mineral fuel (coal) and browning in food material were discussed, especially in relation to the presence of nitrogen in browned products, which was inferred to be derived from amino acids (and related nitrogenous materials) used for synthetic "melanoidins".

Thus Maillard worked energetically on this reaction from 1911 through 1917 with notable results.

Naming the "Maillard Reaction"

Reynolds (23, 24) and Strahlmann (7) have cited Ellis (1959) (25) and Heyns and Paulsen (1960) (26) as the first to call this browning the Maillard reaction. It is certainly true that they published review articles with these words in the title. However, I wondered if any other worker used the term Maillard reaction earlier.

Examination of the subject indexes of Chemical Abstracts (CA) restricted to the terms browning and Maillard reaction gave the results shown in Table I.

Table I. "Browning" and "Maillard reaction" as Index Terms in CA (a)

Vols.	(Years)	Browning (b)			Maillard reaction.		
1-10	(1907-16)	Color, change of			No		
11-20	(1917-26)	"	" "		"		
21-30	(1927-36)	Discoloration*			"		
31-40	(1937-46)	"			"		
41-50	(1947-56)	"	(of foods)	18	Yes		13
51-55	(1957-61)	"	browning	169	"	**	32
56-65	(1962-66)	"	" ***	151	"	**	52
66-75	(1967-71)	"	"	120	"		77
76-85	(1972-76)	Browning****		340	"		157

(a) Number of abstracts under each index term shown.
(b) Browning or substitute index term.
* (See also Color(s); Coloring; Staining; Yellowing.)
** (See also "browning" under Discoloration.)

***Browning reaction. See Maillard reaction; see "browning" under Discoloration.
Discoloration, browning. -- See also Maillard reaction.
****Studies of browning of food and related materials were indexed at this heading.

Thus CA already used the index term Maillard reaction in the 5th Collective Index (1947-56), prior to 1959. There was no index term Maillard reaction up to vol. 43 (1949) of CA. The earliest citations by that term are to the 6 papers (27-32) in CA 44 (1950) and 46-48 (1952-54).

John B. Thompson (27) worked at the Trace Metal Research Laboratories in Chicago, while André Patron (28) was employed at the Institute of Colonial Fruits and Citruses (Institute de Fruits et Agrumes Coloniaux) in Paris, and prepared a review (33) containing the words Maillard reaction in the title. R. Geoffroy (29) of the French College of Milling (Ecole Francaise de Meunerie) in Paris also published a brief review on the Maillard reaction in the cereal industry. H. L. A. Tarr (30) of Pacific Fisheries Station, Vancouver, British Columbia, as well as Hans Wegner (32) of the Research Institute of Starch Fabrication (Forschungsinstitut der Staerkefabrikation) in Berlin, also used this term in the titles of research papers.

P. de Lange (31) of the Central Institute of Food Research (Centr. Inst. Voedingonderzoek), Utrecht, Netherlands reviewed the mechanism of the Maillard reaction, though his title did not show the term.

Thus, I once considered Thompson (1950) and Patron (1950) to be the first namers of the Maillard reaction. However, Barnes and Kaufman (34) of General Foods Corporation, Hoboken, NJ, published a review in 1947, three years earlier, of which the abstract in CA begins thus: "Maillard or browning reaction in foodstuffs is attributed to a reaction between sugars and proteins or other amino bodies," and the review itself repeatedly refers to the Maillard reaction. Later I found that Patron (33) cited a paper by Seaver and Kertesz (1946) (35) with this term in the title.

However, this type of search proved not reliable. In reading the abstract of a paper by Willy Ruckdeschel (36) of the Laboratory of Fermentation Chemistry, Royal Technical College (Koenigliche Technische Hochschule) at Munich I found the words Maillard's reaction in CA more than three times. The original paper by Ruckdeschel (1914) (36) contained the words "die Reaktion Maillards" first and then the words "die Maillardsche Reaktion" four times. Hence, he may be called one of the first namers of this reaction.

It is true that Maillard himself wrote often "my reaction" (ma reaction). It is a very simple but clear conclusion that the first namer of the Maillard reaction was the original author Maillard himself! I believe that no one would protest this deduction. As has already been noted, in 1912 Lintner (6) used the phrase "by the process of Maillard". It is of interest to find two German scientists, Lintner and Ruckdeschel, concerning themselves with a reaction then recently discovered by a French scientist, especially since World War I began in July, 1914.

Chemistry of the Maillard Reaction

Among chemical reviews (22-26, 37-42) the paper of Hodge (38) containing the famous scheme is especially notable. The scheme was considered practically useful and effective even after 25 years (42). The condensation of an amino acid with an aldose to form a Schiff base was first recognized by Maillard (20). The resulting N-substituted aldosylamine is converted to a deoxyketosylamine by the Amadori rearrangement. (In the reaction of an amino acid with a ketose a deoxyaldosylamine is formed by the Heyns rearrangement (24)). The three pathways of Hodge (38) should be augmented to five (41).

Condensation products of triose reductone with glycine, leucine, methionine, and phenylalanine have been characterized (43).

Hashiba, *et al.* (44) isolated six Amadori compounds from

soy sauce and other brewed products. Very recently Hashiba (45) compared the intensity of browning caused by heating sugars with glycine at 120°C for 5 min. It was in the order (highest first): ribose, glucuronic acid, xylose, arabinose, galacturonic acid, galactose, mannose, glucose, and lactose. Reducing power and the amount of glycine consumed were proportional to the intensity of browning, but the amount of Amadori compounds accumulated in the reaction mixture could not be correlated with these. The corresponding Amadori compounds were also compared for browning intensity. A general route

$$\text{Sugar + amino acid} \xrightarrow{\text{I}} \text{Amadori compd} \xrightarrow{\text{II}} \text{brown pigment}$$

has been postulated for the Maillard reaction by Kato (40). It was the purpose of Hashiba's study to find whether differences in browning are due to differences in reaction velocity in step I or II. However, the results were not clear-cut. He considered that the difference of browning among sugars depended on the rate of step I or a combination of rates of both steps I and II, according to the sugar used.

Namiki and Hayashi (46) recently summarized their theory of formation of intermediate free radicals, *N*, *N*'-disubstituted pyrazine cation radicals, in an early stage of the Maillard reaction (cf. 47).

Melanoidins (36, 48-50) are different from melanins, humins, and caramels, but similar to humus (37), according to Maillard (4, 11, 19-21). Kato and Tsuchida (51) studied the possible chemical structure of melanoidins.

Nutritional Aspects of the Maillard Reaction

Some reviews (41, 52-54) are available. The positive aspects are found in the production of desirable flavors and aromas. Fujimaki and Kurata (55) listed aldehydes and pyrazines, volatile compounds produced by heating amino acids with carbonyl compounds, isovaleraldehyde produced by reaction of leucine with carbonyl compounds (aldehydes, 3-deoxyglucosone, xylose, and glucose), and volatile compounds produced by reaction of cysteine with pyruvaldehyde. A recent review (56) cited 127 references including about 30 Japanese ones.

To study the effect of the Maillard reaction on nutritive value of protein, Patton, *et al*. (57) heated purified casein and soybean globulin in 5% glucose solution for 24 hrs at 96.5°C, and found significant losses of lysine, arginine, tryptophan, and histidine (52).

Kawamura, *et al*. (58) reported that the nonreducing sugar level, available lysine, and whiteness decreased in parallel with heating time of defatted soybean flakes at 100 or 120°C. Concentrations of the three oligosaccharides (sucrose, raffinose, and stachyose) tended to decrease. These nonreducing sugars were presumed to undergo hydrolysis upon heating, and reducing sugars formed (fructose, glucose, and galactose) were responsible for the Maillard reaction.

Studies with ^{15}N-nondialyzable melanoidin showed that 76% of the dietary melanoidin was excreted in rats' feces (59). The oral administration of Maillard reaction products caused an increase in the growth of both aerobic and anaerobic lactobacilli in the microflora of rats (60).

Mutagenicity of Maillard Reaction Products

In advanced stages of the Maillard reaction some mutagens might be formed. For example, mutagens appeared when ground beef hamburgers were cooked over 200°C (cf. 41). Reductive intermediates isolated from browning mixtures of triose reductone with guanine and its derivatives also showed mutagenicity (61).

Japanese workers in the National Cancer Center Research Institute published reviews concerning mutagens in heated foods (62-64). Pyrolyzates of proteins, peptides, and amino acids, especially tryptophan and glutamic acid, showed mutagenicity (65-69). Coffee prepared in the usual way for drinking contained some mutagenic substance(s) (70). Black tea and green tea were also mutagenic. However, these favorite beverages were not carcinogenic (64). Nine heterocyclic amines were isolated as mutagenic compounds from pyrolyzates (63): two pyridoindoles, four pyridoimidazoles, two imidazoquinolines, and 2-amino-5-phenylpyridine. None of them have been isolated as products of the Maillard reaction. Moreover, mutagenicity appeared only above 400°, and even more strongly at 500-600°C as the pyrolysis temperature.

Lee, *et al.* (71) carried out long-term (up to 12 months) feeding experiments of browned egg albumin with rats. Several changes were found, but no mutagenic response was observed.

The Maillard Reaction in Vivo

A simple sketch is given along the lines of two recent reviews (72, 73). Borsook, et al. (74) observed the occurrence of Amadori compounds *in vivo*, which were confirmed by Heyns and Paulsen (75) to be 1-deoxy-1-*N*-aminoacyl-D-fructose derivatives. Feeney, *et al.* (76) incubated glucose with egg white at 37°C for a few days and observed the disappearance of glucose. Holmquist and Schroeder (77) first showed that the *N*-terminal valine of Hb-A-1-c was blocked and subsequent researchers showed the occurrence of Amadori rearrangement of an aldosylamine, *N*-(1-deoxyglucosyl)valine. Rahbar (78) discovered an increase in the minor hemoglobin component, Hb-A-1-c, in the blood of diabetic patients.

Tanzer (79) observed the presence of the reduced form of the Amadori compounds formed with hydroxylysine and reducing sugars in aged connective tissues. Mester, *et al.* (80) suggested the formation of Amadori compounds in the blood from glucose and lysine-rich protein or serotonine. Cerami, *et al.*(81) showed that in diabetes mellitus structural proteins of the lens might be affected by high glucose concentration to induce cataract through Amadori-type reaction. It is likely that this kind of study has not been made by Japanese workers.

Maillard Reaction in Relation to Lipids

Lipids upon autoxidation produce carbonyl compounds, which react with amino compounds to form brown high-molecular products. This type of browning has recently been reviewed by Pokorny (82). According to him the first paper on the subject was published by Stansby (83). Fujimoto, *et al.* (84-89) studied this reaction with some fish and reported that most brown pigment of fish muscle was soluble in benzene-methanol. Thus in this case oxidized lipid-protein interactions are more important than the reaction between amino acids and ribose. In fish muscle the browning due to oxidized lipids is accompanied by the Maillard reaction between amino acids and ribose. In this connection it is interesting to note an earlier paper by Tarr (30), who reported that upon heating 1 hr at 120°C halibut browned slightly whereas lingcod darkened markedly and that free ribose was detected in the latter fish muscle.

Maillard reaction products have antioxidant activity (90, 91). Franzke and Iwainsky (92) first reported this activity of melanoidins. Patton (93) noted the contribution of the Maillard reaction in preventing dry milk powder from oxidation. Griffith and Johnson (94) showed that the products from reaction of glucose and glycine exhibited antioxidative properties in model system with lard. The antioxidative properties were attributed to reductones and this was proved by Evans, *et al.* (95). Zipser and Watts (96) reported that sterilization of meat produced antioxidative compounds by the Maillard reaction. Yamaguchi, *et al.* (97) as well as Kirigaya, *et al.* (98) made extensive studies concerning this problem. Eichner (90) studied intermediate reductone-like compounds as antioxidants. Yamaguchi, *et al.* (99) fractionated the reaction products obtained by heating D-xylose and glycine and found strong antioxidative activity in some melanoidin fractions.

Studies on Soy Sauce

Soy sauce is used every day by Japanese people. It is a fermented product made from steamed soybeans, parched wheat, and salt. It has a specially pleasant flavor and a deep but attractive color. It is now known that the color of fresh soy sauce is formed chiefly by the ordinary Maillard reaction, whereas the undesirable dark color of soy sauce stored in contact with atmospheric oxygen is produced not only by the advanced Maillard reaction but also and more pronouncedly by oxidative browning reaction.

The pigments of soy sauce, first studied by Kurono and Katsume (100), are melanoidins, which were reported by Omata, *et al.* (101) to be produced from the reaction of sugars and amino acids. Kato, *et al.* (102) found 3-deoxyglucosone as an intermediate in browning of soy sauce (cf. 103). Oxidative browning (104,105) was again reviewed (44) with emphasis on the browning of Amadori compounds and interaction between melanoidins and iron.

Studies on Dried Milk Products

One of the earlier reviews (31) concerned the Maillard reaction in dried milk during storage. Spray-dried whey has considerable amounts of lactose and protein rich in lysine. Theoretical treatment of the problem in whey powder was the object of recent studies by Labuza and Saltmarch (106, 107).

When the whey powders are stored at a_w (water activity) 0.33, 0.44, and 0.65, protein quality loss and browning extent were greatest not at 0.65 but at 0.44, where amorphous lactose began to shift to the alpha-monohydrate crystalline form with a release of water which mobilized reactants for the Maillard reaction (106). This type of research was continued with special reference to temperature conditions. The temperatures used were fairly constant at 25, 35, and 45°C or fluctuating between 25° and 45°C with alternating 5-day periods at each temperature. Experiments were carried out for 100-200 days. Temperature history, however, did not significantly change reaction mechanisms (107).

Acknowledgments Thanks are due to Prof. George R. Waller for his kind cooperation. Professor Hiromichi Kato of the University of Tokyo and Professor Mitsuo Namiki of Nagoya University provided information on many valuable references. The author is grateful to them and several other colleagues.

Literature cited (The titles of articles are included. CA: Chem. Abstr.)

1. Kawamura, S. Sixty years of the Maillard Reaction. *Shokuhin Kaihatsu* 1972, *7* (12), 64-5.
2. Maillard, L.-C. Action des acides amines sur les sucres: Formation des melanoidines par voie methodique. *C. R. Hebd. Seances Acad. Sci.* 1912, *154*, 66-8.
3. Maillard, L.-C. Condensation des acides amines en presence de la glycerine; Cycloglycylglycine et

polypeptides. C. R. Hebd. Seances Acad. Sci. 1911, 153, 1078-80.
4. Maillard, L.-C. Formation d'humus et de combustibles mineraux sans intervention de l'oxygene atmospherique, des microorganismes, des hautes temperatures, ou des fortes pressions. C. R. Hebd. Seances Acad. Sci. 1912, 155, 1554-6.
5. Ling, A. R. Malting. J. Inst. Brew. 1908, 14, 494-521.
6. Lintner, C. J. Ueber Farbe- und Aromabildung im Darrmalz. Z. Gesamte Brauwes. 1912, 35, 545-8, 553-6.
7. Strahlmann, H. Louis Camille Maillard (1878-1936) und die nach ihm benannte Braeunungsreaktion. Alimenta 1978, 17 (5), 144, 146.
8. Pictet, A.; Chou, T. Q. La formation des bases pyridiques et isoquinoleiques a partir de la caseine. C. R. Hebd. Seances Acad. Sci. 1916, 162, 127-9.
9. Maillard, L.-C. Sur la formation des bases pyridiques a partir des albuminoides. C. R. Hebd. Seances Acad. Sci. 1916, 162, 757-8.
10. Maillard, L.-C. Origines des bases cycliques du goudron de houille. C. R. Hebd. Seances Acad. Sci. 1913, 157, 850-2.
11. Maillard, L.-C. "Genese de Matieres Proteiques et des Matieres Humiques"; Masson et Cie: Paris, 1913; CA 8, 1594.
12. Achard, Ch. Deces de M. Louis Maillard. C. R. Hebd. Soc. Biol. 1963, 122, 347-8.
13. Maillard, L.-C. Synthese des peptides inferieurs par une methode nouvelle et directe, voisine des reaction biologique. C. R. Hebd. Seances Mem. Soc. Biol. 1911, 71, 546-9.
14. Maillard, L.-C. Synthese des polypeptides par action de la glycerine sur le glycocolle. Ann. Chim. (Paris), 1914, (9) 1, 519-78.
15. Maillard, L. C. Synthese des polypeptides par action de la glycerine sur le glycocolle: Etude dynamique. Ann. Chim. (Paris), 1914, (9) 2, 210-68.
16. Maillard, L.-C. Les cyclo-glycyl-glycines: Synthese directe par action de la glycerine sur les acides amines. Ann. Chim. (Paris), 1915, (9) 3, 48-120.
17. Maillard, L.-C. Synthese des cyclo-glycyl-glycine mixtes par action de la glycerine sur les melanges d'acides α-amines. Ann. Chim. (Paris), 1915, (9) 3, 225-52.
18. Maillard, L.-C. Reaction generale des acides amines sur les sucres: Ses consequences biologiques. C. R. Hebd. Seances Mem. Soc. Biol., 1912, 72, 559-601.
19. Maillard, L.-C. Formation des matieres humiques par action de polypeptides sur les sucres. C. R. Hebd. Seances Acad. Sci. 1913, 156, 1159-60.

20. Maillard, L.-C. Syntheses des matieres humiques par action des acides amines sur les sucres reducteurs. Ann. Chim. (Paris), 1916, (9) 5, 258-317.
21. Maillard, L.-C. Identité des matieres humiques de synthese avec les matieres humiques naturelles. Ann. Chim. (Paris) 1917, (9) 7, 113-52.
22. Kato, H. Chemistry of nonenzymatic browning phenomena. "Shokuhin no Henshoku to Sono Kagaku (Food Discoloration and Its Chemistry)"; Nakabayashi, T., Kimura, S.; and Kato, H., Eds.; Korin Shoin: Tokyo, 1967; pp. 223-89.
23. Reynolds, T. H. Chemistry of nonenzymic browning. I. The reaction between aldoses and amines. Adv. Food Res. 1963, 12, 1-52.
24. Reynolds, T. H. Chemistry of nonenzymic browning. II. Adv. Food Res. 1965, 14, 167-283.
25. Ellis, G. P. The Maillard reaction. Adv. Carbohydr. Chem. 1959, 14, 63-134.
26. Heyns, K.; Paulsen, H. Ueber die chemische Grundlagen der Maillard-Reaktion. Wiss. Veroeff. Deutsch. Ges. Ernaehr. 1960, 5, 15-42.
27. Thompson, J. B. A browning reaction involving copper-proteins. "Symposium on Copper Metabolism;" Johns Hopkins Press, 1950, pp. 141-53; CA 47, 11269f.
28. Patron, A. Recherches sur le brunissement nonenzymatique des fruits et des produits de fruits en conserve. Fruits Outre Mer 1950, 5, 201-7; CA 44, 10204d.
29. Geoffrey, R. The Maillard reaction in the cereal industry. Bull. Anc. Eleves Ec. Fr. Meun. 1953, (138), 260; CA 48, 7807a.
30. Tarr, H. L. A. Ribose and the Maillard reaction in fish muscle. Nature 1953, 171, 344-5; CA 47, 5973i.
31. Lange, P. de. Decrease of solubility and nutritive value of milk powder during storage. Conserva 1954, 2, 322-5; CA 48, 7808a.
32. Wegner, H. The Maillard reaction and yellowing of starch sirup. Staerke 1954, 6, 5-10; CA 48, 6150d.
33. Patron, A. La "Reaction de Maillard" et le brunissement non-enzymatique dans les industries alimentaires. Ind. Agric. Aliment. 1951, 68, 251-6; CA 46, 4687f.
34. Barnes, H. K.; Kaufman, C. W. Industrial aspects of browning reaction. Ind. Eng. Chem. 1947, 39, 1167-70.
35. Seaver, J. L.; Kertesz, Z. I. "Browning (Maillard) reaction" in heated solutions of uronic acids. J. Am. Chem. Soc. 1946, 68, 2178-9.
36. Ruckdeschel, W. Ueber Melanoidine und ihr Vorkommen in Darrmalz. Z. Gesamte Brauwes. 1914, 37, 430-2, 437-40.
37. Danehy, J. P.; Pigman, W. W. Reactions between sugars and nitrogenous compounds and their relationship to certain food problems. Adv. Food Res. 1951, 3, 241-90.

38. Hodge, J. E. Chemistry of browning reactions in model systems. J. Agric. Food Chem. 1953, 1, 928-43.
39. Feeney, R. E.; Blankenhorn, G.; Dixon, H. B. F. Carbonyl-amine reactions in protein chemistry. Adv. Protein Chem. 1975, 29, 135-203.
40. Kato, H. Studies on nonenzymic browning of foods. Nippon Nogei Kagaku Kaishi 1968, 42, R9-15.
41. Mauron, J. The Maillard reaction in food; a critical review from the nutritional standpoint. Prog. Food Nutr. Sci. 1981, 5, 5-35.
42. Nursten, H. E. Recent developments in studies of the Maillard reaction. Food Chem. 1981, 6, 263-77.
43. Omura, H.; Inoue, Y.; Eto, M.; Tsen, Y.-K.; Shinohara, K. Reaction products of triose reductone with some amino acids. Kyushu Daigaku Nogakubu Gakugei Zasshi 1974, 29, 61-70.
44. Hashiba, H.; Okuhara, A.; Iguchi, N. Oxygen-dependent browning of soy sauce and some brewed products. Prog. Food Nutr. Sci. 1981, 5, 93-113.
45. Hashiba, H. The browning reaction of Amadori compounds derived from various sugars. Agric. Biol. Chem. 1982, 47, 547-8.
46. Namiki, M.; Hayashi, T. Formation of novel free radical products in an early stage of Maillard reaction. Prog. Food Nutr. Sci. 1981, 5, 81-91.
47. Hayashi, T.; Namiki, M. On the mechanism of free radical formation during browning reaction of sugars with amino compounds. Agric. Biol. Chem. 1981, 45, 933-9.
48. Zikes, H. Ueber Melanoidine. Allg. Z. Bierbrau. Malzfabr. 1915, 43, 57-8.
49. Sattler, L.; Zerban, F. W. Unfermentable reducing substances in molasses -- volatile decomposition products of sugars and their role in melanoidin formation. Ind. Eng. Chem. 1949, 41, 1401-6.
50. Luers, H. The melanoidins. Brew. Dig. 1949, 24 (10), 41-4, 48; CA 44, 794a.
51. Kato, H.; Tsuchida, H. Estimation of melanoidin structure by hydrolysis and oxidation. Prog. Food Nutr. Sci. 1981, 5, 147-56.
52. Patton, A. R. Present status of heat-processing damage to protein foods. Nutr. Rev. 1950, 8, 193-6.
53. Adrian, J. Nutritional and physioloical consequences of the Maillard reaction. World Rev. Nutr. Diet. 1974, 19, 71-122.
54. Dworschak, E. Nonenzymatic browning and its effect on protein nutrition. CRC Crit. Rev. Food Sci. Nutr. 1980, 12, 1-40.
55. Fujimaki, M.; Kurata, T. Aroma of heated foods. Kagaku To Seibutsu 1971, 9, 85-96.
56. Mabrouk, A. F. Flavor of browning reaction products.

"Food Taste Chemistry"; Boudreau, J. C., Ed.; ACS Symp. Ser. 1979, 115, 205-45.
57. Patton, A. R.; Hill, E. G.; Foreman, E. M. Effect of browning on the essential amino acid content of soy glubulin. Science 1948, 108, 659-60; through ref. 52.
58. Kawamura, S.; Kasai, T.; Honda, A. Changes of sugars and decrease in available lysine on autoclaving defatted soybean flakes. Eiyo To Shokuryo 1968, 20, 478-81.
59. Homma, S.; Fujimaki, M. Growth response of rats fed a diet containing nondialyzable melanoidin. Prog. Food Nutr. Sci. 1981, 5, 209-16.
60. Horikoshi, H.; Ohmura, A.; Gomyo, T.; Kuwabara, Y.; Ueda, S. Effects of browning products on the intestinal microflora of the rat. Prog. Food Nutr. Sci. 1981, 5, 223-8.
61. Shinohara, K.; Lee, J.-H.; Tanaka, M.; Murakami, H.; Omura, H. Mutagenicity of intermediates produced in the early stage of browning reaction of triose reductone with nucleic acid related compounds on bacterial tests. Agric. Biol. Chem. 1980, 44, 1737-43.
62. Sugimura, T.; Nagao, M. Mutagenic factors in cooked foods. CRC Crit. Rev. Toxicol. 1979, 7, 189-209.
63. Sugimura, T.; Nagao, M.; Wakabayashi, K. Mutagenic heterocyclic amines in cooked food. Environ. Carcinog. Sel. Methods Anal. 1981, 4, 251-67.
64. Nagao, M. Mutagens in ordinary foods. Hen'igen To Dokusei 1981, 4 (5), 20-31.
65. Nagao, M.; Honda, M.; Seino, Y.; Yahagi, T.; Kawachi, T.; Sugimura, T. Mutagenicities of protein pyrolyzates. Cancer Lett. 1977, 2, 335-40.
66. Matsumoto, T.; Yoshida, D.; Mizusaki, S.; Okamoto, H. Mutagenic activity of amino acid pyrolyzates in Salmonella typhimurium TA 96. Mutat. Res. 1977, 48, 279-86.
67. Matsumoto, T.; Yoshida, D.; Mizusaki, S.; Okamoto, H. Mutagenicities of the pyrolyzates of peptides and proteins. Mutat. Res. 1978, 56, 281-8.
68. Matsukura, H.; Kawachi, T.; Horino, K.; Ohgaki, H.; Sugimura, T.; Takayama, S. Carcinogenicity in mice of mutagenic compounds from a tryptophan pyrolyzate. Science 1981, 213, 346-7.
69. Tsuda, N.; Nagao, M.; Hirayama, T.; Sugimura, T. Nitrite converts 2-amino-α-carboline, an indirect mutagen, into 2-hydroxy-α-carboline, a non-mutagen, and 2-hydroxy-3-nitroso-α-carboline, a direct mutagen. Mutat. Res. 1981, 83, 61-8.
70. Nagao, M.; Takahashi, Y.; Yamanaka, H.; Sugimura, T. Mutagens in coffee and tea. Mutat. Res. 1979, 68, 101-6.
71. Lee, T.-C.; Kimiagar, H.; Pintauro, S. J.; Chichester, C. O. Physiological and safety aspects of Maillard browning of foods. Prog. Food Nutr. Sci. 1981, 5, 243-56.

72. Mester, L.; Szabados, L.; Mester, K.; Yadav, H. Maillard type carbonyl-amine reactions in vivo and their physiological effects. Prog. Food Nutr. Sci. 1981, 5, 295-314.
73. Monnier, V. U.; Stevens, V. J.; Cerami, A. Maillard reactions involving protein and carbohydrates in vivo: relevance to diabetes mellitus and aging. Prog. Food Nutr. Sci. 1981, 5, 315-27.
74. Borsook, H.; Abrams, A.; Lowy, P. H. Fructose-amino acids in liver: stimuli of amino acid incorporation in vitro. J. Biol. Chem. 1955, 215, 111-24; through ref. 72.
75. Heyns, K.; Paulsen, H. Ueber 'Fructose-Aminosaeuren' und 'Glucose-Aminosaeuren' in Leberextracten. Liebigs Ann. Chem. 1959, 622, 160-74; through ref. 72.
76. Feeney, R. E.; Clary, J. J.; Clark, J. R. A reaction between egg white proteins in incubated eggs. Nature (London) 1964, 201, 192-3; through ref. 73.
77. Holmquist, W. R.; Schroeder, W. A. A new N-terminal blocking group involving a Schiff base in hemoglobin A-1-c. Biochemistry 1966, 5, 2489-503; through ref. 72.
78. Rahbar, S. An abnormal hemoglobin in red cells of diabetes. Clin. Chim. Acta 1968, 22, 296-8; through ref. 73.
79. Tanzer, M. L. Cross-linking of collagen. Science 1973, 180, 561-6; through ref. 72.
80. Mester, L.; Kraska, B.; Crisba, J.; Mester, M. Sugar-amine interactions in the blood clotting system and their effects on haemostasis. Proc. Vth Int. Congr. Thromb. Haemostasis 1975, Paris; Int. Soc. Thromb. Haemostasis; through ref. 72.
81. Cerami, A.; Stevens, V. J.; Monnier, V. M. Role of nonenzymatic glycosylation in the development of the sequelae of Diabetes Mellitus. Metabolism 1979, 28, 431-7; through ref. 72.
82. Pokorny, C. Browning from lipid-protein interactions. Prog. Food Nutr. Sci. 1981, 5, 421-8.
83. Stansby, M. E. Oxidative deterioration in fish and fishery products. I. Introduction. Commer. Fish. Rev. 1957, 19, 24-6; through ref. 82.
84. Fujimoto, K.; Maruyama, M.; Kaneda, T. Brown discoloration of fish products. I. Factors affecting the discoloration. Nippon Suisan Gakkaishi 1968, 34, 519-23.
85. Fujimoto, K. Lipid oxidation and oxidized oil stain of aquatic products. Nippon Suisan Gakkaishi 1970, 36, 850-3, 871-2.
86. Fujimoto, K.; Abe, I.; Kaneda, T. Brown discoloration of fish products. II. Effect of several aldehydes, especially azelaaldehydic acid, in the autoxidized oil on discoloration. Nippon Suisan Gakkaishi 1971, 37, 40-3.

87. Fujimoto, K.; Saito, J.; Kaneda, T. Brown discoloration of fish products. III. Effect of ribose on the browning reaction derived from autoxidized oil. Nippon Suisan Gakkaishi 1971, 37, 44-7.
88. Fujimoto, K.; Kandea, T. Brown discoloration of fish products. IV. Nitrogen content of the browning substance. Nippon Suisan Gakkaishi 1973, 39, 179-83.
89. Fujimoto, K.; Kaneda, T. Brown discoloration of fish products. V. Reaction mechanisms in the early stage. Nippon Suisan Gakkaishi 1973, 39, 185-90.
90. Eichner, K. Antioxidative effects of Maillard reaction intermediates. Prog. Food Nutr. Sci. 1981, 5, 441-51.
91. Lingnert, H.; Eriksson, C. E. Antioxidative effect of Maillard reaction products. Prog. Food Nutr. Sci. 1981, 5, 453-66.
92. Franzke, C.; Iwainsky, H. Zur antioxydativen Wirksamkeit der Melanoidine. Dtsch. Lebensm.-Rundsch. 1954, 50, 251-4; through ref. 91.
93. Patton, S. Browning and associated changes in milk and its products: a review. J. Dairy Sci. 1955, 38, 457-478; through ref. 90.
94. Griffith, T.; Johnson, J. A. Relation of the browning reaction to storage stability of sugar cookies. Cereal Chem. 1957, 34, 159-69; through ref. 90.
95. Evans, C. D.; Moser, H. A.; Cooney, P. M.; Hodge, J. E. Amino-hexose-reductones as antioxidants. I. Vegetable oils. J. Am. Oil Chem. Soc. 1958, 35, 84-8; through ref. 90.
96. Zipser, M. W.; Watts, B. M. Lipid oxidation in heat sterilized beef. Food Technol. 1961, 15, 445-7.
97. Yamaguchi, N.; Yokoo, Y.; Fujimaki, M. Studies on antioxidative activities of amino compounds on fats and oils. III. Antioxidative activities of soybean protein hydrolyzates and synergistic effect of hydrolyzate on tocopherol. Nippon Shokuhin Kogyo Gakkaishi 1975, 22, 431-5.
98. Kirigawa, N.; Kato, H.; Fujimaki, M. Studies on antioxidant activity of nonenzymic browning reaction products. III. Fractionation of browning reaction solution between ammonia and D-glucose and antioxidant activity of the resulting fractions. Nippon Nogei Kagaku Kaishi 1971, 45, 292-8.
99. Yamaguchi, N.; Koyama, Y.; Fujimaki, M. Fractionation of antioxidative activity of browning reaction products between D-xylose and glycine. Prog. Food Nutr. Sci. 1981, 5, 429-39.
100. Kurono, K.; Katsume, E. Chemical composition of the pigment of soy sauce. Nippon Nogei Kagaku Kaishi 1927, 3, 594-613.
101. Omata, S.; Ueno, T.; Nakagawa, Y. The pigment components

and their formation in amino-acid seasonings. Hakko Kogaku Zasshi 1956, 34, 166-72.
102. Kato, H.; Yamada, Y.; Izaka, K.; Sakurai, Y. Studies on browning mechanisms of soybean products. I. Separation and identification of 3-deoxyglucosone occurring in soy sauce and miso. Nippon Nogei Kagaku Kaishi 1961, 35, 412-14.
103. Okuhara, A. Browning of soy sauce. Kagaku To Seibutsu 1972, 10, 383-90.
104. Motai, H. Nonenzymatic oxidative browning: Polymerization of melanoidins. Kagaku To Seibutsu 1975, 13, 292-4.
105. Hashiba, H. Oxidative browning of soy sauce. Contribution of Amadori compounds. Kagaku To Seibutsu 1977, 15, 156-8.
106. Saltmarch, R.; Vagnini-Ferrari, M.; Labuza, T. P. Theoretical basis and application of kinetics to browning in spray-dried whey food systems. Prog. Food Nutr. Sci. 1981, 5, 331-44.
107. Labuza, T. P.; Saltmarch, R. Kinetics of browning and protein quality loss in whey powders during steady state and non-steady state storage conditions. J. Food Sci. 1982, 47, 92-6, 113.

RECEIVED January 14, 1983

CHEMICAL ASPECTS

2

A New Mechanism of the Maillard Reaction Involving Sugar Fragmentation and Free Radical Formation

MITSUO NAMIKI and TATEKI HAYASHI

Nagoya University, Department of Food Science and Technology, Nagoya, Japan 464

Analyses of the hyperfine structures of ESR spectra found at an early stage of the Maillard reaction led to the assignment that the radical products are N,N'-disubstituted pyrazine cation radicals. Quantitative product determination at the reaction stage involving the free radicals indicates a sequential formation of glucosylamine, a 2-carbon and other sugar fragmentation products, then certain reducing products and the free radical, the Amadori product, and finally, glucosones. N,N'-dialkylpyrazines, or mixtures of glycolaldehyde with amino compounds, are shown to be highly active in free radical formation as well as in browning. Thus we propose the existence of a new pathway to browning in the Maillard reaction, involving sugar fragmentation and free radical formation prior to the Amadori rearrangement.

The mechanism proposed by Hodge in 1953 (1) for the early stages of the Maillard reaction, involving the Amadori rearrangement as a key step, has been accepted over a quarter of a century as a most apt description. Here, we propose a new mechanism which involves cleavage of the sugar molecule with generation of a highly reactive two-carbon fragment at an early stage of the Maillard reaction, prior to the Amadori rearrangement. We were led to the idea of sugar fragmentation by our finding of the development of a novel free radical product at an early stage in the Maillard reaction, prior to browning (2). Based on the hyperfine structural analyses of ESR spectra of various sugar-amino compound systems, the structures of the radical products were assigned as N,N'-dialkylpyrazine cation radicals (3), and it was assumed that each alkyl-substituted nitrogen originated from the amino compound, and each two-carbon moiety of the pyrazine ring from the sugar fragment. By isolation and identification of glyoxal derivatives from the sugar-amine reaction system (4) we

0097-6156/83/0215-0021$07.50/0

have also established that the two-carbon fragmentation of the sugar molecule occurs prior to Amadori rearrangement. The present paper describes our recent studies on this new mechanism of the Maillard reaction (5,6).

Development of Novel Free Radicals in the Early Stage of the Maillard Reaction

With the single exception of the stable free radical observed in melanoidin prepared from the glycine-glucose reaction (7), there have been no reports of free radical formation in the early stages of the Maillard reaction. We have observed the development of ESR spectra with characteristic hyperfine structures at an early stage in the Maillard reaction (2). Figure 1 shows the changes with time in intensities of the ESR signal with hyperfine structure, the ESR signal with broad singlet, and browning during the reaction of D-glucose with α- or β-alanine in boiling water; the ESR spectra with hyperfine structure are shown for each reaction mixture at the maximum intensity. In the case of glucose-β-alanine the ESR signal with hyperfine structure could be detected as soon as the reaction mixture was heated; the relative intensity increased rapidly during about ten min and then decreased rapidly, with simultaneous loss of hyperfine structure. The disappearance of hyperfine structure in the ESR spectrum was also accompanied by a gradual increase in browning and development of a melanoidin-type ESR signal with broad singlet. We are concerned here only with the ESR spectra with hyperfine structure, due to the new free radical products.

The ESR spectra of reaction mixtures of glucose with α- and β-alanine differed from each other in the complexity of their hyperfine structure, the former being split into 19 lines and the latter into 25 lines. This difference was shown to depend on differences in the structure of amino acid, α- and β-alanine, by comparison with spectra obtained from various sugar-amino acid systems as shown in Table I. All the sugars and their related carbonyl compounds gave essentially the same type of ESR spectrum with a given amino acid, with the exception of glyceraldehyde and dihydroxyacetone. The latter two showed spectra resembling each other and with more complicated hyperfine structures than the others.

The carbonyl compounds most effective in formation of free radicals are also most effective in browning, and the fact that glycolaldehyde has especially high activity in both reactions is particularly significant in this study. Such carbonyl compounds as furfural and crotonaldehyde showed high activity only in browning, and examination of structures of various carbonyl compounds suggest that the presence of an enediol or a potential enediol grouping in the carbonyl compounds is necessary for radical formation.

Investigations of various amino compounds indicated that

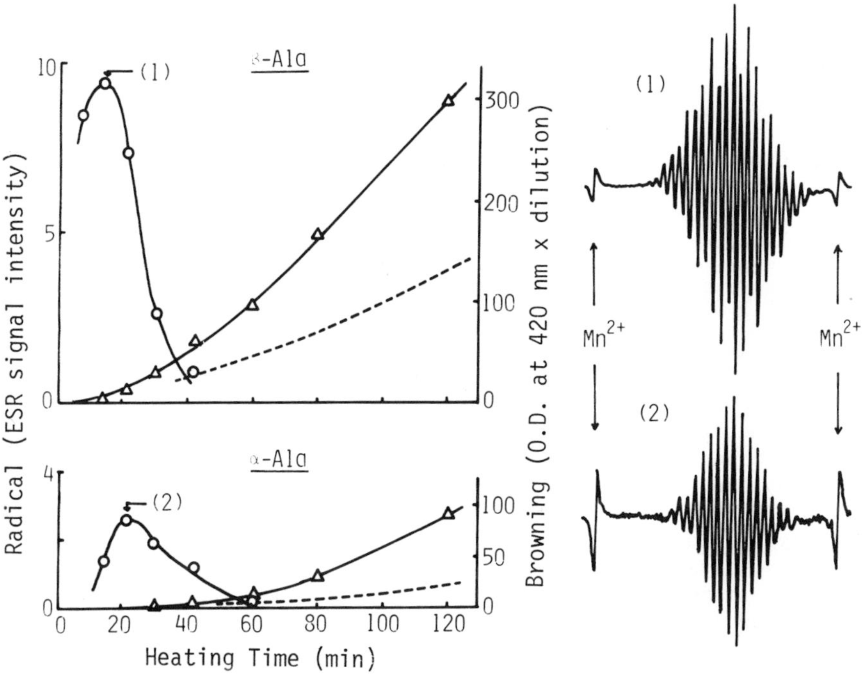

Figure 1. Free radical formation and browning in the reaction of D*-glucose with α-alanine or β-alanine (each 3* M*), and ESR spectra of the reaction mixtures heated in a boiling water bath. Key: ○, ESR signal with hyperfine structure; – – –, ESR signal with broad line; and △, browning.*

Table I

ESR Spectral Data on Free Radicals and Browning in the Reaction of Sugars and Other Carbonyl Compounds with α- or β-Alanine*

	ESR Spectra		Browning
	Splitting Line No.	Intensity	
α-Alanine			
D-Glucose	19	+	+
D-Fructose	19	+	+
D-Arabinose	19	+	++
D-Xylose	19	+	++
D-Ribose	19	+	+++
Glycolaldehyde	19	+++	+++
β-Alanine			
D-Glucose	25	++	++
D-Fructose	25	++	++
D-Arabinose	25	++	++++
D-Xylose	25	++	++++
D-Ribose	25	++	++++
Glyceraldehyde	35 ~	+++	+++++
Dihydroxyacetone	35 ~	+++	+++++
Glycolaldehyde	25	++++	+++++
3-Deoxyglucosone	25	++	++++
5-Hydroxymethyl-furfural			++
Furfural			+++++
Glyoxal	25	++	++++
Crotonaldehyde			++++
Propionaldehyde			+

*Aqueous solutions (each 3 M) were heated in boiling water bath.

ability to develop an ESR spectrum with the characteristic hyperfine structure was only observed in compounds with a primary amino group, although certain amino compounds, such as aniline, cysteine, and ethylenediamine, failed to show the spectrum (3).

Effect of Reaction Conditions on Development and Stability of Free Radicals (2)

pH seemed the most important factor in the Maillard reaction for determining the rates of browning as well as reaction processes. As shown in Fig. 2a, free radical development was observed even at neutral pH, and, like browning, was enhanced markedly by increase in pH, although the ESR signal disappeared rapidly at pH above 11. The free radical formed is known to be fairly stable in weakly acidic reaction mixture, although it is unstable in moderately alkaline solution (Fig. 2b).

It is known that oxygen plays an important role in free radical reactions, and radicals usually disappear rapidly in the presence of oxygen. Interestingly, the free radical formed in the early stages of the Maillard reaction is fairly stable in the reaction mixture, as the ESR signal could be observed in the mixture on heating in an open test tube without exclusion of air, although it was abolished rapidly by bubbling air. Attempts to isolate the radical product are as yet unsuccessful.

Analysis of Hyperfine Structure in ESR Spectra (3)

To elucidate the structure of the free radical products, analyses were made of the hyperfine structures of ESR spectra of various reaction systems of sugar with amino compounds. Figure 3 shows analyses of representative ESR spectra of the reaction mixtures of D-glucose with α-alanine or β-alanine. The hyperfine structures of (A) could be resolved into 8.41 G quintet, 2.86 G quintet and 2.93 G triplet due to two equivalent nitrogens, four equivalent protons and two equivalent protons, respectively. On the other hand, (B) was resolved into the 8.17 G quintet, 2.84 G quintet and 5.36 G quintet, due to two equivalent nitrogens, four equivalent protons and four equivalent protons, respectively, as indicated by the stick diagrams and splitting constants in each figure.

The results of analyses of the ESR spectra of various reaction systems are summarised in Table II. It was shown that all spectra have in common the splittings arising from two equivalent nitrogens (about 8.2 G) and four equivalent protons (about 3.0 G), and additionally from an even number of equivalent protons with different splitting constants. These assignments let to the reasonable assumption that the radical products are N,N'-disubstituted pyrazine cation radical derivatives, as shown in Fig. 4. This assumption was strongly supported by the fact that the hyperfine structure as well as the g-value of the ESR

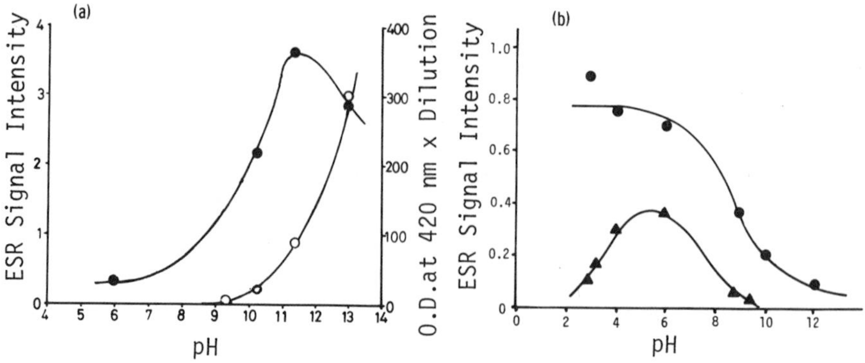

Figure 2. For glucose and β-alanine (each 1.5 M), effects of pH on free radical formation (a) and stability (b). Key to a (heated in boiling water bath): ●, ESR signal; ○, browning. Key to b (at room temperature): ●, 2 min after; ▲, 2 hr after.

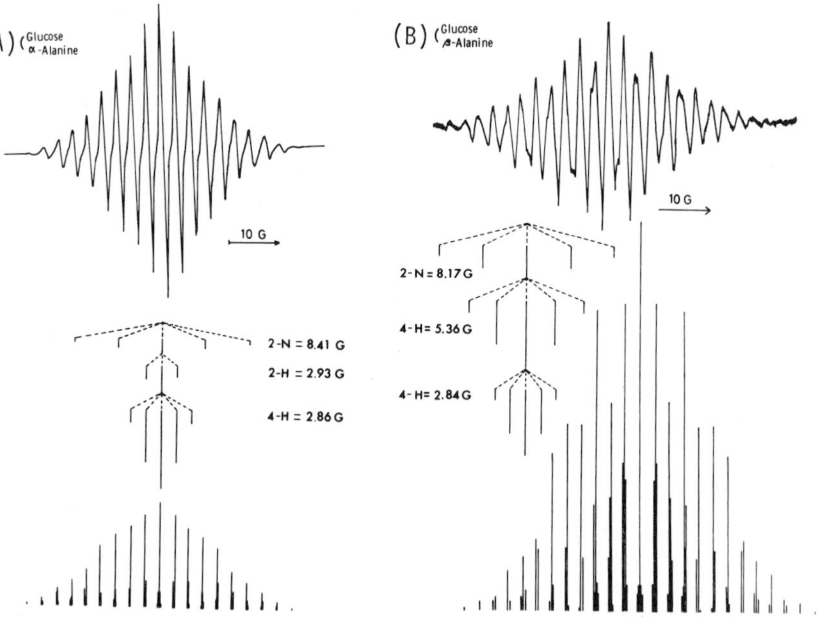

Figure 3. ESR spectra of the reaction mixtures of (A) glucose-α-alanine and (B) glucose-β-alanine. (Reproduced from Ref. 3. Copyright 1977, American Chemical Society.)

Table II

Analysis of Hyperfine Structures of ESR Spectra

Amino acide or Amine	Splitting constants (G)			
	(α-H)		(2-N)	(4-H)
Glycine	4.69	(4H)	8.15	3.04
α-Alanine	2.93	(2H)	8.41	2.86
β-Alanine	5.36	(4H)	8.15	2.84
Valine	1.43	(2H)	8.40	2.88
Phenylalanine	1.31	(2H)	7.99	3.03
Other amino acids*	1.5±0.1	(2H)	8.3±0.15	2.9±0.1
t-Butylamine			8.74	2.74
Methylamine	7.98	(6H)	8.35	2.83
Methyl-d_3-amine	1.25	(6D)	8.38	2.86
Ethylamine	5.37	(4H)	8.35	2.85
N,N'-Diethyl-pyrazinium salt	5.33	(4H)	8.37	2.82

Sugar; D-glucose, D-arabinose, D-xylose or glycolaldehyde.
*; serine, methionine, leucine, isoleucine, tyrosine, arginine.

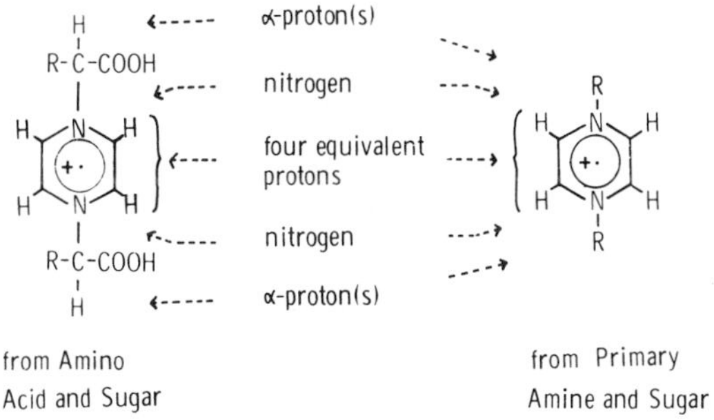

Figure 4. N,N'-*Disubstituted pyrazine cation radicals with assignments for hyperfine structures of the ESR spectra in the Maillard reaction mixtures.*

spectrum of the reaction mixture of ethylamine with D-glucose were in good agreement with those of the ESR spectrum of N,N'-diethylpyrazine cation radical (Fig.5) synthesized according to the method of Curphy and Prasa (8).

Possible Formation Pathways for the Free Radical Products

Despite the fact that a large number of pyrazine derivatives have been identified as Maillard reaction products, there is little information on formation of N,N'-disubstituted pyrazinium salts and on their chemical properties. This is probably due to their instability (8), but in the present work we were able to detect the presence of such pyrazine derivatives by ESR spectrometry. As the pyrazine derivatives proposed have no substituents on the ring carbons, one could propose the following two pathways as plausible formation mechanisms (Fig.6): a) Formation of a two-carbon enaminol product by fragmentation of the sugar or sugar-amine derivative, followed by its dimerization to give the N,N'-disubstituted pyrazine product. b) Bimolecular condensation of the enaminol product of the Amadori rearrangement to form a pyrazine derivative possessing 1,4-diamino residues and 2,5-disugar residues, with subsequent elimination of the substituted sugar residues by C-C bond scission to give the proposed pyrazine product.

Relation of the Free Radical to the Schiff Base, Amadori Compound and 3-Deoxyglucosone (5)

It is of importance to investigate the relation of free radical formation to formation of the main intermediates in the generally accepted scheme of the Maillard reaction (1), especially in connection with the potential route b mentioned above.

Figure 7 shows concentration changes in the main intermediates, and the relative intensities of ESR signals and of browning during the reaction of D-glucose with β-alanine in 0.1 N NaOH solution. The free radical developed rapidly prior to, or simultaneously with, formation of the Amadori product, and then began to decrease after about 10 min, while the Amadori product continued to increase. 3-Deoxyglucosone increased gradually thereafter. These results indicate that the free radical is derived from neither the Amadori product nor, naturally, from 3-deoxyglucosone and that it is produced prior to the Amadori rearrangement.

To confirm these relationships, the ability of these intermediates to generate the free radical was also examined. As shown in Fig. 8, the Amadori product did not provide any free radical on heating alone or with added sugar or amino acid. In contrast to this, glucosyl-β-alanine alone gave the free radical to an extent similar to that in the glucose-β-alanine system. In a separate run, the Schiff base, glucosyl-*n*-butylamine, alone in

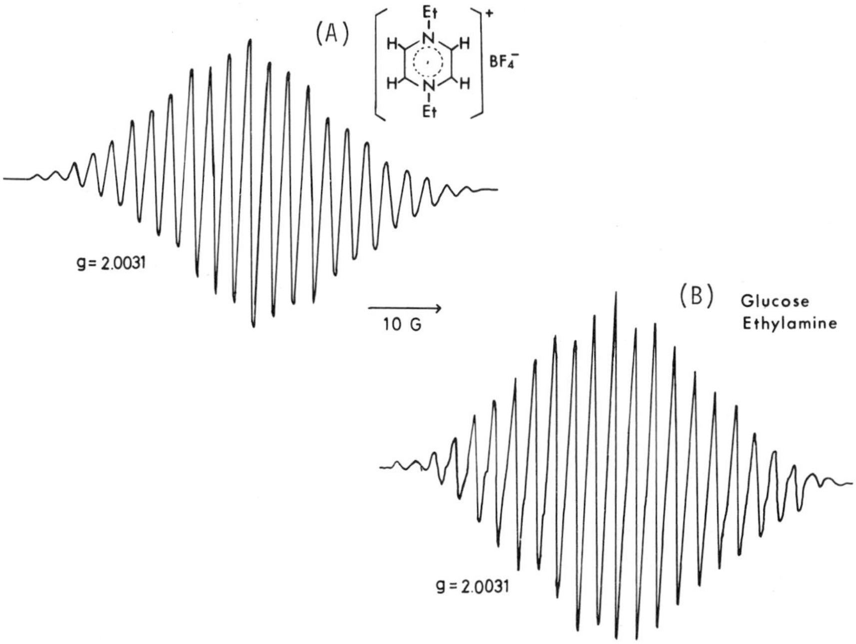

Figure 5. ESR spectra of the reaction mixture of ethylamine with glucose (lower) and of synthesized N,N′*-diethylpyrazine cation radical (upper). (Reproduced from Ref. 3. Copyright 1977, American Chemical Society.)*

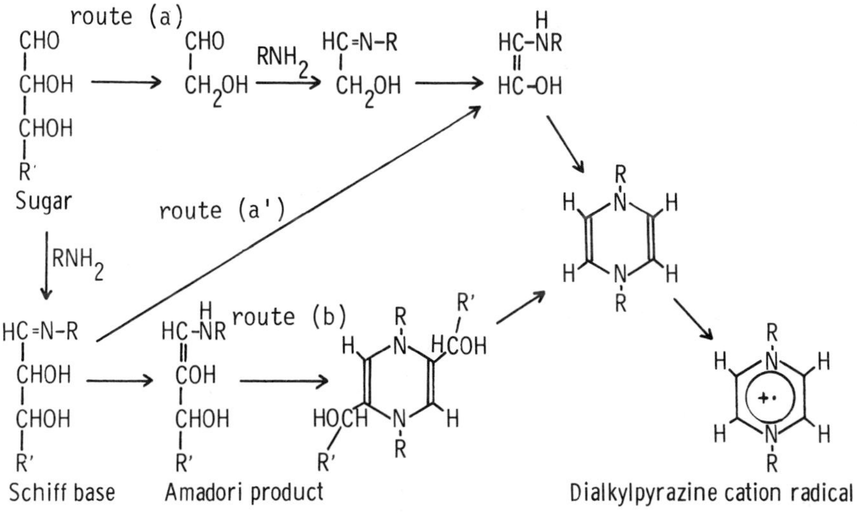

Figure 6. Two possible pathways for radical formation in the reaction of a sugar with an amino compound.

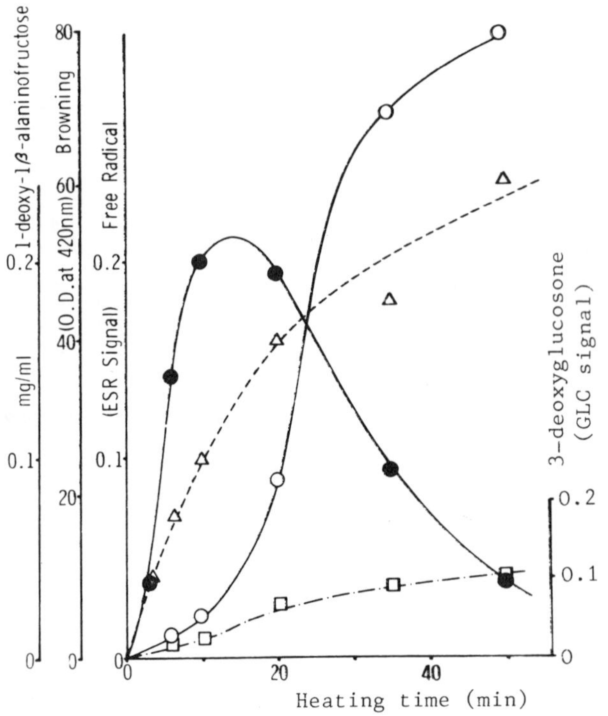

Figure 7. Formation of free radical and other intermediates in the reaction of glucose with β-alanine. The mixture (each 2 M*) with aqueous alkali solution (0.1* N *NaOH) was heated in a boiling water bath. Key:* ●, *free radical;* ○, *browning;* △, *1-β-alanino-1-deoxyfructose; and* □, *3-deoxyglucosone. (Reproduced with permission from Ref. 5.)*

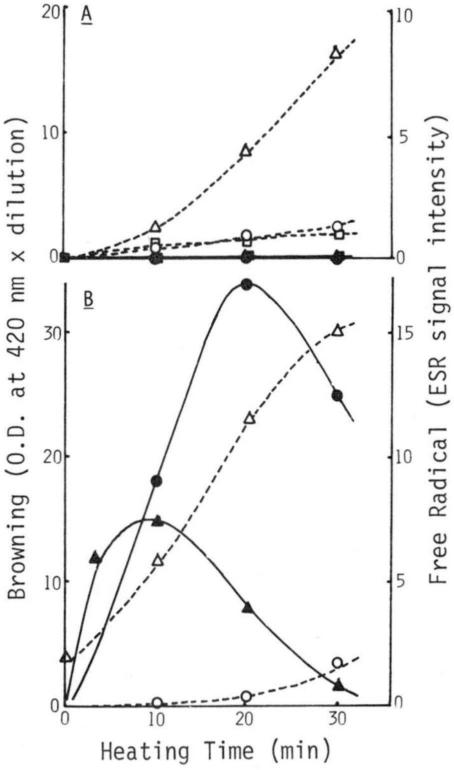

Figure 8. Development of free radicals from intermediate products in the reaction of glucose with β-alanine. Each aqueous solution (each 1 M) was heated in boiling water bath. Key to A for 1-β-alanino-1-deoxyfructose: ●, radical; ○, browning; for 1-β-alanino-1-deoxyfructose + β-alanine: ▲, radical; △, browning; and for 1-β-alanino-1-deoxyfructose + glucose: ■, radical; □, browning. Key to B for glucose + β-alanine: ●, radical; ○, browning; and for glucosyl-β-alanine: ▲, radical; △, browning. (Reproduced with permission from Ref. 5.)

ethanol (1 M), also provided the ESR signal to an extent similar to that in the glucose-*n*-butylamine system.

These results establish that the free radical was produced before the Amadori rearrangement and eliminate the possibility of route b from the enaminol of sugar. Thus, the only remaining possibility is formation by the initial fragmentation of sugar or sugar-amino derivative to give a two-carbon enaminol.

Formation of Fragmentation Products at an Early Stage of the Maillard Reaction

The above results led to the reasonable assumption of sugar fragmentation at an early stage of the Maillard reaction, but prior to this little was known about formation of the two-carbon fragments, though it had been postulated that cleavage would occur at the C-2--C-3 position by a reverse aldol mechanism, catalysed by amino acids in neutral and alkaline solution (1). However, isolation and identification of the products have not been reported. To investigate this fragmentation, we carried out experiments to isolate and identify the two-carbon products produced in the sugar-amine reaction systems.

Mixtures of various sugars with amines were refluxed in ethanol for 10 min. The reddish-brown mixtures were developed on silica gel TLC plates and sprayed with 2,4-dinitrophenylhydrazine (2,4-DNP). As shown in Fig. 9, a particular spot at around R_f 0.7 was observed immediately after the spraying in every case, and an additional spot was also detected in the cases of *n*-butylamine and cyclohexylamine. These spots agreed well in R_f value and characteristic coloration with those of glyoxal and methylglyoxal, respectively.

The substance commonly observed in every reaction systems at around R_f 0.7 which was highly sensitive to 2,4-DNP attracted our particular attention. To isolate this substance, the reaction mixture of glucose with *t*-butylamine was chromatographed on a Bondapak C_{18} column and subjected to preparative TLC. From the fraction with the appropriate R_f, we obtained the 2,4-DNP-positive substance as a purified liquid. GLC analysis of the reduced and acetylated product from the isolated substance showed a main peak identical with that for ethylene glycol diacetate, but apparently differing from that for glycerol triacetate, as shown in Fig. 10. This suggests that the substance was glyoxal or glycolaldehyde, and the former was supported by TLC analyses of the substance itself and of its reaction mixture with 2,4-DNP, carried out in parallel with authentic glyoxal (Fig.10 B). Furthermore, the phenylhydrazone of this substance could be successfully prepared, and its NMR spectral data agreed well with those of authentic glyoxal bisphenylhydrazone. The identification of this substance as glyoxal, however, did not necessarily mean that it was a direct product of the reaction. This is because it was shown that a 2,4-DNP-positive product was certainly present

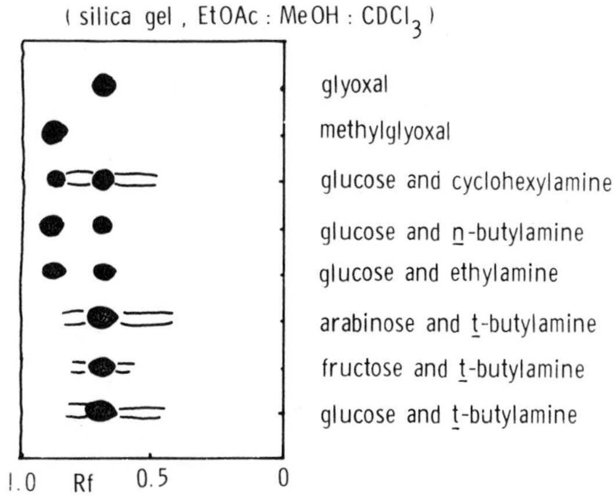

Figure 9. TLC of the reaction mixtures of sugars with amines in ethanol. Mixtures of the materials (each 1 M*) were prepared by refluxing for 40 min (*t*-butylamine system) or 10 min (others), developed with ethyl acetate–methanol–chloroform (5:2:1) on silica gel plate, and sprayed with 2,4-DNP solution (Reproduced with permission from Ref. 4.)*

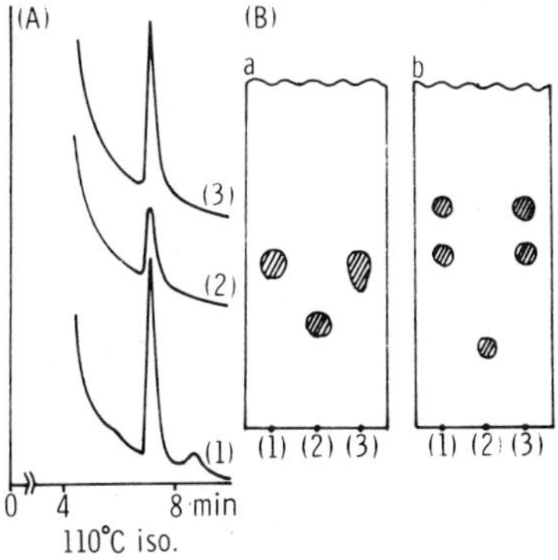

Figure 10. GLC (A) and TLC (B) of the isolated 2,4-DNP-positive substance from the reaction mixture of glucose with t*-butylamine in ethanol. GLC conditions: XE-60 column at 110 °C. In a, 2,4-DNP-positive spots of the reaction mixture were developed with chloroform–ethanol–water (5:2:1). In b, 2,4-DNP derivatives were developed with benzene–hexane–ethyl acetate (5:2:1) on silica gel plate. Key to A and B: (1), isolated substance; (2), glycolaldehyde; (3), glyoxal. (Reproduced with permission from Ref. 4.)*

in the reaction mixture, but this product was easily extractable with ethyl acetate, whereas glyoxal was not. It was further shown that treatment of the reaction mixture by the addition of acid or silica gel powder readily converted the product to glyoxal, which remained in the aqueous layer after extraction with ethyl acetate. It therefore seemed reasonable to assume that the glyoxal detected on TLC was an artifact of hydrolysis of the product in question in the mixture, effected by the silica gel TLC procedure.

In order to isolate the unchanged 2,4-DNP-positive product from the reaction mixture, fractionation by solvent extraction was employed. Chloroform extraction of the reaction mixture of D-glucose with _n_- or _t_-buylamine gave only noncrystalline products, but that with cyclohexylamine yielded a colorless crystalline product. The mass spectrum of this product gave $M^{+\cdot}$ 220, corresponding to the Schiff-type condensation product of two molecules of cyclohexylamine with glyoxal (Fig. 11). The significant peak at _m/z_ 177 can be explained by the fragment $CH_2{=}CH{-}CH{=}N^+{=}CH{-}CH{=}N{-}C_6H_{11}$ resulting from ring fission of the cyclohexyl group, which is frequently seen for cyclohexylamine derivatives (9). Proton NMR of this product showed signals for two cyclohexylamino groups (δ1.30-1.80 and 3.06) and a signal for the two C-H protons of the imine groups (δ7.82). This spectrum, as well as the IR spectrum, agreed well with that of the authentic specimen of glyoxal dicyclohexylimine. The yield of this compound, as estimated by densitometry on the TLC spot, was 12 or 74 mmol from 1 mol glucose, when it was reacted with 1 or 5 mol cyclohexylamine, respectively.

When the isolated or authentic glyoxal dicyclohexylimine was developed on silica gel plate in the same way as the reaction mixture, the only spot produced by 2,4-DNP was that of glyoxal, and heating the plate sprayed with sulfuric acid revealed no additional spot. It was also confirmed that treatment of its solution with acid or silica gel, carried out as was done before, readily hydrolyzed the imine to glyoxal.

These findings confirmed that glyoxal dicyclohexylimine is one of the products from the reaction of glucose with cyclohexylamine in ethanol. Although the formation of similar imines by the reaction of glucose with other alkylamines was not directly established, detection of glyoxal by silica gel TLC, shown in Fig. 9, in all of these cases seems to justify the assumption that the two-carbon diimines are always among the products of this kind of reaction, under the conditions employed.

Fragmentation Products of Sugar in the Reaction Mixture of Sugar with Amine in Aqueous System (10)

When a mixture of glucose and _t_-butylamine, 1 M each in distilled water, was heated in a boiling water bath, the colorless mixture soon became turbid and then turned brown. After

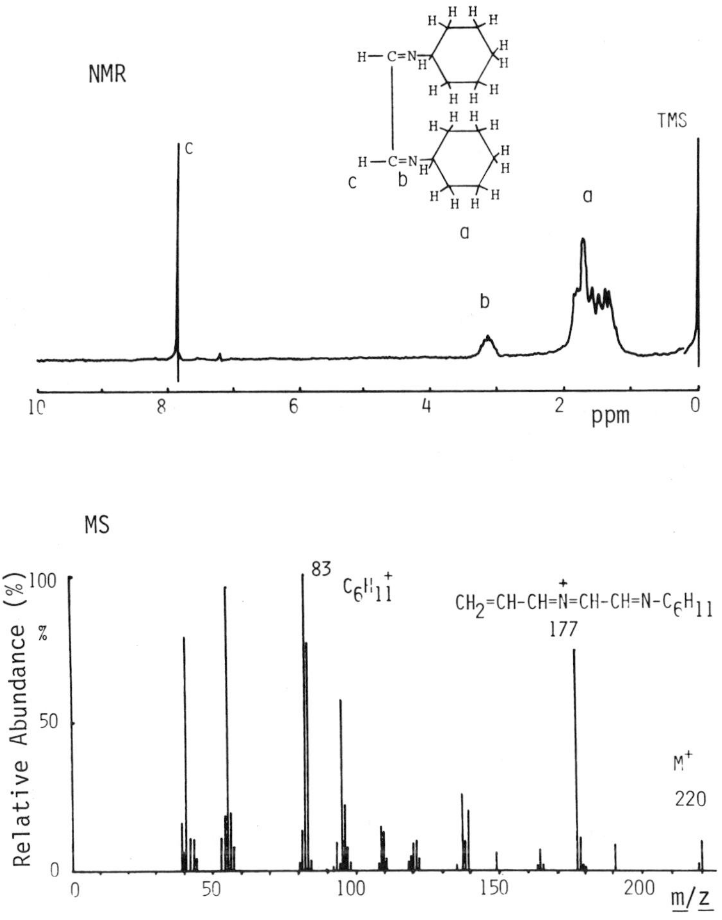

Figure 11. NMR and mass spectra of isolated glyoxal dicyclohexylimine.

about 3 min the reaction mixture was extracted with ether, and the extract was separated by silica gel TLC. As shown in Fig. 12, immediately after spraying with 2,4-DNP two spots were observed at R_f values corresponding to those of authentic glyoxal and methylglyoxal. Proton NMR of the ether-free extract in $CDCl_3$ showed signals for the protons assignable to glyoxal di-*n*-butylimine and methylglyoxal di-*n*-butylimine as indicated in Fig. 12.

It is therefore clear that fragmentation of the sugar to two- or three-carbon products occurred readily at a very early stage in the Maillard reaction in aqueous systems.

Relation of Sugar Fragmentation to Formation of Free Radicals and the Main Intermediates (4)

To elucidate the relation between formation of the identified two-carbon product, the free radical and the main intermediates of the early stages of the Maillard reaction, the progress of reaction of D-glucose with *t*-butylamine in ethanol was followed by simultaneous analyses with respects to glycosylamine, glyoxal dialkylimine, 3-deoxyglucosone, the ESR signal, and the degree of browning. The results are shown in Figs. 13 and 14. It is apparent that fragmentation occurred after formation of glycosylamine, but prior to formation of the free radical and 3-deoxyglucosone. Formation of fragmentation products from 3-deoxyglucosone has been postulated, but these results demonstrate that the fragmentation in question is quite independent of 3-deoxyglucosone formation.

Formation Pathway of the Two-carbon Fragmentation Products

Figure 15 shows the pathway proposed for the formation of the isolated two-carbon product (glyoxal diimine derivative). The initial two-carbon fragment might be a glycolaldehyde alkylimine derived from the C-C bond scission of glycosylamino compound by a reverse-aldol-type reaction. However, it will be easily oxidized to a glyoxal monoimine derivative, and will subsequently give the glyoxal diimine derivative, which is the product isolated from the reaction mixture. Glyoxal is an artifact of the isolation and identification procedures.

Formation of Four-carbon Product by the Reaction of Sugar with Amine (10)

Formation of mainly two-carbon fragmentation products of sugars at an early stage of the Maillard reaction implies the presence of residual product(s) of the fragmentation in the reaction mixture. To demonstrate this, the reaction mixture of D-glucose with *n*-butylamine in ethanol was treated at the initial stage with $NaBH_4$ and then acetylated for GLC analysis. As shown

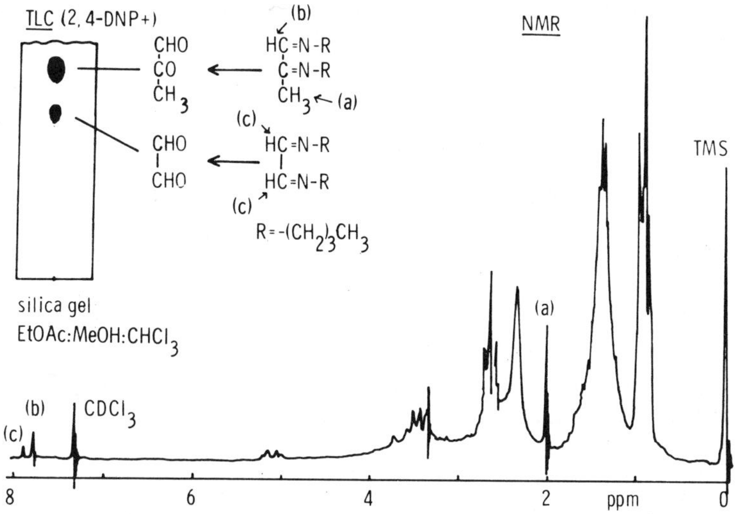

Figure 12. TLC and NMR spectra of the ether extract of the reaction mixture of glucose and n-*butylamine in aqueous system (each 1* M*), heated for 3 min in boiling water bath. TLC: silica gel, developed with ethyl acetate–methanol–chloroform (5:2:1) and sprayed with 2,4-DNP (6).*

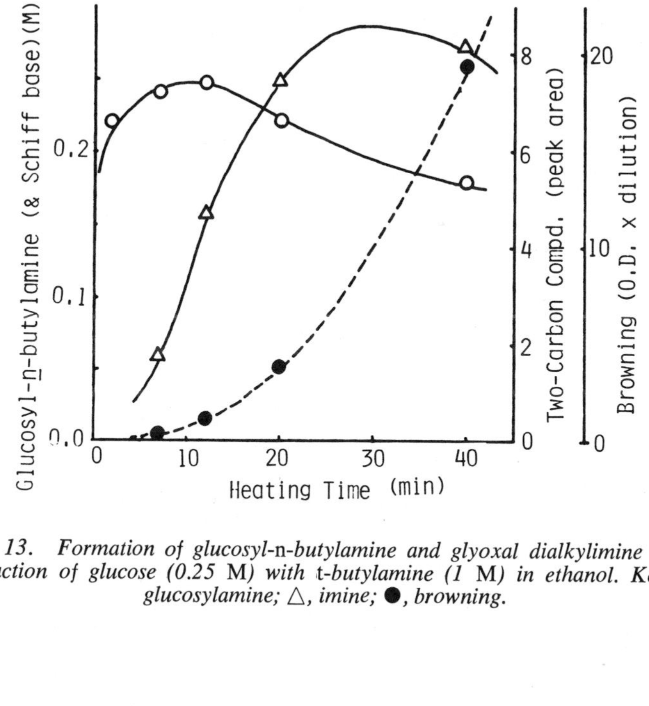

*Figure 13. Formation of glucosyl-*n-*butylamine and glyoxal dialkylimine during the reaction of glucose (0.25* M*) with* t-*butylamine (1* M*) in ethanol. Key: ○, glucosylamine; △, imine; ●, browning.*

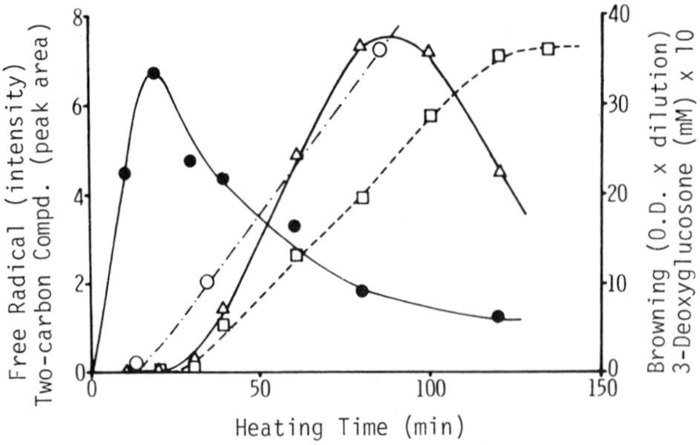

Figure 14. Formation in glyoxal dialkylimine, 3-deoxyglucosone, and the free radical from glucose with t-*butylamine (each 1* M*) in ethanol. Key: ●, glyoxal dialkylimine; ○, 3-deoxyglucosone; △, free radical; □, browning (absorbancy at 420 nm). (Reproduced with permission from Ref. 4.)*

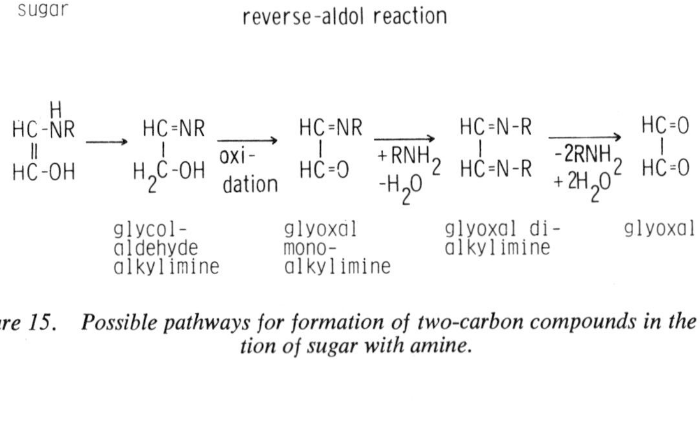

Figure 15. Possible pathways for formation of two-carbon compounds in the reaction of sugar with amine.

in Fig. 16A, the presence of a C_4 product in the reaction mixture, probably erythrose, was shown by identity of the Rt of the GLC peak with that of authentic erythritol acetate.

Changes in the amounts of the C_4 and C_2 products measured by GLC and TLC, respectively, are shown in Fig. 16B. The C_4 product was produced at a very early stage of the reaction and increased rapidly within about 10 min, and then decreased gradually. Formation and increase of the C_2 fragment product occurred at somewhat after this, but this may be partly due to lower sensitivity in measurement of the C_2 product by TLC, coloration with 2,4-DNP followed by densitometry.

Formation of the Free Radical Product from the Two-carbon Fragmentation Products (4)

Although glycolaldehyde alkylimines were proposed as initial fragmentation products, they will be easily oxidized to glyoxal monoalkylimine, and will subsequently give rise to glyoxal dialkylimines. These processes suggest two different pathways for free radical product formation in the reaction system. The first is the bimolecular ring formation from the enaminol form of the glycolaldehyde alkylimine, followed by oxidative formation of the free radical product. The second is the formation of *N,N'*-dialkylpyrazinium ions from glyoxal monoalkylimine followed by reduction to the free radical product.

The respective intermediates in these pathways (glycolaldehyde alkylimine and glyoxal monoalkylimine) may be formed by reaction of glycolaldehyde or glyoxal with amino compound. Then, radical formation and browning were therefore compared between these two systems. As shown in Fig. 17, the glycolaldehyde system showed much faster and stronger radical formation and browning than the glyoxal system; in particular, the latter gave only negligibly weak ESR signal. These results indicate that the former is by far the predominant intermediate.

Formation of Reducing Substances and Its Relation to Free Radical Formation (5)

When a reaction mixture of glucose with *t*-butylamine in ethanol was refluxed for 50 min and subjected to sampled DC polarography, there was observed a group of anodic waves consisting of one or two components, which were better resolved by DP polarography. The height of the sampled DC polarographic current at +0.10 V increased steadily with time, after an initial induction period, as shown in Fig. 18. The reducing ability of the reaction mixture was also followed by the use of Tillman's reagent (2,6-dichlorophenol-indophenol sodium), and gave essentially the same result. Interestingly, the ESR signal measured simultaneously indicated a similar increase during the reaction, although it began to decrease past 80 min. These

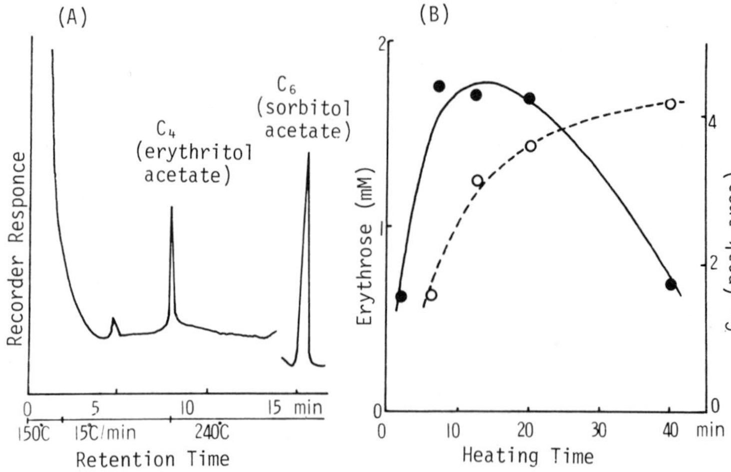

Figure 16. Formation of C_4 product by the reaction of glucose (0.5 M*) and* n*-butylamine (2* M*) in ethanol system. (A): GLC of the reduced and acetylated reaction mixture. Key to (B): ●, C_4 product; ○, glyoxal dialkylimine (6).*

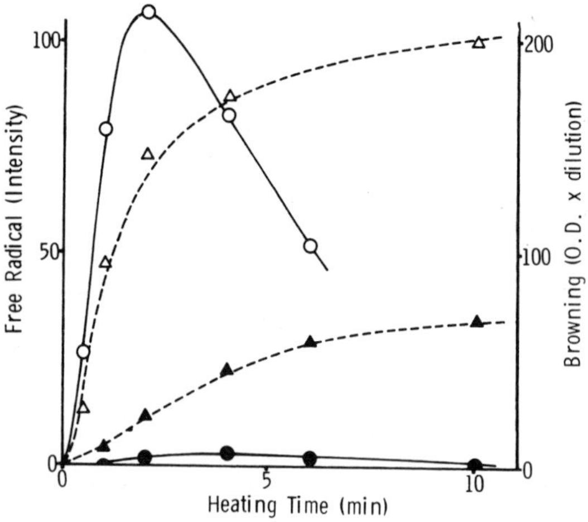

Figure 17. Free radical formation and browning in the reaction of two-carbon aldehydes with β-alanine (each 1 M*) in water. The mixtures were heated in a boiling water bath. Key for glycolaldehyde: ○, radical; △, browning. Key for glyoxal: ●, radical; ▲, browning. (Reproduced with permission from Ref. 5.)*

results suggest that free radical formation is closely connected with increase in the reducing power of the reaction mixture, i. e., a particular intermediate is formed during an induction period in the initial stage and is then converted by reduction to the radical as the reducing ability of the reaction mixture begins to increase. If this is true, addition of a suitable reductant to the reaction mixture in the induction period should produce the free radical; the following experiment was carried out to verify this idea.

Radical Precursor in the Reaction Mixture (5,6)

A portion of the reaction mixture of glucose with t-butylamine was taken at the induction period of the reaction (after about 15 min), where no detectable amounts of free radical were present and no reducing ability was observed. When an aqueous solution of ascorbic acid was added to the mixture, an intense ESR signal appeared as shown in Fig. 19, and increased gradually with time at room temperature. The hyperfine structure of the ESR spectrum agreed well with that observed in the reaction of glucose with t-butylamine. The appearance of free radical on addition of ascorbic acid to the reaction mixture was similarly observed in the system of glucose-n-butylamine in ethanol, and moreover, though to a lesser extent, in the case of glucose-β-alanine heated for 5 min in a boiling water bath.

These facts show that there exists an intermediate product in the reaction mixture, which can easily give the free radical product on mild reduction: This intermediate is tentatively termed the radical precursor.

The amount of this radical precursor in the reaction mixture of glucose with t-butylamine was estimated from intensity of the ESR signal that appeared 30 min after ascorbic acid was added to the mixture. The increase in ESR signal during the reaction is shown in Fig. 20. The radical precursor was observed at a very early stage of the reaction and increased rapidly for about 25 min, then decreased along with the development of the free radical. This pattern is very similar to that observed in the formation of glyoxal dialkylimine as shown in Fig. 18. The similarity of these changes in the precursor and in glyoxal dialkylimine seemed at first to suggest that the latter may be the precursor. However, this assumption is unacceptable since glyoxal dialkylimine is known not to give the free radical by itself.

As mentioned previously, we have found that certain product(s) could be extracted with ether from the early stages of the reaction between glucose and n-butylamine in aqueous system, and have assigned glyoxal dialkylimine as the main product. Recently, it has been shown that the ether-extractable substance is not itself a radical product but when it is dissolved in an aqueous solution of ascorbic acid or acidic ferrocyanide, an

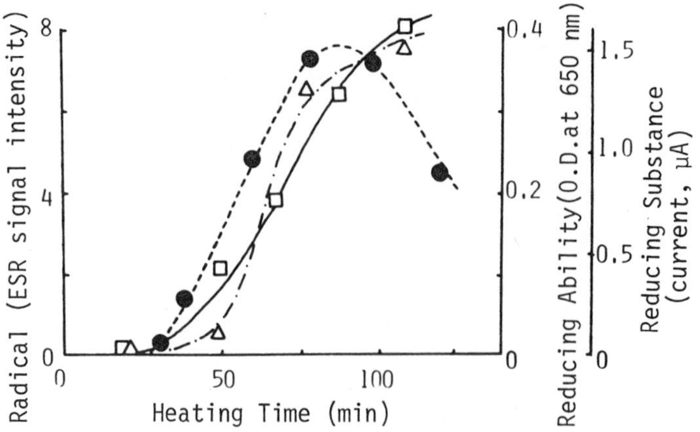

Figure 18. Formation of free radical and reducing substances in the reaction of glucose with t-*butylamine (each 1* M*) in ethanol. Key:* ●*, free radical;* □*, reducing substances (determined by polarography);* △*, reducing ability (determined with Tillman's reagent).*

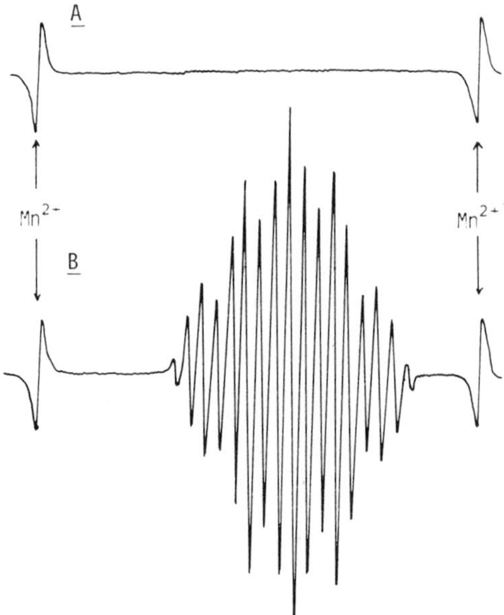

Figure 19. ESR spectrum (B) produced by the addition of ascorbic acid to the reaction mixture of glucose with t-*butylamine (each 1* M*) in ethanol refluxed for 15 min. A: reaction mixture. B: reaction mixture (0.5 mL) kept at room temperature for 15 min after the addition of ascorbic acid (0.1 g/mL, 0.05 mL). (Reproduced with permission from Ref. 5.)*

intense ESR signal was developed and its hyperfine structure agreed well with that of the glucose-*n*-butylamine reaction mixture. These results reasonably lead to the conclusion that the radical precursor termed above is glyoxal dialkylimine, and in acidic condition it will easily give glyoxal monoalkylimine derivative, and subsequent reduction will provide glycolaldehyde monoalkylimine as a very active product to the radical formation.

Reaction of N,N'-diethylpyrazinium Salt

As described above, the free radical products are considered as *N,N'*-disubstituted pyrazine cation radical products. However, formation of such pyrazine derivatives in the Maillard reaction is unknown, and little work has been done on the synthesis of *N,N'*-diethylpyrazinium salt (8), the one-electron oxidation product of the *N,N'*-diethylpyrazine cation radical. The dialkylpyrazinium compounds are known to be very unstable; they are readily decomposed to give polymer products of unknown structure. The diethyl compound is assumed to give important information on the role of free radical products in browning in the Maillard reaction. Free radical formation and browning of the synthesized *N,N'*-diethylpyrazinium salt were therefore examined in comparison with the similar reactions of ethylamine with glycolaldehyde or glucose systems. As shown in Fig. 21, the salt is very unstable and on dissolution in pH 6.0 buffer solution gave instantaneous browning as well as free radical formation. Heating enhanced the browning, and its increase was far more rapid than those observed in the sugar-amine reaction mixtures. An intense ESR signal was observed immediately after the dissolution and then decreased with rapid increase in browning with heating time.

It was thus shown that *N,N'*-disubstituted pyrazinium compounds are very likely among the active intermediates of the browning reaction.

Conclusion

Formation of novel free radical products at an early stage of the Maillard reaction was demonstrated by use of ESR spectrometry. Analyses of the hyperfine structures for various sugar-amino compound systems led to the conclusion that the radical products are *N,N'*-disubstituted pyrazine cation radicals. These new pyrazine derivatives are assumed to be formed by bimolecular condensation of a two-carbon enaminol compound involving the amino reactant residue. The presence of such a two-carbon product in an early stage reaction mixture of sugar with amine was demonstrated by isolation and identification of glyoxal dialkylimine by use of TLC, GLC, NMR, MS and IR.

Determination of product concentrations during the early stages of the Maillard reaction indicated a sequential reaction

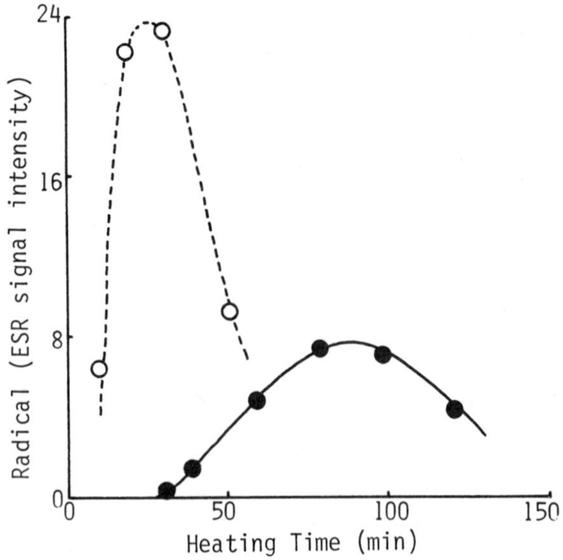

Figure 20. *Formation of radical precursor by the reaction of glucose with* t-butyl*amine (each 1* M*) in ethanol. Key:* ○, *radical precursor;* ●, *free radical. (Reproduced with permission from Ref. 5.)*

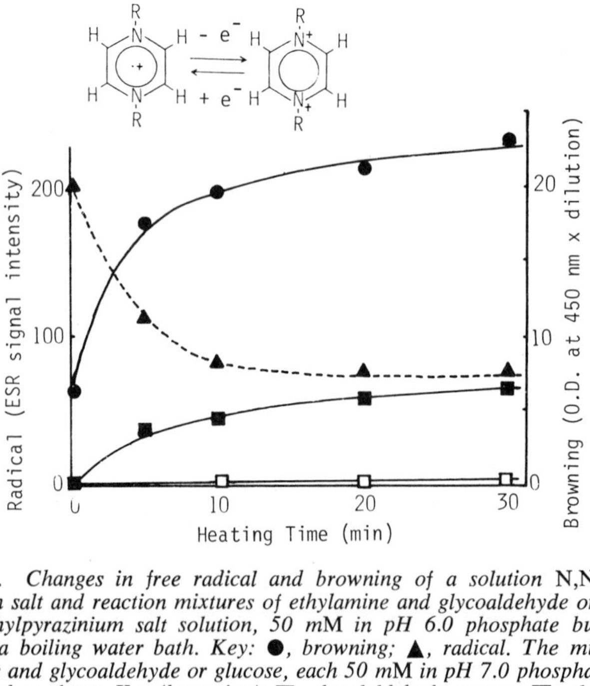

Figure 21. *Changes in free radical and browning of a solution* N,N′*-diethylpyrazinium salt and reaction mixtures of ethylamine and glycoaldehyde or glucose.* N,N′*-Diethylpyrazinium salt solution, 50 m*M *in pH 6.0 phosphate buffer, was heated in a boiling water bath. Key:* ●, *browning;* ▲, *radical. The mixtures of ethylamine and glycoaldehyde or glucose, each 50 m*M *in pH 7.0 phosphate buffer, were heated as above. Key (browning):* ■, *glycolaldehyde system;* □, *glucose system. (Reproduced with permission from Ref. 5.)*

process, _i.e._, initial formation of glucosylamino compound, followed by fragmentation to C_2 and C_4 products, and subsequently almost simultaneous formation of the reducing substance, free radical and Amadori product and, finally, formation of glucosones.

It was shown that both a _N,N_'-dialkylpyrazinium salt and a mixture of glycolaldehyde with an amino compound are highly active in free radical formation as well as in browning.

Thus, it is proposed that there exists a new pathway to browning in the Maillard reaction, involving sugar fragmentation and free radical formation at an early stage prior to the Amadori rearrangement. The summarized scheme is shown in Fig. 22. This reaction pathway seems very interesting not only from the viewpoint of basic food chemistry but also from an industrial viewpoint, because this knowledge might prove useful in countering such an early stage browning of foods and beverages.

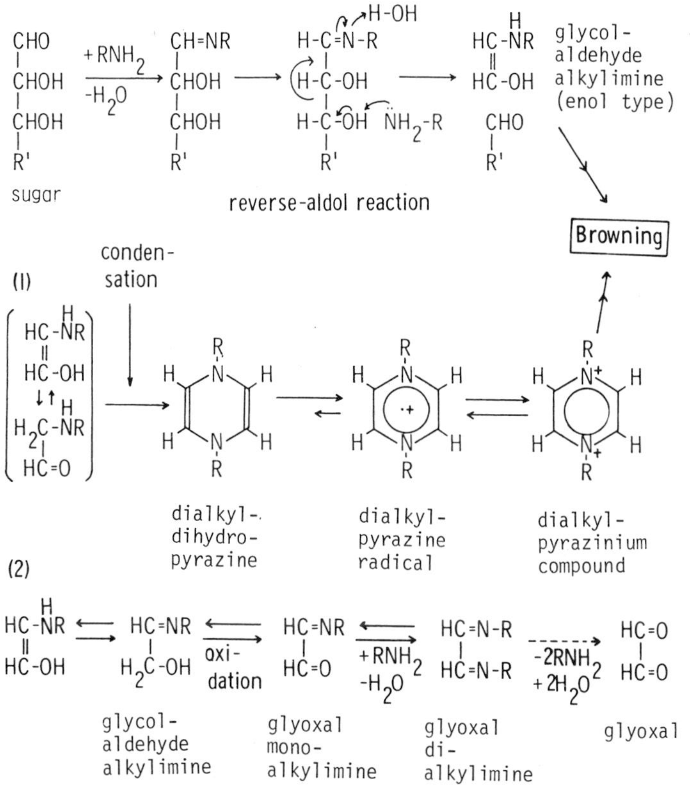

Figure 22. A possible pathway for formation of the free radical product and browning in the reaction of sugar with amino compound. (Reproduced with permission from Ref. 5.)

Literature Cited

1. Hodge, J. E. J. Agric. Food Chem. 1953, 1, 928.
2. Namiki, M.; Hayashi, T. J. Agric. Food Chem. 1947, 23, 487.
3. Hayashi, T.; Ohta, Y.; Namiki, M. J. Agric. Food Chem. 1977, 25, 1282.
4. Hayashi, T.; Namiki, M. Agric. Biol. Chem. 1980, 44, 2575.
5. Hayashi, T.; Namiki, M. Agric. Biol. Chem. 1981, 45, 933.
6. Hayashi, T.; Mase, S.; Namiki, M. in preparation.
7. Mitsuda, H.; Yasumoto, K.; Yokoyama, K. Agric. Biol. Chem. 1965, 29, 751.
8. Curphy, T. J.; Prasad, K. S. J. Org. Chem. 1972, 37, 2259.
9. Budzikiewicz, H.; Djerassi, C.; Williams, D. H. "Mass Spectrometry of Organic Compounds", Holden-Day, Inc., London, 1967; p 304.
10. Hayashi, T.; Mase, S.; Namiki, M. in preparation.

RECEIVED October 5, 1982

3

Analytical Use of Fluorescence-Producing Reactions of Lipid- and Carbohydrate-Derived Carbonyl Groups with Amine End Groups of Polyamide Powder

W. L. PORTER, E. D. BLACK, A. M. DROLET, and J. G. KAPSALIS
U.S. Army Natick Research and Development Laboratories, Natick, MA 01760

Carbonyl compounds arising from lipid oxidation and reducing carbohydrates both produce fluorescent compounds on polyamide powder coated on plastic or glass, as conventionally used in TLC. Lipid oxidative fluorescence has an excitation maximum at 356 nm and emission at 422 nm. It may be produced in either the liquid or vapor phase of wet or dry oxidizing lipid. Fluorescence from reducing sugars can be distinguished by its excitation at 362 nm and emission at 430 nm, and, unlike lipid-derived fluorescence, is greatly reduced if measured in the vapor phase. The plate fluorescence may be reproducibly measured using a solid sample holder. The fluorescence from lipid oxidation is presumed to arise from polymer-bound amino-imino-propene compounds resulting from the reaction of malondialdehyde and the known amine end groups of the polyamide. The pH dependence, excitation and emission wavelengths, and other chemical evidence strengthen this hypothesis. The nature of the fluorescent compounds produced by the reaction of reducing sugars with amine end groups on polyamide is not known. The sugar reaction is pseudo-first order with a high activation energy. Relative reaction rates are triose > pentose > hexose. Low water activity inhibits the reaction. Alkali quenches the fluorescence and acid restores it. Plate fluorescence declines at the onset of visible browning. The lipid oxidative fluorescence can be used to monitor abuse of frying oils and to evaluate antioxidants in both emulsions and dry systems. The sugar-derived fluorescence can be used in an accelerated evaluation of browning potential in foods.

The general nature of the fluorophore in oxidative lipid-amine browning is known, but that for the sugar-amine browning reaction is not. To study this, we have used a fluorescence-producing reaction that we have found between the volatiles from oxidizing lipids and polyamide powder in thin films on solid supports of plastic or glass, the conventional TLC display (1). Reducing sugars coated on the powder or in ambient solution produce a similar fluorescence, which in all cases can be ascribed to reaction with the amine end groups of the polyamide. This fluorescence of compounds formed on polyamide can be measured by solid sample fluorescence spectrophotometry. In the work reported here we have used the oxidative reaction to measure the abuse state of dry oils and relative effectiveness of antioxidants in lecithin microdispersions. The sugar-amine reaction has been used to assess relative reaction rates of various sugars with amines on a solid polymer structure like that of wool.

The oxidative fluorescence measured has many features suggestive of that resulting from the combination of malonaldehyde and primary amines. The sugar-amine fluorescence is very similar in many ways, and the precise geometry of the reacting amine substrate may give some insights as to the fluorophore.

Background

The malonaldehyde-amine reaction has been extensively studied by Tappel and co-workers (2), by Csallany et al. (3), by Privett and co-workers (4), Knook et al. (5), Buttkus and Bose (6), as well as many others. As shown in Figure 1, malonaldehyde may react with a primary amine to produce an initial ene-amine Schiff base (7), then by further condensation with another amine, an amino-imino-propene compound. Heat and acid are required in the usual model reaction. The model compounds (lysine with malonaldehyde, for example) fluoresce between 430 and 470 nm when excited between 350 and 360 nm (8). These are typical of the "aging" pigments, the organic solvent-soluble lipofuscin in oxidizing tissue.

The production of fluorescence in the sugar-amine reaction has been treated by Hodge (9, 10), Adhikari and Tappel (11), Dillard and Tappel (12), Acharya and Manning (13), Monnier and Cerami (14), and many others. In particular, sugar-derived fluorescence on solid supports has been studied for wool and amino-cellulose (15, 16) and for soluble polymers on polylysine (17), and polyvinylamine (18). In no case has the fluorophore been identified. We felt that the solid polymers so far studied do not have a precise geometry of amine group distribution. Because of the crystalline nature of polyamides such as nylon or Perlon, the distribution therein is more predictable. In addition, the availability of uniform thin layers permits solid sample spectrofluorometry. Since the fluorogens have been shown to be precursors of the brown pigments (17, 19) and are therefore an early warning of lysine loss, their definition and measurement seem important.

Characteristics of the Polyamide Substrate

The typical polyamide powder, a poly-ε-caprolactam, is formed by heating aqueous solutions of ε-caprolactam in a sealed tube at 223°C (20) (Figure 2). The product is variously called nylon-6 or Perlon (Figure 3) (21). The highly ordered crystalline structure is typically composed of linear bundles of about 100 residues with a molecular weight of about 10,000 daltons. Hydrogen bonding insures that the many amide bonds are in registry in the same plane transverse to the bundle (21). The powder is neutral in pH and appears to be zwitterionic at end-group locations. Eighty-five to 90% of the amine and carboxylate ions are available for titration, although apparent pK values are shifted toward the extremes of the pH scale from the values for the monomer ε-aminocaproic acid (20). The titration does not obey the simple Henderson-Hasselbach equation. These effects are most plausibly explained as due to the electrostatic effects of fields of closely adjacent zwitterionic polyelectrolytes. We have demonstrated the presence of available amine groups in many ways. The reagent Fluoram™, fluorescamine or 4-phenylspiro[furan-2(3*H*)-1'-phthalan]-3,3'-dione, adds to primary amines to give an intense yellow-green fluorescence, excitation at 390 nm, emission at 475 to 490 nm. Polyamide powder, but not silica or Whatman filter paper, shows this reaction intensely. We have found that reducing sugars, but not sucrose, and polyamide produce fluorescence with heat and the proper water activity as expected from the potential Maillard reaction. *p*-Benzoquinones and dehydroascorbic acid on polyamide show the characteristic intense pink color of the known amine adducts of these compounds (22). The benzoquinone reaction quantitatively removes the alkali-titrable amine groups which are protonated at neutral pH (20).

The existence of an amino-iminopropene reaction product on this solid substrate might place rather precise limits on the spacing of the available amine groups.

Materials and Methods

Fluorescence was measured for polyamide reaction products with three types of materials: 1) products of oxidation of a dry oil, 2) products of oxidation of microdisperse lecithin, and 3) sugars.

Dry Oil Oxidation Fluorescence may be measured within or over an oil or within or over an emulsion of lipid in water providing the pH is below 5.5. When plate fluorescence is to be measured, solid sample fluorescence spectrophotometry is necessary (23). A Hitachi Solid Sample Holder Attachment for Model MPF-2A Hitachi-Perkin Elmer Fluorescence Spectrophotometer (Figure 4) was

$$O = CHCH = CHOH + RNH_2 \xrightarrow{-H_2O} O = CHCH = CHNHR$$

MALONALDEHYDE	N - HEXYL-	ENAMINE
(ENOL FORM)	AMINE OR	
	α - N - ACETYL-	
	LYSINE	

$$O = CHCH = CHNHR + RNH_2 \xrightarrow{-H_2O} RN = CHCH = CHNHR$$

ENAMINE N,N' - DISUBSTITUTED
 1 - AMINO - 3 - IMINOPROPENE

WHERE
$R = (CH_2)_5CH_3$

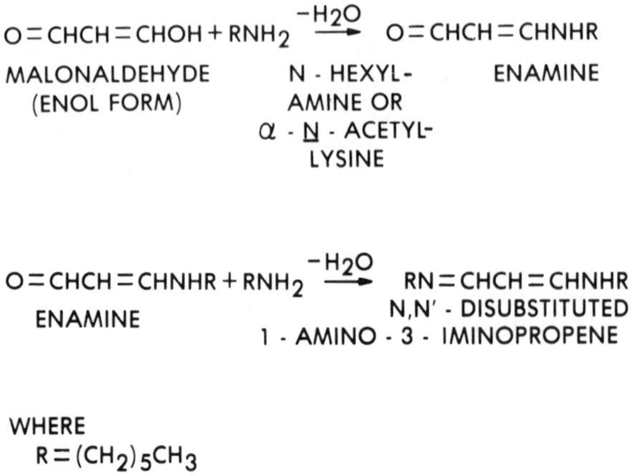

*Figure 1. Malonaldehyde–amine reactions. (Reproduced from Ref. 8. Copyright
1969, American Chemical Society.)*

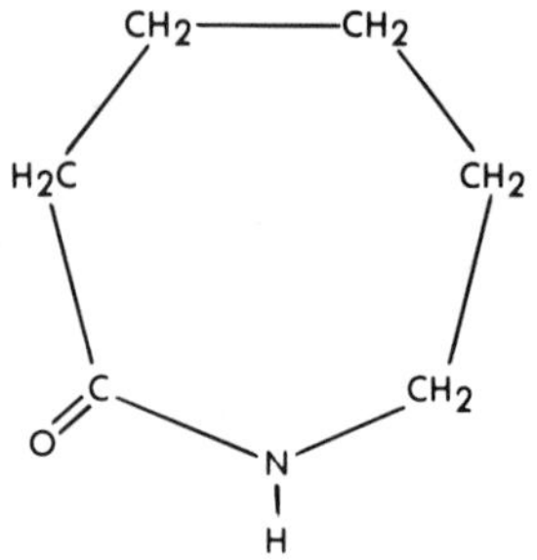

Figure 2. ε-Caprolactam.

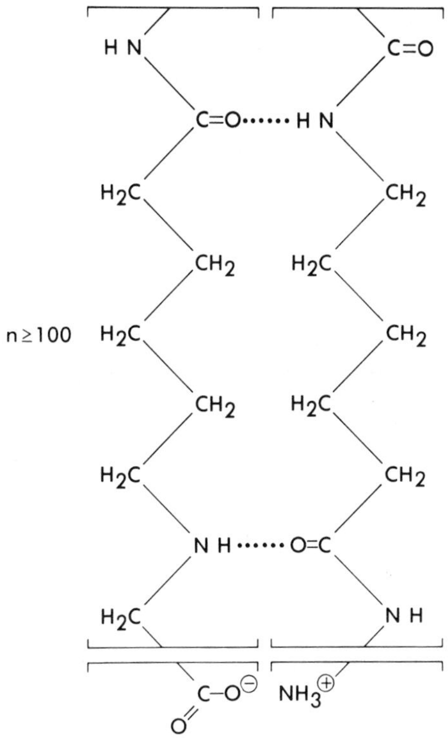

Figure 3. Polyamide-6 (nylon-6, Perlon, or polycaprolactam).

used. Unlike solution fluorescence spectrophotometry, solid sample measurement demands precise positioning. We used thin plastic spacer frames to achieve it. We used the following settings: Excitation wavelength 360 nm; emission wavelength 430 nm; excitation slit width 2 nm; emission slit width 4 nm; filter 39; sensitivity variable 2-4. Calibration using plates spotted with 30 µl quinine sulphate solution (1 µg/ml in 0.1N H_2SO_4) compared to the relative fluorescence intensity of known model compounds (8) suggested a sensitivity of 10 ng of the malonaldehyde-amine moiety.

For vapor phase measurement of dry oil oxidation, the plate assembly shown in Figure 5 is used. The lower glass or terephthalate plastic plate holds a disk of polyamide (1.2 x 0.1 cm) on which 1 µl of a suitable oil (linoleic acid, methyl linoleate, cottonseed oil) is deposited from a micropipette. A Buna "O" ring of 1.5-mm thickness is placed surrounding the spot and a similar plate with a polyamide disk is inverted over the ring forming a tight seal when clamped with "C" clamps. The plate assembly is placed in a 65°C draft oven with the receiver plate bottommost. It is disassembled, the receiver plate removed, and the fluorescence measured.

Vapor Phase over Oxidizing Lipid Emulsions For practical vapor phase measurement of malonaldehyde over oxidizing lipid emulsions, the pH must be below 5.5, since the pK of malonaldehyde is 4.65 (24). We have used a sonicated aqueous dispersion (3 g/l) of either crude soybean lecithin (35% acetone solubles, mostly triglyceride) or acetone-extracted lecithin (3-5% acetone solubles) procured from Ross and Rowe, Inc., an Archer, Daniels, Midland subsidiary. The dispersion medium is de-ionized water containing 0.013M phosphate buffer. Sonication of the crude (plastic) lecithin is done at maximum power for 20 min on a Biosonik BP-III Ultrasonic System, Bronwill Scientific. Sonication is carried out under a stream of nitrogen, with the vessel suspended in an ice-salt bath. The stripped lecithin ("Arlec"), which is in dry granular form, takes only 5 min sonication to produce a clear, opalescent microdispersion with no particles visible in an optical microscope (1250 X). The crude lecithin microdispersion, which was largely used for the work herein reported, is cloudy gray and has spherical particles sized from 0.1 to 1.2 µm, as measured on the optical microscope. The most frequently occurring particle size is about 0.1 µm. The crude lecithin (Yelkin DS) has been double-bleached in factory processing with hydrogen peroxide and benzoyl peroxide. Ultraviolet spectra of the phosphatide microdispersions prior to adding hematin showed no measurable conjugated diene absorption at 233 nm. Sonicated microdispersions are prepared either on the day of an experiment, or the day before, being stored at 40°C in the dark.

Thin-layer chromatography of the crude lecithin on silica gel using chloroform revealed two main spots, the phosphatides at the origin and the triglycerides at 0.67 (sterol esters and pigments

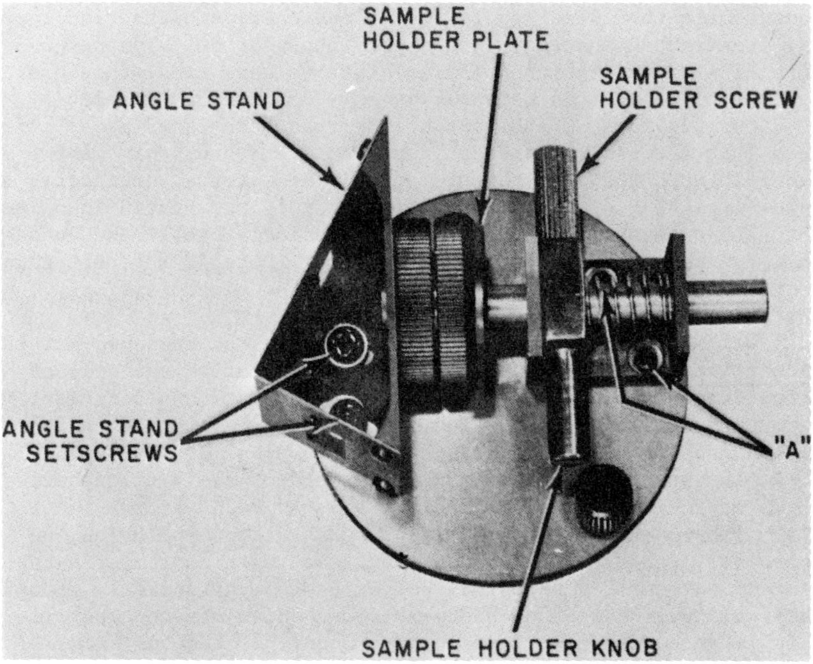

Figure 4. Solid-sample holder attachment.

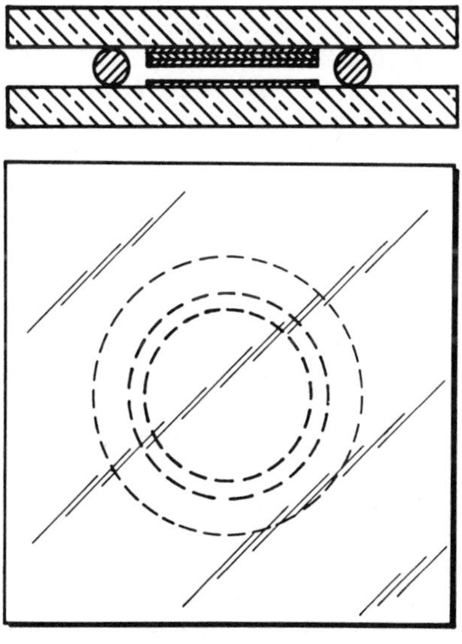

Figure 5. Plate assembly for oxidative fluorescence assay.

contaminate the latter spot). Complete acetone extraction removes all these contaminants to leave only phosphatide. In the commercial stripped lecithin (Arlec) that we have used, this process is incomplete and the material is only 95% acetone insoluble.

Our previous GC analyses of the transmethylated phosphatides from this material, commercial soybean "lecithin," have shown 54% linoleic acid and 5% linolenic acid. The material designated soybean "lecithin" of course contains not only phosphatidylcholine, but also substantial amounts of phosphatidylethanolamine, phosphatidylinositol, and phosphatidic acid. Both of the latter would confer a substantial negative charge to the microdisperse particles.

We have used both the commercial crude and stripped materials without further purification in tests of relative antioxidant effectiveness because they effectively simulate two different food exposures of phosphatide. The crude material simulates a typical emulsified oil (35% triglyceride), as in a baking shortening, while the stripped powder in microdispersion (3-5% triglyceride) is an approximate membrane model (25).

Although the crude material has 35% acetone solubles, we found only traces of compounds that react with the Emmerie-Engel reagent, a test for potential antioxidants, particularly tocopherols. In particular, no spot was found for alpha-tocopherol. This is to be expected, since the product has been peroxide treated in bleaching.

For emulsion tests, polyamide-coated terephthalate plastic plates are attached powder-side down by double-faced transparent tape strips to the undersurface of the lid of 9 x 1 cm Pyrex Petri dishes. The plates, 2 x 3 cm, are cut from standard 20 x 20 cm polyamide-terephthalate plates used for thin-layer chromatography. They are Polygram^{R} Polyamide-6 UV_{254}, procured from Macherey-Nagel and Co. through Brinkmann Instruments, Inc., Westbury, N.Y. As indicated, they contain a fluorophore (zinc silicate) activated by short-wave UV, but not active in the 360-nm range used herein. For the usual test, 25 ml of microdispersion is placed in the dish about 4 mm in depth.

Tests of relative effectiveness of antioxidants are conducted with hematin acceleration. Hematin is procured from Calbiochem-Behring Corporation. It is used as received, since the UV spectrum and TLC behavior indicate purity. In the standard preparation, 75 mg of hematin is dissolved in 75 ml deionized water with 8 drops of 10% KOH, and brought to a volume of 100 ml. Two ml of this preparation are added to 50 ml of the microdispersion at zero time to give a phosphatide hematin ratio of 100/1. Final pH is 5.5 to 5.6.

Antioxidants are tested for purity by melting point and TLC on heat-activated silica using solvent systems 1) chloroform, 2) chloroform/methanol, 19/1, 3) chloroform/methanol/acetic acid, 19/1/0.1. Antioxidants are added at 0.1% by weight of dispersed lipid. In a typical test, 50 ml of the lecithin microdispersion are treated with 0.5 ml ethanol containing 0.15 mg antioxidant. Controls are 50 ml dispersion and 0.5 ml ethanol. After addition, the dispersions are bubbled with glass-filtered air for thirty minutes. Two ml of hematin solution are added at zero time to each 50-ml portion, and 25 ml of the mixture placed in each Petri dish. The covered dishes, with attached polyamide strip, are placed in a 65°C draft oven and sampled at 30-min intervals by removing the dish to a room temperature cupboard, substituting a labeled Petri lid for the lid with polyamide strip, and measuring the accumulated fluorescent material in the solid-sample fluorescence spectrophotometer, a process taking 5 to 8 min.

Sugar-amine Browning Fluorescence development from sugar-amine browning is measured in covered 9 x 1 cm Pyrex Petri dishes, much as in the method used for oxidation of emulsions. A 2 x 3 cm polyamide plate is attached, powder-side down, to the undersurface of the Petri lid, after sugar saturation. For the latter, the plate is soaked for five minutes in the appropriate concentration (often 0.02 M) of the sample sugar, removed, edge-drained on filter paper briefly, and dried under nitrogen. The lid with the plate is placed over 25 ml of a salt solution or activated silica of appropriate water activity, and after ten min equilibration the assembly is placed in an 80°C ($\pm$2) draft oven. The very thin polyamide layers (100 μm), the short distance to the solution (0.5 cm), and the relatively large volume of solution make short equilibration adequate. The fluorescence of the cover plate is measured at 30-min intervals as described above. Concentration and type of sugar, temperature, and water activity can be varied.

Results

Characteristics of the Fluorescence Spectra from Oxidation Products Figure 6 shows a typical fluorescence spectrum produced on polyamide powder exposed to the vapors of a dry film of linoleic acid oxidizing at 70°C. Figure 7 shows a similar spectrum of polyamide exposed to the vapors of oxidizing aqueous soy lecithin microdispersions at pH 5.6 and 65°C. Figure 8 shows similar spectra obtained after polyamide exposure to vapors of oxidizing packaged potato chips and freeze-dried carrots. Polyunsaturated fatty acids, their esters and triglycerides produced similar spectra. The similarities are obvious. In all cases, excitation maximum is at 356 nm and emission at about 422 nm. The blank shown is the spectrum from a polyamide plate exposed in a similar situation but without lipid. It shows the residual of the scatter peak at 360 nm which is not removed by the 39 filter. There is also a pattern of diffraction peaks produced by the polyamide (and, indeed, by any fine powder-coated surface like silica TLC plates). The wavelengths of this diffraction pattern are excitation wavelength-dependent, unlike the situation in normal fluorescence, a given peak of which is conservative in wavelength with changes in

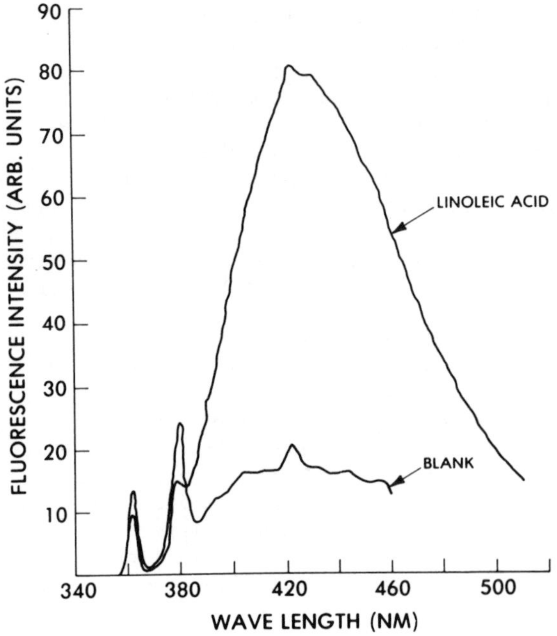

Figure 6. Fluorescence emission spectrum for polyamide on plastic facing oxidizing linoleic acid. Conditions: 69 °C for 20 h.

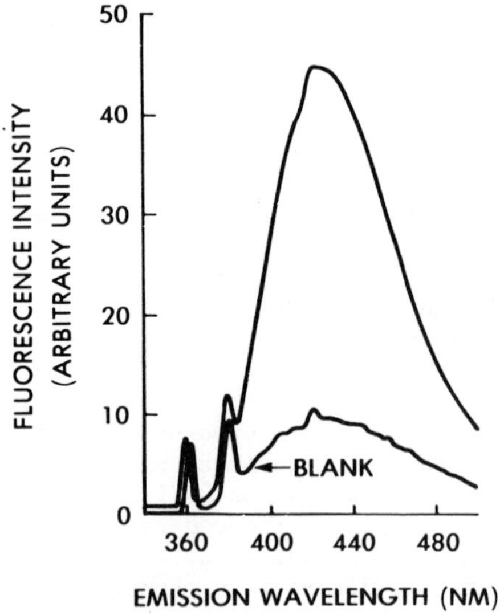

Figure 7. Oxidative polyamide fluorescence emission spectrum from oxidized soy lecithin liposomes (excitation wavelength, 360 nm).

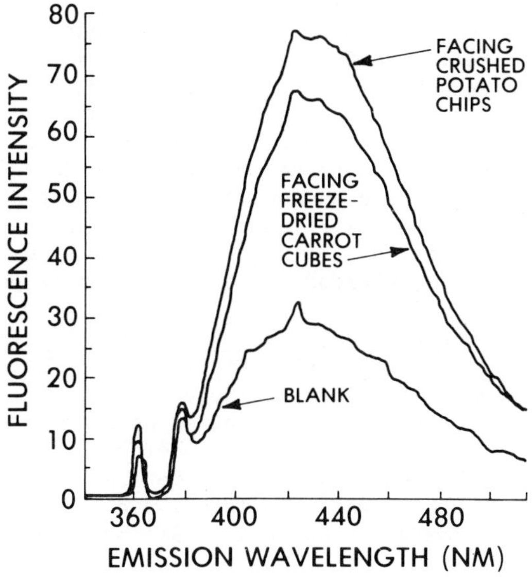

Figure 8. Oxidative polyamide fluorescence emission spectra in packaged foods.

excitation wavelength. As can be seen, the increase in oxidative fluorescence gradually obliterates this pattern.

Measurement of the Spectral Intensity For measurement, we use the residual 360 nm peak as internal reference (23). The fluorescence intensity is recorded in the standard method at 430 nm. Fluorescence index is defined by us as the ratio I_{430}/I_{360}. This is measured at every sampling period for both the blank, the control without antioxidant, and the antioxidant-treated sample. Relative mean deviation of replicate measurements (repositioning the plate) is $\pm 5\%$.

When fluorescence index in an oxidizing system is plotted against time, there is always a typical oxidative induction period whether the system does not contain antioxidant (Figure 9, linoleic acid) or has endogenous or added antioxidant (Figure 9, cottonseed oil; Figure 10, lecithin plus BHA). The antioxidant-dependent induction periods are, of course, more pronounced and much longer. Pro-oxidants like cobalt (dry system) and hematin (wet system) accelerate the reaction, and antioxidants inhibit or slow the development of fluorescence. Fatty acids and esters with only two methylene-interrupted double bonds (linoleic acid) produce fluorescence, at a lower rate than the linolenate of soybean phosphatides. In addition, the fluorescence is somewhat broadened to the greenish yellow, whereas the soy emission is pure blue.

Chemical Characteristics of the Oxidative Fluorescence-Producing Reaction and Presumptive Implication of Malonaldehyde The reaction producing measurable fluorescence on polyamide powder can be carried out in the gas phase by exposure to the volatiles from oxidizing polyunsaturated lipids, whether acids, esters, triglycerides, or phosphatides. It can be also carried out within either the oil phase or a water emulsion, and, as noted above, it can be followed in the vapor above an emulsion, providing the pH is 5.5 or below.

If polyamide is exposed to an authentic malonaldehyde source, i.e., vapors from acidified 1,1,3,3-tetraethoxypropane, intense fluorescence excited at 360 nm is produced, initially in the 430 nm range, and broadening rapidly to 440-450 nm.

Neither silica nor Whatman filter paper, so exposed, develops fluorescence. As we have indicated, polyunsaturated lipids produce the same fluorescence on polyamide, but not on silica or filter paper.

In addition, in our hands in the most sensitive fluorescence test for malonaldehyde (acidified, moist p-aminobenzoic acid, PABA; 26) authentic malonaldehyde from the tetraethoxypropane produces a characteristic intense yellowish-green fluorescence with excitation at 360 nm and emission at 450-460 nm. Oxidizing lipids of the type used in our system show the same fluorescence with PABA.

Moreover, the presumptive evidence for malonaldehyde participation is strengthened by the fact that a polyamide strip

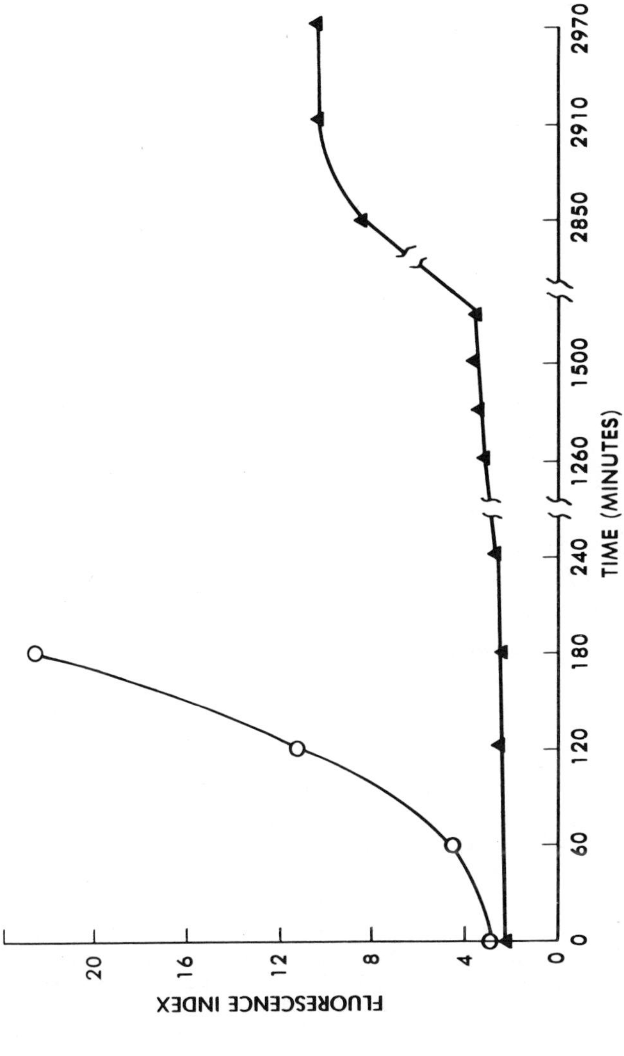

Figure 9. Oxidative polyamide fluorescence production over oils. Key: ○, linoleic acid; ▲, cottonseed oil.

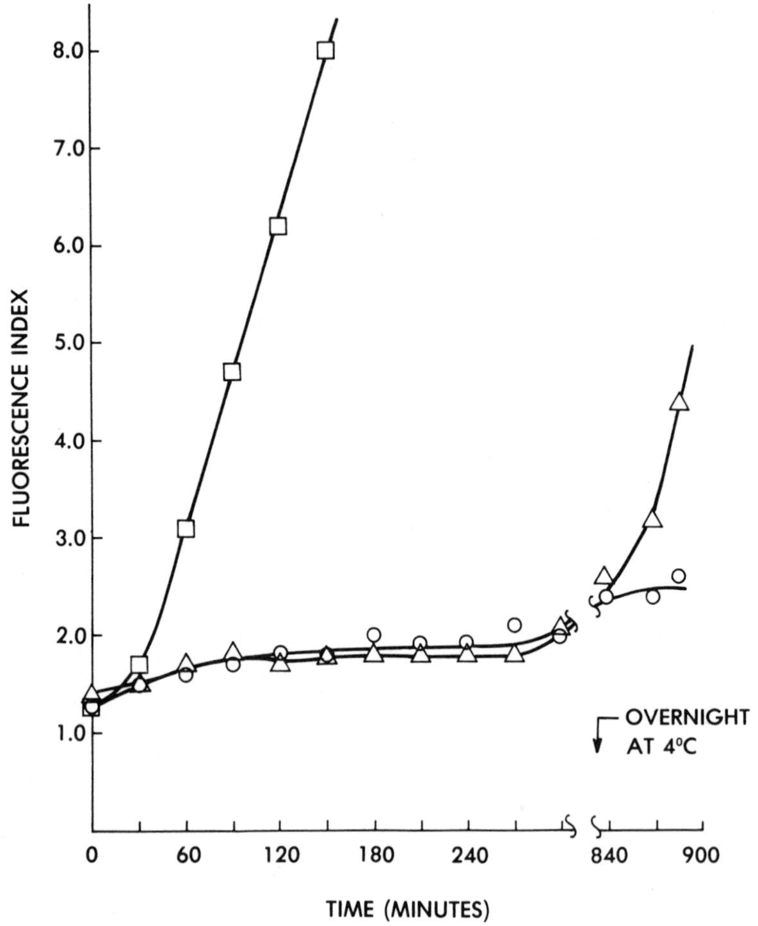

Figure 10. Antioxidant evaluation by oxidative polyamide fluorescence from soy lecithin liposomes. Key: ○, blank; □, lecithin plus hematin; △, lecithin–hematin plus 0.1% BHA.

immersed 12 h at 65°C in a microdispersion of soybean lecithin prepared in borate buffer, pH 9.1, shows strong fluorescence, whereas a plate in the vapor space above shows little to none. After adjustment of the same oxidized lecithin microdispersion to pH 2.6 and holding for three h at 65°C, the vapor space plate shows strong fluorescence. This pH-dependent shift in development of fluorescence on polyamide in the vapor space over an oxidizing emulsion seems strong evidence for malonaldehyde involvement, since the writer knows of no other carbonyl compound with such dependence that could be expected from lipid oxidation.

Finally, the polyamide oxidative fluorescence is unmistakably quenched when the fluorescing plate is treated with sodium methoxide or 1N sodium hydroxide, dried, and viewed under long-wave UV light, a test developed by Malshet, et al. (27). Methanol or water alone has no effect. Strong acid restores the fluorescence. We were not able to quantitate this quenching and restoration, because of the risk of damage to the solid sample holder.

Characteristics of the Fluorescence Spectra from Sugar-amine Browning Figure 11 shows a typical fluorescence spectrum produced on a polyamide plate coated with a 0.2 M solution of the reducing sugar fructose, dried, and heated at 80°C at water activity 1.0. Excitation maximum is at 362 nm and emission maximum is at 430 nm. (Adhikari and Tappel (11) found excitation and emission maxima of 350 nm and 430 nm in browned glucose/glycine; Monnier and Cerami (14) found 360 nm and 430 nm in glucose-browned lens protein.) There is a close similarity to the spectra from oxidizing lipid (Figures 6, 7, 8). Sugar-amine fluorescence is quantitated in the same manner as that from oxidizing lipid (see above). In each case the blank fluorescence index is subtracted from the sample index to yield change in fluorescence index (ΔFI). As heating progresses, the spectrum broadens and the peak shifts toward 440-450 nm in a manner similar to that from oxidizing lipids. The onset of pronounced visual browning (tan color) is accompanied by a pronounced decrease in fluorescence intensity. The form of the initial spectrum is similar for trioses, pentoses, and hexoses, although the spectra of the more rapidly reacting trioses and pentoses broaden to longer wavelength much earlier at a given temperature and water activity.

Reaction Kinetics of Sugar-amine Fluorescence Figure 12 shows a typical plot of fluorescence index against time for polyamide coated with 0.02 M glyceraldehyde, xylose, and arabinose and held at water activity 1.0 and 80°C. The kinetics of fluorescence development is, of course, only an apparent one, since concentration is undefined. There is clearly in the early stages, however, a quasi-first order decrease from an initially nearly linear, rapid reaction with no induction period, unlike lipid oxidation. The reaction rather rapidly approaches a limit, presumably by saturating the surface amine groups. The maximum attained change

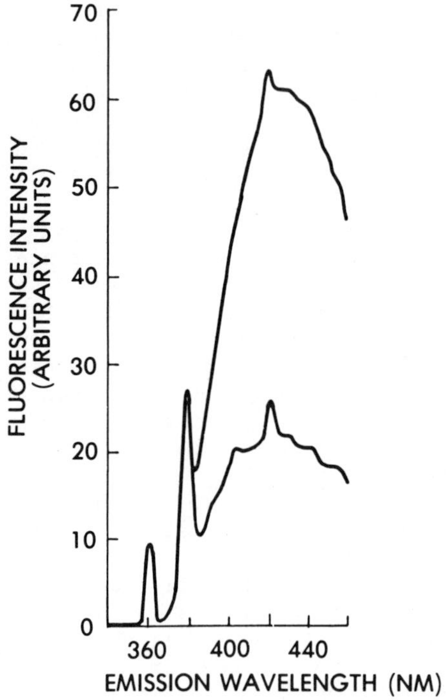

Figure 11. Polyamide fluorescence emission spectrum from fructose-amine browning (excitation wavelength, 360 nm). Condition: Fructose 0.2 M, *80 °C.*

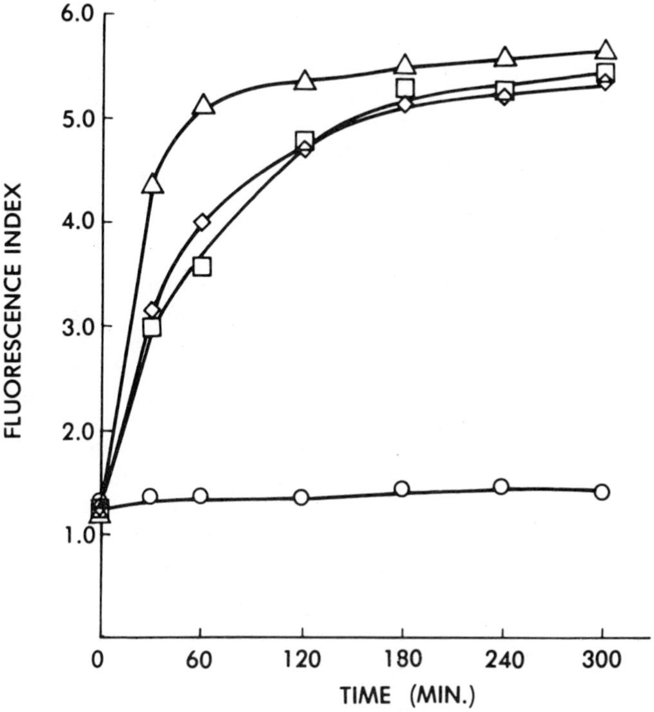

Figure 12. Polyamide fluorescence production in sugar–amine browning. Conditions: Water activity 1.0, 80 °C, sugar, 0.02 M *and dipped and dried plate. Key:* ○*, blank;* △*, glyceraldehyde;* ◇*, xylose; and* □*, arabinose.*

in fluorescence index (ΔFI_{max}) for sugar-amine browning is much less than that for lipid oxidation. ΔFI_{max} for sugar-amine browning is rarely more than 5, whereas it can be as much as 15 in lipid oxidation. In addition, in sugar-amine browning, as mentioned above, at the time of intense visual browning, fluorescence index begins a pronounced decline; this is rarely seen in lipid oxidation.

Effect of Sugar Type on Rate of Fluorescence Development

Because of the quasi-linear initial rate, relative initial rate is defined in this study as FI/hour expressed in arbitrary units. This is obtained from the 30-min observation. Maximum ΔFI (ΔFI_{max}) is the maximum excess over blank at any time in the experimental run for a given sugar.

We have studied the effect of sugar type and water activity on rate of fluorescence development. As expected, rate increases with surface concentration on the polyamide, but since this parameter is not well defined in our system, we have not studied its effects in detail. The effect of temperature has also not been studied.

Table I shows relative rates of fluorescence development for a triose, pentoses, and hexoses at 80°C, at water activity 1.0. The relative rate order is triose>pentose>hexose. Nonreducing sugars (e.g. sucrose, trehalose, raffinose) do not develop fluorescence without a long lag to permit hydrolysis. The expected large rate difference--fructose>>glucose--occasioned by the greater percentage of acyclic carbonyl group in fructose (0.7 versus 0.002; 28) was not found. It may be that the restricted mobility of the amine group in the crystalline polyamide favors the aldose over the less electrophilic ketose sugar (29).

Table I
Rate of Polyamide Fluorescence Development with Reducing Sugars over Water*

Sugar	Initial Rate ΔFI/60 Min	ΔFI 240 Min
Glyceraldehyde	6.4	4.5
Arabinose	3.6	4.0
Ribose	5.0	4.0
Xylose	4.2, 3.3	4.0, 3.6
Dextrose	1.7, 1.6	3.4, -
Fructose	2.0, 2.0	3.2, 2.4
Mannose	2.8	4.0

*T = 80°C ±2. Plate immersed in 0.02 M sugar and dried, suspended over water (water activity = 1.0).

Table II shows similar relative rates and ΔFI_{max} measured over saturated aqueous NaCl (water activity 0.76). Limited studies were also made over saturated KI solutions (water activity 0.60). Rate was somewhat greater at the higher water activity, but there was not sufficient difference to warrant further comparative study. The same relative rate order for sugar type occurs over NaCl as over water. Table III reports similar studies over activated silica. The experimental scheme does not permit rigid water activity control at near zero activity, but it is clear that rates are much lower and that the same relative rate order for pentose and hexose sugars is preserved, although the differences are much less pronounced. The triose is relatively less reactive here.

Table II
Rate of Polyamide Fluorescence Development with Reducing Sugars over Saturated NaCl*

Sugar	Initial Rate ΔFI/60 Min	ΔFI 240 Min
Glyceraldehyde	4.8	3.2
Arabinose	3.6	3.4
Ribose	4.5	3.9
Xylose	3.6, 4.1	- , 3.9
Dextrose	2.8, 2.3, 1.7	4.2 -
Fructose	1.9, 1.7	2.5, -
Mannose	2.4	4.4

*T = 80°C ±2. Plate immersed in 0.02 M sugar and dried, suspended over saturated NaCl (water activity 0.76).

Table III
Rate of Polyamide Fluorescence Development with Reducing Sugars over Activated Silica*

Sugar	Initial Rate ΔFI/60 Min	ΔFI 240 Min
Glyceraldehyde	1.4	2.7
Arabinose	1.9	2.1
Ribose	1.7	2.1
Xylose	1.7, 1.6	2.3, 1.4
Dextrose	1.3, 1.3	2.0, 2.4
Fructose	1.3	2.0
Mannose	1.1	2.0

*T = 80°C ±2. Plate immersed in 0.02 M sugar and dried, suspended over activated silica (water activity~0.0).

Table IV summarizes the effects of water activity. In this system, there does not appear to be the usual biphasic dependence on water activity. Fluorescence production rises monotonically with water activity to saturation. In a freeze-dried immobilized system, Eichner (30) has found a similar dependence for initial rate of reducing power development.

Table IV
Effect of Sugar Type and Water Activity

Sugar	Means of Initial Rates (ΔFI/60 Min) Over Activated Silica	Over Sat. NaCl	Over Water
Triose	1.4*	4.8*	6.4*
Pentose	1.77 $\pm$ 0.12**	4.00 $\pm$ 0.46**	4.13 $\pm$ 0.76**
Hexose	1.23 $\pm$ 0.12**	2.17 $\pm$ 0.32****	2.17 $\pm$ 0.57***

*N = 1
**N = 4
***N = 5
****N = 6

Chemical Characteristics of the Sugar-amine Fluorescence and Implications as to Fluorophore Nature The near, but not complete, similarity of the polyamide fluorescence spectra from oxidation and from sugar-amine browning (excitation and emission wavelengths, general shape of the curve) prompts speculation as to the similarity of the fluorophores, whether in solution (11) or on polyamide. For both oxidative and sugar-amine browning, Adhikari and Tappel found partial fluorescence quenching in alkali and reversible restoration on neutralization. We have found the same phenomena for both the oxidative and sugar-amine polyamide fluorescence spectra. Adhikari and Tappel suggest an imine-ene-ol structure to accommodate their sugar-amine data, which also include a europium ion chelation effect. The structure proposed by Adhikari and Tappel is a tautomer of the initial Schiff base of the malonaldehyde with an amino acid, a structure which is not fluorescent (Figure 13, II), and which is hydrolyzed easily by acid (7); only the amino-iminopropene product fluoresces (Figure 13, I).

(I) HO-C(=O)-CH_2-N=CH-CH=C-NH

(II) HO-C(=O)-CH_2-N=CH-CH=C(-OH)

Figure 13. Possible structures of the fluorophore in sugar–amine browning. (Reproduced from Ref. 11. Copyright 1972, Institute of Food Technologists.)

We have found, as did Burton et al. (16) with wool, and Hannan and Lea (17) with polylysine, that the fluorescent material is firmly bound to the polyamide and cannot be removed by any usual combination of organic or organic-aqueous solvents, including boiling 1N HCl. In addition, rather strong oxidizing agents such as hydrogen peroxide (30%) and 0.1% $FeCl_3$ do not irreversibly quench the fluorescence after washing of the surface. These data are not consistent with the presence of a simple low-molecular-weight water-soluble compound.

Antioxidant Effect of Sugar-amine Products on Polyamide We have also found that the fluorescent polyamide contains acid-ferricyanide-reducing power (31) and antioxidant activity for linoleic acid (30). The latter was tested by a simple variant of the procedure outlined above, depositing 1 μl of linoleic acid on a plate previously made fluorescent by brief heating with a 0.2 M xylose coating. Oxidative fluorescence development in a fresh polyamide cover plate is greatly delayed in contrast with a similar fresh plate facing linoleic acid deposited on either Whatman filter paper or a fresh, unbrowned plate. In this case, of course, the fluorescence on the receiver plate arises from malonaldehyde from the oxidizing linoleic acid.

Like the fluorescence, the reducing power and antioxidant activity are not decreased by washing the plate with the usual organic or organic-aqueous solvent combinations. The substance responsible is bound to the polyamide surface.

Possible Chemical Nature of the Sugar-amine Fluorophore In sugar-amine browning, the relative amount of the fluorophore is usually low. Adhikari and Tappel found 130 μmoles per mole of glucose based on an assumption of an amino-imino-propene moiety (11). The relative amount of acid reducing power is usually much higher (28). But of course it cannot be assumed, no matter how simple the system, that the polymer-bound fluorophore and reducing power reside in the same molecule.

The polyamide fluorophore has characteristic properties. The bound nature of the fluorophore and its lack of ready hydrolysis, the quenching by alkali and its total reversal by acid, the strong basicity thus suggested for the fluorophore (sodium carbonate does not quench as greatly), the lack of quenching by oxidizing agents, and the very marked similarity of the excitation and emission spectra from lipid oxidation and sugar-amine browning, are consistent with a possible fluorophore of a carbohydrate-substituted amino-iminopropene type (Figure 13A), a vinylog of an amidinium compound similar to that for the malonaldehyde-adduct in oxidative browning. Such compounds are strongly basic (32). This compound would exist at the usual pH range as the symmetric and resonant cation, alkali titration to the free base destroying much of the resonance. Such a compound is clearly a crosslinking entity, to the extent of its concentration. If it were a part of a small pool of a rapidly reacting obligate intermediate, it might be important in crosslinking.

In this regard, Baltes, et al. (33), using p-chloroaniline as a slow-reacting primary amine with D-glucose, have detected in the refractory melanoidin fraction largely fragments derived from diamine substitution on a three-carbon fragment, the quinoline analogs of an amino-iminopropene skeleton.

The total lack of permanent quenching by oxidizing agents dictates that the fluorophore is not part of a resonant structure that includes the reducing moiety.

Applications of Oxidative and Maillard Polyamide Fluorescence

We have used the polyamide fluorescence reaction in accelerated shelf life tests of fats and oils, either with the neat oils or with plates pre-dipped in cobalt chloride for acceleration.

We described (34) use of the test in abused frying oils. We have patented an application to in-package detection of oxidative quality loss. We have studied in this manner the relative effectiveness of antioxidants in a thin layer of dry oil.

We reported (35) on a fluorescence test of the relative effectiveness of antioxidants in oxidizing soy lecithin liposomes (sonicated microdispersions).

We have reported here on comparison of the sugar-amine browning fluorescence with oxidative lipid-derived fluorescence to bring out the similarities and differences of the two emissions. The sugar-amine browning fluorescence can also be studied within an antioxidant-containing medium to assess browning potential of the system.

Summary and Conclusions

Polyamide microcrystalline powders form measurable polymer-bound fluorescent reaction products with malonaldehyde from oxidizing lipids and with reducing sugars. The compounds form on the terminal amine groups which appear to exist in zwitterionic fields with carboxylate anions, as revealed by titration with acid, alkali, or benzoquinones.

Similarities in the spectrofluorometric and chemical behavior of the malonaldehyde and Maillard adducts and the constraints of the rather precise geometry of the available amine groups in microcrystalline polyamide are consistent with a possible carbohydrate-substituted amino-iminopropene derivative for the sugar-amine fluorophore. The malonaldehyde product with amino acids and amines has been demonstrated by others to have an amino-iminopropene moiety.

Analytical applications of the malonaldehyde and Maillard fluorescent product formation to assessment of abuse status of oxidizing oils, relative effectiveness of antioxidants, and sugar-amine browning potential are available.

Literature Cited

1. Porter, W. L.; Black, E. D.; Drolet, A. Abstract No. 2, Div. Agricultural and Food Chem. 182d Ann. Meet. Am. Chem. Soc., New York, N.Y., August, 1981.

2. Trombly, R.; Tappel, A. L. Lipids 1975, 10, 441.
3. Csallany, A. S.; Manwaring, J. D. Abstract No. 227, J. Am. Oil Chemists Soc. 1980, 57, 115A.
4. Shimasaki, H.; Veta, N.; Privett, O. S. Lipids 1980, 15, 236.
5. Knook, D. L.; van Bezooijen, C. F. A.; Sleyster, E. C. Abs. No. 311, J. Am. Oil Chemists Soc. 1980, 57, 115A.
6. Buttkus, H.; Bose, R. J. J. Am. Oil Chemists Soc. 1972, 49, 440.
7. Crawford, D. L.; Yu, T. C.; Sinnhuber, R. O. J. Agric. Food Chem. 1966, 14, 182.
8. Chio, K. S.; Tappel, A. L. Biochemistry 1969, 8, 2821, 2827.
9. Hodge, J. E. "Food Chemistry"; Fennema, O. R., Ed.; Marcel Dekker, Inc. N.Y., 1976; p 41.
10. Hodge, J. E. J. Agric. Food Chem. 1953, 1, 928.
11. Adhikari, H. R.; Tappel, A. L. J. Food Sci. 1973, 38, 486.
12. Dillard, C. J.; Tappel, A. L. J. Agric. Food Chem. 1976, 24, 74.
13. Acharya, A. S.; Manning, J. M. J. Biol. Chem. 1980, 255, 7218.
14. Monnier, V. M.; Cerami, A. Science 1981, 211, 491.
15. Burton, H. S.; McWeeny, D. J.; Biltcliffe, D. O. J. Sci. Food Agric. 1963, 14, 911.
16. Burton, H. S.; McWeeny, D. J.; Pandhi, P. N.; Biltcliffe, D. O. Nature 1962, 196, 948.
17. Hannan, R. S.; Lea, C. H. Biochim.Biophys. Acta. 1952, 9, 293.
18. Micheel, F.; Büning, R. Chem. Ber. 1957, 90, 1606.
19. Overby, L. R.; Frost, D. V. J. Nutr. 1952, 46, 539.
20. Mathieson, A. R.; Whewell, C. S.; Williams, P. F. J. Appl. Polymer Sci. 1964, 8, 2009.
21. Kohan, M. I. "Nylon Plastics"; John Wiley & Sons, Inc.: New York, N.Y., 1973.
22. Thompson, R. H. "Naturally Occurring Quinones" 2d Edition; Academic Press, Inc., New York, N.Y., 1971.
23. Sawicki, E.; Stanley, T. W.; Johnson, H. Microchem. J. 1964, 8, 257.
24. Kwon, T.; Watts, B. M. J. Food Sci. 1964, 29, 294.
25. Kaschnitz, R. M.; Hatefi, Y. Arch. Biochem. Biophys. 1975, 171, 292.
26. Sawicki, E.; Stanley, T. W.; Johnson, H. Anal. Chem. 1963, 35, 199.
27. Malshet, V. G.; Tappel, A. L.; Burns, V. M. Lipids 1974, 9, 330.
28. Hayward, L. D.; Angyal, S. J. Carbohydr. Res. 1977, 53, 13.
29. Bunn, H. F.; Higgins, P. J. Science 1981, 213, 222.
30. Eichner, K. "Autoxidation in Food and Biological Systems"; Simic, M.; Karel, M., Eds.: Plenum Press: New York, N.Y.,1980; p 367.
31. Crowe, L. K.; Jenness, R.; Coulter, S. T. J. Dairy Sci. 1948, 31, 595.
32. Euler, von H.; Eistert, B. "Chemie und Biochemie der Reduktone und Reduktonate"; F. Enke Verlag: Stuttgart, Germany, 1957.

33. Baltes, W.; Franke, K.; Hörtig, W.; Otto, R; Lessig, U. "Maillard Reactions in Food"; Eriksson, C., Ed.: Pergamon Press: New York, N.Y., 1981; p 137.
34. Porter, W. L.; Wetherby, A. M.; Kapsalis, J. G. Abstract No. 497, J. Am. Oil Chemists Soc. 1980, 57, 115A.
35. Porter, W. L. U.S. Patent 4,253,848, Mar. 3, 1981.

RECEIVED November 4, 1982

4

Strecker Degradation Products from (1-^{13}C)-D-Glucose and Glycine

TOMAS NYHAMMAR, KJELL OLSSON, and PER-ÅKE PERNEMALM[1]
Swedish University of Agricultural Sciences, Department of Chemistry and Molecular Biology, S-750 07 Uppsala, Sweden

The effect of pH, reactant ratio and reaction time on the yields of 5-methylpyrrole-2-carboxaldehyde (4), methyl pyrrol-2-yl ketone (5), 6-methyl-3-pyridinol (6) and 2-methyl-3-pyridinol (7) in the reaction of D-glucose with glycine at 100°C in aqueous solution has been studied. The reaction of the potential intermediate 3-deoxy-D-*erythro*-hexosulose with glycine was investigated in a similar way. The reaction of (1-^{13}C)-D-glucose with glycine yielded (^{13}CHO)-4, (5-^{13}C)-5, (2-^{13}C)-6 and (6-^{13}C)-7. The results support previously proposed routes to 4 and 6 but disqualify those to 5 and 7. Based on the smooth formation of 7 from 2-deoxy-D-*arabino*-hexose (9) and ammonia, a new route to 5 and 7 through an enamine derived from 9 is proposed. This route was also supported by the formation of 2,3-dideoxy-D-*erythro*-hexose from 3-deoxy-D-*ribo*-hexose and glycine.

In the Strecker degradation, an α-amino acid suffers oxidative decarboxylation to an aldehyde. The amino group is transferred to a carbonyl compound, which is thereby reduced. It has been generally accepted (1, 2) that the carbonyl compound may be an α-dicarbonyl compound (Figure 1a) or its vinylogues but not, *e.g.*, an acyloin (Figure 1b). Compounds 1-3 or their enol forms are important intermediates in the dehydration of hexoses (3). In Maillard reactions of hexoses, the Strecker degradation may therefore be induced by any of 1-3 or by an α-dicarbonyl compound formed through further dehydration of 1-3. Cyclization of the resulting amino sugar or some of its dehydration products yields heterocyclic compounds, formally derived from a deoxy sugar and ammonia.

[1] Current address: Pharmacia Fine Chemicals AB, P.O. Box 175, S-751 04 Uppsala, Sweden

0097-6156/83/0215-0071$06.00/0

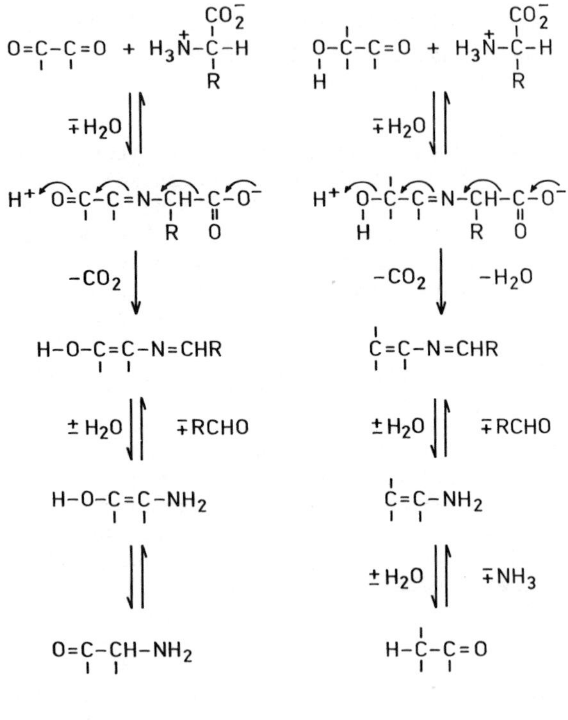

Figure 1. Strecker degradation induced by (a) an α-dicarbonyl compound and (b) an acyloin.

Thus, in 1977, Shaw and Berry (4) obtained the 2-acylpyrroles 4 and 5 by refluxing a slightly acidic aqueous solution of D-fructose and DL-alanine. As shown in Figure 2, both products were believed to form through 1 and to contain C-6 of the fructose in the methyl group. In 1978, the pyrroles 4 and 5 as well as the 3-pyridinols 6 and 7 were obtained under similar conditions from D-glucose and glycine at our department (5). As shown in Figure 3, compounds 4 and 6 were believed to form through 1 and to contain C-6 of the glucose in the methyl group. However, 5 and 7 were thought to form via 2 (or 3) and to contain C-1 of the glucose in the methyl group. The proposed routes to 4 differ only as to whether 1 is further dehydrated or not before the Strecker degradation. The routes to 5 are entirely different, however. The glucose-glycine reaction has therefore been reinvestigated using glucose labelled with 90 % ^{13}C at C-1. The behavior of a few potential intermediates under similar reaction conditions has also been studied.

The glucose-glycine reaction

Experiments with unlabelled glucose To make maximum use of the expensive labelled compound, we first studied the influence of the reaction conditions on the yields of 4-7, using ordinary glucose. Samples were withdrawn from each reaction mixture at suitable time intervals and processed according to the scheme in Figure 4. The resulting Extract 1 contained all of the relatively lipophilic pyrroles (4 and 5) and some 5-(hydroxymethyl)-2-furaldehyde (8), while Extract 3 contained all of the weakly basic pyridinols (6 and 7). The yields of 4-7 were determined by GC analysis of Extracts 1 and 3 on 3 % NPGS (Neopentyl Glycol Succinate).

Preliminary experiments showed that a considerable excess of glycine was required to minimize the formation of 8 and other caramellization products. Aqueous solutions 0.17 M in glucose and 3.4 M in glycine were then brought to a pH ranging from 2.0 to 8.0 and refluxed for 72 h. The yields of 4-7 at initial pH 2.0, 3.0, and 6.0 are shown in Figure 5. Whereas the yields of the pyridinols increased steadily with time, those of the pyrroles reached a maximum and then declined. This was not due to conversion of pyrroles into pyridinols, as shown by similar experiments with 4 or 5 instead of glucose. The predominance of 4 and 6 at pH 2.0, and of 5 and 7 at pH 6.0, seemed to support the scheme in Figure 3, since the importance of 2 and 3 relative to 1 as intermediates in the dehydration of hexoses is considered to increase with pH (3).

Experiments with labelled glucose The experiment at initial pH 3.0 was repeated with 540 mg of (1-^{13}C)-D-glucose and interrupted after refluxing for 13 h. Those at initial pH 2.0 and 6.0 were repeated on a three times smaller scale and interrupted

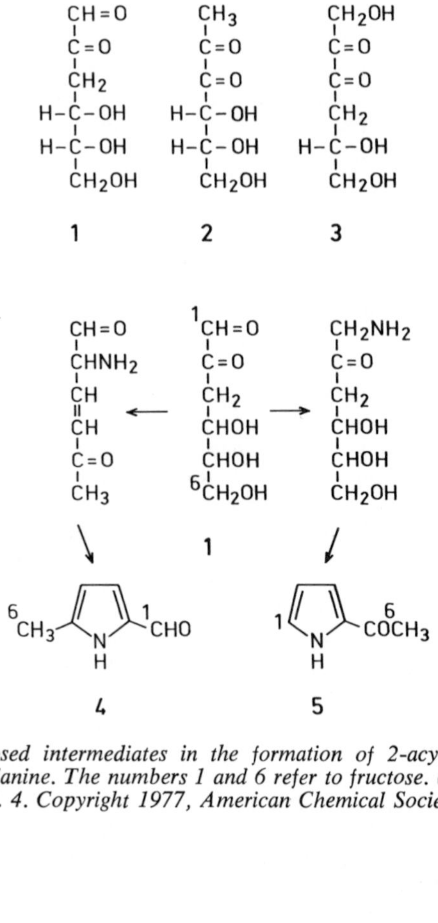

Figure 2. Proposed intermediates in the formation of 2-acylpyrroles from D-fructose and DL-alanine. The numbers 1 and 6 refer to fructose. (Reproduced from Ref. 4. Copyright 1977, American Chemical Society.)

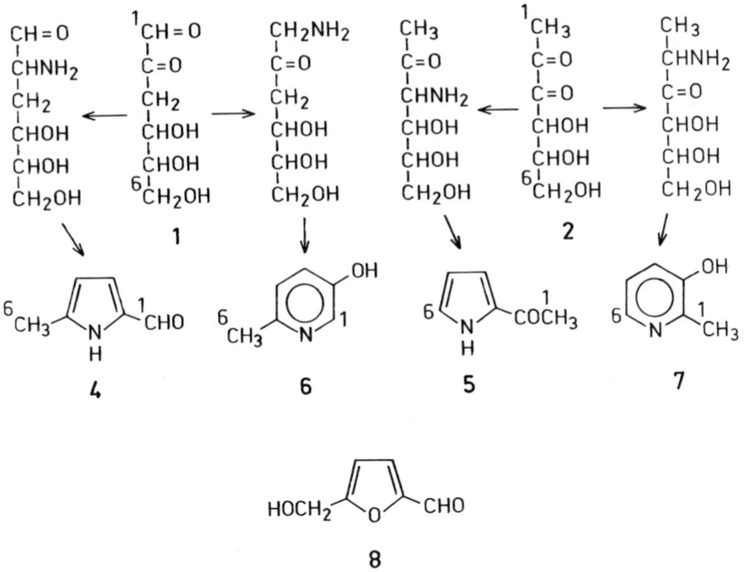

Figure 3. *Proposed intermediates in the formation of 2-acylpyrroles and 3-pyridinols from D-glucose and glycine (5). (Numbers 1 and 6 within the structures refer to glucose.)*

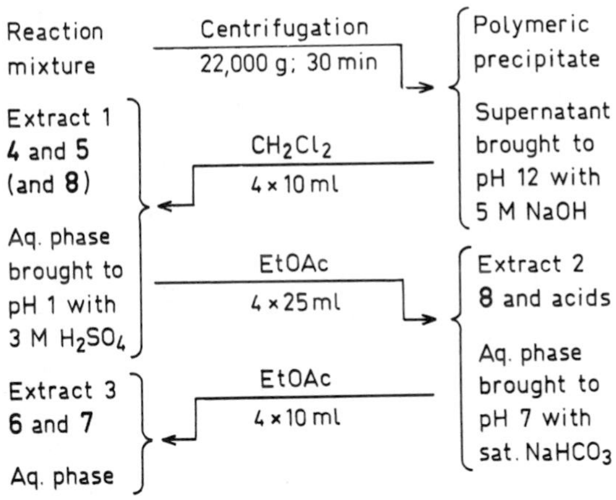

Figure 4. *General scheme for separation of 2-acylpyrroles (Extract 1) and 3-pyridinols (Extract 3) from a reaction mixture.*

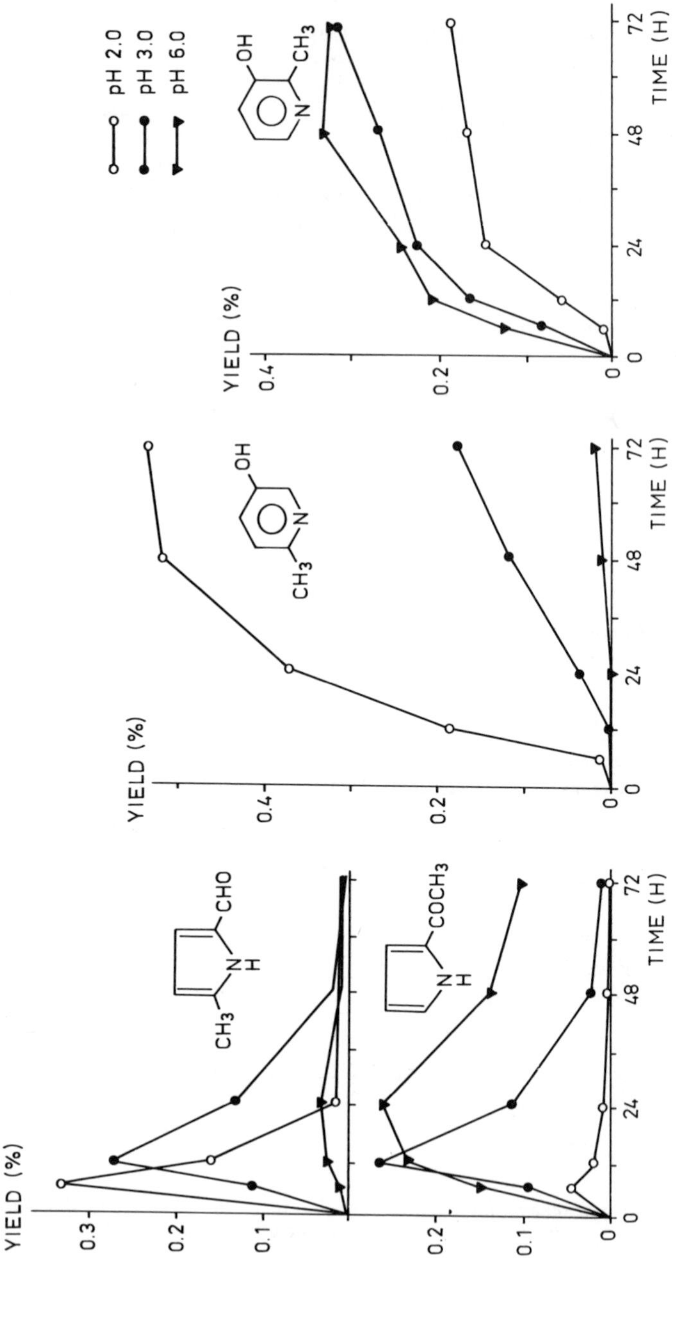

Figure 5. *Yields of compounds 4–7 on refluxing an aqueous solution of 0.17* M D-*glucose and 3.4* M *glycine.*

after 48 and 24 h, respectively. The whole reaction mixtures were processed according to Figure 4. Each extract 1 or 3 was dried and evaporated. The residues were dissolved in ($^{2}H_4$)methanol. The solutions were analyzed with ^{13}C-NMR spectrometry, those from the pH 3.0 experiment also with GC/MS. The spectra were compared with those of unlabelled 4-7.

In the "off-resonance" ^{13}C-NMR spectrum of each unlabelled compound, the high-field quartet was, of course, due to the methyl carbon. Being linked to nitrogen, the other terminal atom of the C_6 chain corresponded to the doublet at lowest field. Both signals were very well separated from those due to the four internal atoms of the chain. The positions of the signals from the terminal carbons in the proton-decoupled spectra are shown in Figure 6. Clearly, separation of 4 from 5 or 6 from 7 was not required to interpret the proton-decoupled spectra of Extracts 1 and 3, containing the labelled compounds. These spectra showed only the low-field signals in Figure 6. Taking into account the moderate noise level, the extent (90 %) of labelling, and the relative intensities in the spectra of the unlabelled compounds, we estimated that ≤ 2 % of the 6 and ≤ 6 % of the 7, which formed at initial pH 2.0, contained C-1 of glucose in the methyl group. At initial pH 3.0, this limit was 6 % for 4 and 5, and 2 % for 7. At initial pH 6.0, the limits 6 % for 5 and 9 % for 7 were estimated.

The fragmentation of 2-acylpyrroles (6) and 3-pyridinols (7) on electron impact (EI) has been elucidated and is shown for 4-7 in Figure 7. The two numbers under each ion are the *m/z* values expected with the ^{13}C label at the alternative positions, the latter number within parantheses corresponding to labelled methyl carbon. Where the two *m/z* values differed, the peak at the first one always predominated in the EI mass spectra of the labelled 4-7 obtained on GC/MS analysis of Extracts 1 and 3. This was in complete accordance with the ^{13}C-NMR spectrometric results, but for several reasons the mass spectrometric results were less accurate and did not exclude up to 15 % of the label in the respective methyl groups. Most important, the spectra showed clusters of peaks at consecutive *m/z* values, including those of interest. The ^{13}C content was calculated by means of a computer program based on the somewhat doubtful assumption that ions in the same cluster differed only in hydrogen content and/or isotopic composition.

The results obtained with the labelled glucose disqualify our routes to 5 and 7 but are consistent with those to 4 and 6 (Figure 3) and with the routes to 4 and 5 proposed by Shaw and Berry (Figure 2).

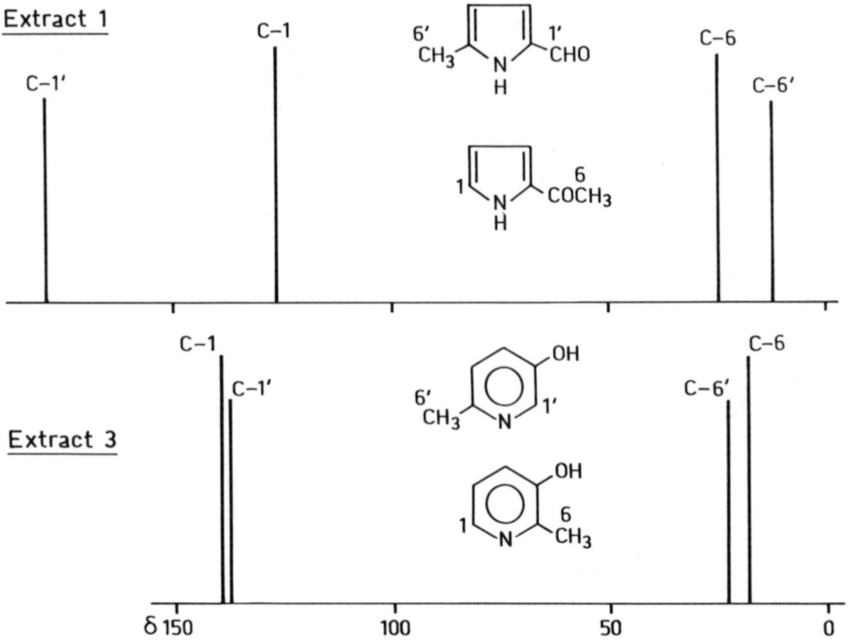

Figure 6. Positions of ^{13}C-NMR signals from terminal atoms of the C_6 chain in compounds 4–7. (Numbers 1 and 6 refer to glucose.)

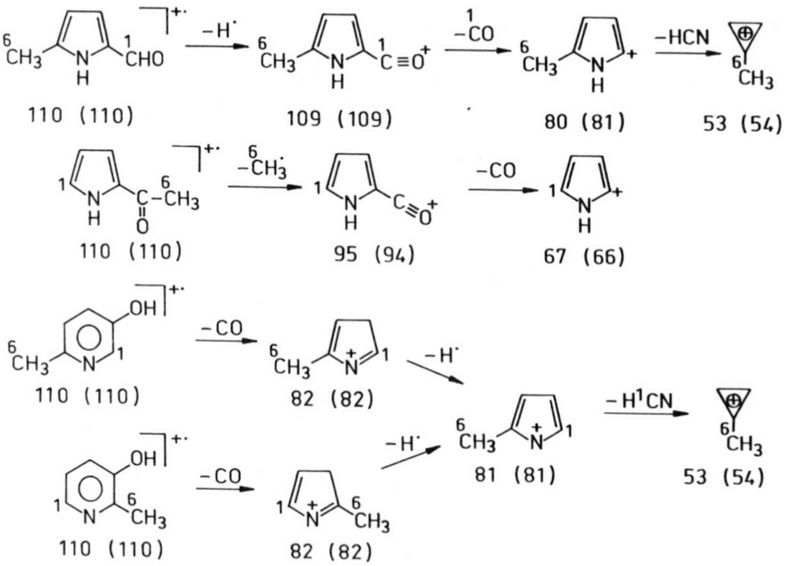

Figure 7. Fragmentation of compounds 4–7 on electron impact. Two m/z values are given under each ion. The first value predominates in the observed mass spectrum and corresponds to the 1-^{13}C-labeled ion. The second value, in parentheses, corresponds to the 6-^{13}C-labeled ion (6, 7). (Numbers 1 and 6 within the structures refer to glucose.)

Related reactions

Experiment with 3-deoxy-D-erythro-hexosulose (1). The experiment with unlabelled glucose at initial pH 3.0 was repeated after replacing the glucose by an equimolar amount of 1, prepared via its bis(benzoylhydrazone) (8). The yields of 4-7 in the two experiments are compared in Figure 8. This further supports the proposed routes to 4 and 6 but also shows that 1 is not an important intermediate in the formation of 5 or 7. Thus, the route to 5 proposed by Shaw and Berry (Figure 2) must also be discarded.

It should be noted in 5 and 7 that no hetero atom is linked to the carbon now known to originate from C-2 of glucose. This is hard to explain with an intermediate such as 1-3 where C-2 is oxidized. However, without intermediates such as 1-3, the Strecker degradation cannot follow the accepted course (Figure 1a). To explain the elimination of the hydroxyl group at C-2 of glucose, we therefore propose that the alternative route in Figure 1b is followed when 5 or 7 is formed, as shown in Figure 9. The end product from glucose in Figure 1b would be 2-deoxy-D-arabino-hexose (9). However, attempts to detect 9 during the glucose-glycine reaction were unsuccessful, although several methods of sugar analysis were tried. A possible reason may be that the enamine precursor of 9 is dehydrated through β-elimination much more rapidly than it is hydrolyzed to 9. In Figure 9, a "vinylogous" Amadori rearrangement is assumed to follow this β-elimination.

Experiments with deoxyhexoses That the reaction steps after the Strecker degradation in Figure 9 are indeed fast, was indicated by the following experiment. An aqueous solution 0.17 M in (commercial) 9 and 3.4 M in ammonium acetate was brought to pH 6.0, refluxed for 3 h and processed according to Figure 4. Gas chromatographic analysis of Extracts 1 and 3 showed a 1.0 % yield of 5 and a 19 % yield of 7. This may be compared with the respective maximum yields 0.27 and 0.34 % in the much slower glucose-glycine reaction (Figure 5).

To support the mechanism in Figure 1b, the reaction of 3-deoxy-D-ribo-hexose (10), prepared via 3-deoxy-3-iodo-1,2:5,6-di-O-isopropylidene-α-D-allofuranose (9), with glycine was investigated. The end product from 10 in Figure 1b would be 2,3-dideoxy-D-erythro-hexose (11). Since β-elimination from its enamine precursor is not possible, 11 should accumulate, if the unconventional Strecker degradation in Figure 1b takes place. An aqueous solution 0.17 M in 10 and 0.85 M in glycine (pH 6.0) was refluxed. Samples were withdrawn after 3, 6 and 12 h, and evaporated. The sugars in each residue were converted to per-O-acetylaldononitriles (10), which were analyzed by GC and GC/MS using a glass capillary column coated with CP Sil 5. One of the GC peaks was assigned to the nitrile derivative of 11 on the basis of the EI mass spectrum (11). Still more convincing was the spectrum recorded

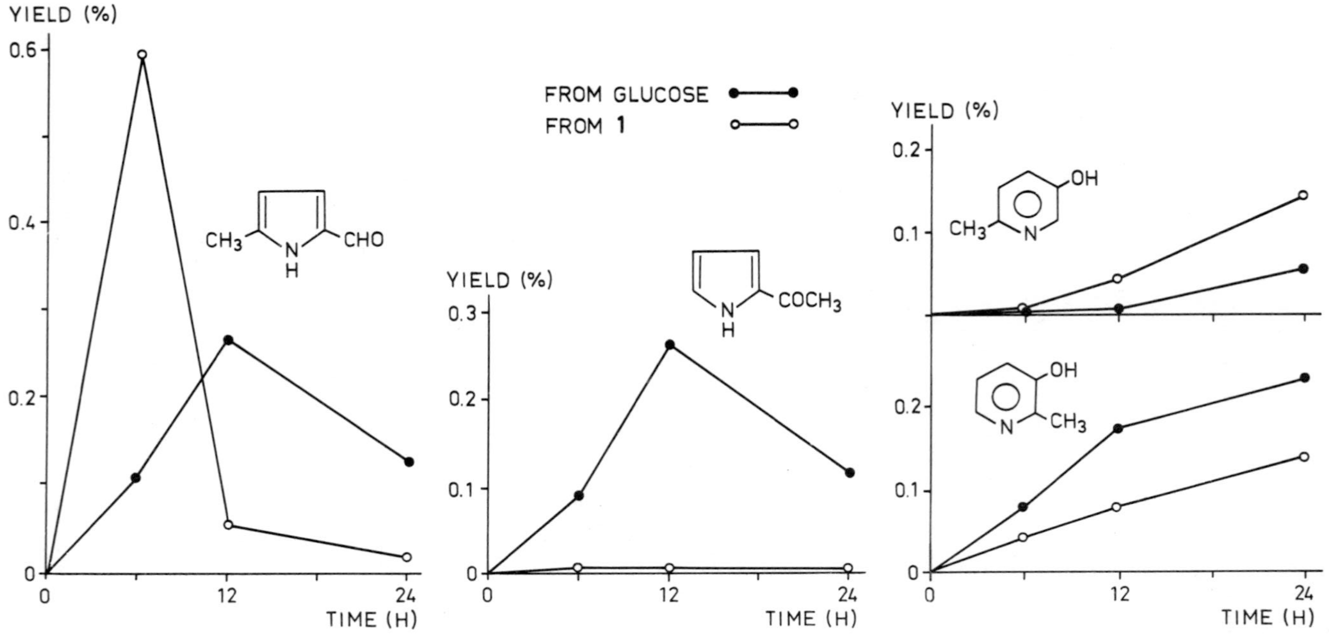

Figure 8. Yields of compounds 4–7 on refluxing an aqueous solution of 0.17 M *D-glucose or 3-deoxy-D-*erythro*-hexosulose (1) and 3.4* M *glycine (pH 3.0).*

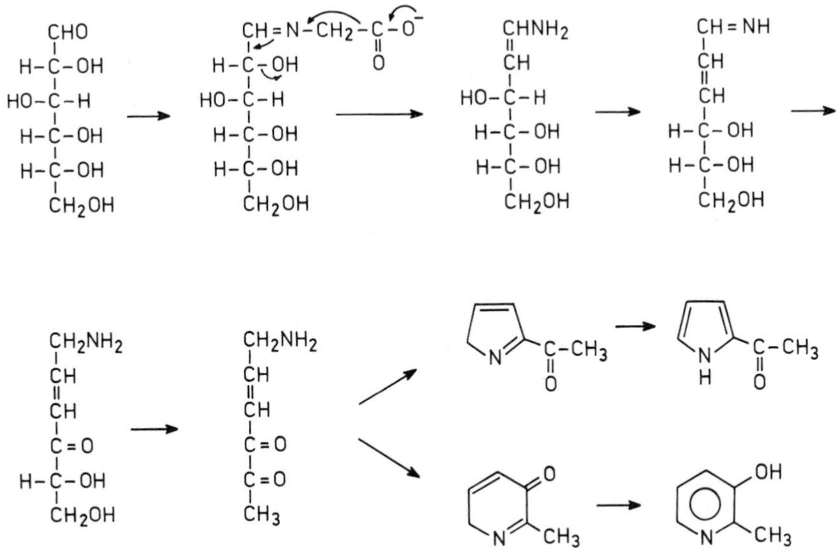

Figure 9. Proposed route to compounds 5 and 7 from D-*glucose and glycine.*

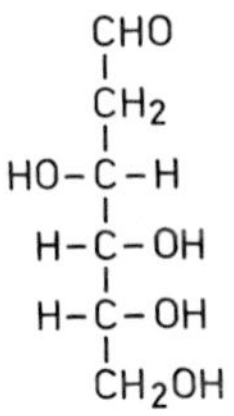

9

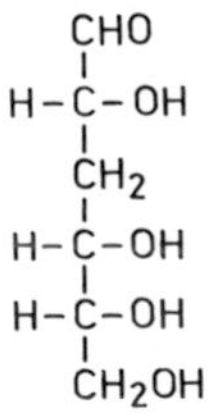

10

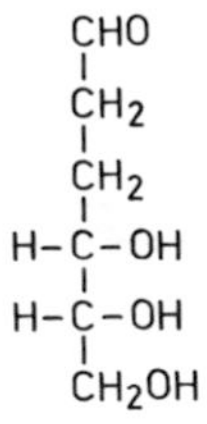

11

with chemical ionization (10), using ammonia as reaction gas. The intensity of $(M + H)^+$ at *m/z* 272 was *ca.* 50 % relative to the base peak, which was due to $(M + H + NH_3)^+$ (*m/z* 289). If there was no difference in GC response factor between the derivatives of **10** and **11**, the maximum yield of **11** was 4.7 % and was obtained after 6 h. At that time, 27 % of **10** remained.

It remains to be explained how **5** is formed from fructose and alanine (4). Perhaps a Heyns rearrangement will give rise to 2-[(1-carboxyethyl)imino]-2-deoxy-D-glucose, where the (protonated) alanine residue might be eliminated as proposed in Figure 9 for the hydroxyl group at C-2 of glucose. It may be noted that 2-amino-2-deoxy-D-glucose is unstable in slightly acidic solution, particularly in the presence of glycine (12).

Acknowledgements

We thank Professor Olof Theander for his kind interest in the present work, Mr Rolf Andersson and Mr Suresh Gohil for recording the spectra, and the Swedish Board for Technical Development for financial support.

Literature Cited

1. Schönberg, A.; Moubacher, R. *Chem. Rev.* 1952, *50*, 261.
2. Mauron, J. *Prog. Food Nutr. Sci.* 1981, *5*, 5.
3. Feather, M. S.; Harris, J. F. *Adv. Carbohydr. Chem. Biochem.* 1973, *28*, 161.
4. Shaw, P. E.; Berry, R. E. *J. Agric. Food Chem.* 1977, *25*, 641.
5. Olsson, K.; Pernemalm, P.-Å.; Theander, O. *Acta Chem. Scand. Ser. B* 1978, *32*, 249.
6. Budzikiewicz, H.; Djerassi, C.; Williams, D. H. "Mass Spectrometry of Organic Compounds"; Holden-Day: San Francisco, CA, 1967; p. 602.
7. Undheim, K.; Hurum, T. *Acta Chem. Scand.* 1972, *26*, 2075.
8. El Khadem, H.; Horton, D.; Meshreki, M. H.; Nashed, M. A. *Carbohydr. Res.* 1971, *17*, 183.
9. Garegg, P. J.; Samuelsson, B. *J. Chem. Soc. Perkin Trans. 1* 1980, 2866.
10. Li, B. W.; Cochran, T. W.; Vercellotti, J. R. *Carbohydr. Res.* 1977, *59*, 567.
11. Dmitriev, B. A.; Backinowsky, L. V.; Chizhov, O. S.; Zolotarev, B. M.; Kochetkov, N. K. *Carbohydr. Res.* 1971, *19*, 432.
12. Feather, M. S. *Prog. Food Nutr. Sci.* 1981, *5*, 37.

RECEIVED October 5, 1982

5

Nitrite Interactions in Model Maillard Browning Systems

GERALD F. RUSSELL
University of California, Department of Food Science and Technology, Davis, CA 95616

The Maillard reaction is well recognized in food chemistry and by carbohydrate and protein chemists. Initial condensation reactions of glucose and amino acids lead to the formation of intermediates such as the Amadori compounds and their rearrangement products. Secondary amines in these products are thus potential sites for reactions with nitrite to form nitrosamines and, as such, may be of significance as genotoxic agents. Further Maillard reactions yield volatile productrs which may have nitrosation potential. Recent evidence has shown that formation of compounds formed from Maillard type reactions are biologically active in bacterial mutagen assays. Heterocyclic compounds such as thiazolidines can react with nitrous acid to form nitrosothiazolidines which also show genotoxicity in bacterial assays. A discussion of studies from model system reactions and formation of toxic products from reaction with nitrite is presented.

Many reports in the literature have identified volatile and nonvolatile products formed from Maillard browning reactions. Several reports have shown the formation of 1,3-thiazoles, 1,3-thiazolines, and 1,3-thiazolidines in the early stages of browning reactions from fragmentations which lead to the formation of these small heterocyclic molecules. Recently, some of these compounds have shown mutagenic activity in bacterial systems such as the Ames mutagen assay and hence are of interest to food toxicologists as genotoxic agents. Also, there is evidence that at least one Amadori compound, which when nitrosated, has strong mutagenic activity. This paper will review these recent developments as an overview of the the role of nitrite in Maillard reactions from the standpoint of their toxicological significance.

We first became interested in this area from studying volatiles produced in model Maillard systems. Shibamoto and Russell (1,2) studied model systems such as glucose/hydrogen sulfide/ammonia and reported volatile profiles which contained many of the

0097-6156/83/0215-0083$06.00/0

volatiles reported in cooked meat products. Certain volatile fractions produced from three model browning systems of glucose/cysteamine, acetaldehyde/cysteamine, and glyoxal/cysteamine displayed mutagenic activity in the Ames assay (3). Of particular interest were the heterocyclic thiazoles and their presumed thiazoline and thiazolidine precursors, since purified synthetic nitroso-thiazolidines were shown to exhibit varying levels of mutagenic response (4).

Amadori compounds (N-substituted-1-amino-1-deoxy-2-ketoses) are potential precursors to the formation of many of these heterocyclic volatile products. The secondary nitrogen in most Amadori compounds is weakly basic and is therefore a likely site for rapid nitrosation reactions via normal reactions with nitrous acid, under mildly acidic conditions. However, purified Amadori compounds are usually obtained only after tedious isolation procedures are invoked to separate them from the complex mixtures of typical Maillard browning systems. Takeoka et al. (5) reported high performance liquid chromatographic (HPLC) procedures to separate Amadori compounds in highly purified form on a wide variety of columns, both of hydrophilic and hydrophobic nature. They were able to thus demonstrate that reaction products could be followed for kinetic measurements as well as to ensure purity of isolated products.

Takeoka et al. (5) also reported methods for derivatizing both aliphatic and aromatic Amadori compounds as *p*-nitrobenzyloxyamine (PNBO) derivatives to allow facile UV detection in the picomolar range for HPLC separations. They reported in the same paper a simple method for derivatizing the Amadori compounds to allow gas chromatographic/mass spectrometric (GC/MS) separation and identification of highly purified Amadori compounds.

Only recently have *N*-nitroso Amadori compounds been characterized chemically. The first description of an *N*-nitroso derivative of an Amadori compound reported the formation of 1-deoxy-1-(*N*-nitroso-3,4-xylidino)-D-fructose to confirm that a secondary amino group had been formed in an Amadori compound (6). Coughlin et al. (7) and Heyns et al. (8) described the formation of nitrosated Amadori compounds. Since Amadori compounds are weakly basic secondary amines and occur widely in Maillard browned foods and beverages (5) and unburned tobacco (9), the genotoxic potential of these compounds is of interest.

Procedures

Analytical Instrumentation A Hewlett Packard model 5880-A series gas chromatograph equipped with flame ionization detection (FID) was fitted with an 18 m x 0.25 mm id glass capillary column coated with OV-101. Reaction mixtures and standards were injected through a stainless steel injector splitter with a split ratio of 100:1. The column oven was linearly programmed from 90 to 200°C at 2 or 5°/min. The injector and detector temperatures were

250°C. The gas flow rates were: nitrogen carrier gas -- 0.5 mL/min (linear velocity of 15 cm/sec), nitrogen make-up gas -- 50 mL/min, hydrogen -- 40 mL/min, and compressed air -- 240 mL/min.

A DuPont model 21-492B GC/MS with an Incos data system was used to confirm the presence of thiazolidine and nitrosated thiazolidine. A 30 m x 0.24 mm id glass capillary column coated with SP-2250 was used for GC/MS analyses with a 30:1 split ratio (with the larger fraction vented to atmosphere through a charcoal trap). Electron impact ionization was used at 70 eV. The data system generated total ion current chromatograms and recorded mass spectra.

A Perkin Elmer model 257 infrared spectrophotometer with a beam condenser and normal slit program was used to identify some reaction products. Infrared (IR) spectra of the compounds were obtained from neat liquids between KBr crystals.

Preparation of Standards and Model System Reaction Mixtures N-Nitroso-1,3-thiazolidine was prepared by direct nitrosation of 1,3-thiazolidine and also from reaction of cysteamine/formaldehyde/nitrite. Details for the experimental procedures involved in the preparation and work up of these standards and mixtures will be presented elsewhere (10). Quantitative determination of 1,3-thiazolidine and N-nitroso-1,3-thiazoline were obtained for seven reaction mixtures (10):

(1) Thiazolidine (0.5M) heated in aqueous solution (pH 5.5, 60°C, 12 h).
(2) Thiazolidine (0.2M) heated in aqueous solution (pH 4.5, 60°C, 12 h).
(3) Thiazolidine (0.2M) heated in aqueous solution with 0.4M nitrite (pH 4.5, 60°C, 2 h).
(4) Cysteamine (0.5M) and formaldehyde (0.5M) aqueous solution heated under nitrogen (pH 1.8, 100°C, 12 h).
(5) Cysteamine (0.5M) and formaldehyde (0.5M) aqueous solution heated under nitrogen (pH 4.5, 100°C, 12 h).
(6) Cysteamine (0.5M), formaldehyde (0.5), and nitrite (0.5M) aqueous solution heated under nitrogen (pH 4.5, 100°C, 2 h).
(7) Cysteamine (0.5M), formaldehyde (0.5M), and nitrite (0.5M) aqueous solution heated under nitrogen (ph 4.5, 60°C, 2 h).

Nitrosation of Fructose-L-tryptophan The Amadori compound fructose-L-tryptophan (fru-trp), was synthesized as described by Sgarbieri et al. (11). Purity was checked by the methods given by Takeoka et al. (5). An aqueous solution of reactants containing fru-trp (10 micromol/mL) and sodium nitrite (60 micromol/mL) was adjusted to pH 4.0 and incubated in a closed vial at 37°C for 24 h as described by Coughlin et al. (7). The methods of Yahagi et al. (12) were used to modify the Ames assay in which *S. typhimurium* strains TA98 and TA100 and portions of the nitrosated fru-trp were preincubated in liquid suspension with and without S-9 mix for 20 min at 37°C prior to being poured onto a plate.

Results and Discussion

GC/MS Analyses Confirmation of the identity of peaks on chromatograms was based largely upon GC/MS data. Dichloromethane extracts of appropriate compounds or reaction mixtures were chromatographed as described under procedures. Mass spectra for the major peaks obtained from nitrosation mixtures showed a prominent molecular ion [$M^{+\bullet}$] of m/z 118, corresponding to the molecular weight of N-nitroso-1,3-thiazolidine. A base peak at m/z 88 corresponding to the loss of NO and a prominent m/z 60 peak were also present. This type of fragmentation is consistent with that shown by Vestling and Ogren (13). In addition, high resolution mass spectrometry on the main peak gave a $M^{+\bullet}$ of m/z 118.0211 (calc. = 118.02). Mass spectra of the minor peaks from these reaction mixtures matched those of commercially available 1,3-thiazolidine.

The infrared spectra of reaction product thiazolidines matched the standard Sadler Reseach Laboratory spectra, with a band at 3270 cm^{-1} consistent with that for an NH group. IR spectra of the nitrosated products in a KBr pellet showed no peak in the NH region, but did show prominent N=O absorption at 1420 cm^{-1}, supporting the presence of nitrosated thiazolidines.

GC/FID Analyses Chromatograms for all seven reaction mixtures were obtained on an 18-m OV-101 open-tubular glass capillary column as described under procedures. Oven temperatures were programmed linearly from 90 to 200°C at 2°C/min for mixtures 1, 2, 4, and 5, and at 5°C/min for mixtures 3, 6, and 7.

The chromatogram for mixture 1 showed several chromatographic peaks with one peak at retention time t_R 3.50 (min) identified as 1,3-thiazolidine. The thiazolidine peak represented approximately 20% of the total integrated peak area. There were several other peaks which were probably formed from degradation reactions from the thiazolidine.

The chromatogram for mixture 2 showed very similar results with the thiazolidine peak representing approximately 19% of the total integrated area. These results along with others from similar experiments suggest that heating time, thiazolidine concentration, and pH adjustment prior to heating did not greatly alter the degradation pattern of the heated thiazolidine.

The chromatogram of mixture 3 showed a major peak with t_R 9.29 which accounted for 98% of the total integrated area; this peak was identified by GC/MS as N-nitroso-1,3-thiazolidine. A minor peak with t_R 3.63 representing 0.45% of the total integrated area was identified as 1,3-thiazolidine.

The chromatogram of mixture 4 showed a major peak at t_R 3.39 which was identified as 1,3-thiazolidine and was essentially the only peak on the chromatogram, other than solvent impurities.

The chromatogram of solution 5 showed a minor peak at t_R 3.87 corresponding to 1,3-thiazolidine and 6% of the total integrated area. The chromatographic profile was akin to that of mixtures 1 and 2, indicating that the higher pH (4.5-5.5 compared to 1.8 in mixture 4) facilitated rapid degradation of the thiazolidine. Ratner and Clarke (14) studied the reaction of cysteine and formaldehyde and noted the rapid reaction rate increase with increasing pH. It is possible that rapid formation of the thiazolidine at the higher pH permitted rapid degradation to occur while the slower formation at low pH was not conducive to further degradation.

The chromatogram of mixture 6 showed a major peak with t_R 8.40, identified as N-nitroso-1,3-thiazolidine, which represented 97% of the total integrated area. A minor peak for 1,3-thiazolidine was seen at t_R 3.21.

The chromatogram of mixture 7 showed a pattern akin to that for mixtures 3 and 6. The nitrosothiazolidine at t_R represented 96% of the total integrated area, with a trace peak at t_R representing unreacted thiazolidine.

The three chromatograms for mixtures 3, 6, and 7 had similar patterns with no indication of thiazolidine degradation products. Presumably, the nitrosation was faster than degradative reactions of the thiazolidine precursor.

Mirvish (15) showed that weak bases are more easily N-nitrosated than strong bases under mildly acidic conditions. Since the secondary amine nitrogen on the 1,3-thiazolidine ring is much less basic (pK_a 6.3) than in the pyrrolidine ring (pK_a 11.3), the formation of N-nitroso-1,3-thiazoline should be favorable and hence would be predicted to occur in products such as bacon where N-nitrosopyrrolidine is of major concern.

It was demonstrated by Coughlin (16) that formaldehyde catalyzed the N-nitrosation of 1,3-thiazolidine at pH 7. Therefore, it is possible that HCHO catalyzed the formation of N-nitroso-1,3-thiazolidine in these reactions, presumably following ring closure.

The modified Ames mutagen assay was used to investigate the mutagenic potential of nitrosated fru-trp, a fru-trp control, and a nitrite control incubated in solutions under mild conditions (37°C, pH 4, 24 h). Figure 1 shows dose response curves obtained with *S. typhimurium* TA98 and TA100. Nitrosated fru-trp demonstrated a positive dose-response relationship with both tester strains in the absence of S-9 mix and a somewhat reduced dose-response relationship in its presence. The fru-trp and nitrite controls did not manifest mutagenic properties in either tester strain.

The positive dose-response results of the mutagen assays are of interest since both strains showed the nitrosated product to be direct-acting mutagens for both TA98 and TA100 strains. Most N-nitroso compounds cause mutations by base-pair substitution only

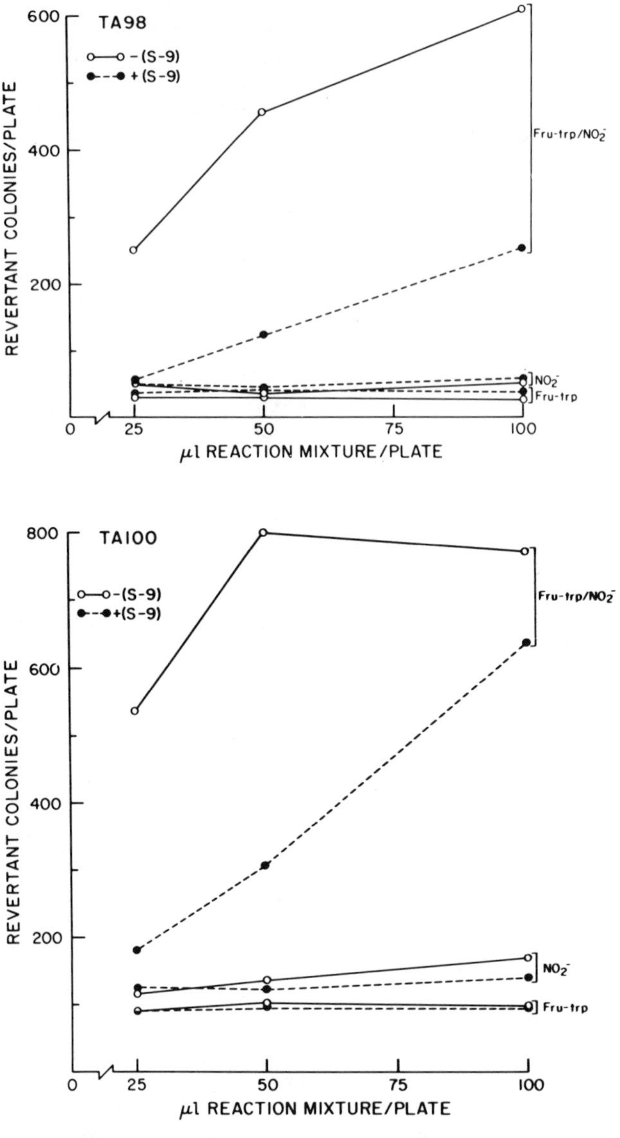

Figure 1. Ames assay dose–response curves of the mutagenic effects of 25-, 50-, and 100-μL aliquots of nitrosated fru-trp reaction mixture and controls on S. typhimurium *strains TA98 (top) and TA100 (bottom). Bacteria and test mixtures were preincubated with and without S-9 mix for 20 min at 37 °C. Background revertant levels were not subtracted from the data points shown.*

with only a few reported to cause frameshift mutations (17). The planar aromatic ring of fru-trp may contribute to its frameshift mutagenic activity as a result of intercalation with DNA. Similarly, most N-nitrosamines require S-9 activation to effect a mutagenic response in the Ames system; the direct-acting properties of fru-trp are unusual, but not unique.

The reaction between fru-trp and nitrite under mildly acidic conditions could lead to a mixture of mono- and di-nitroso derivatives since both the secondary nitrogen and indole ring nitrogen can be nitrosated. Nakai et al. (18) demonstrated the rapid room-temperature nitrosation of the indole nitrogen of N-acetyl-L-tryptophan using excess nitrite at pH 4. A characteristic shift in UV absorption maximum (18) was observed with our reaction product; the incubated fru-trp control showed an absorption maximum at 277 nm, while the nitrosated fru-trp mixture showed a complete shift in the absorption maximum to 257 nm which suggested complete nitrosation of the indole nitrogen in the presence of excess nitrite. Amino acid autoanalysis showed no unreacted fru-trp and the mixture gave a negative ninhydrin reaction indicating little if any unreacted secondary nitrogen. Since Amadori-type products can easily react through further steps in browning reactions, further work is required to isolate and identify the exact mutagenic product(s). Professor Heyns' group is actively looking at these products and has identified similar questions that need to be further addressed before the active components can be clearly identified (19).

Literature Cited

1. Shibamoto, T.; Russell, G.F. J. Agric. Food Chem. 1976 24, 843.
2. Shibamoto, T.; Russell, G.F. J. Agric. Food Chem. 1977, 25, 109.
3. Sakaguchi, M.; Shibamoto, T. J. Agric. Food Chem. 1978, 26, 1179.
4. Sekizawa, J.; Shibamoto, T. J. Agric. Food Chem. 1980, 28, 781.
5. Takeoka, G.R.; Coughlin, J.R.; Russell, G.F. In "Liquid Chromatographic Analysis of Food and Beverages"; Vol. 1; Academic Press: New York, 1979; p. 179.
6. Zaugg, H.E. J. Org. Chem. 1961, 26, 2718.
7. Coughlin, J.R.; Russell, G.F.; Wei, C.I.; Hsieh, D.P.H. Toxicol. Appl. Pharmacol. 1979, 48, A45.
8. Heyns, K.; Roper, H.; Roper, S.; Meyer, B. Angew. Chem. Int. Ed. Eng. 1979, 18, 878.
9. Schmeltz, I.; Hoffman, D. Chem. Rev. 1977, 77, 295.
10. Assadi, F.; Coughlin, J.R.; Russell, G.F. (in preparation).
11. Sgarbieri, V.C.; Amaya, J.; Tanaka, M.; Chichester, C.O. J. Nutr. 1973, 103, 657.
12. Yahagi, T.; Nagao, M.; Seino, Y.; Matsushima, T.; Sugimura, T.; Okada, M. Mutation Res. 1977, 48, 121.

13. Vestling, M.M.; Ogren, R.L. J. Heterocyclic Chem. 1975, 12, 243.
14. Ratner, S.; Clarke, H.T. J. Am. Chem. Soc. 1037, 59, 200.
15. Mirvish, S.S. Toxicol. Appl. Pharmacol. 1975, 31, 325.
16. Coughlin, J.R. Formation of N-Nitrosamines from Maillard Browning Reaction Products in the Presence of Nitrite. 1979. Dissertation, Univ. of California, Davis.
17. Olajos, E.J. Ecotoxicol. Environ. Safety 1972, 1, 175.
18. Nakai, H.; Cassens, R.G.; Greaser, M.L.; Woolford, G. J. Food Sci. 1978, 43, 399.
19. Erbersdobler, H.F. 1982. Personal communication.

RECEIVED January 18, 1983

6

Studies on the Color Development in Stored Plantation White Sugars

HUNG-TSAI CHENG, WAI-FUK LIN, and CHUNG-REN WANG
Taiwan Sugar Research Institute, 54 Sheng Chan Road, Tainan, Taiwan, Republic of China

Crude plantation white sugars from sugar cane, manufactured by a carbonation or a sulphitation process, developed color during storage leading to degradation in their color grade. This development of color or poor keeping quality was more marked in carbonation than in sulphitation sugars. Spectrophotometric and chromatographic studies indicated that the color-bearing compounds responsible for the sugar quality changes were humic acids, caramel, 5-hydroxymethylfurfural, and melanoidins. In carbonation, a high percentage of reducing sugar destruction in the highly alkaline condition of the first carbonation stage is believed responsible for the formation of color-bearing compounds of the 5-hydroxymethylfurfural and humic acid categories, and these reducing sugar degradation products play an important role on the melanoidin and caramel formation in sugar crystals during storage. Carbonates evidently catalyze the caramelization much faster than sulphites, leading to faster development of color in carbonation sugars. Sulphitation inhibited the melanoidin formation, presumably by blocking the carbonyl function. Lowering of the level of carbonates by replacing the second carbonation stage with sulphitation or phosphatation, and improvement of the first carbonation technique for reducing sugar destruction, were recommended to improve the keeping quality of the carbonation sugars.

The development of brown color in sugar during storage is one of the oldest problems in the sugar industry. In Taiwan, about 400,000 tons of plantation white sugar from sugar cane are

0097-6156/83/0215-0091$06.00/0

annually produced in seven carbonation factories and two sulphitation factories, during the period of November-May. The sugar is stored in bags in warehouses at ambient conditions, i.e., the relative humidity (RH) is from 65 to 90% and the temperature is between 5 and 40°C. Under these conditions of storage, the plantation white sugar deteriorates in the sense that some coloring matter is produced in the sugar crystals, leading to degradation in their color grade. This deterioration in color on storage is more marked in carbonation sugar than in sulphitation sugar, as shown in Figure 1.

Though much knowledge concerning the coloration of sugar products was obtained from studies involving alkali and heat treatment of sugars and the reactions of amino compounds and sugars (1,2), understanding of the nature of the coloring matter formed during storage is still very limited. Most investigations have concerned the browning of raw sugar. Tsuchida (3) reported that a direct relationship exists between darkening of such sugar on storage and nitrogen content, which is probably related to the Maillard reaction between the amino acids and degradation products of reducing sugars. Chen (4) concluded that a sugar with high original color increases more than a sugar with low original color. Monterde (5) stated that no relationship exists between darkening and bacterial content. According to Cortis-Jones (6), color development during raw sugar storage depends on original color and the ambient conditions. Ramon Samaniego (7) concluded that coloring matter present in raw sugars are simply caramelization products which interact with each other to give rise to the dark color developed during storage. Gillett (2) reported that the color development of sugar is dependent on several factors such as variety of cane, soil and season.

Plantation white sugars contain ash, reducing sugars, and some amino acids. These may interact during storage to give rise to colored products. This study was conducted to (a) develop a simple method of isolating sugar colorants, (b) determine the possible causes of color development during storage, and (c) find methods to improve the keeping quality of carbonation sugar in particular. The present communication reports a summary of these studies.

Experimental Materials and Methods

Investigation of factors affecting the browning occurring during storage of plantation white sugar. Samples of sugar from the sulphitation process and the carbonation process were stored at constant temperature of 70°C and RH 60% for 90 h. Changes in terms of color variation or development under such conditions were found to be equal to those of sugar stored at about 20°C and RH 80% for one year (4).

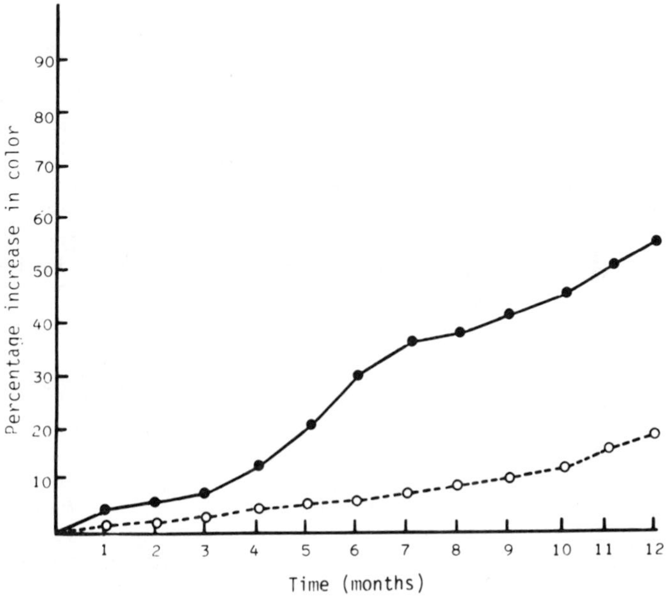

Figure 1. Development of color in carbonation (●) and sulfitation (○) sugars during storage.

Four plantation white sugar samples from the same area exhibiting various degrees of browning were collected for this experiment as follows:

Samples	Source	Color Value (Lumetron) B	Color Value (Lumetron) A	Whiteness B	Whiteness A
1	Sulphitation	60.08	69.75	0.723	0.675
2	Sulphitation	62.14	71.44	0.650	0.512
3	Carbonation	57.96	92.55	0.705	0.390
4	Carbonation	61.70	94.18	0.665	0.310

B: Before storage A: After storage

These were analyzed for ash, reducing sugar, pol, starch, protein and moisture to determine if any relationship exists between these factors and the color development of the sugar. Color values were determined with a Lumetron Spectrophotometer at 420 nm. Ash, reducing sugar, pol, and moisture were determined in accordance with the Official Methods of ICUMSA (8). "Pol" is the value determined by direct or single polarization of the normal weight solution of a sugar product in a saccharimeter; the figure is used as the concentration of sucrose present in the solution. Protein was determined colorimetrically by Lowry's Folin-phenol method using bovine albumin (Sigma Chemical Co.) as a standard (9). Starch was determined colorimetrically by Whistler's iodine method using amylose from corn starch (G.R. TCI Chemical Co.) as a standard (10).

Isolation of coloring matter produced in plantation white sugar during storage Fresh and stored plantation white sugar samples from the carbonation process and the sulphitation process were compared. A rapid method for isolating colorants from sugar samples was developed. A prepacked mini-column of SEP-PAK C_{18} of I.D. 0.8 x 1.0 cm was used. Colorants of a sugar sample were isolated from sucrose in minutes by three simple steps. First, sample after dilution with buffer of pH 2.5 was pumped through the SEP-PAK C_{18} column. About 70% of the colorants were adsorbed on the column while sugar molecules passed through. Second, residual sugar in the column was washed by pumping through portions of the same buffer. Finally, the adsorbed colorants were eluted with 2 ml of 50% aqueous CH_3CN containing 0.01 N NaOH, 1 ml of 50% CH_3CN containing 0.02 N HCℓ, and 2 ml of 0.1 M Tris buffer of pH 7.5 successively. The concentrated sugar-free colorant solution was analyzed by high performance gel permeation chromatography (HPGPC), HPLC, UV, and IR.

Nature of the coloring matter produced in plantation white sugar during storage The liquid chromatograph used (Water Associates) consisted of a Model 6000A high pressure solvent delivery system, a Model 440 dual wavelength detector, and a Model U6K injector. The column used for fractionation was Bondagel E-125/3.9 mm I.D. × 30 cm length. It is an ether-bonded silica gel column with nominal molecular weight separation range of 2000 to 50,000. The operating conditions for HPGPC were as follows:

Solvent: 0.1 M Tris buffer of pH 7.5
Flow rate: 0.4 ml/min
Injection volume: 30 μℓ
Detection: UV detector with 436 nm wavelength filter
Sensitivity: 0.05 aufs (absorbance units full scale)
Chart Speed: 1.0 cm/min

Hitachi Model 200-20 double-beam spectrophotometer with wavelength scanning was used for spectrophotometric study. The scanning covered the range 220 to 500 nm.

Results and Discussion

It has been observed that the color development of plantation white sugar during storage is more marked in carbonation sugar than in sulfitation sugar, as shown in Figure 1. We have observed that (1) Sugars produced in the early periods of the campaign (November-December) or the end of the campaign (April-May) developed color faster than those manufactured in the periods of midseason (January-March), (2) The higher the temperature, the faster is the color development; (3) Color development occurred even when sugar was kept out of contact with air and in the dark. To help explain these phenomena, chemical analyses were made to examine the relationship of constituents and the color development in the sugar crystals. Four plantation white sugar samples from the same area exhibiting various degrees of browning were analyzed. The results are shown in Table 1. There are no correlations between the factors studied (reducing sugar, pol, starch, protein and moisture) and the color development. The ash content of sugars varied between 0.037% and 0.071%. It is interesting to note that the sugars of darker color development had the higher ash content. Ash constituents present in the crystals catalyze the development of color in sugars during storage.

In order to understand the mechanism of color formation or the chemistry of the colorant itself in sugar products, a rapid chromatographic method for isolation of colorants from sugar samples was developed. The mini-column adopted, a SEP-PAK C_{18} cartridge, is able to isolate colorants from sugar products of very low color. By this technique, highly concentrated and sugar-free color solutions can be obtained from 100 g of sugar sample; sufficient colorants were recovered on the SEP-PAK C_{18} column for HPGPC studies by treating 100 g of sugar samples. This high efficiency of color concentration and sugar removal of the cartridge is a consequence of its small column size and well placed nonpolar packing. The small size of the cartridge permits rapid operation, and only a small amount of eluent is needed for releasing the adsorbed colorants. Consequently, colorant is concentrated after elution. The nonpolar packing of the cartridge can retain organics of moderate to low polarity disolved in a polar moving phase, yet reject very polar compounds such as sugar completely. Because little sugar colorant is very polar,

Table 1. Chemical Analyses of Sugars

Sugar sample	Ash %	Reducing Sugar %	Pol %	Starch ppm	Protein ppm	Moisture %	Color (Lumetron)
1A	0.037	0.032	99.81	20	38	0.035	69.75
1B	0.042	0.038	99.82	28	43	0.044	60.08
2A	0.048	0.041	99.80	30	30	0.036	71.44
2B	0.045	0.036	99.83	36	45	0.023	62.14
3A	0.054	0.044	99.81	20	52	0.032	92.55
3B	0.067	0.045	99.82	24	68	0.031	57.96
4A	0.062	0.039	99.83	44	29	0.037	94.18
4B	0.071	0.046	99.84	50	47	0.038	61.70

significant amounts of colorants can be isolated free from sugar. Thus analysis of colorant by methods such as HPGPC or TLC can be conducted with higher resolution and lesser difficulty. HPGPC fractionation of sugar colorant is compared with that of colored sugar itself in Figure 2. This suggests that the principal peak of colorant is indeed present in the original sugar, and that the colorants adsorbed by the SEP-PAK C_{18} cartridge possessed the same range of molecular size and polarity as the original sugar solution. Hence the treatment is considered nonselective and gentle for isolating colorants. Figure 2 also illustrates the adverse effect of sugar itself, and the difficulty of fractionating very light-colored samples, even when monitoring is in the UV region of the spectrum. The application of the SEP-PAK C_{18} cartridge as a tool for sugar colorant isolation gives excellent efficiency and simplicity.

Fractionation of sugar colorants based on molecular size is usually conducted with Sephadex gel. Many investigators (11,12) have employed this method for sugar colorant studies. Usually three to four peaks are obtained after at least 2 h of separation. In this experiment, HPGPC was adopted for the fractionation of isolated colorants. Repeatability and efficiency of HPGPC were good. Fractionating one sample into six peaks detected at 436 nm took only 20-40 min. HPGPC colorant molecular profiles of fresh and stored plantation white sugar samples from the carbonation process and the sulphitation process showed significant differences, as shown in Figure 3. Sulfitation sugar contained more high-molecular-weight colorants, and less low-molecular-weight colorants, than carbonation sugar. Fractions (2-3 ml and 4.5-5.5 ml) collected and analyzed by UV absorption spectra also showed differences as shown in Figure 4. The carbonation sugar colorant of the 4.5-5.5-ml fraction had an absorption maximum at 283 nm, and a stronger total absorbance around 280 nm. After storage, the low-molecular-weight colorants of carbonation sugar had become high-molecular-weight colorants, as shown in Figure 5. Compared with Figure 3, these data indicated that the low-molecular-weight colorants undergo slow polymerization leading to the development of color in sugar crystals during storage, particularly caramelization. The calcium carbonate present in the crystals catalyzes the caramelization of the reducing sugar. UV spectra of the colorant fractions (2.5-3.5 ml) from HPGPC for sulfitation sugar and carbonation sugar after storage (Figure 6) show carbonation sugar colorant had stronger absorption at 280-285 nm, and more 5-HMF, caramel, humic acids and melanoidins; further studies will be done to characterize these compounds. UV absorption maxima of these "compounds" are at 265 nm (humic acid), 282 nm (caramel), 285 nm (5-HMF), and 300 nm (melanoidins), as shown in Figure 7. In carbonation, a high percentage of reducing sugar destruction in the highly alkaline condition of the first carbonation stage helped cause formation of color-bearing compounds of the 5-HMF and humic categories, and these reducing sugar degradation products play an

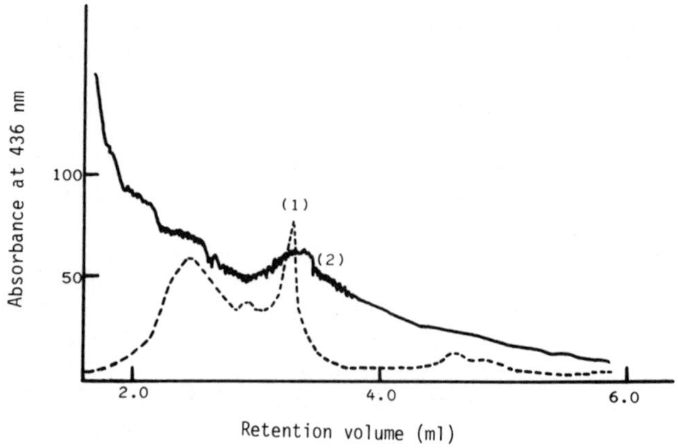

Figure 2. HPGPC fractionation of sugar (—) and sugar colorant (– – –).

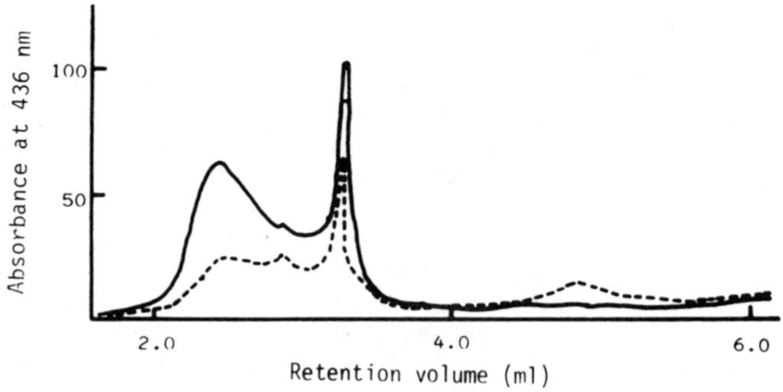

Figure 3. HPGPC colorant molecular profiles of sulfitation sugar (—) and carbonation sugar (– – –) before storage.

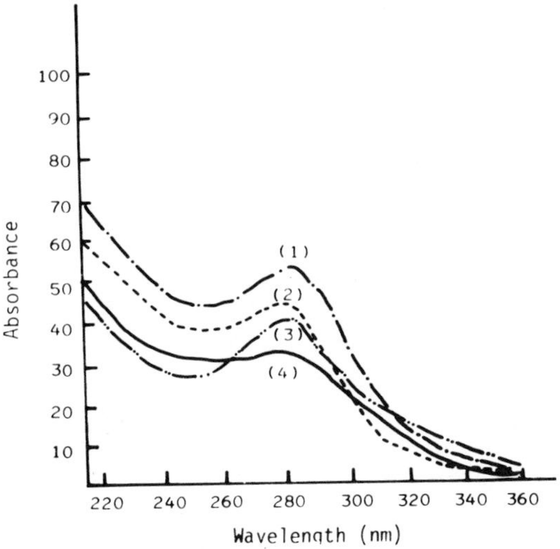

Figure 4. UV spectra of colorant fractions from HPGPC for fresh sulfitation sugar and carbonation sugar. Key for 2–3-mL fraction: — • —, sulfitation sugar colorant; – – –, carbonation sugar colorant. Key for 4.5–5.5-mL fraction: – • • –, carbonation sugar colorant; —, sulfitation sugar colorant.

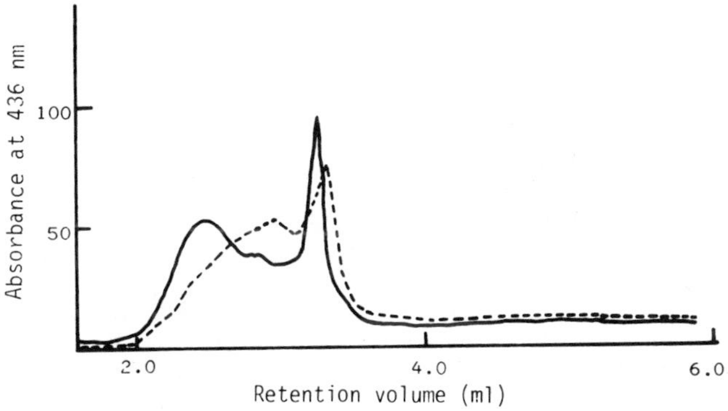

Figure 5. HPGPC colorant molecular profiles of sulfitation sugar (—) and carbonation sugar (– – –) after storage.

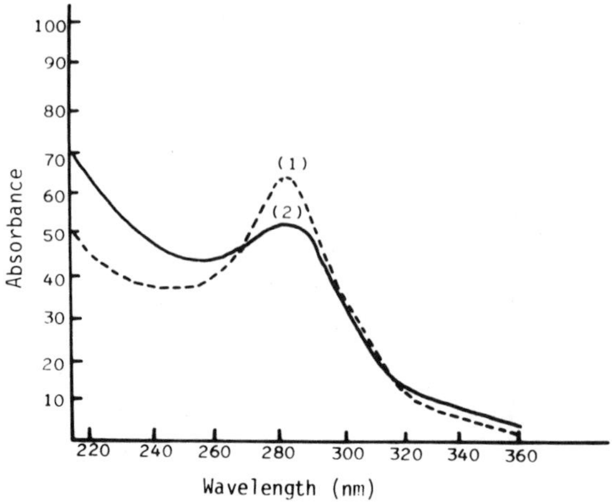

Figure 6. UV spectra of colorant fraction (2.5–3.5-mL) from HPGPC for sulfitation sugar (—) and carbonation sugar (- - -) after storage.

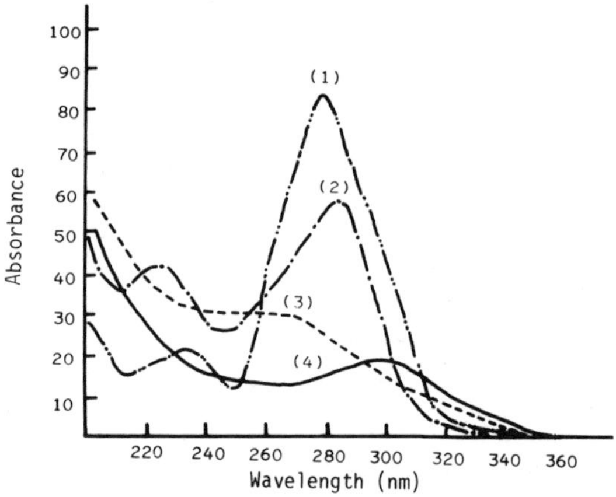

Figure 7. UV spectra of main colorants. Key: – • • –, 5-HMF; — • —, caramel; – – –, humic acids; and —, melanoidins.

important role in melanoidin and caramel formation in the sugar crystal during storage. They are formed by slow polymerization of reducing sugars present as impurities along with ash constituents, imbedded between the crystal layers. The nature of the ash constituents played an important role in the rate of deterioration of the sugar. Carbonates evidently catalyze the caramelization much faster than sulphites, leading to faster development of color in carbonation sugars. Sulphitation inhibited the melanoidin formation, presumably by blocking the carbonyl function. Reduction of carbonate content by replacing the second carbonation stage with sulphitation or phosphatation, and improvement of first carbonation technique so as to minimize reducing sugar destruction, were recommended to improve the keeping quality of the carbonation sugars.

Conclusions

A rapid method developed for sugar colorant isolation by adsorption column chromatography has been proven to possess excellent efficiency and simplicity. The mini-column adopted, a SEP-PAK C_{18} cartridge, is able to isolate colorants from crystal sugar of very low color. Colorants adsorbed by the column cover the same range of molecular size and polarity as the colorants before isolation. By using this technique, highly concentrated and sugar-free color solution can be obtained simultaneously from sugar products. For fractionation of the isolated colorants, a HPLC system for gel permeation chromatography was developed. The efficiency is considered higher than for the conventional gel filtration technique.

Spectrophotometric and chromatographic studies indicated that the color-bearing compounds responsible for the sugar color development were 5-HMF, caramel, humic acids, and melanoidins. Reduction of these compounds by improving first carbonation technique to minimize reducing sugar destruction would improve color development of the carbonation sugar.

Literature Cited

1. Keely, F.H.C.; Brown, D. W. "Sugar Technology Reviews"; Hutson, M.; McGinnis, R. A., Eds.; Elsevier: Amsterdam, 1978; Vol. 6, p. 1-41.
2. Gillett, T. R. "Principles of Sugar Technology"; Honig, P., Ed.; Elsevier: Amsterdam, 1953; Vol. 1, p. 230-234.
3. Tsuchida, H.; Komoto, M. Proc. Res. Soc. Japan Sugar Refin. Technol. 1969, 21, p. 89-94.
4. Chen, W. Proc. 14th Congr. Intern. Soc. Sugar Cane Technol. 1971, p. 1564-1568.
5. Monterde, J.; Ruso, R.; Fajardo, R. Bol. Cienc. Tech. Univ. Central Las Villas 1970, 5 (6), 27-41.
6. Cortis-Jones, B. Intern. Sugar J. 73 (7),219.

7. Ramon, S.; Salahuddin, S. Proc. 15th Congr. Intern. Soc. Sugar Cane Technol. 1974, p. 1412-1424.
8. Payne, J. H. "Sugar Cane Factory Analytical Control"; Elsevier: Amsterdam, 1969, p. 632-636.
9. Lowry, O. H.; Rosebrough, N. J.; Farr, A. L.; Randall, R. L. J. Biol. Chem. 1951, 193, 265.
10. Colburn, C. R.; Schoch, T. J. "Methods in Carbohydrate Chemistry"; Whistler, R. L., Ed.; Academic, New York, 1964, Vol. 4, p. 157-160.
11. Tu, J. C.; Kondo, A.; Sloane, G. E. Proc. 15th Congr. Intern. Soc. Sugar Cane Technol. 1974, p. 1393-1401.
12. Linecar, D.; Paton, N. H.; Smith, P. Proc. 1978 Tech. Sess. Cane Sugar Refin. Res. 1979, p. 81-91.

RECEIVED November 3, 1982

7

Colored Compounds Formed by the Interaction of Glycine and Xylose

H. E. NURSTEN and ROSEMARY O'REILLY
University of Reading, Department of Food Science, London Road, Reading, RG1 5AQ, United Kingdom

Colored compounds were isolated and characterised from the reaction products of xylose (1M) and glycine (1M), refluxed for 2 h, initial pH 6. The solubles in light petroleum ether (b.p. 60-80°) were fractionated into at least 13 peaks by semipreparative reverse-phase HPLC, using a water-methanol gradient. The peaks were collected and freed from adjacent peaks and shoulders by further HPLC, before examination by mass spectrometry and electronic absorption spectroscopy. Compounds of the following molecular masses were progressively eluted on reverse-phase HPLC: 114, 346, 178, 194, 194, 194, 192, 272, 272, 371, 257, 326, 342 daltons. With the exception of the first, 4-hydroxy-5-methyl-3(2H)-furanone, all compounds were colored. The seventh compound dominated the chromatogram and was identified as 2-furfurylidene-4-hydroxy-5-methyl-3(2H)-furanone. Its derivative with a further furfurylidene substituent on its methyl group (270 daltons) was not found to be present.

Color production is the primary characteristic of the Maillard reaction, yet surprisingly little is known about any chromophores present (1). In view of the labile nature of at least some of the browning products, rapid separation of these complex mixtures with minimal exposure to heat and air is necessary. High-performance liquid chromatography promised to provide almost the ideal answer.

High-performance liquid chromatography has been used for the separation of Amadori products (2) and for the isolation of 3,8-dihydroxy-2-methylchromone from the products of xylose degradation at 100° (3). In 1981 (4) we reported an HPLC separation scheme for the colored products of the xylose/glycine reaction.

0097-6156/83/0215-0103$06.00/0

To make the initial work relatively simple, we chose to investigate the least polar material formed in that system, that soluble in light petroleum, b.p. 60–80°.

Experimental

Preparation of browning product mixtures Xylose (15 g) and glycine (7.5 g) were dissolved in Sørensen's phosphate buffer (5) (1/15M pH 8.2, 100 ml) and heated under reflux for 2 h. On 250-fold dilution with water, the absorbance at 450 nm was 0.75 ± 0.05. The preparation was also carried out in the absence of buffer, initial pH 6.0. Larger amounts were prepared on 5 and 10 times this scale.

Solvent extraction Immediately after preparation, the browning product mixture was continuously extracted first with light petroleum, b.p. 60–80°, and then with ethyl ether. About 2 h were required for each stage before no further color was extracted. In each case the solvent was removed under vacuum $< 60°$.

HPLC The sample was injected on to an ACS Model LC750 chromatograph with dual pumps (Applied Chromatography Systems Ltd, Luton, Beds, U.K.) by means of a loop injection valve, Model 7120 (Rheodyne Inc., Berkeley, CA 94710, U.S.A.) A single-channel, variable-wavelength UV detector with a 1-cm path length (8-μl volume) quartz flowcell was used, Model CE212 (Cecil Instruments Ltd, Cambridge, U.K.) The chart recorder, Model BS600 (Bryans Southern Instruments Ltd, Surrey, U.K.) was preset at 10 mV full scale deflection. For semipreparative HPLC, a 50-μl loop was made from AISI Type 316 stainless steel tubing.

The HPLC was carried out at room temperature on a 0.8 i.d. x 25 cm stainless steel column, laboratory-packed with Spherisorb ODS (Phase Separations Ltd, Clwyd, U.K.), a reverse-phase material.

Samples were injected as concentrated solutions in methanol. The initial composition of the solvent was 0% methanol/100% water, but was changed at 5%/min using a linear gradient program, the solvent composition reaching 100% methanol in 20 min. Methanol of HPLC grade (Rathburn Chemical Ltd, Peebleshire, U.K.) was used with solvent flow rate of 6 ml/min. With the recorder set at 0.2 absorbance units full scale, the respective detection wavelengths and sample loadings were 450 nm (50 μl and 20 μl) and 260 nm (2 μl). Assessment of peak purity using 260 nm takes advantage of the increased sensitivity and also detects colorless but UV-absorbing compounds.

TLC Separation by TLC was performed on silica layers (BDH Chemicals Ltd, Poole, Dorset, U.K.). Separated colored components were detected visually.

Electronic absorption spectroscopy Samples were examined in methanol or ethanol on a double-beam spectrophotometer, Model SP800B (Pye Unicam Instruments Ltd, Cambridge, U.K.)

Infrared spectroscopy KBr discs (3-mm) were made with a micropelleting kit (Model 198854) and mounted in a micropellet holder (Model 195465), used in conjunction with a C621 beam condenser (all ancillary equipment by Beckmann RIIC Ltd, Purley, Surrey, U.K.). Spectra were obtained on a Pye Unicam SP200G spectrometer.

PMR All PMR spectra were recorded at 100 MHz, using a Varian HA-100 instrument (Varian Associates Ltd, Walton-on-Thames, Surrey, U.K.), chloroform-d as solvent and tetramethylsilane as internal standard (NMR Ltd, High Wycombe, Bucks, U.K.), with a normal sweep width of 1000 Hz and sweep time of 500 s. Fourier transform PMR spectra were recorded using a Bruker WH 300 instrument (Bruker Instruments Inc., Manning Park, Billerica, MA 01821, USA) and methanol-d_4 as solvent (NMR Ltd).

Mass spectrometry Low resolution spectra were determined with an MS9 instrument (AEI Ltd, Urmston, Manchester, U.K.) at 70 eV. A probe was used, normally at 100°. Molecular ions were also examined under high resolution. For A178, high resolution data were obtained using the ARC/FRI Mass Spectrometry Service (Food Research Institute, Norwich, Norfolk, U.K.) and perfluorokerosine as reference.

Results

Here we are solely concerned with the light petroleum solubles (designated Fraction A), derived from a xylose/glycine model system. Fraction A was obtained as a yellow solid in 0.05% yield. Analysis by HPLC at 450 nm showed it to consist of at least 25 components (Figure 1), and examination at 315 and 260 nm further increased the number detected.

Reproducibility of HPLC separations requires care. Use of a 0 to 100% methanol gradient helped, as did purging the column between injections with methanol and then with water, for 5 min each. Although Fraction A was stable in methanol at -20° without change in HPLC profile, it did alter after more than a day at room temperature.

Thirteen of the components of Fraction A were isolated by semipreparative HPLC, using multiple injection of methanolic solutions. Two of the components, A114 and A192 (the numbers represent the molecular weights as determined by mass spectrometry), were present in far larger amounts than the others. To isolate sufficient material of the others, the column had to be overloaded with respect to the major components. Even so, not enough material for mass spectrometry was obtainable for some of the components. To purify compounds producing individual peaks, they were subjected to HPLC, sometimes twice more. Each step caused losses of 30-50%. The process for A178 is shown in Figure 2.

Table I shows the amounts obtained of the different components and the identification techniques which it was feasible to apply to them in consequence. All compounds isolated (except A114) were colored. With the exception of A192 and A178, the amounts of samples are estimated, based on peak height, using detection at 450 nm and assuming absorptivities equal to that of A192.

Data for A114 and A192 are summarized in Tables II and III, alongside data for 4-hydroxy-5-methyl-3(2*H*)-furanone and 2-furfurylidene-4-hydroxy-5-methyl-3(2*H*)-furanone, respectively. A178 is a new colored compound, the data for which are summarized in Table IV. Both HPLC and MS data for the 10 other colored compounds isolated are summarized in Table V.

Discussion

The discussion is in three parts: (a) the 4 compounds for which detailed analytical data have been obtained, (b) compounds with MS data only, and (c) the examination of Fraction A for the presence of LS270.

(a) Compounds A114, A192, A178 and A326

A114, $C_5H_6O_3$ This colorless compound is identified as 4-hydroxy-5-methyl-3(2*H*)-furanone. The data for the compound isolated are compared with those for 4-hydroxy-5-methyl-3(2*H*)-furanone in Table II. 4-Hydroxy-5-methyl-3(2*H*)-furanone has been previously isolated from several foods (12), including raspberries (14), beef broth (7, 14) and shoyu (14), and from the reaction products of primary amines (11) or methylammonium acetate (8) with pentoses. The almost superimposable IR spectra are particularly noteworthy.

A192, $C_{10}H_8O_4$ The data for A192 match closely the data for 2-furfurylidene-4-hydroxy-5-methyl-3(2*H*)-furanone (see Table III). The two have identical HPLC retention and TLC R_fs, and closely matching mass, PMR, IR, and electronic absorption spectra. They have identical melting points, a 1:1 mixture melting sharply at the same temperature. There can be no doubt about their identity. 2-Furfurylidene-4-hydroxy-5-methyl-3(2*H*)-furanone has previously been isolated from the products of

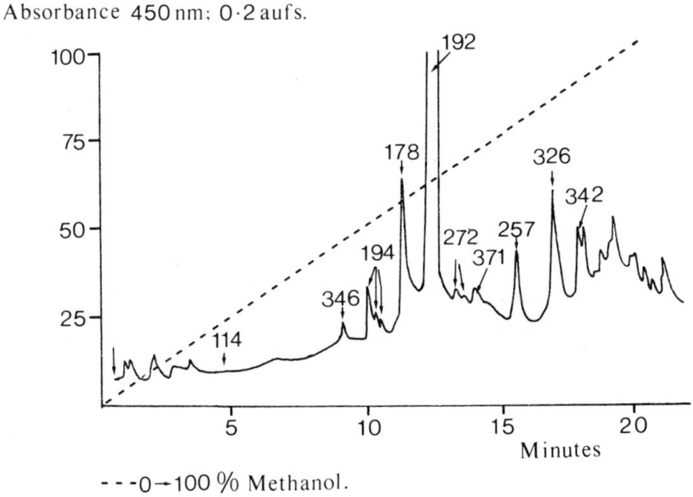

Figure 1. Separation of Fraction A by HPLC.

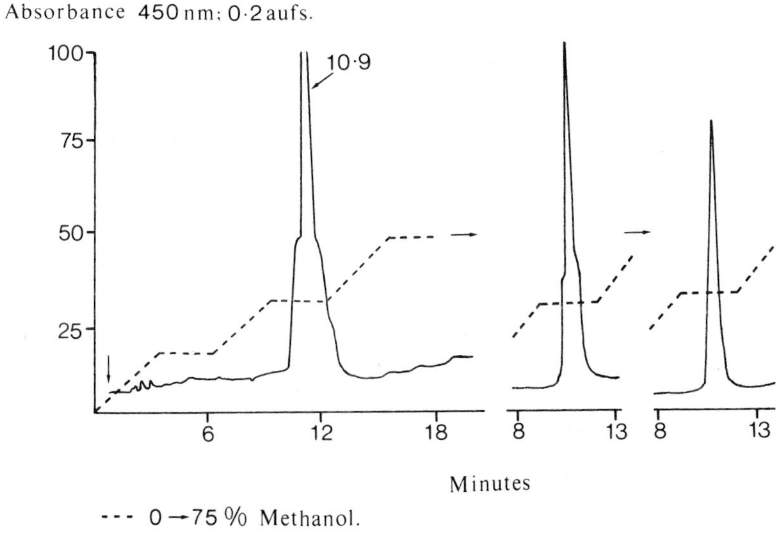

Figure 2. Purification of A178 in three stages by HPLC.

TABLE I

TECHNIQUES APPLIED TO THE SEPARATION AND ANALYSIS OF COMPONENTS OF FRACTION A

Component Molecular Weight (daltons)	Approximate Amount Obtained (mg)	Techniques							
		TLC	HPLC	M.p.	MS resolution low	MS resolution high	IR	PMR	EAS[†]
114	-	*	*		*	*	*		*
192	5	*	*	*	*	*	*	*	*
178	0.5		*		*	*		*	*
326	0.5		*		*	*			*
346	0.1		*		*				
194a	0.2		*		*				
194b	0.1		*		*				
194c	0.05		*		*				
272a	0.05		*		*				
272b	0.05		*		*				
371	0.1		*		*				
257	0.2		*		*				
342	0.2		*		*				

† EAS - Electronic absorption spectroscopy

TABLE II

COMPARISON OF DATA FOR A114 WITH 4-Hydroxy-5-methyl-3(2H)-furanone

	A114	4-Hydroxy-5-methyl-3(2H)-furanone	
Mass spectral ions (m/z) (% relative abundance)	$M^{+\cdot}$ 114(72)	114(60)	(6)
	43(100)	43(100)	
	55(8)	55(10)	
	42(6)	42(9)	
	29(6)	29(9)	
	44(5)	44(31)	
	115(5)	115(3)	
	71(4)	71(3)	
	58(4)	58(4)	
Accurate mass	114.032	114.032	(7)
	Calculated for $C_5H_6O_3$	114.0317	

Infrared peaks cm^{-1}

A114	Reference data (7)	A114	Ref. data (7)
3195 (broad)	3195 (broad)	1315	1315
2950	2950	1195	1195
2920	2920	1143	1140
2850	2850	1000	1000
1745	~1740	958	955
1705	1705	920	920
1695	1695	883	885
1600-1650	1600-1650	745	745
1450	1450	698	700
1405	1405	660	660
1365	1365		

Electronic absorption, λ max, nm

288-290	287 (7)
(ethanol)	(water)

TABLE III

COMPARISON OF DATA FOR A192 WITH

2-Furfurylidene-4-hydroxy-5-methyl-3(2H)-furanone

	A192	2-Furfurylidene-4-hydroxy-5-methyl-3(2H)-furanone Experimental	Literature (8)
HPLC retention, min	12.4	12.4	-
TLC (silica) R_f			
Methyl acetate : water, 8:1	0.72	0.72	-
Ethyl acetate : toluene, 1:1	0.44	0.44	0.5
Melting point, ^{o}C	172	172	172
	yellow needles	yellow needles	
Mass spectral ions m/z (% relative abundance)	$M^{+\cdot}$ 192(100)	$M^{+\cdot}$ 192(100)	$M^{+\cdot}$ 192(84)
	121(90)	121(78)	121(100)
	65(14)	65(14)	65(19)
	43(46)	43(33)	43(42)
	52(18)	52(8)	
	193(11)	193(11)	
	39(11)	39(8)	
	51(11)	51(5)	
Accurate mass	192.042	Not determined	Not given

Calculated for $C_{10}H_8O_4$ 192.042

	A192	2-Furfurylidene-4-hydroxy-5-methyl-3(2H)-furanone	
		Experimental	Literature (8)
PMR	2.38(sCH_3) 3H	2.36(sCH_3) 3H	2.44(sCH_3) 3H
	6.58(mCH) 1H	6.58(mCH) 1H	6.59(mCH) 1H
	6.75(sCH) 1H	6.75(sCH) 1H	6.80(sCH) 1H
	7.00(dCH) 1H J = 4Hz	7.02(dCH) 1H J = 4Hz	7.05(dCH) 1H J = 4Hz
	7.58(dCH) 1H J = 2Hz	7.59(dCH) 1H J = 2Hz	7.61(dCH) 1H J = 2Hz
	OH group not located	OH group not located	OH located by acetylation
IR	3150–3200	3150–3200	
cm^{-1}	1690	1690	1695
	1605	1605	1605
	1475	1475	
	1325	1325	
	960	960	
EAS			
EtOH λ_{max}	355 nm	355 nm	Not given
log ϵ	4.2	4.2	
H_2O	366 nm	366 nm	365 nm

TABLE IV
DATA FOR A178

HPLC retention: 11.2 cm.
MS: $M^{+\cdot}$178(100), 150(43), 107(29), 31(26), 163(25), 79(23), 43(23), 39(21), 160(20), 121 m/z (19% relative abundance).
Accurate molecular mass, 178.06352; calculated for $C_{10}H_{10}O_3$, 178.062989 u.
High-resolution mass spectral data for fragment ions:

Observed ion mass	ion formula	Observed ion mass	ion formula
163.0708	$C_9H_7O_3$	83.0474	C_5H_7O
150.0661	$C_9H_{10}O_2$	82.0781	C_6H_{10}
149.0274	$C_8H_5O_3$	82.0394	C_5H_6O
135.0428	$C_8H_7O_2$	81.0674	C_6H_9
131.0132	$C_8H_3O_2$	81.0318	C_5H_5O
122.1099	C_9H_{14}	80.0611	C_6H_8
122.0704	$C_8H_{10}O$	79.0536	C_6H_7
121.0987	C_9H_{13}	78.0434	C_6H_6
	C_8H_9O	77.0373	C_6H_5
108.0949	C_8H_{12}	77.0211	$C_2H_5O_3$
108.0548	C_7H_8O	75.0458	C_3H_7O
107.0842	C_8H_{11}	75.0254	C_6H_3
107.0520	C_7H_7O	69.0705	C_5H_9
107.0461	C_7H_7O	69.0330	C_4H_5O
96.0955	C_7H_{12}	68.0654	C_5H_8
96.0580	C_6H_8O	67.0557	C_5H_7
95.0892	C_7H_{11}	66.0505	C_5H_6
95.0498	C_6H_7O	65.0404	C_5H_5
94.0814	C_7H_{10}	63.0240	C_5H_3
94.0440	C_6H_6O	61.0317	$C_2H_5O_2$
94.0000	$C_5H_2O_2$	60.0234	$C_2H_4O_2$
93.0735	C_7H_9	56.0641	C_4H_8
91.0572	C_7H_7	55.0570	C_4H_7
91.0407	$C_3H_7O_3$	55.0198	C_3H_3O
83.0835	C_6H_{11}		

Table IV. Continued

With the exception of m/z 163, all observed masses lie within ± 0.005 u. of those calculated.

λ_{max}^{MeOH}, 255 nm, 290 nm, tailing broadly into the visible region and responsible for the yellow color.

FT PMR: aromatic region

Chemical Shift (ppm)	Coupling constant J (Hz)	Multiplicity	Relative area	Assignment (9,10) 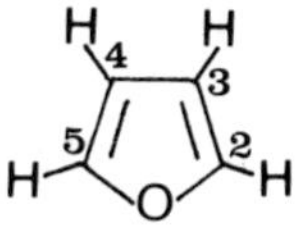
7.40-7.41	1.5	Quad.	1	C-2
6.33-6.35	1.8	Quad.	1	C-3
6.19-6.20	3	Doublet	1	C-4

Insufficient material for clean spectrum in the aliphatic region.

TABLE V
MASS SPECTRAL AND HPLC DATA
FOR 10 OTHER COMPONENTS OF FRACTION A

Component	HPLC retention (min)	Mass spectrum: $M^{+\bullet}$ and 10 most prominent peaks, m/z (% relative abundance)
A346	8.8	$M^{+\bullet}$ 346(22), 43(100), 29(80), 31(33), 81(32), 60(32), 41(30), 55(23), 45(20), 57(17)
A194a	10.0	$M^{+\bullet}$ 194(28), 31(100), 81(93), 29(83), 44(84), 149(79), 63(66), 91(49), 39(34), 41(34), 55(34)
A194b	10.3	$M^{+\bullet}$ 194(62), 81(100), 43(42), 55(16), 94(14), 53(12), 192(11), 121(11), 57(9), 60(8) M* observed 159.6, $(176)^2/194 = 159.7$. 194→176 + 18 M* observed 117.5, $(151)^2/194 = 117.5$. 194→151 + 43
A194c	10.5	$M^{+\bullet}$ 194(10), 29(100), 43(52), 81(48), 31(48), 136(43), 108(39), 39(33), 52(28), 91(24), 27(22), 41(22)
A272a	13.1	$M^{+\bullet}$ 272(8), 81(100), 43(57), 121(48), 192(48), 160(24), 147(21), 271(20), 95(19.5), 134(19)
A272b	13.6	$M^{+\bullet}$ 272(2), 192(100), 28(100), 43(90), 206(85), 121(83), 177(79), 135(79), 81(66), 258(29), 244(29)
A371	14.2	$M^{+\bullet}$ 371(1), 177(100), 81(100), 43(64), 258(55.5), 255(54), 206(47), 256(45.5), 135(44.5), 228(44.5), 192(42)
A257	15.5	$M^{+\bullet}$ 257(5), 45(100), 43(32.5), 29(23), 75(23), 192(18), 149(18), 121(18), 81(13), 76(13), 55(12)
A326	16.7	$M^{+\bullet}$ 326(100), 325(69), 77(36), 43(33), 149(25), 57(23), 170(20), 81(20), 233(19), 55(19) M* observed 198.3, $(215)^2/233 = 198.4$. 233→215 + 18 Accurate mass observed = 326.07134, calculated for $C_{20}H_{10}N_2O_3$ = 326.069137 $C_{18}H_{14}O_6$ = 326.079030
A342	18.1	$M^{+\bullet}$ 342(100), 77(37.5), 341(31), 109(29), 43(27), 60(26), 249(26), 149(24), 343(24), 233(22.5)

interaction of isopropylammonium acetate with xylose (13) and synthesized by the reaction of 4-hydroxy-5-methyl-3(2H)-furanone with 2-furaldehyde (8). It would be useful to confirm the presence of the OH group by acetylation as Ledl and Severin have done.

A178, $C_{10}H_{10}O_3$ A178 is a new colored compound. The molecular formula was derived by high resolution mass spectrometry (Table IV). Further support comes from the following observations:

(1) The M+1 peak is consistent with the presence of 10 carbons (1.05/9.05 x 100 = 11.6%/1.1% = 10.5).

(2) It is likely to be formed from 2 x 5 carbon moieties.

(3) A178 would be expected to possess some similarities to A192, both coming from Fraction A and having close HPLC retention times.

(4) High resolution MS data gave ion fragments which can only come from $C_{10}H_{10}O_3$.

Important losses from the molecular ion are at M-15, M-18, M-28, and M-43.

The formulae for the neutral fragments lost in transitions of metastable ions and confirmed by high resolution MS data build up into a scheme (Figure 3). Most metastable ions appeared within 0.1% of the calculated value, but two, at m/z 58.5 and 89.3, did not (calc. for m/z 150→94, 58.9; 163→121, 89.8).

Successive losses of 28 daltons are due to the expulsion of CO in three stages by two parallel but staggered routes.

The presence of a monosubstituted furan ring at C-5 is evident from the aromatic region of the PMR spectrum. A cyclic structure is favored by the high stability of the molecular ion; a straight chain would require a terminal carbonyl group, and the PMR spectrum does not show an aldehydic proton. $C_{10}H_{10}O_3$ has 6 rings and/or double bonds, and the conjugated unsaturation of the 2-furyl group needs to be extended to account for the pinkish-yellow color.

The metastable ion at m/z 93.8 is of particular structural significance, involving m/z 122, $C_8H_{10}O$→m/z 107, C_7H_7O. This shows a methyl group to be present in a group of eight carbon atoms, having lost 2 mol. CO in successive stages. The furyl ring can supply only one of these CO molecules, so the rest of the molecule must provide at least one and probably both. In the latter case, $C_8H_{10}O$ still contains a furyl ring. The ion at m/z 43 is likely to be $CH_3C{\equiv}\overset{+}{O}$, which implies that the methyl group must be next to an oxygen-carrying C. Such reasoning leads to 1 as the most probable structure for A178. The evidence is not unequivocal, and problems that remain to be resolved are:

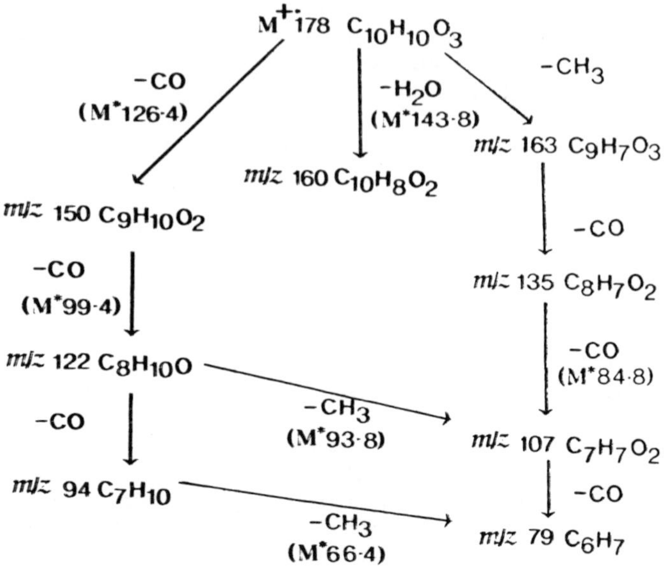

Figure 3. Mass spectral fragmentation of A178 based on high resolution mass spectral data and observed metastable ions.

High resolution data showed no listing at m/z *160, but such an ion was clearly observed in the low resolution spectrum and is implied by the metastable ion.*

(a) the absence of a fourth H(C-6) in the aromatic region of the PMR,

(b) the hydrogen-transfer necessary to liberate C-9 as CO from 1.

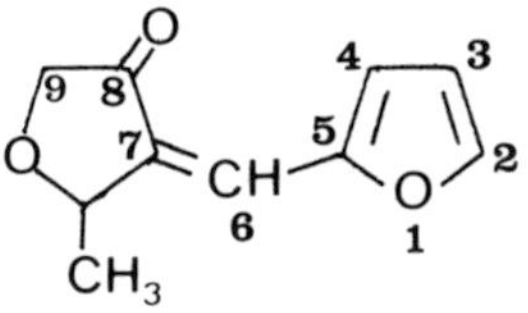

1

A326 Methanolic solutions of A326 are dark brownish-orange. Two formulae, $C_{20}H_{10}N_2O_3$ and $C_{18}H_{14}O_6$, fall just outside the normal ±5 p.p.m. tolerance limits, but the former lies closer. In addition, the compound possesses a multiple of 5 carbon atoms. It has featureless absorption in the UV region, and in this respect resembles the total browning product.

(b) Compounds with MS data only Some generalizations can usefully be made about these spectra:

(i) Some fragments occur in several of them, at m/z 39, 43, 55, 81, 121, 149, and 192.

(ii) The ion at m/z 81 is the most striking of these and is in fact present in all the spectra. It is most likely due to a 2-furfuryl group.

(iii) Because the ion at m/z 121 is very prominent in the spectrum of A192 and of LS270, it can be attributed to structure 2.

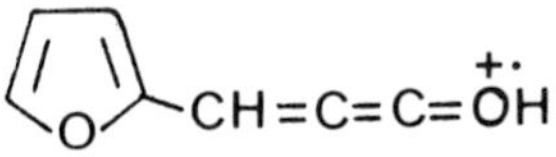

2

It is present in spectra of A194b, A272a, A272b, A371, and A256, but also in A178, where high resolution mass spectrometry gave a different composition, C_8H_9O.

(iv) The ion at m/z 135 could well be 2 with a 5-methyl group, since it is very prominent in spectra of appropriate compounds synthesized by Ledl and Severin (8). It is present in A272b and A371.

(v) The ion at m/z 43 is likely to be due to the acetylium ion-radical, $CH_3C \equiv \overset{+\cdot}{O}$.

(vi) The ion at m/z 192 is the base peak of A272b and important in spectra of A272a, A371, A256, and A326. Its structure is likely to correspond to A192.

(vii) The spectra for A326 and A342 are noteworthy in having their molecular ions as base peaks, as do A178 and A192.

It is difficult to assemble this information into a coherent picture at this stage. Further data are required, but it will need considerable effort to obtain them because of poor yields.

(c) Examination of Fraction A for LS270 A key compound identified in this study was 2-furfurylidene-4-hydroxy-5-methyl-3(2*H*)-furanone. Its methyl group is sufficiently activated to condense with 2-furaldehyde in a second stage, which Ledl and Severin (8) have demonstrated synthetically (Figure 4).

Figure 5 shows the HPLC separation of Fraction A spiked with LS270. The peak at 16.9 min in Fraction A (unspiked) was collected, but mass spectrometry gave its molecular weight as 326 daltons. No fragments characteristic of LS270 were present in its mass spectrum.

Whereas quite large amounts of free 4-hydroxy-5-methyl-3(2*H*)-furanone were found, 2-furaldehyde was not evident in Fraction A, both with normal column loading and with overloading. The amount of 2-furaldehyde formed appears to limit the formation of LS270 and even of A192. As long as 4-hydroxy-5-methyl-3(2*H*)-furanone is present in excess of 2-furaldehyde, A192 is likely to be formed in preference to LS270. 2-Furaldehyde is probably involved in yet other reactions, limiting even further the amount of A192 formed. Formation of 2-furaldehyde is favored by low pH, and thus, provided sufficient furanone is produced, LS270 is more likely to be found in model systems with a low starting pH, but no evidence for its presence was obtained with our model systems, starting pH 6.0.

Conclusion

Use of HPLC has permitted the separation and isolation of several colored Maillard products, illustrating at the same time the complexity of the mixtures formed even from relatively simple reactants. Thirteen compounds from Fraction A were isolated, their molecular weights being, in order of elution: 114, 346, 194, 194, 194, 178, 192, 272, 272, 371, 257, 326, and 342. With the exception of the first compound, all are colored.

Four compounds were characterized in more detail, and two of them had previously been associated with browning, namely, 4-hydroxy-5-methyl-3(2*H*)-furanone and 2-furfurylidene-4-hydroxy-5-methyl-3(2*H*)-furanone. A178 is a new yellow compound; it probably has structure **1**, which remains to be confirmed by synthesis.

The dark brownish-orange A326 has the formula $C_{20}H_{10}N_2O_3$.

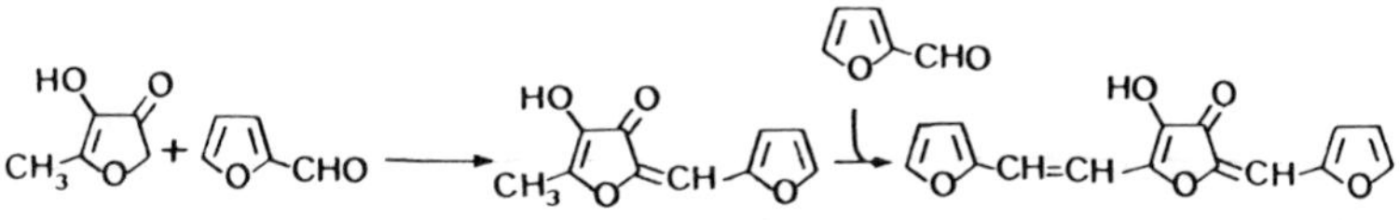

Mol. wt. daltons	114 colorless		192 yellow	270 purplish red
HPLC cm.	4·4	7·6	12·4	16·9

Figure 4. Ledl and Severin's route to colored compounds from pentoses (8).

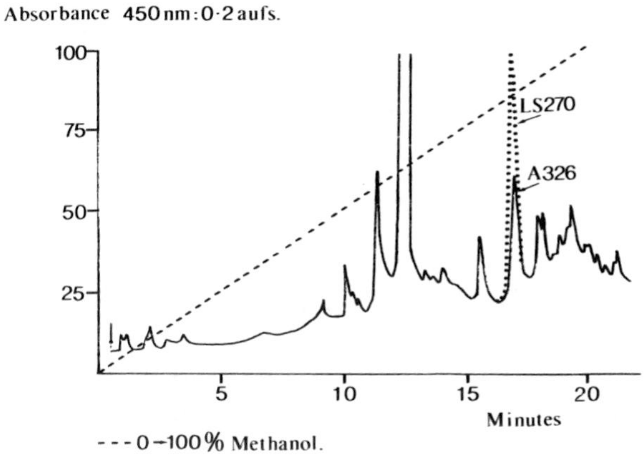

Figure 5. HPLC separation of Fraction A spiked with LS270.

Of the remaining nine compounds, structural information is limited to that derivable from low-resolution mass spectra, which indicate the presence of furfuryl groups in all of them. For progress to be made, high resolution mass spectra would be helpful, but scale-up of preparation would be required. Synthesis of appropriate model compounds for comparison and automation of analysis by combined LC/MS (15, 16) are two promising approaches.

Acknowledgments

We are grateful to Dr. F. Ledl for samples of the synthetic compounds LS192 and LS270, to Lady Richards for FT-PMR, and to Dr. D. Manning for mass spectra.

Literature cited

1. Nursten, H.E. Food Chem. 1981, 6, 263-277.
2. Takeoka, G.R.; Coughlin, J.R.; Russell, G.F. "Liquid Chromatographic Analysis of Food and Beverages"; Charalambous, G., Ed.; Academic Press Inc.: New York, 1979; Vol. 1, 179-214.
3. Lindgren, G.; Pernemalm, P.-A. J. Liq. Chromatog. 1980, 3, 1737-1742.
4. O'Reilly, R. Progr. Food Nutr. Sci. 1981, 5, 477-481.
5. "Documenta Geigy, Scientific Tables"; Diem, K., Ed.; Geigy Pharmaceutical Company Ltd: Manchester, U.K., 6th ed., 1968.
6. Mass Spectrometry Data Centre. "Eight-Peak Index of Mass Spectra"; A.W.R.E.: Aldermaston, Berkshire, U.K., 1975.
7. Tonsbeek, C.; Plancken, A.; Weerdhof, T. J. Agr. Food Chem. 1968, 16, 1016-1021.
8. Ledl, F.; Severin, T. Z. Lebensm.-Unters. Forsch. 1978, 167, 410-413.
9. Williams, D.H.; Fleming, I. "Spectroscopic Methods in Organic Chemistry", 2nd ed.; McGraw-Hill: U.K., 1973.
10. Dyke, S.F.; Floyd, A.J.; Sainsbury, M.; Theobald, R.S. "Organic Spectroscopy", 2nd ed.; Longmans: New York, 1978.
11. Severin, T.; Seilmeier, W. Z. Lebensm.-Unters. Forsch. 1967, 134, 230-232
12. Ruiter, A. Lebensm.-Wiss. Technol. 1973, 6, 142-146.
13. Severin, T.; Kronig, U. Chem. Mikrobiol. Technol. Lebensm. 1972, 1, 156-157.
14. van Straten, S.; de Beauveser, J.C.; Visscher, C.A., Ed. "Volatile Compounds in Food", 4th ed.; TNO Division for Nutrition and Food Research: Zeist, Netherlands, 1977, and supplements.
15. Throck Watson, J. "Biochemical Applications of Mass Spectrometry, 1st Supplementary Vol."; Waller, G.R.; Dermer, O.C., ed.: Wiley-Interscience: New York, 1980, 3-11.

16. McLafferty, F.W. "Biochemical Applications of Mass Spectrometry, 1st Supplementary Vol."; Waller, G.R.; Dermer, O.C., ed.; Wiley-Interscience: New York, 1980, 1159-1167.

RECEIVED December 14, 1982

FLAVORS, TASTES, AND ODORS

8

Conditions for the Synthesis of Antioxidative Arginine–Xylose Maillard Reaction Products

G. R. WALLER, R. W. BECKEL, and B. O. ADELEYE
Oklahoma State University, Oklahoma Agricultural Experiment Station, Department of Biochemistry, Stillwater, OK 74078

An antioxidative Maillard reaction product (AX) can be formed from arginine and xylose. Optimum results were obtained by refluxing 1M arginine-HCl with 1M xylose in water at 100° C for 10–20 h at initial pH of approximately 5. Tris buffer and pressures up to 5 bar using N_2, O_2, and air had negligible effect on yield of antioxidative activity but the use of 1:1 pyridine/water as a reaction medium produced a 2-fold increase in this activity.

A considerable amount of research pertaining to the Maillard reaction has revolved around the conditions of the reaction (1, 2, 3). However, very little attention has been directed toward the conditions for maximizing the yield of antioxidative products. The studies that have been conducted have focused primarily on very specific reactants, as does this study, and although some generalizations can be drawn about reaction parameters, most reactant combinations used so far have only been treated with a single set of reaction conditions.

By increasing the initial pH from 4 to 9 in a solution of glycine and glucose, Kirigaya, et al. (4) found that the yield of undialyzable antioxidative product was increased. Similar increases were also obtained with arginine and glucose. In addition, Kirigaya, et al. reported that increasing the ratio of glycine to xylose increased the yield of antioxidants.

In one of the most extensive studies to date, Tomita (5) found that phosphate buffer more effectively enhanced the yield of antioxidative activity produced from tryptophan and glucose than either veronal or borate buffer. The optimum concentration of phosphate buffer was 0.1 M and the optimum initial pH was 9. He also reported that an increase in antioxidative activity was directly associated with an increase in the molar ratio of sugar to amino acid and was

0097-6156/83/0215-0125$06.00/0

favored by high total reactant concentrations and reaction temperature (140° C). Eichner (6) indicated that low water activity favors the production of antioxidative browning intermediates more than it does the formation of higher-molecular-weight melanoidins. This implies that the yield of higher-molecular-weight antioxidants would be favored by moderate to high reactant concentrations.

Recently, Lingnert and Eriksson (7) reported the effect of varying the amino acid-sugar combinations on the yield of antioxidative activity. Histidine with glucose as well as arginine with xylose produced high levels of antioxidative Maillard reaction products (MRP'S). The effects of initial pH, reaction time, and molar ratio of amino acid to sugar on the yield of antioxidative activity from the histidine-glucose reaction were also reported. A high yield was favored by refluxing a 2:1 molar ratio of histidine to glucose for 20 h at a pH initially between 7.0 and 9.0.

Although Lingnert and Eriksson obtained highly antioxidative products in the arginine-xylose reaction, no attempt was made to define the optimum conditions. In light of this, Foster (8) established that maximum antioxidative activity could be generated by refluxing for three h a 2:1 arginine-xylose mixture buffered at pH 8.0 with 0.1 M phosphate. At least some of the antioxidative activity was clearly associated with a high-molecular-weight fraction. No other reaction parameters were explored in Foster's study. Consequently, the purpose of this present investigation was to discover the optimum reaction conditions under which arginine and xylose would yield, following partial purification, active products. Although many of the antioxidants presently used by the food industry are effective in preventing rancidity, their safety is often questioned by the consumer. Most diets contain MRP's so the possibility of utilizing these as "natural" preservatives is attractive and should not be overlooked.

Materials and Methods: Conditions for Synthesis

Synthesis of MRP All reactions were conducted using the appropriate amounts of L-arginine hydrochloride (Sigma grade, Sigma Chemical Co., St. Louis) and D(+)-xylose (grade II, Sigma) in 50 ml of distilled water; the solutions were refluxed for varying times. All samples produced were stored at -20°C until assayed for antioxidative activity.

Tris Buffer A 1.0 M solution of Tris (tris(hydroxymethyl)aminomethane) (Sigma) was adjusted to a pH of either 7.0 or 8.0 with KOH and conc. HCl.

Organic Additives Pyridine (Reagent grade ACS) was obtained from Eastman Kodak (Rochester, New York). Geraniol was supplied by Chemicals Procurement Laboratories (College

Point, New York) and 1-nonanol was purchased from K and K Laboratories (Jamaica, New York). The other organic solvents were of HPLC grade, from J. T. Baker and Company (Phillipsburg, New Jersey).

Pressure Reactions The Parr pressure reaction apparatus (Parr Instrument Co., Moline, Illinois) was used for all pressure and gas studies. A specially designed Teflon stopper was used in place of the original rubber stopper so that the commercial antioxidants in the rubber would not contaminate the reaction mixture. Standard 99.9% oxygen, lamp-grade nitrogen, and Type-1 grade-E breathing air were obtained from Sooner Supplies (Shawnee, Oklahoma).

Assays for Antioxidative Activity Spectrophotometric and polarographic (9) assays were utilized to monitor levels of antioxidative activity. In the spectrophotometric assay, a 2-ml portion of linoleic acid emulsion (9) was placed in a test tube along with 10 µl of a sample to be assayed, and from this, 200 µl were withdrawn and mixed with 2 ml of 100% methanol and 6 ml of 60% methanol-water. The remainder of the 2 ml of emulsion and sample mixture was then incubated at 37°C for 15 to 20 h. Meanwhile, the absorbance of the methanol solution to which the 200 µl of sample had been added was measured at 234 nm. After incubation, the absorbance of 200 µl of the emulsion and sample mixture was also measured at 234 nm in the same manner. Antioxidative activity (A.O.A.) was calculated using the equation:

$$A.O.A. = \frac{\Delta A_{234c} - \Delta A_{234s}}{\Delta A_{234c}}$$

where ΔA_{234c} is the difference in absorbance between a fresh and an incubated control, and ΔA_{234s} in the analogous change in absorbance of the sample.

The polarographic assay was carried out by adding 4 ml of the linoleic acid emulsion to 100-200 µl of the sample and then measuring the length of time required for 50% of the oxygen to be consumed after the addition of 0.2 ml of a solution of hemin catalyst (prepared by dissolving 5 µmol of bovine hemin (Sigma Chemical Co.) in 500 ml of a one-to-one mixture of 0.02 M potassium phosphate buffer (pH 7.0) and 95% ethanol). Oxygen consumption was monitored using a YSI Model 53 Oxygen Monitor (Yellow Springs Instrument Company, Yellow Springs, Ohio). The polarographic assay is calculated using the equation:

$$A.O.A. = \frac{t_s - t_1}{t_1}$$

where t_s is the time required for 50% of the oxygen to be consumed in the emulsion containing the sample being assayed, and t_1 is the corresponding time for the emulsion to which no sample has been added.

Purification of AX Aliquots (15-ml) of crude reaction material were dialyzed against 250 ml of triply distilled, degassed, N_2-saturated water by use of Spectrapor 6 tubing (43 mm x 10 m, MW cutoff nominal at 1000). The inner contents (retentate) and 250 ml of the outer liquid (dialyzate) were separately lyophilized, weighed, and stored in desiccators at 20 C prior to being assayed.

Results

Effect of Time Figure 1 reflects the change in pH with time, and Figure 2 reflects the change in antioxidative activity of crude product with time, both the spectrophotometric and polarographic assays being used. The pH of the crude reaction mixture reached a constant minimum value after two to five h. The time required to produce maximum antioxidative activity varied between 10 and 20 h depending on which assay was used to monitor it. This difference was not surprising, however, and reflected a problem commonly encountered when autoxidation is measured with more than one assay system. The assays in this case were applied to the linoleic acid emulsion at different stages of autoxidation and, therefore, exhibited different sensitivities to equal amounts of antioxidants.

Effect of Tris Buffer Figures 3 and 4 illustrate the changes in pH and antioxidative activity with time of three reaction systems in which 1.0 M Tris was used as a buffer. No significant difference in antioxidant yield could be observed between the buffered and nonbuffered systems, and all three systems were characterized by a rapid drop in pH during the initial three h of the reaction. Within five h, however, the pH reached a minimum and was constant for the remainder of the reaction.

Effect of Initial pH Figure 5 depicts the change in pH with time and Figure 6 shows the antioxidative activity after 20 h of refluxing of six nonbuffered systems having different initial pH values. In all six reactions, the pH attained a minimum and constant value within five h. The two reaction mixtures characterized by initial pH values of 5.0 and 7.0 generated the maximum amount of activity; in these the pH reached a final value of 3.5 and 4.0 respectively.

Effect of Molar Ratio of Reactants Several mixtures containing different molar ratios of arginine to xylose were refluxed in 25 ml of nonbuffered water for 20 h. The total concentration in all cases was 3.0 M. Figure 7 is a plot of the final activity versus the molar ratio of arginine to xylose, and clearly indicates that the 1:1 ratio was the superior combination. In addition, the bell-shaped

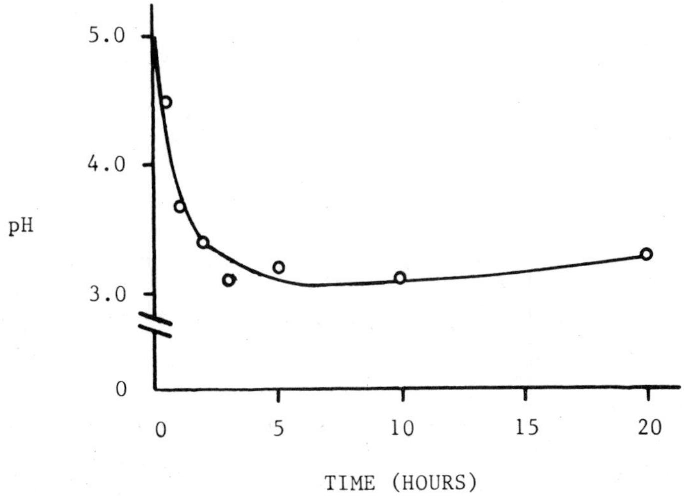

Figure 1. Average change in pH with time of the unbuffered arginine–xylose reaction mixture.

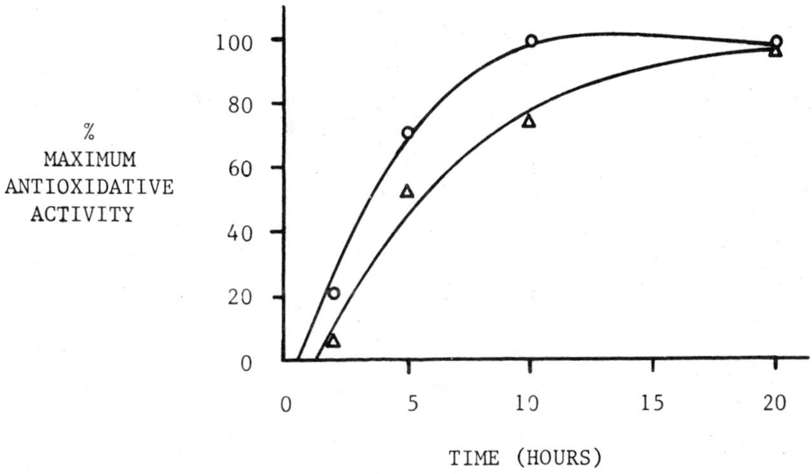

Figure 2. Average change in percent maximum antioxidative activity with time of the nonbuffered arginine–xylose reaction mixture. Key: ○, spectrophotometric assay; △, oxygen electrode assay.

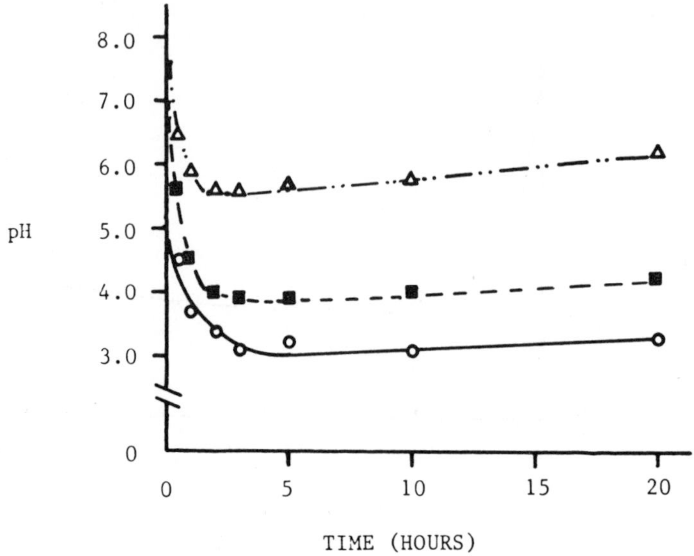

Figure 3. Average change in pH with time of the nonbuffered, Tris 7.0, and Tris 8.0 arginine–xylose reaction mixtures. Key: ○, *nonbuffered;* ■, *Tris 7.0;* △, *Tris 8.0.*

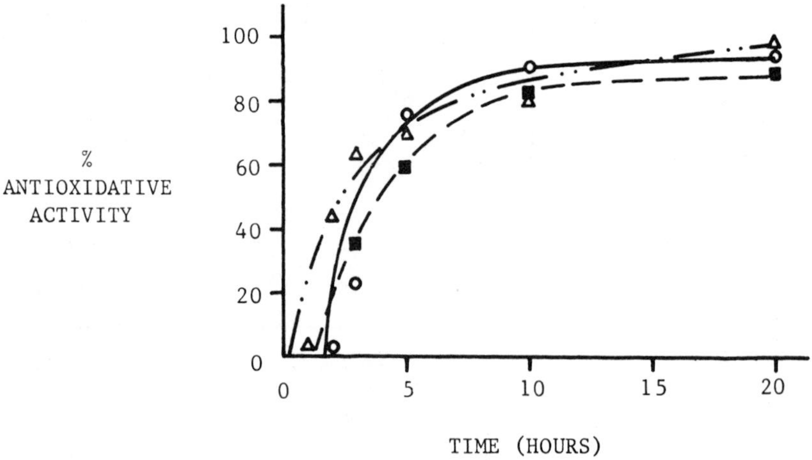

Figure 4. Change in percent antioxidative activity with time of the nonbuffered, Tris 7.0, and Tris 8.0 arginine–xylose reaction. Key: ○, *nonbuffered;* ■, *Tris 7.9;* △, *Tris 8.0.*

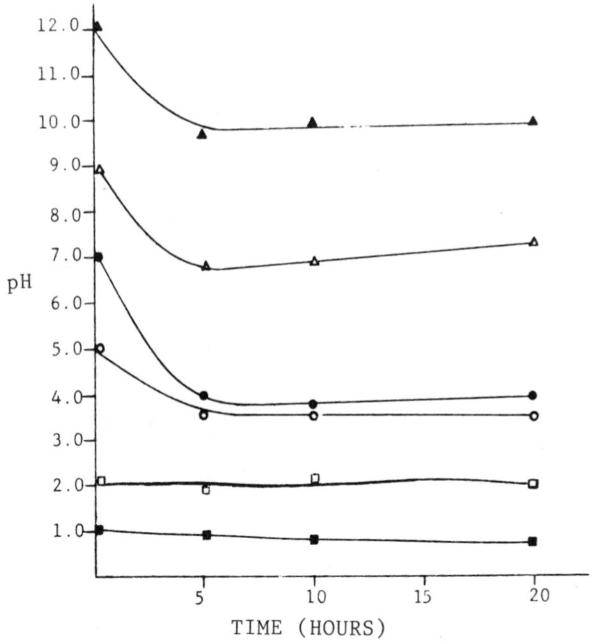

Figure 5. Change in pH with time of six nonbuffered arginine–xylose systems possessing different initial pH values.

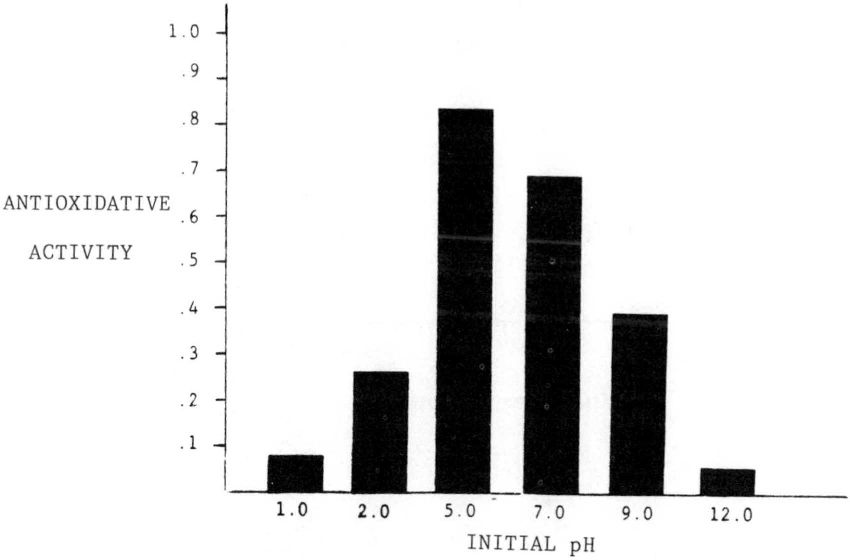

Figure 6. Antioxidative activity produced after 20 h in six nonbuffered arginine-xylose systems possessing different initial pH values.

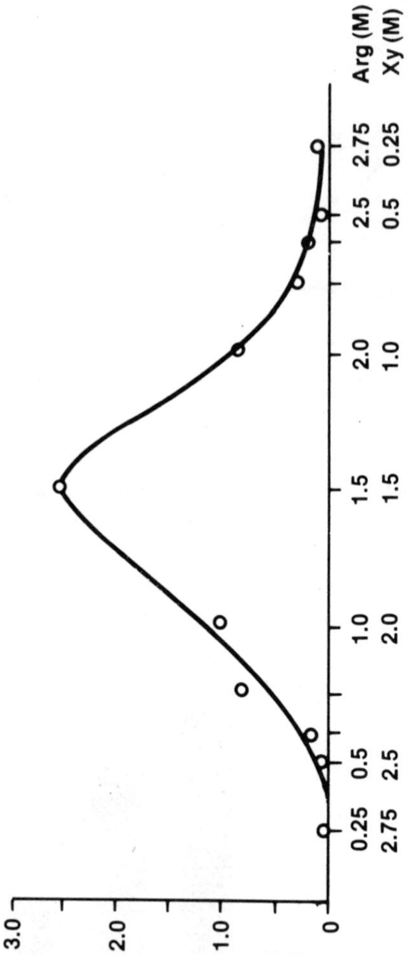

Figure 7. Antioxidative activity versus molar ratio of arginine to xylose. 100 μL of 1:100 dilution of crude.

distribution of activity suggested the possibility of predicting the outcome of heating several combinations or arginine and xylose.

Effect of Organic Additives To determine if other organic compounds could be used to enhance the yield of antioxidative activity, mixtures were made containing 0.1 mole of arginine, 0.05 mole of xylose, and one of the following: 50 ml 95% ethanol, 50 ml acetone, 45 ml 1-pentanol, 9 ml benzyl alcohol, 1 ml 1-nonanol, 1 ml geraniol, 50 ml tetrahydrofuran (THF), 50 ml carbon tetrachloride (CCl_4), and 50 ml pyridine. The volumes were adjusted to 100 ml with distilled water and the solutions refluxed for five h. The final relative activities of these systems can be seen in Figure 8. Only the pyridine mixture displayed a marked increase in activity compared to the control, even though pyridine by itself was shown to be pro-oxidative. It seemed plausible that pyridine perhaps by combining with other antioxidants produced in the reaction might enhance the effectiveness of these compounds. An equivalent amount of pyridine, therefore, was added to a control sample just prior to performing the assay and the results showed the sample to be also pro-oxidative in activity. Equal weights of lyophilized control and lyophilized arginine-xylose-pyridine (AXP) reaction mixture were then assayed. The dry AXP was first dissolved in 1.0 N acetic acid and then lyophilized in order to discard the pyridine. The pyridine system still displayed an antioxidative activity twice that of the control.

In order to obtain a clearer perspective of the difference between the pyridine-influenced reaction and the control, both mixtures were refluxed for seventy h, and aliquots were withdrawn at various times so that the change in activity could be monitored. Figure 9 clearly shows that throughout the entire time, the pyridine system generated twice the activity of the control and did not reach a maximum until after fifty h. Finally, Figure 10 shows that twice as much control was needed in order to produce an antioxidative effect equivalent to that of the pyridine system. Thus, pyridine may have served as a catalyst or it may have been incorporated into new antioxidative compounds, though the latter is unlikely, considering the low reactivity of pyridine. It should be noted that the pH of this reaction dropped from an initial value of 5.6 to a final value of 5.2. In light of this, it was first considered that the enhanced activity was perhaps due solely to the buffering effect of the pyridine. However, this seemed unlikely since no appreciable differences in antioxidant production could be obtained in reactions containing no pyridine and running at pH's between 3.5 and 6.5 (see Figs. 3 & 4). Since the unbuffered water reaction ran at a pH between 3.5 and 4.0, it was doubtful that the increase in antioxidative effect generated by the pyridine

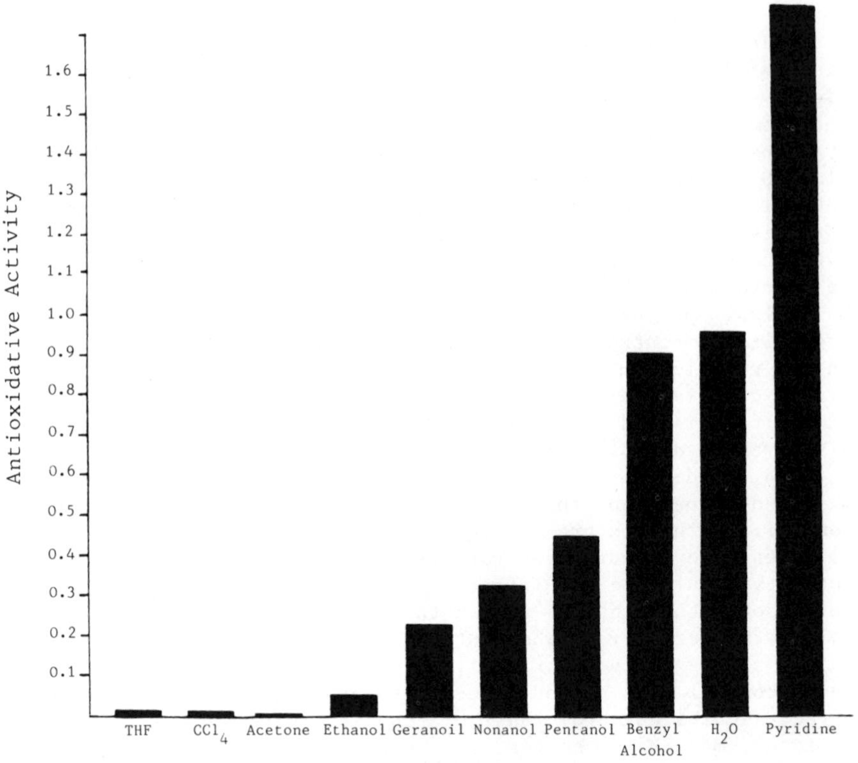

Figure 8. A comparison of the effects of different organic compounds on the production of antioxidative activity in the arginine–xylose reaction.

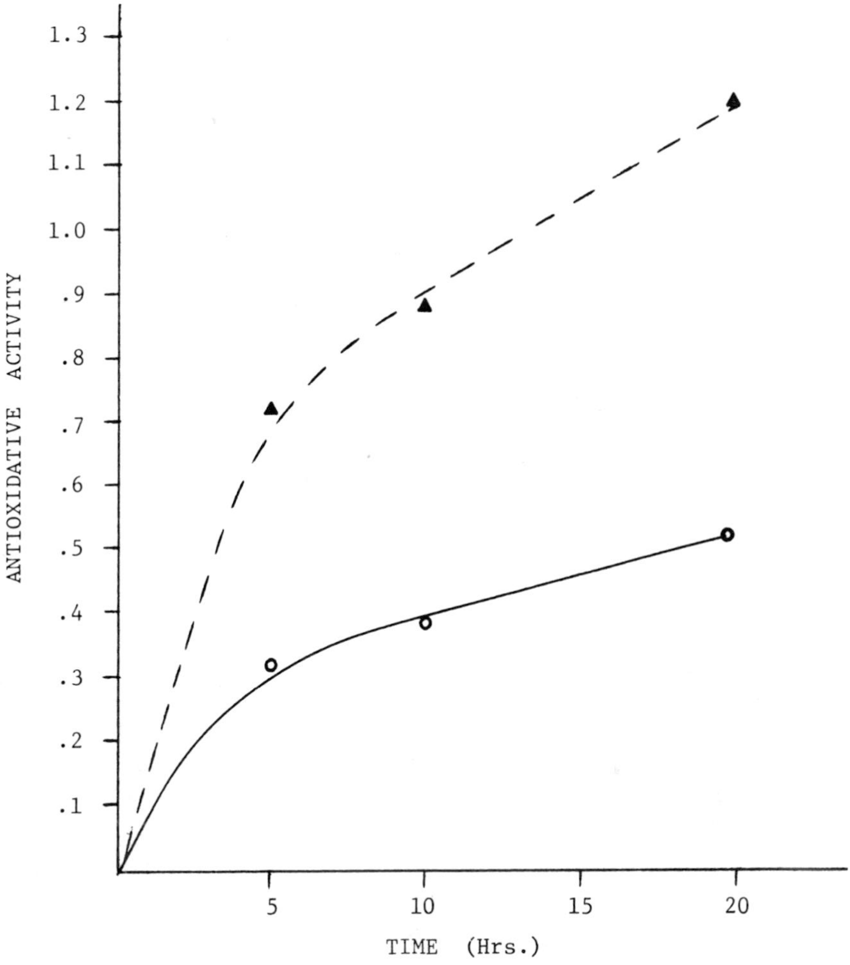

Figure 9. A comparison of the change with time in antioxidative activity produced by arginine–xylose in a water system and a pyridine–water system. Key: ○, water system; ▲, pyridine–water system.

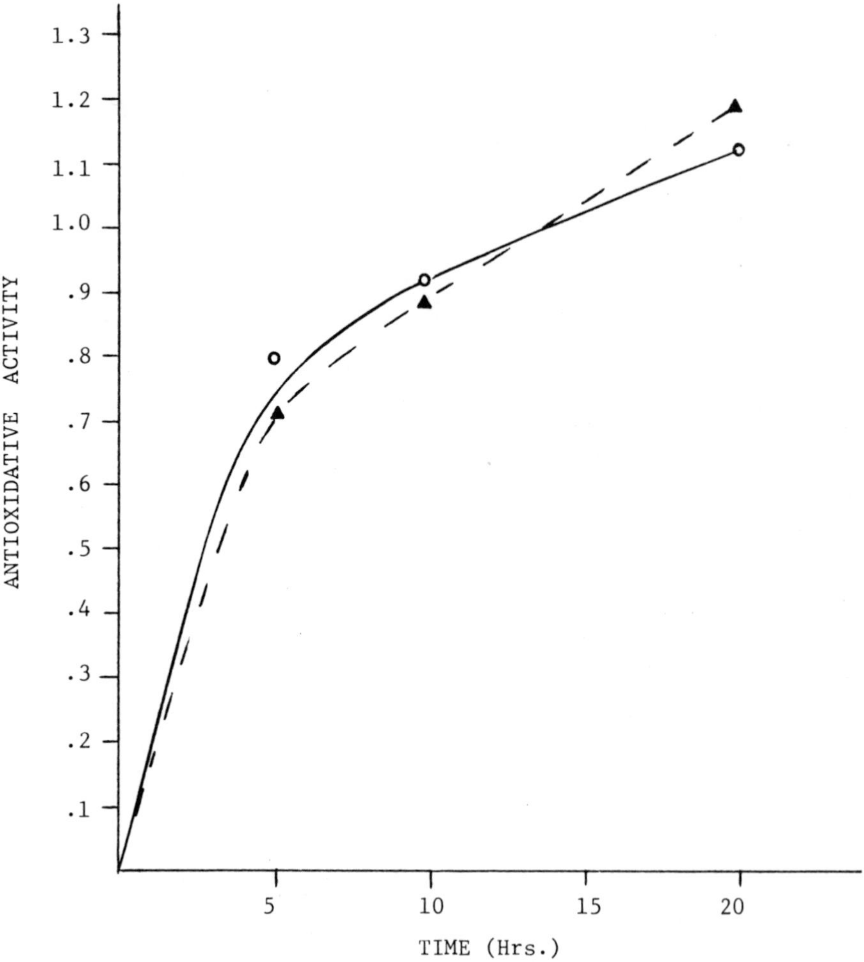

Figure 10. A comparison of the change with time in antioxidative activity produced by arginine–xylose in 2 μL of the water system and 1 μL of the pyridine–water system. Key: ○, water system; ▲, pyridine–water system.

system was entirely due to the influence of pH. Both systems were reacting within a pH range that results in very little difference in antioxidant production.

To gain an insight into the composition of the reaction products, a C, H, O, N analysis was performed by Galbraith Laboratories, Tennessee on samples prepared by lyophilization; the results are displayed in Table 1.

Table 1

Elemental Composition of AXC and AXP

% Total Dry Weight

	C	N	O	H	Total
AXC	37.1	20.7	22.7	7.2	87.7
AXP	40.9	17.7	24.4	7.1	90.1

Clearly, the AXP contained more carbon and oxygen and less nitrogen and hydrogen than did the AXC. If 95-97% of the total reaction products were nonantioxidative and if the antioxidants present in the AXC mixture contained a greater percentage of carbon and oxygen than nitrogen and hydrogen, then an increase in their yield would have been reflected by an increase in the carbon and oxygen content relative to nitrogen. This was the case when pyridine was used in the reaction, thus giving credence to its role as a catalyst. If the pyridine had participated in the reaction and in doing so generated a new antioxidative compound then it is unlikely that the new product would contain a ratio of elements consistent with the observed analysis.

Effect of Pressure and of Gas Used Solutions containing 30 ml of nonbuffered arginine-xylose were allowed to react in a 50-ml Parr vessel for 10 h at 100°C either at one or five bars of pressure of air, N_2, or O_2. Prior to starting the reaction, air, nitrogen, or oxygen was bubbled through the solution for one min.

Comparison of the antioxidative activity of equal amounts of product obtained by repetitive dialysis yielded the results shown in Table 2.

Table 2

Relative Antioxidative Activities produced from 2:1 Ratio of Arginine and Xylose under 1 and 5 Bars of Air, Oxygen, and Nitrogen

	No. of Dialyses	1 Bar			5 Bars		
		Air	Oxygen	Nitrogen	Air	Oxygen	Nitrogen
Dialysate	1	0.25	0.02	0.32	0.21	-0.22	0.51
	2	1.49	0.07	1.07	0.25	-0.11	0.79
	3	0.39	0.00	2.99	0.35	-0.09	1.17
	4	---	0.05	1.00	---	0.00	0.58
Retentate	1	1.45	0.01	3.23	0.30	-0.02	1.44
	2	3.67	-0.12	8.94	0.21	-0.02	1.90
	3	0.89	-0.01	0.89	0.35	0.00	1.06
	4	0.99	-0.13	0.64	0.36	0.07	0.88

Both retentates and dialyzates showed a higher antioxidative activity for the samples prepared under 1 bar of pressure of air and nitrogen than under 5 bars except for retentates 3 and 4 under nitrogen, in which the antioxidative activity for 5 bars was slightly higher. Generally, the antioxidative activity in samples prepared under 1 and 5 bars of pressure of oxygen was significantly lower. Except for samples heated under oxygen, the antioxidative activity was at its highest in the second retentate. When the crude samples were dialyzed more than twice, the antioxidative activity in the dialysates increased with decreasing activity in the retentates. This showed that increasing repetitive dialysis resulted in more diffusable antioxidative substances that crossed the 1,000 molecular weight cutoff tubing into the dialysate. The structures and the molecular weights have not been determined.

Discussion

Depending on which assay was used, the optimum time for producing antioxidative activity was between 10 and 20 h. However, in either case, the major part of the yield had been formed within 10 h. If time is the critical factor when producing antioxidants from this system, the increase in yield from 10 to 20 h would probably not be substantial enough to justify the doubling of reaction time.

The presence of Tris exerted no influence on the amount of final activity produced. Only 1.0 M Tris was used and it is conceivable that any enhancement of activity resulting from the buffering could have been negated by an inhibitory action of the Tris, though this was unlikely since Tris is an amino compound which if anything would have increased the rate. A more extensive survey of other buffers including phosphate might identify some that could increase the yield of antioxidant, but the increase would have to be substantial in order to outweigh the added expense of the buffer.

Maximum activity was produced with an initial pH of 5.0 and, in fact, the results in Figure 6 indicated that a wide range between 3.0 and 7.0 could be employed to obtain acceptable yields. However, this is based partly on extrapolation since no data for a pH of 3.0 were obtained.

Although the 1:1 molar ration produced a higher yield of activity, unpublished results obtained in this laboratory have shown that the majority of the activity is lost upon dialysis. This implies that the increase in activity with the 1:1 system was the result of an increase in the yield of low-molecular-weight antioxidants at the possible cost of higher antioxidants.

Just as certain nitrogenous compounds are known to increase the rate of browning, the results with pyridine indicated that this may also be true for the production of antioxidants. The present study revealed that use of pyridine doubled the yield of activity. No attempts, however, were made to isolate and purify the antioxidative components, and therefore, no conclusion can be made as to whether pyridine was acting merely as a catalyst or if it was involved in producing a different and stronger antioxidant.

Conclusions

This study has defined a few of the optimum conditions, but, the effects of other buffers and of the total concentrations of reactants need to be investigated. In addition, the influence of other base catalysts merits further study. The mechanisms by which pyridine enhances the yield is not understood, but the use of isotopically labeled pyridine might provide evidence as to whether or not pyridine is incorporated into a new antioxidant. Another approach would be to carry out some experiments with a pH-stat to control pH at the same level that pyridine maintains. Finally, caution should be exercised in the interpretation of the results of this study. Except for the 1:1 molar ratio products, no attempt was made to associate molecular weights with antioxidative activity, and therefore, little can be said about the conditions for producing a maximum yield of high-molecular-weight antioxidants. Even though it would be tempting to do so, it is clear from the dialysis results that more than assumptions must be made before defining the optimum conditions for producing the antioxidative compounds.

Acknowledgments

The authors thank Otis C. Dermer for critically reviewing and editing this manuscript. Portions of this work were taken from the M.S. thesis of Ronald W. Beckel. This research was supported in part from the National Science Foundation in the form of a research grant No. INT-8018220.

Literature Cited

1. Smirnov, V. A.; Geispits, K. A. Biokhimiya 1956, 21, 633-635; Biochemistry (translation), 21, 655-657.
2. Kliely, R. J.; Nowlin, A. C.; Moriarty, J. H. Cereal Sci. Today 1960, 5, 273-274.
3. Heyns, K.; Klier, M. Carbohydrate Res. 1968, 6, 436-448.
4. Kirigaya, N; Kato, H.; Fujimaki, M. Nippon Nogei Kagaku Kaishi 1969, 43, 484-491; CA (1970) 72:11493a; BA (1970) 51:18420.
5. Tomita, Y. Kogashima Daigaku Nogakubu Gakujutsu Hokoku 1971, 21, 161-170; CA (1972) 76:98098e.
6. Eichner, K. "Water Relations of Food"; Duckworth, R. B., eds.; Academic Press, London, New York, 1975; pp. 417-437.
7. Lingnert, H.; Eriksson, C. E. J. Food Process. Pres. 1980, 4, 161-172.
8. Foster, R. C. "Preliminary isolation and characterization of an antioxidative component of the Maillard reaction between arginine and xylose"; M. S. Thesis, Oklahoma State University, Stillwater, Oklahoma, 1980; 72 pp.
9. Lingnert, H.; Vallentin, K.; Eriksson, C. E. J. Food Process. Pres. 1979, 3, 87-103.

RECEIVED November 22, 1982

9

The Variety of Odors Produced in Maillard Model Systems and How They Are Influenced by Reaction Conditions

M. J. LANE[1] and H. E. NURSTEN[2]

The University, Procter Department of Food and Leather Science,
Leeds, LS2 9JT, United Kingdom

The odors of more than four hundred model systems drawn from twenty-one amino acids and eight sugars, treated under different conditions of temperature and humidity with the inclusion of selected third components, were evaluated by one assessor only. Organoleptic assessment of odors produced by a full range of common amino acids in "dry" 1:1 admixture with glucose was carried out by immersion of an ignition tube containing the reaction mixture into a heated Wood's metal bath. The investigator assessed variations in aroma produced by: (a) changes in heating temperature from 100 to 220°C at selected time intervals up to thirty minutes, (b) different levels of relative humidity of the reactants, (c) changes in the sugar used, and (d) adding third components, such as powdered cellulose and soluble starch and two nitrogen sources, asparagine and glutamine. The initial experiments, the heating of glucose-amino acid mixtures at selected temperatures, proved of greatest interest, many aromas being reported for the first time. However, as the experiments progressed from variations in time and temperature only to variations of water content, change of sugar and inclusion of third components, the aromas emerged as basically the same for each particular amino acid.

[1] Current address: Quaker Oats Ltd., Bridge Rd., Southall, Middlesex
[2] Current address: University of Reading, Department of Food Science, Whiteknights, Reading, Berkshire

0097-6156/83/0215-0141$06.00/0

When our interest in this area was aroused in 1969, a survey of previous work showed that, although many aqueous mixtures had been studied, attention had been accorded to very few systems of very low water content. In fact, at that time no "dry", two-component system had been examined with respect to the volatiles produced, a surprising omission considering the importance of roasting and grilling.

Since then, some important work along these lines has been carried out (1), culminating in the recent paper by de Rijke, et al. (2), in which a sulfur-containing amino acid, methionine or cysteine, was heated under "dry" conditions with one of six sugars in turn for 15 min at 190-200° and 2-3 Torr., the volatiles being collected in cold traps and examined.

Our approach was different, ambitious, yet simple. The odors of more than 400 model systems drawn from 21 amino acids and 8 sugars, treated under different conditions of temperature and humidity with the inclusion of selected third components, were evaluated by one assessor only.

Experimental

Reagents. Analar amino acids and sugars (B.D.H. Chemicals Ltd) were used. Thin layer chromatography revealed no impurities in the reagents as delivered, and no further purification was carried out. The reagents were kept over phosphorus pentoxide in a vacuum desiccator for at least three days before use.

Two-component systems (a) Glucose-amino acid systems Finely ground, equimolar mixture (0.3 g) was placed into a small test tube, a glass spacer added to hold down the solid in case of frothing and also to increase the contact of the mixture with the outside wall (see Figure 1), and the tube closed with an aluminum cap (Oxoid Ltd) to lessen the escape of volatiles. The test tube was partially immersed in a glass dish of molten Wood's metal, kept at 180 (±2)° on a magnetic hot-plate stirrer. At regular intervals the test tube was withdrawn to smell the volatiles produced. Glucose was heated with the following amino acids: ala, arg, asn, asp, cys, cys_2, glu, gln, gly, his, ile, leu, lys, met, phe, pro, ser, thr, try, tyr, and val, his and lys being in the form of their monohydrochlorides. The amino acids and glucose were also heated alone.

The aromas were normally assessed after ½, 1, 2, 3, and 4 min. The maximum yield (0.5%) of the interesting 2-(5-hydroxymethyl-2-formyl-1-pyrrolyl) propionic lactone has been shown to occur from a mixture of glucose and ala in about 3 min under "dry" conditions at 200° (3). The full results are available (4) and a set obtained at 180±2° is given in Table I as an example.

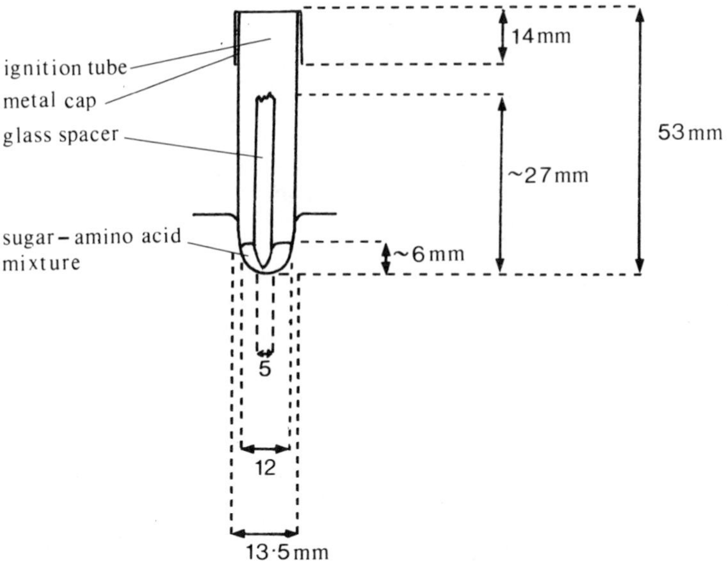

Figure 1. Heating tube.

TABLE I

EXAMPLE OF AROMA ASSESSMENT OF 1:1 GLUCOSE : AMINO ACID MIXTURE AT 180°

	glucose-cysteine		cysteine alone
30, 60 s	strongly puffed wheat	30 s	very slight H_2S smell
2 min	crusty, burnt	1, 2 min	slightly sulfurous
3 min	burnt bread, with underlying puffed wheat or over-roasted meat (cf. meaty at 150°) (5)	3, 4 min	unpleasant sulfurous, some puffed wheat

(b) Temperature as a variable The experiments were repeated at 100, 140, and 220 (±2)° for up to 30, 7, and 1½ min, respectively.

(c) Humidity as a variable Henceforth, to save time, the number of amino acids investigated was cut from 21 to 6: cys, ile, lys, ser, thr, and tyr, giving a selection of interesting aromas, as well as of structures. The glucose-amino acid mixtures were conditioned in desiccators over one of 5 appropriate saturated solutions or solid P_2O_5, nominally at 98-0% r.h. Heating was for up to 5 min at 140±2°.

(d) Sugar as a variable The six amino acids were heated individually with each of the sugars, fructose, galactose, maltose, ribose, sucrose, and xylose, and with ascorbic acid. Heating was for up to 5 min at 140±2°. Because of the high reactivity of the two pentoses, their mixtures were also heated for up to 30 min at 100±2°.

Three-component systems (a) A nitrogen source, asn or gln Pyrazines have been recognized for some years as important odorants, and the availability of additional nitrogen was expected to promote their formation. Both asn and gln when heated alone produce ammonia, so 1:1:1 mol. mixtures were heated, each serving as nitrogen source in turn.

(b) Asparagine with change of sugar At this stage, the effect of asn on the lys mixtures considered most interesting had already been investigated. Eight other two-component systems exhibit distinct and interesting aromas (see Table II), and they were now heated in presence of an equimolar amount of asn. A number of related systems was also heated.

(c) Cellulose Mixtures of half quantities of the six 1:1 glucose:amino acid mixtures with, in turn, 0, 1, 3, and 10 50-mg portions of cellulose were heated at 140±2°, as were a few related systems.

(d) Soluble starch Four mixtures, parallelling those containing cellulose, were heated at 140±2°.

(e) Polymerization-blocking components The aldol condensation is an important reaction in the Maillard network (see Figure 2). Not all the reactions can occur in all cases. Equilibrium 1 cannot occur where the carbonyl compound lacks α-hydrogen atoms, i.e., it belongs to what we designate Group 1 (Figure 3). Vinylogues of α-hydrogens may well need to be considered. In general, β-elimination next leads to the α,β-unsaturated aldehyde by Reaction 3. This, the first "irreversible" step, is not possible where the aldol lacks

TABLE II

TWO-COMPONENT SYSTEMS GIVING DISTINCT AND INTERESTING AROMAS

Amino acid	Sugar	Temperature °C	Aroma
cys	ribose	100	roast beef
cys	ascorbic acid	140	chicken
ile	ascorbic acid	140	celery
ile	glucose	100	celery
lys HCl	glucose	140	bread
ser	glucose	100	chocolate
thr	ascorbic acid	140	beef extract, meaty
tyr	glucose	100	chocolate
tyr	ascorbic acid	140	chocolate

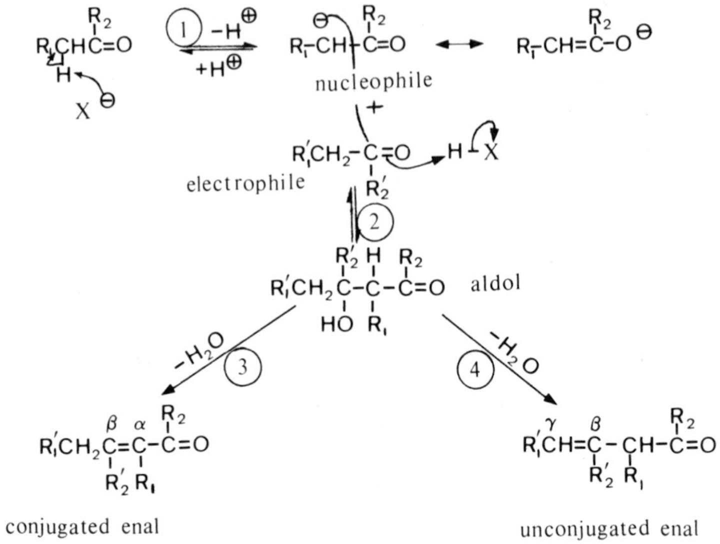

Figure 2. Aldol condensation.

α-hydrogens. Such aldols are formed from nucleophilic aldehydes having only one α-hydrogen, which we designate Group 2 (Figure 4). These aldols may dehydrate by the less favored Reaction 4, giving the β,γ-unsaturated aldehyde, provided the electrophilic aldehyde did not belong to Group 1.

Thus, the aldol route will be blocked, or partially blocked, if the only available aldehydes have fewer than two α-hydrogens. Hence, inclusion of "blocking" aldehydes of Group 1 or 2 in a browning reaction mixture should increase the proportion of free aldehyde produced and generally lower the chain length of aldol polymers formed. Overall, an increase in aroma volatiles should be observed, except with aldehydes of such low boiling point as to be able to escape physically and for which the aldol condensation acts as a restraint.

In consequence, furfural (Group 1) was added to two glucose: amino acid mixtures in the molar proportion 0.1:1:1, followed by heating at 140±2° (see Table III).

Comparison of the difference between (1) and (2) with that between (3) and (4) should show the effect of the greater degree of blocking of mixture 1, through a higher final furfural concentration. Furfural can only be involved as the electrophile and, in mixture 1, reaction with 2-methylbutanal (from ile, Group 2) cannot proceed even via the less favored β,γ-elimination.

Comparison of (5) with (6a) + (6b) (heated in separate tubes but smelled together) should show similarly the extent to which furfural can affect a normal aldehyde. Comparison of (7) with (8a) + (8b) and of (9) with (10a) + (10b) should show little difference because of considerable survival of volatile aldehydes in presence of a blocking Group 2 aldehyde (from ile).

Results and Discussion

(a) Glucose-amino acid systems at different temperatures
To reduce the mass of data to manageable proportions, most of the aroma descriptors were placed into 14 groups (Table IV), some others were also placed in these groups (for example, pork crackling, lamb, and meat extract were included in meaty), and some were omitted (for example, rubbery, earthy, peaty). Two descriptors, toast and potato crisps, appear in two groups. This grouping allowed tables to be constructed showing the incidence of each type of aroma. The simplest example was the floral aroma (Table V). Only 1 amino acid gave rise to it, both with glucose and alone. It appears at almost all temperatures, there being 3 occurrences on heating with glucose and 4 on heating alone. The other extreme is presented by Group 14, which each of the 21 amino acids exhibited under at least one of the conditions, there being 55 occurrences with glucose and 63 without, out of a maximum of 84 in each case. More interesting than either are Groups 2-4 (Tables VI-VIII).

formaldehyde	HCHO
α-methylacrolein	$CH_2{=}C(CH_3)-CHO$
2-furaldehyde	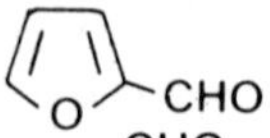
3-furaldehyde	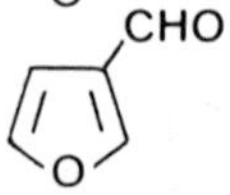
benzaldehyde	Ph–CHO

Figure 3. Examples of Group 1 aldehydes (those lacking α-hydrogens).

2-methylbutanal	$CH_3CH_2CH(CH_3)-CHO$
2-methylpropanal	$CH_3CH(CH_3)-CHO$
β-angelica lactone	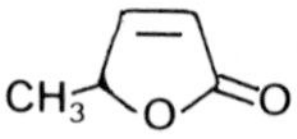
2-hydroxypropanal	$CH_3CHOH\ CHO$

Figure 4. Examples of Group 2 aldehydes (those having only one α-hydrogen).

TABLE III

BLOCKED AND PARTIALLY BLOCKED SYSTEMS INVESTIGATED

1.	Glucose + ile + furfural
2.	Glucose + ile
3.	Glucose + tyr + furfural
4.	Glucose + tyr
5.	Pentanal + furfural
6a.	Pentanal and 6b. Furfural
7.	Glucose + ile + pentanal
8a.	Glucose + ile and 8b. Pentanal
9.	Glucose (X2) + ile + tyr
10a.	Glucose + ile and 10b. Glucose + tyr

TABLE IV

GROUPING OF AROMA DESCRIPTORS USED

Group	Descriptor(s)
1.	Sweet, boiled sugar, caramel, toffee
2.	Chocolate, cocoa
3.	Bread, crusty, biscuits, cakes, toast
4.	Meaty, beefy
5.	Potato, potato skins, potato crisps
6.	Fruity, aromatic ester
7.	Celery, chicory, leeks, Brussels sprouts, turnips
8.	Puffed wheat, Sugar Puffs
9.	Nutty
10.	Floral
11.	Ammoniacal
12.	Unpleasant, 'caused coughing'
13.	Aldehydic
14.	Burnt, charred, scorched, acrid, potato crisps, toast, smoky

TABLE V

INCIDENCE OF FLORAL AROMA (GROUP 10)

Amino acid	With glucose (temperatures, °C)				Alone			
Phe	100	140		220	100	140	180	220
1 (1 + 1)	1	1	0	1	1	1	1	1
Total	3				4			

TABLE VI

INCIDENCE OF CHOCOLATE, COCOA AROMAS (GROUP 2)

Amino acid	With glucose (temperatures, ^{o}C)				Alone			
Arg	100		180	220				
Asn	100			220				
Asp	100		180	220				
Glu			180	220				
Gln	100	140	180	220				
Ile	100			220				
Leu	100					140		
Phe	100	140	180	220				
Ser	100	140	180					
Thr			180	220				
Try			180	220				
Tyr	100	140	180	220				
Val	100	140	180					
13 (13 + 1)	10	5	10	10	0	1	0	0
Total	35				1			

TABLE VII

INCIDENCE OF BREAD, CRUSTY, BISCUITS, CAKES, TOAST AROMAS (GROUP 3)

Amino acid	With glucose (temperatures, °C)				Alone			
Arg	100	140						
Asn			180	220				
Cys			180					
Cys_2				220				
Gln		140	180	220				
Gly				220				
His	100	140	180					
Lys	100	140	180	220				
Pro	100							
Ser	100		180	220				
Thr		140						
Tyr				220				
12 (12 + 0)	5	5	6	7	0	0	0	0
Total	23				0			

TABLE VIII

INCIDENCE OF MEATY, BEEFY AROMAS (GROUP 4)

Amino acid	With glucose (temperatures, °C)				Alone			
Asn						140	180	
Cys	100	140	180			140	180	
Cys_2		140	180					
Glu							180	
Gly					100	140		
Ile					100			
Ser					100	140	180	220
Thr	100				100	140	180	220
8 (3 + 7)	2	2	2	0	4	5	5	2
Total		6				16		

Based on all the data available, the following comments can usefully be made:

1. Almost without exception, there is no one temperature at which the aromas start or cease to occur, although in several of the groups the number of occurrences rises above 100° and then falls above 180°.

2. Aromas of Groups 1-3 and 9 are produced almost exclusively by amino acid mixed with sugar, whereas others, of Groups 4, 7, and 11, show a greater occurrence when amino acids are heated alone. Noteworthy is the fact that cys_2 only gave a meaty aroma in presence of glucose, presumably because of reduction to cys, which gave a meaty aroma with and without glucose (Table VIII).

3. Generally, the aromas become stronger and more unpleasant, aldehydic, and burnt with increased temperature and time.

4. The aromas obtained from some amino acids possess a consistent note throughout the temperature range investigated (Table IX).

5. On the other hand, each amino acid gives rise to more than one characteristic note over the temperature range used, ile appearing in no less than six groups.

(b) Variation in water content A limited range of amino acids was investigated, no interesting new aromas being produced, and little or no pattern of change becoming apparent.

(c) Other sugars in place of glucose In general, the results at 140° did not show much divergence from the aromas produced with glucose, but they do call for some comments:

(i) Cys + ascorbic acid gave rise to the only mention of chicken meat aroma so far.

(ii) Lys + ribose provided a definite mention of toast.

(iii) None of the glucose replacements gave rise to a chocolate aroma with ser.

(iv) Thr + ribose and thr + xylose gave an almond and a marzipan aroma, respectively.

(v) Whereas mixtures of thr with maltose, ribose, sucrose, and ascorbic acid elicited a meaty note, those with glucose, galactose, fructose, and xylose did not.

(vi) Tyr mixtures gave a consistent chocolate aroma, except with maltose and ribose.

(vii) The high reactivity of ribose and xylose shows itself in their ability to produce aromas at 140° equivalent to those obtained with glucose mixtures at about 200°. At 100°, the pentose mixtures gave less burnt and therefore more interesting results.

TABLE IX

THE AROMAS OBTAINED FROM SOME AMINO ACIDS AND POSSESSING A CONSISTENT ELEMENT THROUGHOUT THE TEMPERATURE RANGE USED

System	Aroma
Ala + glucose	caramel
Ala alone	burnt
Arg + glucose	aldehydic
Asn alone	ammoniacal
Cys + glucose	puffed wheat, Sugar Puffs
Cys_2 + glucose	puffed wheat, Sugar Puffs
Gln + glucose	chocolate
Gln alone	ammoniacal
Gly + glucose	caramel and burnt
Gly alone	burnt
Ile alone	fruity
Lys + glucose	bread, cakes, etc.
Met + glucose	potatoes
Phe + glucose	chocolate
Phe alone	floral
Pro + glucose	nutty
Ser alone	meaty
Thr + glucose	burnt
Thr alone	meaty
Tyr + glucose	chocolate
Val + glucose	aldehydic

(d) Three-component systems with asparagine and glutamine No evidence of increased pyrazine formation was detected, the overall effect being a sweetening of the aromas. With two-component systems, a meaty aroma was evident, possibly just a reflection of ammonia production. In conjunction with sugars other than glucose, asn did not increase or improve the aroma.

(e) Three-component systems with cellulose and starch Inclusion of powdered, polymeric material was considered likely to affect the browning reaction physically, by providing additional surfaces, which could constitute the significant difference between vegetable material and model mixtures. Little evidence of any such effect was found.

(f) Systems with polymerization-blocking components The relatively unsophisticated comparisons made provided no evidence for the effects sought.

Conclusions

The results of this survey of the aromas produced over time by heating glucose-amino acid mixtures at a series of temperatures in the range 100-220° proved of great interest. Many mixtures were heated in the "dry" state for the first time. Some produced the expected result, for example, methionine and phenylalanine led to potato and to floral aromas, respectively. Others were unexpected, for example, the large number of amino acids that was capable of producing chocolate aroma under one or other set of conditions.

Variables other than time and temperature, i.e., moisture content, nature of sugar, and inclusion of third components, asparagine, glutamine, cellulose, starch, and blocking agents, did not lead to important differences. The number of aromas produced still depended basically on the amino acid primarily involved. Yet the range of aromas produced in the cooking of foods is much wider than this, and so the factors responsible for this diversity remain to be identified. The ideas presented here have only been tested in a preliminary way and, even though this initial attempt to verify them has proved largely negative, we intend to follow them up at the first opportunity.

Acknowledgments

Financial support from Bush Boake Allen Ltd is gratefully acknowledged, as is Dr. T. A. Rohan's initiation and encouragement of this work and Dr. A. Montgomery's advice.

Literature cited

1. Mabrouk, A.F. ACS Symp. Ser. 1979, 115, 205-245.
2. de Rijke, D.; van Dost, J.M.; Boelens, H. "Flavour '81"; Schreier, P. Ed.; W. de Gruyter: Berlin, 1981; 416-431.
3. Shigamatsu, H.; Kurata, T.; Kato, H.; Fujimaki, M. Agric. Biol. Chem. 1971, 35, 2097-2105.
4. Lane, M.J. Ph.D. Thesis, Leeds University, 1978, 245 pp.
5. Kiely, P.J.; Nowlin, A.C.; Moriarty, J.H. Cereal Sci. Today 1960, 5, 273-4.

RECEIVED December 14, 1982

10

Characteristics of Some New Flavoring Materials Produced by the Maillard Reaction

E. DWORSCHÁK and V. TARJÁN
Institute of Nutrition, H-1097 Gyáli út 3/a, Budapest, Hungary

S. TUROS
Kanold AB, Göteborg, Sweden

Milk-crumbs and soya-crumbs are new aroma sources produced by the Maillard reaction and intended to be used in the food industry. During the production of the crumbs thermal reactions take place in the material (e.g. milk) and the added glucose and amino acid constituents. Fructose-lysine and lactulose-lysine, biologically active Amadori compounds, are formed in high levels in the milk-crumb, but much less in the soya-crumb, caused by lysine degradation. Extracts from the crumbs show only a low mutagenic activity measured by Ames test. Soya-crumb has a significant antioxidant effect; for this reason its possible role in the food industry is discussed.

The KANOLD AB Company has developed a series of new materials, produced from mixtures of skimmed milk, glucose, and lysine (or other basic amino acids) by drying on a roller at a temperature of 125°C (1). Instead of skimmed milk, soya flour can be used as basic material. The new products, named milk- and soya-crumbs respectively, have a very pleasant odor resembling that of caramel. The yellow to deep-brown coloration and the aroma characteristics are attributed to the Maillard reaction. According to experiments carried out in the confectionery industry, up to 50% of cocoa powder could be replaced by the crumbs without any significant changes in flavor, appearance, or consistency of food products, e.g. in chocolates.

Our aim was to investigate some chemical properties as well as the possible mutagenic effects of the new materials. We determined the content of furosine, which is an amino acid derivative formed by acid hydrolysis of fructose-lysine or lactulose-lysine (2) (Fig. 1).

0097-6156/83/0215-0159$06.00/0

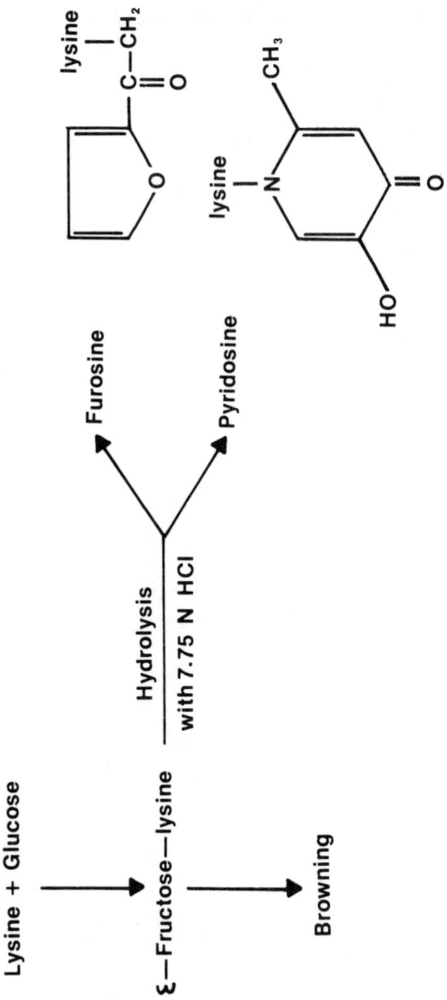

Figure 1. Formation of furosine and pyridosine by acid hydrolysis.

Experimental

Material and Supplies Milk-crumb and soya-crumb were prepared by the KANOLD AB Co., Göteborg, Sweden.

N-α-acetyl-L-lysine. A Grade, Calbiochem.

Tween 20 and 80 (sorbitan, polyoxyethylated, stearic ester), BDH Chemicals Ltd.

Hemoglobin, horse, pure. REANAL, Hungary.

Sunflower oil, Ph. Hg. VI purity.

BHT (3,5-ditert-butyl-4-hydroxytoluene), A grade REANAL

Preparation of furosine solution:
N-α-acetyl-L-lysine (0.5 mmole) and glucose (1.0 mmole) are dissolved in 1 ml 0.1 M K_2HPO_4 buffer solution. The solution is heated in a closed vessel for 5 h at 100°C. Then the solution is made 6 N with HCl and heated at 105° for 24 h. The hydrolysate is evaporated in vacuo. A solution of the residue in 20 ml water can be used for the TLC production of furosine spots. For quantitative evaluation the same absorbance was assumed for the ninhydrin reaction product both arginine and furosine.

Methods

Determination of Furosine For the determination of furosine the food crumbs were hydrolyzed with 6 N HCl for 24 h at 105° C in a closed glass vessel. The 0.5% hydrolysate was chromatographed on a cation exchange resin-coated thin layer plate (sodium cycle FIXION 50 X 8 plates, CHINOIN Pharmaceutical Works, Budapest) developed by Devenyi, et al. (3). For the separation of furosine a 0.12 M citrate buffer (sodium concentration 400 meq/l, pH = 6.0) was used. The chromatogram was developed at 50° C. After the drying of the plate, the chromatogram was sprayed with 0.4% ninhydrin in acetone. Furosine spots were evaluated with a VIS-UV-2 Chromatogram Analyzer, FARRAND Optical Co., Inc., N.Y.

Test for Mutagenic Activity Milk-crumbs and soya-crumbs extracted in two different ways were tested for possible mutagenicity by the Ames test.

A portion of the crumbs was extracted with a 5-fold amount of 5% Tween 80 aqueous solution.

In another extraction, a 5-g sample was kept in 60 ml refluxing ethanol for one h. The filtered ethanol was distilled from the dissolved fatty material at a normal pressure. The distillate was used for the Ames test.

In the Ames test the following *Salmonella typhimurium* strains were used: TA 1535 and TA 100, which are sensitive to mutagens causing base-pair substitutions, and TA 1537, TA 1538, and TA 98, sensitive to mutagens causing frame-shift mutations.

Manipulation of the test strains, and preparation of the microsomic fraction of the Arochlor 1254 pretreated R-Amsterdam male rats, were carried out as described by Ames, et al. (4). The control mutagen was 2-aminofluorene in a 0.1 mg/plate dose.

Determination of Antioxidant Effect The measurement was based on consumption of dissolved oxygen by a sunflower oil emulsion in a closed system with or without the presence of antioxidant material.

The emulsion was freshly prepared (5) from 10 ml sunflower oil, 2 ml 30% aqueous solution of Tween 20, and 88 ml 0.1 M phosphate buffer at various pH values. This mixture was homogenized in a BIOMIX-mixer (LABOR MIM, Budapest) for 5 min.

To 50 ml emulsion was added 0.1 g finely ground crumb sample, and the mixture homogenized with a magnetic stirrer. The oxygen content of the emulsion was evaluated with a RADELKIS (Budapest) pO_2 -pCO_2 analyzer using an OP-9263-type oxygen membrane electrode. From the moment of adding 5 ml 0.14% hemoglobin solution to the stirred emulsion, the time was measured with a stopwatch. The time required for 50% reduction of the dissolved oxygen concentration was recorded.

The antioxidant effect (A.E.) is calculated from the equation (5)

$$A.E. = (T_a - T_c)\ T_c$$

where T_a is time elapsed for 50% reduction of the available gaseous oxygen in the sample containing the antioxidative material, and T_c the corresponding time interval in the control, without crumb sample.

For comparison, a strong antioxidative material used in the food industry, 10^{-5} mM BHT in 0.1 ml sunflower oil, was added to a portion of the emulsion.

Results and Discussion

Table I shows the furosine content of the milk-crumb and soya-crumb samples in comparison to other food protein sources thermally processed. Milk-crumb has the highest furosine content, probably due to the free lysine and glucose added in the production; moreover, the milk protein has also a considerable level of protein-bound lysine.

Data from Table I point out that other milk-based products are likely to form Amadori products, in this case lactulose-lysine and fructose-lysine, even under moderate thermal treatments. The furosine content, originating from the acid hydrolysis of Amadori products, was high in a whipping agent made from sugar and Na caseinate. It is surprising, however, that cheeses have also relatively high furosine levels.

Table I

Furosine content of milk and soya-crumb, and of other foods and basic materials

	Furosine, mg/100 g protein
Milk-crumb	4000
Soya-crumb	150
Milk treated at ultra-high temperature	180
Condensed sweet milk	400
Infant formulae (EGYT products)	110-230
Whipping agent (from sodium caseinate)	3000
Whole milk powder	70
Gruyére cheese	210
Port salut cheese	390
Roasted coffee	200
Nescafe	350
UNIBIT Textured Soya Concentrate S 202/24 (PURINA)	20

Soya products contain much less reducing carbohydrate and also lower bound lysine than those from milk; moreover, only small amounts of free lysine and glucose were added in the production of soya-crumb. The smaller furosine values from soya products can be explained by these facts. The effect of intensive heat treatment on the formation of furosine is demonstrated by roasted coffee.

Erbersdobler, et al. (6) found karyocytomegaly in the kidneys of rats fed 16200 to 23300 ppm fructose-lysine in the diet. The diet did not contain lysinoalanine (Fig. 2) but the renal damage was similar to that found after lysinoalanine administration by other authors. The details of this question are reviewed elsewhere (7). Erbersdobler's report has not been confirmed with new data by others.

Among the values given in the Table I only milk crumb and the whipping agent have furosine levels that can be compared to the fructose-lysine (or lactulose-lysine) content used in Erbersdobler's experiment.

The test for mutagenic activity of the detergent (Tween) extract did not induce any revertants; all plate counts were in the range of spontaneous mutation rate, with and without S-9 mix. From the ethanol distillate we got a positive response demonstrated in Table II. In the case of milk-crumb the numbers of induced revertants were about a hundredfold those of the spontaneously reverted colonies with the most sensitive strains (TA 98, 100) only without metabolic activation. In the same experiment the test strain TA 100 responded to the soya-crumb sample, also only without metabolic activation.

Thus, according to our results, the crumbs have only slight mutagenic activity as compared to other heat-treated and Maillard products described in the literature (8).

Our experiments on the antioxidative effect of soya-crumbs are summarized in Table III. Surprisingly milk-crumb had no effect at any pH value, although the opposite could be expected from its high level of Amadori products.

Soya-crumb has a significant antioxidative effect in the sunflower oil emulsion system, although less than BHT in the concentration generally used in the food industry (Table III). The antioxidative effect is higher in the acid region with a maximum observed between pH values of 4 and 5.

Our results confirm the antioxidative effect of some Maillard products. The low toxicity of these products may permit using them in certain food products (e.g. in sausages) as mild antioxidants of natural origin (5).

Conclusions

Intensive search has been started recently to find basic materials of moderate price to be used in the food industry.

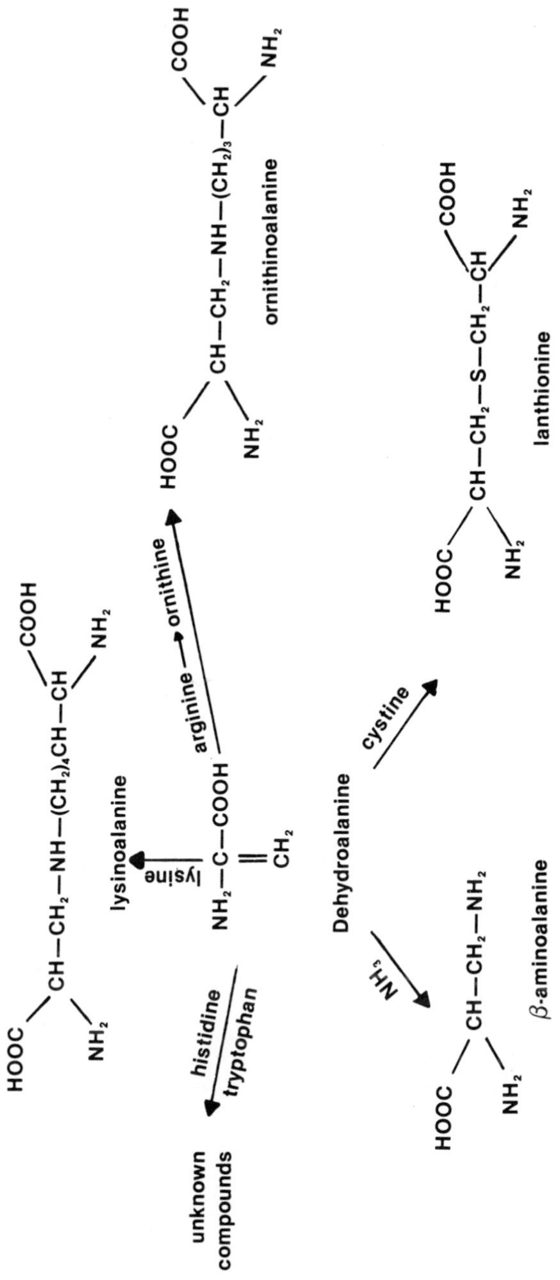

Figure 2. Formation of lysinoalanine and other amino acid derivatives from proteins.

Table II

Number of revertant colonies produced by ethanol extracts of milk crumb and soya crumb

Materials mg/plate	S-9	*Salmonella typhimurium* TA				
		1535	1537	1538	98	100
(spontaneous reversion)	+	4	5	6	20	29
	-	4	3	5	13	22
2-Amino-fluorene 0.1 mg	+	8	14	130	1100	160
	-	8	4	6	66	13
Soya-crumb 62.5 mg	+	7	1	0	18	5
	-	0	0	0	0	2700
Milk-crumb 62.5 mg	+	6	7	5	7	3
	-	0	0	0	1200	2200

The column S-9 shows whether the experiment was carried out with (+) or without (-) metabolic activation.

Table III

Antioxidative effect (A.E.) of soya-crumb at different pH values

Soya-crumb, 0.1 g	A.E.
pH = 7.47	0.84
5.92	1.30
5.03	1.83
4.15	1.90
3.20	1.45
2.25	1.23
BHT 10^{-5} mmole	
pH = 7.47	2.18

Milk-crumbs and soya-crumbs belong to this group. Their favorable organoleptic characters and good adaptability to further technological processes have been established, but more scientific investigations are still needed. The analysis of the aroma constituents of the crumbs is a task for the future.

The potential toxicity of the Amadori products (e.g. fructose-lysine) is still uncertain. The high levels of furosine in the milk-crumb add to the need to settle this question.

Our experiments on the mutagenicity of these materials revealed that a water extract of the crumbs did not give a positive Ames test. On the other hand the ethanol extract after distillation produced reverted colonies in the range of the positive control (2-AF) with the most sensitive strains, but only without metabolic activation. The Maillard products responsible for mutations seemed to be metabolized in the living organism. In view of our experiments and the present international evaluation on heat-processed foods, the crumbs present no more detrimental risk to human health than other heat-processed foods, when consumed.

The model experiments on the antioxidative effect of crumbs revealed the potential ability of the soya-crumb to prevent lipid oxidation in certain foods. The mild acid medium of meat products seems to be favorable for fulfilling this hope. However, further investigations in food products are necessary to obtain a decisive answer. As the case of milk crumb shows, a Maillard product is not necessarily an antioxidant.

Literature Cited

1. Kanold AB Co., Swedish Patent Application 10, 811/72.
2. Finot, P. A.; Bricout, J.; Viani, R.; Mauron, J. Experientia 1968, 24, 1097.
3. Devenyi, T.; Hazai, I.; Ferenczi, S.; Bati, J. Acta Biochem. Biophys. Acad. Sci. Hung. 1971, 6, 385.
4. Ames, B. N.; McCann, J.; Yamasaki, E. Mutat. Res. 1976, 31, 347.
5. Lingnert, H. "Antioxidative effect of Maillard reaction products". ISBN 91-7290-079-2 (SIK-Rapport), Göteborg, 1979.
6. Ebersdobler, H. F.; Brandt, A.; Scharrer, E.; von Wangenheim, B. Prog. Food Nutr. 1981, 5, 257.
7. Dworschak, E. CRC Crit. Rev. Food Sci. Nutr. 1980, 13, 1.
8. Sugimura, T. CRC Crit. Rev. Toxicol. 1979, 6, 189.

RECEIVED November 22, 1982

11

The Maillard Reaction and Meat Flavor

MILTON E. BAILEY
University of Missouri, Department of Food Science and Nutrition, Columbia, MO 65211

Many desirable meat flavor volatiles are synthesized by heating water-soluble precursors such as amino acids and carbohydrates. These latter constituents interact to form intermediates which are converted to meat flavor compounds by oxidation, decarboxylation, condensation and cyclization. O-, N-, and S-heterocyclics including furans, furanones, pyrazines, thiophenes, thiazoles, thiazolines and cyclic polysulfides contribute significantly to the overall desirable aroma impression of meat. The Maillard reaction, including formation of Strecker aldehydes, hydrogen sulfide and ammonia, is important in the mechanism of formation of these compounds.

Raw meat has little desirable flavor, but each type of meat has a characteristic flavor due to the animal species and the temperature and type of cooking. Both water-soluble and lipid-soluble fractions contribute to meat flavor and the water-soluble components include precursors which upon heating are converted to volatile compounds described as "meaty."

There are two approaches to studying meat flavor: one is concerned with identification of nonvolatile precursors of flavor components and the other involves isolation and identification of volatile flavor molecules. Water-soluble meat flavor precursors encompass a number of different organic classes of compounds, including: nucleic acids, nucleotides, nucleosides, peptides, amino acids, free sugars, sugar amines, glycogen, and amines. Bender and co-workers (1,2) were perhaps the first to thoroughly examine contents of water-soluble extracts of boiled beef, which they believed contained precursors of meaty flavor. These investigators reported the presence of many low-molecular-weight nitrogen compounds. Kramlich and Pearson (3) reported that pressed fluid from raw beef produced meat flavor upon heating.

Hornstein and Crowe (4) found that flavor precursors of meat were extractable with cold water and demonstrated that lyophilized

0097-6156/83/0215-0169$06.00/0

diffusate from cold water extracts of raw beef and pork produced meaty odors upon heating. Wood (5) demonstrated that heating at 100°C resulted in increase in inorganic phosphate and a decrease in organic phosphate and reducing substances. This was accompanied by pronounced meat odor. Heating mixtures of amino acids in the absence of reducing sugars produced neither browning nor meaty flavor. To obtain meaty aromas synthetically, a mixture of all compounds identified in beef extract was required. Wood concluded, "it is shown beyond reasonable doubt that development of the brown color and meaty flavor characteristics of these extracts is a result of the Maillard reaction."

Macy, et al. (6-10) studied the water-soluble precursors of beef, pork and lamb including the dialyzable low-molecular-weight constituents, amino acids, carbohydrates, nucleotides, and nucleosides. They studied the influence of heating on these constituents. The predominant amines in the dialyzable diffusate were alanine, anserine-carnosine and taurine, and these decreased considerably during heating. Other amino acids decreasing during heating included methylhistidine, isoleucine, leucine, methionine, cystine, serine, lysine, glycine, and glutamic acid.

Ribose-5-phosphate, ribose, glucose and fructose decreased extensively in concentration during heating at 100°C for one h. Ribose and ribose-5-phosphate decreased most readily under these conditions. Similar results were published by Wasserman and Spinelli (11), who examined the amino acids and sugars in beef diffusate after heating for one h at 125°C. They also studied changes in amino acids and sugars during heating in model systems. Wasserman (12) also concluded that Maillard browning was important for the formation of desirable flavor compounds during heating of meat, but that the relationship between compounds formed by this reaction and meat flavor was unknown.

These statements concerning the Maillard reaction are associated with the formation of 1-amino-1-deoxy-2-ketose addition products during heating of reducing sugars and amines. These addition products can then degrade by several pathways to form flavor compounds. The addition product for one pathway enolizes at C-2 - C-3 and eliminates the amine from C-1 to form methyl dicarbonyl intermediates (13,14), which further react to give fission products such as C-methyl reductones, keto aldehydes, dicarbonyl compounds and reductones (15). Important volatile flavor producing compounds include pyruvaldehyde, diacetyl and hydroxyacetone. Another pathway begins with 1,2-enediol form of the addition product with subsequent deamination at C-1 yielding 3-deoxyhexosone (16). Dehydration results in the formation of important flavor compounds such as 2-furaldehydes (furfural), a formylpyrrole, and similar compounds. Another pathway possibly beginning with the 2,3 enediol involves the enolic form of a 1-deoxy-2,3-dicarbonyl intermediate or an equilibrium enolic form which leads to formation of flavor compounds such as methylfuranones, isomaltol, and maltol (17). The acidity of the reaction medium and the basicity of the amine constituting the addition product (Amadori compound) are important variables for the formation of these various flavor volatiles.

Two branches of the above reaction pathways provide active reagents for the degradation of α-amino acids to aldehydes and ketones of one less carbon atom (Strecker degradation), which is another arm of the Maillard reaction. Strecker aldehydes from these reactions are important flavor compounds (18).

Further evidence of the importance of the Maillard reaction in the formation of volatile flavorants from meat precursors is gleaned by examining ingredients in reaction mixtures patented as synthetic meat constituents. Ching (19) examined 128 patents of meat flavor and found that 55 specified use of both amino acids and sugars. Amino acids were the predominant components of these mixtures, and among the amino acids, sulfur-containing cysteine and/or cystine were used in 39 mixtures. Glutamic acid was also used in 39 such mixtures.

In their recent comprehensive review of natural and synthetic meat flavors, MacLeod and Seyyedain-Ardebili (20) listed 80 patents describing "reaction products" procedures that produced meat-like flavors upon heating. Approximately one-half of these precursor mixtures included amino acids and reducing sugars. Most of the mixtures described in patented procedures for synthetic meat flavor are modeled after ingredients found in the water-soluble dialyzable fraction of fresh meat. These constituents serve as reagents for Maillard reactions.

Many investigations have been made of the volatile flavor compounds resulting from meat cookery or heating of mixtures simulating meat flavor precursors in model systems. These have been reviewed by Herz and Chang (21), Dwivedi (22), Ching (19) and MacLeod and Seyyedain-Ardebili (20). It has been estimated that approximately 600 volatiles have been identified from meat or simulated meat precursors (23). These volatiles are not only from pre-formed volatiles in meat but largely from changes of precursors by oxidation, decarboxylation, fragmentation, recombination, rearrangement, condensation, and cyclization.

Meat aroma is not the result of one chemical constituent but the sum of the sensory effects of many of these volatiles. Over 90% of the volume of volatile constituents from freshly roasted beef is from lipid, but approximately 40 percent of the volatiles from the aqueous fraction are thought to be heterocyclic compounds, many resulting from Maillard reaction products or their interactions with other ingredients.

Heterocyclic compounds as meat volatiles have been reviewed recently by Ohloff and Flament (23) and by Shibamoto (24). More comprehensive coverage of these constituents as food flavorants is currently being published (Vernin, 25).

Heterocyclic compounds contribute significantly to the overall aroma impression of meat. They include O-, N-, and S-heterocyclic structures. Meat flavor heterocyclics include furans, furanones, pyrazines, thiophenes, thiazoles, thiazolines, oxazolines and cyclic polysulfides. These compounds can all be formed by Maillard type interactions.

Furans and Furanones

Heterocyclics, particularly the S-containing ones, are extremely important contributors to "meaty" flavors. Ching (19) identified 11 furans and 7 furanones from reaction mixtures containing precursors responsible for beef flavor. Furfural derivatives were obtained by heating several reaction mixtures including low-molecular-weight water-extractable components from beef and simple amino acid-sugar mixtures. 4-Hydroxy-2,5-dimethyl-3(2H)-furanone, 4-hydroxy-5-methyl-3-(2H)-furanone, and four similar compounds were identified by Ching (19) from beef by dialyzing water-soluble materials and concentrating the diffusible solutes. The diffusates were then heated with or without sugars at 130°C for 2 h. Furanones can be formed by Amadori compound pathways (17, 18). Some of these compounds were originally isolated and identified from beef by Tonsbeek, et al. (26), who stated they might be formed through interaction of ribose-5-phosphate and taurine during heating. Hexoses form 5-methylfufurals and 4-hydroxy-2,5-dimethyl-3(2H) furanone, while pentoses produce furfural and 4-hydroxy-5-methyl-3(2H) furanone. Hicks and Feather (27) and Hicks, et al. (28) synthesized the latter compound by heating amines with xylose, ribose, ribose-5-phosphate, or gluconic acid.

Although exact mechanisms have not been described for the formation of other furanoid compounds through amine-carbohydrate interactions, probably many of the 32 furans described by Ohloff and Flament (23) from meat aroma mixtures are from this source. Shibamoto (24) described many of the same components from mixtures producing meat odors.

Herz and Chang (21) examined several furan compounds which had a wide variety of aromas, but none of them were meaty. Furans that do not contain sulfur are usually fruity, nutty, and caramel-like in odor. The furanones described above have burnt pineapple and roasted chicory odors, but these contribute to overall flavor impression of meat and important N and S meat flavor compounds might be formed from them during cooking.

Thiophenes

MacLeod and Seyyedain-Ardebili (20) listed 36 thiophene derivatives as having been found during various investigations of meat or meat constituents. Ching (19) found 18 and Shibamoto (24) listed 29.

Thiophenes are extremely important in flavor and are responsible for the mild sulfurous odor of cooked meat. Numerous other thiophenes have been identified during heating of meat or meat constituents. The sulfur in thiophene may be derived from amino acids (cysteine, cystine, methionine) or from vitamin B_1.

Probably the most important reactant in the formation of volatile meat flavor compounds is hydrogen sulfide. It can be formed by several pathways during meat cookery, but one mechanism is Strecker degradation of cysteine in the presence of a diketone as established by Kobayashi and Fujimaki (29). The cysteine condenses with the diketone and the product in turn decarboxylates to amino carbonyl compounds that can be degraded to hydrogen sulfide, ammonia and acetaldehyde. These become very reactive volatiles for the formation of many flavor compounds in meat and other foods.

Shibamoto (30) formed α-thiophenecarboxaldehyde by reacting furfural and hydrogen disulfide. This means that there probably was exchange between the S and O in the furan ring during heating.

van den Ouweland and Peer (31) made a significant contribution to the elucidation of meat flavor when they demonstrated that 4-hydroxy-5-methyl-3(2H)-furanone formed a number of "meat-like" mercapto-substituted furan and thiophene derivatives when heated in the presence of hydrogen sulfide. These authors postulated that the initial stage in forming the thiophene derivatives involves a partial substitution of the ring oxygen by sulfur to give the thio analog. Compounds having a meaty odor included 3-mercapto-2-methyl-2,3-dihydrothiophene, 3-mercapto-4-hydroxy-2-methyl-2,3-dihydrothiophene, 4-mercapto-2-methyltetrahydrothiophene, and 3-mercapto-2-methyltetrahydrothiophene. The proposed (31,32) reactions for the formation of some of these "meaty" compounds is summarized in Figure 1. Although these authors assumed that the dihydrofuranone was derived from ribose-5-phosphate via a dephosphorylation-dehydration reaction (31), there is evidence which indicates that the furanone can be formed by a reaction of aldose sugars with amines to produce Amadori products which subsequently dehydrate with amine elimination (17,28,33). The amine could be taurine, which is in ample supply in muscle (7) and was mentioned by Tonsbeek, et al. (26) in their original discovery of these compounds. None of these compounds have been identified from meat extracts, but Ching (19) identified 4-mercapto-2-methyl-4,5-dihydrothiophene from synthetic beef diffusate. The mercaptothiophenes are undoubtedly formed at low concentrations in meat.

Evers, et al. (34) identified several S-substituted furans having meaty aroma including 3-mercapto-2-methylfuran and 3-mercapto-2,5-dimethylfuran from Maillard reaction mixtures. These compounds were readily oxidized to sulfides, some of which retained meaty odors. All furans having the sulfur atom bound to the β-carbon had meaty aromas, whereas those with sulfur bound to the α-carbon had hydrogen sulfide-like odors.

Maillard reaction products formed by interaction of reducing sugar and amino acids such as α-dicarbonyl compounds, aldehydes, hydrogen sulfide, and ammonia can react further to form derivatives that have been identified from meat or its components during heating. Important reviews of sulfur compounds that might be produced by these reactions have been published by Schutte (35) and

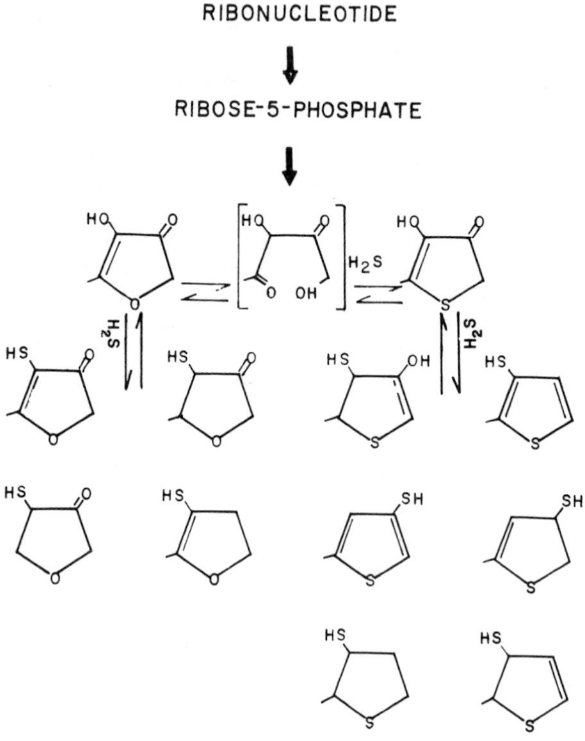

*Figure 1. Reaction of 4-hydroxy-5-methyl-3(2*H*)-furanone with hydrogen sulfide to form mercaptofurans and mercaptothiophenes having beef-like odors (*31*).*

by Takken, et al. (36). Some of the important sulfur-containing compounds include thiazoles, thiazolines, and polysulfide heterocyclics.

Thiazoles and Thiazolines

Takken, et al. (36) found that these compounds could be formed by combining butane-2,3-dione (diacetyl), pentane-2,3-dione, or pyruvaldehyde with acetaldehyde, hydrogen sulfide, and ammonia. Some of the reaction pathways suggested for thiazole and thiazoline formation for these reaction ingredients are outlined in Figure 2. Thiazoles have also been identified in volatiles resulting from heating cysteine and cystine with glucose and pyruvaldehyde at 160°C (37) and by Mulders (38) who heated a similar system at 125°C and pH 5.6 for 24 h. Possible pathways for these reactions were discussed by Schutte (35). Corresponding thiazolines and thiazoles frequently coexist due to the dehydration of thiazolines during heating (39).

MacLeod and Seyyedain-Ardebili (20) listed 12 thiazoles and thiazolines that had been identified in beef samples by different investigators. Some patented compounds listed as having meaty flavor are 4-methyl-5-(2-hydroxyethyl)thiazole, 4-ethyl-5-(3-acetoxypropyl)thiazole, 2-acetyl-2-thiazoline and 2-acetyl-5-propyl-2-thiazoline. Ching (19) identified 12 thiazole derivatives in her studies of the volatile produced by heating components of meat. Important thiazoles identified in these studies included 2 and 4-acetyl derivatives. It was postulated by Tonsbeek, et al. (40) that 2-acetyl-2-thiazoline is formed by Strecker degradation of cysteine with methylglyoxal or similar compounds followed by oxidative cyclization as shown in Figure 3. This compound was isolated from 225 kg of cooked beef and had a strong odor of freshly baked bread. Other thiazoles have been isolated from meat or meat constituents which have undergone Maillard-type reactions (Wilson, et al., 41).

Polysulfide Heterocyclics

Wilson, et al. (41) also confirmed the presence of polysulfur heterocyclics in meat including thialdine (5,6-dihydro-2,4,6-trimethyl-1,3,5-dithiazine) and trithioacetone (2,2,4,4,6,6-hexamethyl-1,3,5-trithiane). Wilson (42) later discussed the possible routes of formation of some of these compounds from cysteine. Thialdine was found by Brinkman, et al. (43) in the headspace volatiles of beef broth. These workers also identified 3,5-dimethyl-1,2,4-trithiolane from the same source. Both cis and trans isomers of this compound had previously been identified as flavor components of boiled beef by Chang, et al. (44) and Herz (45).

Chicken has also been found to be a source of polysulfide heterocyclics. Pippen and Mecchi (46) theorized that 2,4,6-tri-

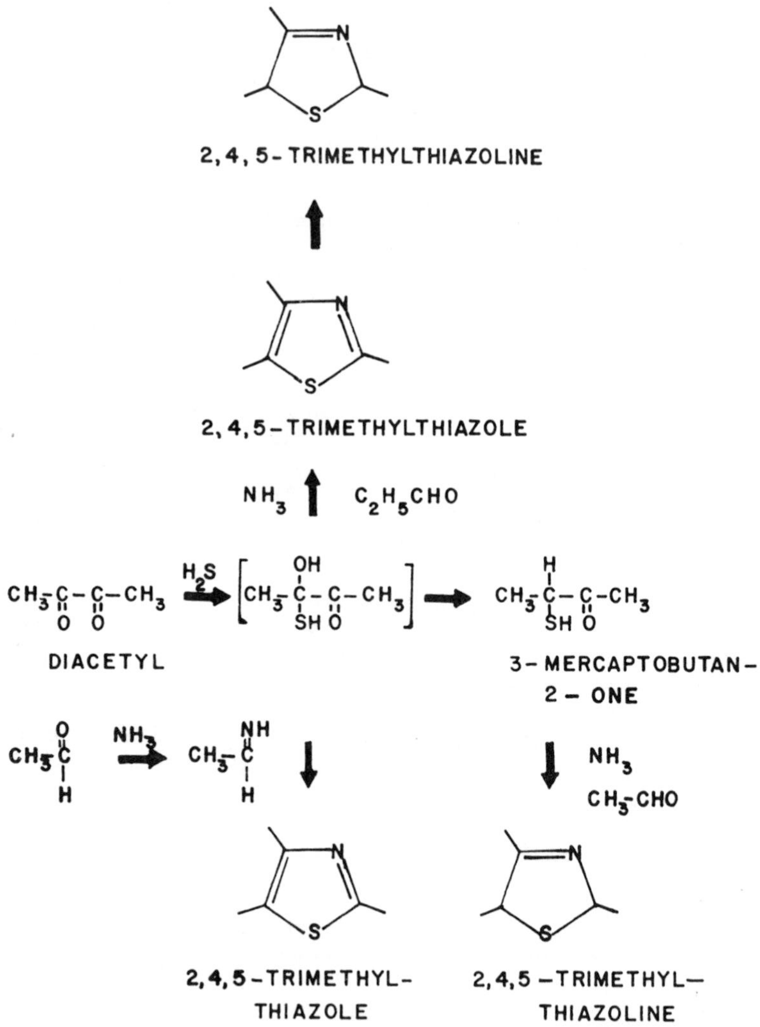

Figure 2. Reaction pathways for formation of thiazoles and thiazolines by heating Maillard degradation products of meat precursors (36).

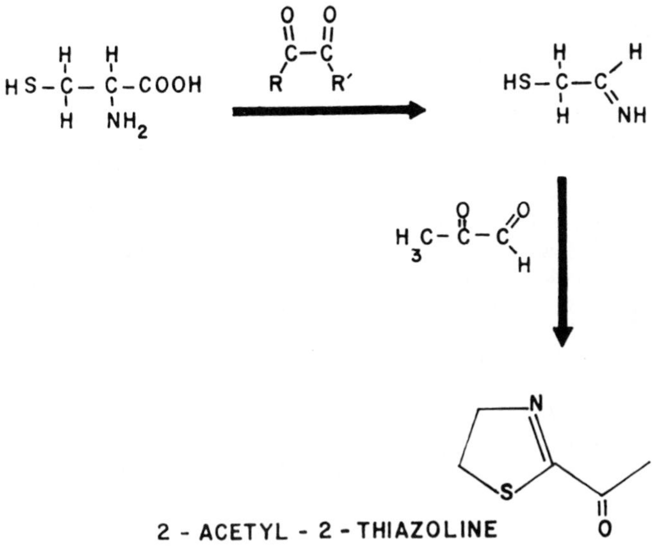

Figure 3. Formation of 2-acetyl-2-thiazoline by Strecker degradation of cysteine. (Reproduced from Ref. 40. Copyright 1971, American Chemical Society.)

methyl-1,3,5-trithiane might be formed in chicken by indirect combinations of hydrogen sulfide and acetaldehyde in fat. The same heterocyclic was listed as a flavor volatile in chicken by Wilson and Katz (47) in their review of chicken flavor. These same workers (48) also identified this compound in lean ground beef.

The formation of 5,6-dihydro-2,4,6-trimethyl-1,3,5-dithiazine, 2,4,6-trimethyl-1,3,5-trithiane, and 3,5-dimethyl-1,2,4-trithiolane by heating of acetaldehyde, hydrogen sulfide, and ammonia was outlined by Takken and coworkers (36) and is summarized in Figure 4. Under oxidative conditions, dialkyltrithiolanes are formed; at low pH there is conversion to trialkyltrithianes; at elevated temperature isomerization into trisulfides occurs, which compounds disproportionate into di and tetrasulfides; and in the presence of ammonia, dithiazines are formed. These compounds and the conditions for their formation are of extreme importance for the production of desirable meat flavors.

1-(Methylthio)ethanethiol

This important flavor compound was identified in the headspace volatiles of beef broth by Brinkman, et al. (43) and although it has the odor of fresh onions, it is believed to contribute to the flavor of meat. This compound can be formed quite easily from Strecker degradation products. Schutte and Koenders (49) concluded that the most probable precursors for its formation were ethanal, methanethiol and hydrogen sulfide. As shown in Figure 5, these immediate precursors are generated from alanine, methionine and cysteine in the presence of a Strecker degradation dicarbonyl compound such as pyruvaldehyde. These same precursors could also interact under similar conditions to give dimethyl disulfide and 3,5-dimethyl-1,2,4-trithiolane previously discussed.

Pyrazines

The formation of pyrazines in foods has been reviewed extensively by Mega and Sizer (50). Temperature and pH are very important factors in the formation of specific pyrazines. Forty-two pyrazines have been identified in meat from various sources by these authors. MacLeod and Seyyedain-Ardebili (20) listed 49 pyrazines found in beef by various investigators. Ching (19) identified 28 pyrazines in her studies of sugar-amine reactions simulating beef flavor.

Several mechanisms have been reported for pyrazine formation by Maillard reactions (21,52,53). The carbon skeletons of pyrazines come from α-dicarbonyl (Strecker) compounds which can react with ammonia to produce α-amino ketones as described by Flament, et al. (54) which condense by dehydration and oxidize to pyrazines (Figure 6), or the dicarbonyl compounds can initiate Strecker degradation of amino acids to form α-amino ketones which are hydrolyzed to carbonyl amines, condensed and are oxidized to substituted

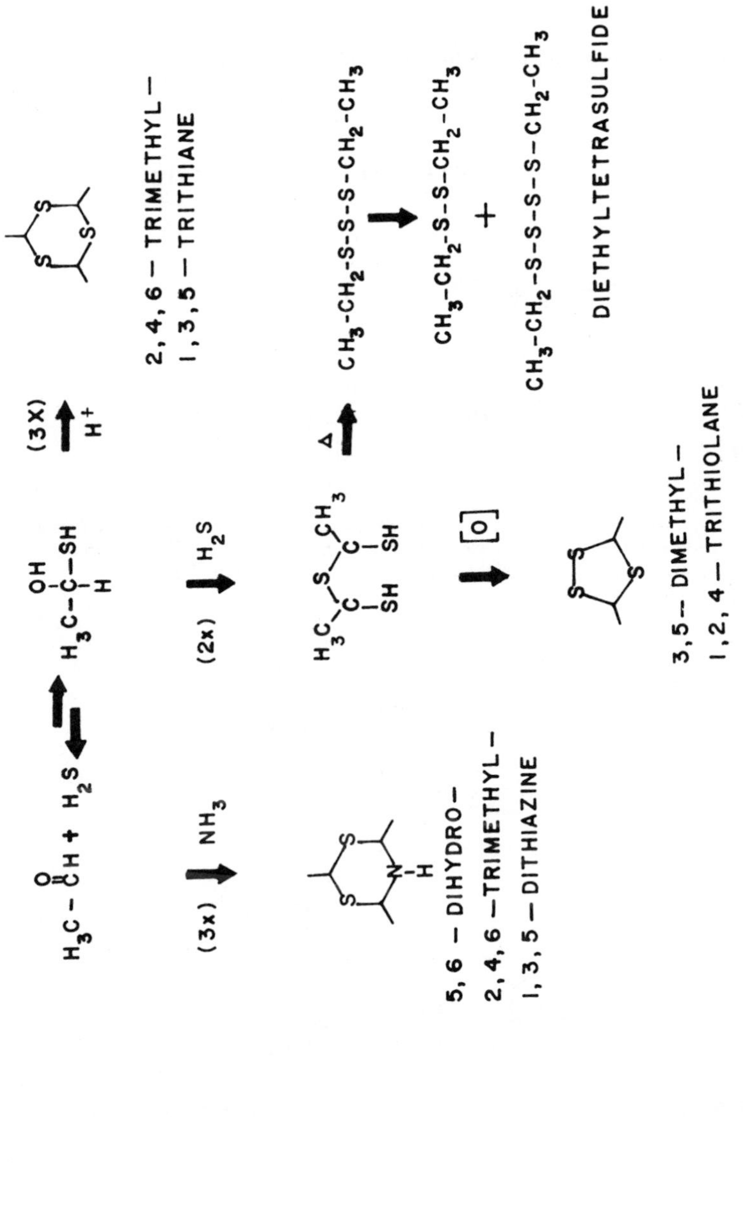

Figure 4. Formation of important meat flavor polysulfide heterocyclics by heating acetaldehyde and hydrogen sulfide (36).

ALANINE ➡ ETHANAL

CYSTEINE ➡ HYDROGEN SULFIDE

METHIONINE ➡ METHIONAL ➡ METHANETHIOL

I-(METHYLTHIO)ETHANETHIOL

$$H_3C-\overset{O}{\underset{H}{C}} + HS-CH_3 \rightleftharpoons H_3C-\overset{OH}{\underset{H}{C}}-S-CH_3$$

$$H_3C-\overset{OH}{\underset{H}{C}}-S-CH_3 + SH^{\ominus} \longrightarrow H_3C-\overset{H}{\underset{SH}{C}}-S-CH_3 + OH^{\ominus}$$

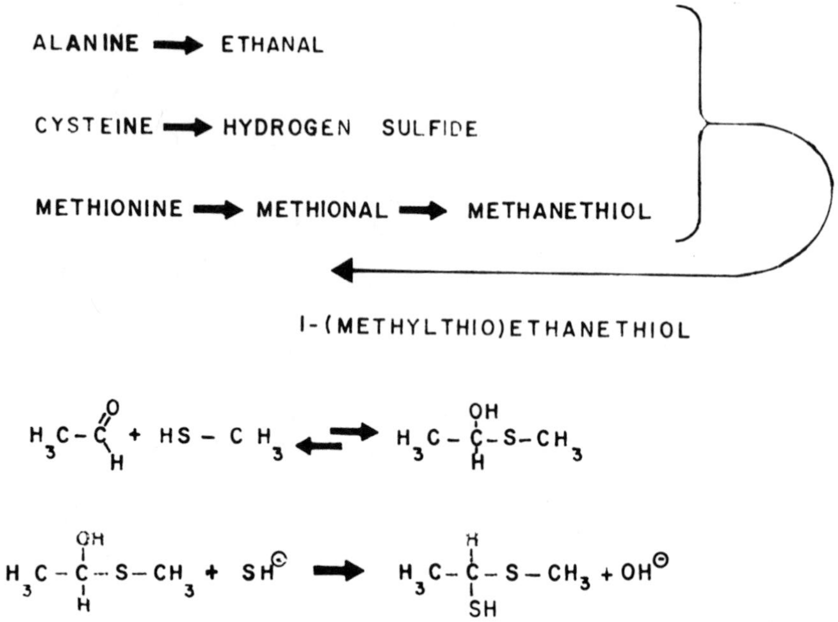

Figure 5. Formation of secondary reaction products involved in production of 1-(methylthio)ethanethiol by Strecker degradation of amino acids. (Reproduced from Ref. 49. Copyright 1972, American Chemical Society.)

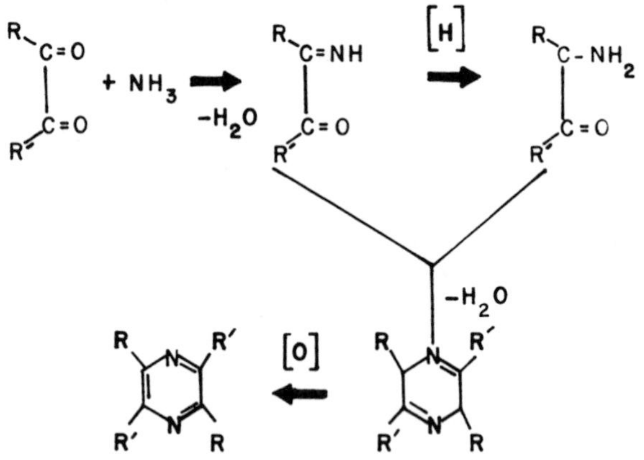

Figure 6. Interaction of Maillard reaction products to form pyrazines (54).

pyrazines. The latter mechanism was proposed by Hodge and Osman (55).

Some meat flavor-contributing pyrazines patented for use for synthetic purposes are methylpyrazine, 5,7-dihydro-5,7-dimethylfuro-(3,4-b)pyrazine, and 5-methyl-(7H)-cyclopenta[b]pyrazine (20).

Present Outlook

Even though many compounds discussed in the above presentation are thought to be important in meat flavor, a delicate blend of these compounds and other ingredients at the appropriate concentration is needed to synthesize acceptable flavor. In view of the possible instability of the flavor compounds themselves, precursors that supply the precise mixture of volatiles upon heating will be needed. Attempts have already been made to use this approach as judged by the numerous patented mixtures of precursors listed in the literature. More effort should be given to the quantitative aspects of meat flavor production and work must be continued on the qualitative aspects of the volatiles and the appropriate Maillard reaction precursors chosen.

One method presently being used to determine the contribution of volatile compounds to food flavor is to combine sensory analyses with GC/MS fractionation technics. Using these combined procedures, Persson and coworkers (56-59) were able to correlate acceptability of canned beef with levels of different compounds in the product. Pearson and coworkers (60,61,62) have used a similar method to optimize the ingredients responsible for desirable beef aroma. In their most recent publication (62) they report that heterocyclic compounds containing oxygen, nitrogen and sulfur were the most important ingredients for beef aroma. Hydrophilic compounds contributed more to meat aroma than did hydrophobic compounds, verifying the importance of water-soluble compounds in meat flavor. Important mixtures of ingredients were found to be hydrogen sulfide with 2-acetyl-3-methylpyrazine and hydrogen sulfide with 4-hydroxy-5-methyl-3(2H)-furanone. This approach will ultimately be useful in solving the problem of producing desirable meat flavors from purified precursors and certainly lends support for the importance of the Maillard reaction in the formation of these flavors.

Literature Cited

1. Wood, T.; Bender, A. E. Biochem. J. 1957, 67, 366.
2. Bender, A. E.; Wood, T.; Palgrave, J. A. J. Sci. Food Agric. 1958, 9, 812.
3. Kramlich, W. E.; Pearson, A. M. Food Res. 1958, 23, 567.
4. Hornstein, I.; Crowe, P. F. J. Agr. Food Chem. 1960, 8, 494.
5. Wood, T. J. Sci. Food Agr. 1961, 12, 61.
6. Macy, R. L., Jr.; Naumann, H. D.; Bailey, M. E. J. Food Sci. 1968, 33, 53.

7. Macy, R. L., Jr.; Naumann, H. D.; Bailey, M. E. J. Food Sci. 1964, 29, 142.
8. Macy, R. L., Jr.; Naumann, H. D.; Bailey, M. E. J. Food Sci. 1970, 35, 78.
9. Macy, R. L., Jr.; Naumann, H. D.; Bailey, M. E. J. Food Sci. 1970, 35, 81.
10. Macy, R. L., Jr.; Naumann, H. D.; Bailey, M. E. J. Food Sci. 1970, 35, 83.
11. Wasserman, A. E.; Spinelli, A. M. J. Food Sci. 1970, 35, 328.
12. Wasserman, A. E. J.Agric.Food Chem. 1972, 20, 737.
13. Hodge, J. E. J. Agric. Food Chem. 1953, 1, 928.
14. Hodge, J. E.; Fisher, B. E.; Nelson, E. C. Am. Soc. Brew. Proc. 1963, 84.
15. Hodge, J. E."Chemistry and Physiology of Flavors"; Schulz, H. W.; Day, E. A.; Libbey, L. M., Eds.; AVI Publishing Co.: Westport, Conn., 1967; p. 465-491.
16. Anet, E. F. L. Adv. Carbohyd. Chem. 1964, 19, 131.
17. Feather, M. S. "Maillard Reactions in Food"; Erikson, C., Ed.; Pergamon Press: Oxford, 1981; p. 37-45.
18. Hodge, J. E.; Mills, F. D.; Fisher, B. E. Cereal Sci. Today 1972, 17, 34.
19. Ching, J.C-Y. "Volatile Flavor Compounds From Beef and Beef Constituents", Ph.D. Dissertation, University of Missouri, Columbia, MO, 1979.
20. MacLeod, G.; Seyyedain-Ardebili, M. CRC Crit. Rev. Food Sci. Nutr. 1981, 14, 309.
21. Herz, K. O.; Chang, S. S. Adv. Food Res. 1970, 18, 1.
22. Dwivedi, B. K. CRC Crit. Rev. Food Sci. Nutr. 1975, 5, 487.
23. Ohloff, G.; Flament, I. Prog. Chem. Org. Nat. Prod. 1978, 36, 231.
24. Shibamoto, T. J. Agric. Food Chem. 1980, 28, 237.
25. Vernin, G. "Heterocyclic Compounds and Their Precursors in Food Aroma," 1982; Ellis Harwood Ltd.:
26. Tonsbeek, C. H. T. J. Agric. Food Chem. 1968, 16, 1016.
27. Hicks, K. B.; Feather, M. S. Carbohyd. Res. 1977, 54, 209.
28. Hicks, K. B.; Harris, D. W.; Feather, M. S.; Loeppky, R. N. J. Agric. Food Chem. 1974, 22, 724.
29. Kobayashi, N.; Fujimaki, M. Agric. Biol. Chem. 1965, 29, 698.
30. Shibamoto, T. J. Agric. Food Chem. 1977, 25, 206.
31. van den Ouweland, G. A. M.; Peer, H. G. J. Agric. Food Chem. 1975, 23, 501.
32. van den Ouweland, G. A. M.; Olsman, H.; Peer, H. G. "Agricultural and Food Chemistry: Past, Present, Future"; Teranishi, R., Ed.; AVI Publishing Co.: Westport, Conn., 1978; p. 292-314.
33. Mills, F. D.; Hodge, J. E. Carbohyd. Res. 1976, 51, 9.
34. Evers, W. J.; Heinsohn, H. H., Jr.; Mayers, B. J.; Sanderson, A. "Phenolic, Sulfur, and Nitrogen Compounds in Food Flavors"; Charalambous, G.; Katz, I., Eds.; ACS Symposium Series 26: Washington, D. C., 1976; p. 184-193.

35. Schutte, L. "Fenaroli's Handbook of Flavor Ingredients", 2nd Ed., Vol. I; Furia, T. E.; Bellanca, N., Eds.; CRC Press: Cleveland, OH, 1975; p. 132-183.
36. Takken, H. J.; van der Linde, L. M.; Devalois, D. J.; van Dort, H. M.; Boelens, M. "Phenolic, Sulfur and Nitrogen Compounds in Food Flavors"; Charalambous, G.; Katz, I., Eds.; ACS Symposium Series 26, Washington, D. C., 1976; p. 114-121.
37. Kato, S.; Kurato, T.; Fujimaki, M. Agric. Biol. Chem. 1973, 37, 539.
38. Mulders, E. J. Z. LebensmUnters-Forsch. 1973, 152, 193.
39. Asinger, T.; Thiel, M.; Dathe, W.; Hampel, O.; Mutag, E.; Plaschil, E.; Schroeder, C. Lichigs Annalen Chem. 1960, 639, 146.
40. Tonsbeek, C. H. T.; Copier, H.; Plancken, A. J. J. Agric. Food Chem. 1971, 19, 1014.
41. Wilson, R. A.; Mussinan, C. J.; Katz, I.; Sanderson, A. J. Agric. Food Chem. 1973, 21, 873.
42. Wilson, R. A. J. Agric. Food Chem. 1975, 23, 1032.
43. Brinkman, H. W.; Copier, H.; deLeuw, J. J. M.; Tgan, S. B. J. Agric. Food Chem. 1972, 20, 177.
44. Chang, S. S. ; Hirai, C.; Reddy, B. R.; Herz, K. O.; Kato, A.; Simpa, G. Chem. Ind., London; 1968, 1639.
45. Herz, K. O. "A Study of the Nature of Boiled Beef Flavor", Ph.D. Thesis, Rutgers State University, New Brunswick, N. J. 1968.
46. Pippen, E. L.; Mecchi, E. P. J. Food Sci. 1969, 34, 443.
47. Wilson, R. A.; Katz, I. J. Agric. Food Chem. 1972, 20, 741.
48. Wilson, R. A.; Mussinan, C. J.; Katz, I.; Sanderson, A. J. Agric. Food Chem. 1973, 21, 873.
49. Schutte, L.; Koenders, E. B. J. Agric. Food Chem. 1972, 20, 181.
50. Maga, J. A.; Sizer, C. E. "Fenaroli's Handbook of Flavor Ingredients"; 2nd Ed., Vol. I; Furia, T. E.; Bellanca, N., Eds.; CRC Press, Inc.: Cleveland, Ohio, 1975; p. 47-131.
51. Newell, J. A.; Mason, M. E.; Matlock, R. S. J. Agr. Food Chem. 1967, 15, 767.
52. Rizzi, G. P. J. Agric. Food Chem. 1972, 20, 1081.
53. Koehler, P. E. and Odell, G. V. J. Agric. Food Chem. 1970, 18, 895.
54. Flament, I.; Kohler, M.; Aschiero, R. Helv. Chim. Acta 1976, 59, 2308.
55. Hodge, J. E.; Osman, E. M. "Principles of Food Science"; Part I, Food Chemistry; Fennema, O. R., Ed.; Marcel Dekker, Inc.: New York, 1976; p. 41-138.
56. Persson, T.; von Sydow, E. J. Food Sci. 1974, 39, 406.
57. Persson, T.; van Sydow, E. J. Food Sci. 1974, 39, 537.
58. Persson, T.; van Sydow, E. J.; Akesson, C. J. Food Sci. 1973, 38, 377.
59. von Sydow, E. Food Technol. 1971, 25, 40.
60. Pearson, A. M.; Baten, W. D.; Goembel, A. J.; Spooner, M. E. Food Technol. 1962, 16, 137.

61. Hsieh, Y.P.C.; Pearson. A. M.; Magee, W. T. J. Food Sci. 1980, 45, 1125.
62. Bodrero, K. O.; Pearson, A. M.; Magee, W. T. J. Food Sci. 1981, 46, 26.

RECEIVED October 5, 1982

12

Sensory Properties of Volatile Maillard Reaction Products and Related Compounds

A Literature Review

SUSAN FORS

SIK–The Swedish Food Institute, Box 5401, S-402 29 Göteborg, Sweden

Volatile compounds formed by non-enzymatic browning reactions are of great importance for the sensory properties of heat-treated foods.

Literature information about the sensory properties for nearly 450 Maillard reaction products has been compiled in a survey. It includes qualitative aroma and flavor descriptions as well as sensory threshold values in different media for the compounds, classified according to their chemical structure.

In recent decades a large number of papers and reviews about research on food aroma and flavor have been published, and this research field continues to expand. The expansion is reflected in the rapid increase in the number of volatile compounds identified in various foods.

Non-enzymatic browning reactions play a central role in the formation of food aroma and flavor, especially in heat-treated foods. The purpose of this work is to present sensory data, scattered in the literature, for volatile non-enzymatic browning reaction products and related compounds. The compilation has no pretensions to completeness and only a small part of the extensive patent literature has been covered. Anyhow, it is felt that a compilation of this kind, which has not been available hitherto, would be useful to workers in the field.

The majority of the compounds in this compilation are Maillard reaction products and likewise recognized as important aroma and flavor substances in foods.

The sensory properties, presented in tabular form, include the following

- The qualitative odor and/or flavor description of the compound. The concentration may have a strong influence on the odor or flavor quality; one compound strongly diluted will not have the sensory characteristics of the same compound in a more concentrated form.

0097-6156/83/0215-0185$21.20/0

- The threshold values determined in different liquids (water, oil, orangejuice, beer, sugar solution etc.) or less frequently, in air. The majority of the threshold values given are detection rather than recognition thresholds, although in many cases the type of threshold determined is not indicated in the original paper.

The threshold values given by different investigators vary considerably depending on the choice of solvent(s), the method used for the determination, and the purity of the compound. Therefore, the same aroma or flavor compound may have as many threshold values as there are investigators. Care must be exercised when threshold values from the literature are used, and the mentioned sources of discrepancies must always be borne in mind. Odor threshold values, expressed as concentration in the gaseous phase, have the distinct advantage of being independent of the solvent used.

Major classes of flavor compounds in general

The volatile compounds formed by the Maillard reaction are only one group of flavor compounds in foods. Schutte (1) presents a brief summary of the major classes and their modes of formation from precursors. Some of them can be formed by different pathways. An example is the furans, which can be formed by non-enzymatic browning reaction but also by biotransformation.

Classification of Maillard reaction products

Since this compilation concerns volatile Maillard reaction products, a brief presentation of different types of substances in this group is justified. The classification system given by Nursten (1980-1981) (2) has been a valuable tool. The volatiles may be classified into three groups

1) 'Simple' sugar dehydration/fragmentation products:
   Furans
   Pyrones
   Cyclopentenes
   Carbonyl compounds
   Acids
2) 'Simple' amino acid degradation products:
   Aldehydes
   Sulfur compounds
3) Volatiles produced by further interactions:
   Pyrroles
   Pyridines
   Imidazoles
   Pyrazines
   Oxazoles
   Thiazoles
   Compounds ex aldol condensations

As seen above, the volatile Maillard compounds have very diverse chemical structures. The most important groups will now be presented one by one.

The importance of heterocyclic compounds

Among the known constituents of food aroma, heterocyclic compounds deserve particular attention. These are formed in large numbers during preparation of the food, e.g., by cooking, baking or roasting. Owing to their characteristic odors, heterocycles contribute significantly to the flavor of processed foods. These heterocyclics do not always arise by non-enzymatic Maillard reactions; enzymatic (including microbiological) processes are also important routes of formation. Degradation by pyrolysis of certain compounds (sugars, amino acids, thiamin, trigonelline, etc.), also releases aroma compounds (3).

Among the heterocyclics there is one group which will be thoroughly examined: the pyrazines. The compounds belonging to this family play a very important role as contributors of desirable food flavor properties. Structurally, pyrazines are heterocyclic nitrogen compounds and their formation is a quite complicated process. Maga and Sizer (4) present a summary of these formation pathways.

Non-heterocyclic compounds

There are other compounds than heterocyclics that also deserve attention: Sulfur compounds and other classes of aliphatics.

Schutte (1) lists substances containing sulfur and their mode of formation. (Note that thiazoles, thiophenes, and other cyclic compounds can be classified also in the preceding group, the heterocyclics).

Other groups of aliphatic substances (aldehydes, ketones, esters, etc.) may also contribute to agreeable food sensations.

Aroma and flavor properties of important substance groups

This chapter gives brief general descriptions of the sensory properties of the most important substance groups. It is intended as an introduction to the more detailed sensory information in the following tables.

Pyranones, furanones, and related compounds (6, 7). Structurally these substances are generally cyclic ethers, mainly furanoid compounds. They are found in condensates from carbohydrates that have been subjected to browning reactions.

As a rule furan derivatives are considered important aroma constituents from a sensory point of view. They are mainly associated with sweet, fruity, nutty or caramel-like odor-impressions. The furans have no meaty characteristics, but it seems possible that they contribute to the overall odor of broiled and roasted meat.

Furan derivatives with several functional groups have increased odor intensity as compared with lower homologues.

Pyrroles (5 - 8). About fifty members of this group have been detected in various food stuffs, though it appears that pyrroles are not present in fresh, raw foods. Pyrroles have not received much attention as flavor-contributing components, but they seem to contribute an unddesirable odor to cooked meat.

Pyridines (6, 9). Pyridines have been found in coffee, barley and roasted lamb. Their importance as aroma constituents is limited.

Pyrazines (4, 6, 10, 11). The pyrazines constitute a very important class among flavor compounds. They have been identified in various food systems, and they are associated with pleasant and desirable food flavor properties. As a rule, the alkyl derivatives produce roasted--nutlike sensory impressions. The acetylpyrazines also have an essential place among flavoring agents. They have a characteristic roasted note, reminiscent of popcorn.

Sulfur compounds Sulfur-containing volatiles contribute to both pleasant and unpleasant overall flavors in many foods.

• Thiols (7, 12)

The traditional name for this group is mercaptans; nowadays, the term thiols is more common. They have been identified in more than sixty different foodstuffs.

Maga (12) gives an explanation for their importance as aroma and flavor components:

a) they have objectionable sensory properties (although there are exceptions)
b) the thresholds for most thiols are in the low parts-per-million range or lower.

• Thiophenes (13)

The majority of the thiophenes has been identified as constituents of meat-based products. This group of flavor compounds is relatively new, therefore sensory data are somewhat lacking.

• Oxazoles and Oxazolines (14, 15)

Their presence has only been reported in a limited number of foods: coffee, cocoa, meat products, barley and soy sauce. Very few sensory properties have been reported for oxazoles and oxazolines.

• Thiazoles (6, 16)

Thiazoles are in a certain way unique; they contain a heterocyclic ring containing both nitrogen and sulfur. Most of the thiazoles have been isolated from coffee and meat. The alkyl derivatives are most common; they generally have green, nutty, and vegetable-like odors.

Sensory data for compounds

Explanatory notes

Notation, symbols and abbreviations used in the following tables:

Names

The nomenclature is intended to follow IUPAC 1979 Rules, "Nomenclature of Organic Chemistry".

Synonyms

Synonyms are at times given, especially when accepted trivial names exist and/or to avoid doubt.

Descriptions

The sensory properties of each compounds are described.

F signifies Flavor, i.e. odor and taste together (the senses of smell and taste), which is the case when the item is taken into the mouth.

O signifies Odor, i.e. a property perceived through the sense of smell only.

Leaving out one or both of these letters signifies lack of such information, for example: green note, characteristic.

Threshold values

These values are usually evaluated through the sense of smell; they are marked with O = Odor threshold value. Some values are evaluated through the sense of taste;

T = Taste threshold value
F = Flavor threshold value, i.e. odor and taste together.

The solvent used is mentioned after the threshold value:

Examples:

O	0.05 ppb/water	the odor threshold of an aroma compound is measured (dissolved) in water
T	150 ppm/water	the taste threshold of a compound in beer
*	For further values,	see references Nos. 23, 141, and 142.

Remarks

Under this heading, place(s) of occurrence of each compound is given.

F for Foods, i.e. the compounds has been identified in a food or beverage, such as meat, vegetables, coffee, cocoa, and tea.

M for Model System, i.e. the compound has been identified in a non-food milieu. Example: glucose-cysteine model system.

S for Synthetic, i.e. the compound has been synthesized.

FURANS

NAME / (SYNONYM)	R_1	R_2	R_3	R_4
Furan				
2-Methylfuran	CH_3			
2-Ethylfuran	C_2H_5			
2-Propylfuran	$\underline{n}$-C_3H_7			
2-Butylfuran	$\underline{n}$-C_4H_9			
2-Pentylfuran	$\underline{n}$-C_5H_{11}			
2-Vinylfuran (2-Furylethylene)	$CH{=}CH_2$			
2-Acetylfuran (2-Furyl methyl ketone)	$COCH_3$			
2-Furyl-2-propanone (2-Furfuryl methyl ketone)	CH_2COCH_3			
2-Furfuryl methyl ether (2-(Methoxymethyl)furan)	CH_2OCH_3			

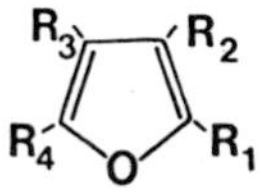

DESCRIPTION	THRESHOLD VALUE	REM.	REF
O: Peculiar spice-smoky, slightly cinnamon-like		F,M,S,	17
O: Ethereal			18
O of GC eluates: Sickly, nasty			19
	O 4.5 ppm/water		20
O: Ethereal		F,M,S,	18
O of GC eluates: Sickly			19
	O 3.5 ppm/water		20
	O 27 ppm/oil		21
O: Powerful sweet, burnt; when dilute, warm, sweet F: Caffeine-like		F,S	17
	O 8 ppm/oil		21
	O 6 ppm/oil	F	21
O: Weak, noncharacteristic		F	22
	O 10 ppm/oil		21
O: Fruity		F,S	23
O of GC eluates: Green, sweet pungent			19
O: Sweet, pungent F in vegetable oil at concentration up to 10 ppm: Beany, grassy			6
	O 6 ppb/water		24
	O 2 ppm/oil		21
O: Phenolic, coffee-grounds		F	22
	O 1 ppm/oil		21
O: Pleasant, ketonic F: Burning, sweetish		F,M	18
O: Powerful balsamic-sweet with a tobacco-like, almost narcotic pungence. Floral undertones of balsamic-cinnamic character			17
F in beer: Almonds, rubber, burnt/ /phenolic, pyrazole	80 ppm/beer		25
	T 110 ppm/orange juice		26
	F 80 ppm/water		27
O: Mild, sweet, fruity-caramellic, somewhat spicy F: Sweet, fruity-spicy, slightly nut-like		F,S	17
F: Suggestive of radish			23
O: Rum-like			6
O: Mustard-like		F	22

Continued on next page

FURANS (cont.)

NAME / (SYNONYM)	R_1	R_2	R_3	R_4
2-Furfuryl methyl sulfide	CH_2SCH_3			
2-Furfuryl acetate	CH_2OCOCH_3			
2-Furfuryl propionate	$CH_2OCOC_2H_5$			
2-Ethyl furoate (2-Furancarboxylic acid, ethyl ester)	$COOC_2H_5$			
2-Furoic acid (2-Furancarboxylic acid)	COOH			
2-Furfuryl alcohol (2-Furylmethanol, 2-(Hydroxymethyl)furan)	CH_2OH			
2-(Hydroxyacetyl)furan (2-Furyl hydroxymethyl ketone)	$COCH_2OH$			
2-Propionylfuran (Ethyl 2-furyl ketone)	COC_2H_5			
2-Furaldehyde (Furfural, 2-Furancarbaldehyde)	CHO			
2,5-Dimethylfuran	CH_3			CH_3

DESCRIPTION	THRESHOLD VALUE	REM.	REF
O: Coffee-like		F,M	28
O: Nut-like		F	29
O: Bitter, nut-like		F	22
O: Burnt, buttery, vanilla-like		F	6
O: Stinging F: Sour		F,M,S	18
O: Practically odorless F: Clear acid, mildly caramellic note			17
O: "Characteristic" F: Bitter		F,M,S	18
O: Very mild, warm-oily, "burnt" F: Warm, slightly caramellic, in higher concentration burning, yet somewhat creamy			17
O: Mild sweet			22
F in beer: Sugar cane, woody	>3000 ppm/beer		25
	T 30 ppm/orange juice		26
	F 5 ppm/water		27
O: Burnt			30
O: None F: Burning, sweetish		F,M	18
	T >200 ppm/ orange juice		26
O: Sweet, caramellic		F,M	28
O: Pungent, but sweet, bread-like caramellic, cinnamon-almond-like of poor tenacity F: Sweet bread-like, caramellic in proper dilution		F,M,S	17
O: "Characteristic" F: Sweet			18
F: Bitter			29
F in beer: Paper, husk	150 ppm/beer		25
	F 5 ppm/water		27
	T 80 ppm/orange juice		26
	O 3000 ppb/water		24
O: Ethereal		F,M	28

Continued on next page

FURANS (cont.)

NAME / (SYNONYM)	R_1	R_2	R_3	R_4
5-Methyl-2-furaldehyde (5-Methylfurfural)	CHO			CH_3
5-Hydroxymethyl-2-furaldehyde (HMF, 5-(Hydroxymethyl)furfural)	CHO			CH_2OH
2,5-Diformylfuran (2,5-Furandicarbaldehyde)	CHO			CHO
2-Acetyl-3-hydroxyfuran (Isomaltol)	$COCH_3$	OH		
2-Acetyl-5-methylfuran	$COCH_3$			CH_3
2-Butyl-5-methylfuran	$\underline{n}$-C_4H_9			CH_3
2-(2-Hydroxybutyl)-5-methylfuran	$CH_2CH(OH)C_2H_5$			CH_3
2-(1,2-Dioxopropyl)-5-methylfuran	$COCOCH_3$			CH_3
2-(2,3-Dioxobutyl)-5-methylfuran	$CH_2COCOCH_3$			CH_3
2-Cyano-5-methylfuran (5-Methyl-2-furonitrile)	CN			CH_3

DESCRIPTION	THRESHOLD VALUE	REM.	REF
O: Sweet-spicy, warm and slightly caramellic F: Sweet-caramellic, warm		F,M,S	17 "
O: Burnt, caramel-like, slightly meaty			6
F in beer: Almonds, burnt/phenolic, pyrazole	20 ppm/beer		25
	T 10 ppm/orange juice		26
O: None F: Bitter, astringent		F,M,S	18
O: Warm-herbaceous, winy-ethereal remotely resembling Hungarian chamomile (matricaria oil). A natural sweetness common; similarity to hay, caramel, tobacco, etc. often perceptible F: Sweet, herbaceous-hay-like, mildly tobacco-like			17
F: Bitter			29
F in beer: Aldehyde, stale, vegetable oil	1000 ppm/beer		25
	T >200 ppm/orange juice		26
	F 100 ppm/water		27
	F >100 ppm/water	M,S	27
O: Burnt, pungent, fruity F: Sour, sweet, fruity		F,M,S	18
O: Caramellic-sweet, but rather pungent, of good tenacity F: First sour, then sweet, caramellic--fruity, bread-like, depending upon the concentration used. Sour taste mostly noticed in high concentration of the material.			17
O: Caramel-like		F,M	28
O: Green		S	31
O: Green		S	31
O: Caramel	F 2 ppm/water	M,S	27
O: Like roasted rye bread		F	22
O: Nut-like, bitter almond-like		F	22

Continued on next page

FURANS (cont.)

NAME / (SYNONYM)	R_1	R_2	R_3	R_4
2-Methyl-3-furanthiol (5-Mercapto-2-methylfuran)	CH_3	SH		
2-Methyl-4-furanthiol (4-Mercapto-2-methylfuran)	CH_3		SH	
2,5-Dimethyl-3-furanthiol (3-Mercapto-2,5-dimethylfuran)	CH_3	SH		CH_3
S-Methyl 2-thiofuroate (2-Thiofuroic acid, methyl ester)	$COSCH_3$			
5-Methylthio-2-furaldehyde (5-Methylthiofurfural)	CHO			SCH_3
2-(Mercaptomethyl)furan (2-Furylmethanethiol, Furfuryl mercaptan)	CH_2SH			
S-2-Furfuryl thioacetate	CH_2SCOCH_3			
2-Methyl-5-(methylthio)furan (2-Methyl-5-furyl methyl sulfide)	CH_3			SCH_3
2- ((Methyldithio)methyl)furan (Furfuryl methyl disulfide)	CH_2SSCH_3			
2-Furfurylfuran	CH_2-furan			
Difurfuryl sulfide	CH_2SCH_2-furan			

DESCRIPTION	THRESHOLD VALUE	REM.	REF
F: Sweetish, fried meat, beef broth		S	32
O: Green, meaty, herbaceous		S	31
F: Strongly roasted meat		S	32
O: Mercaptan-like		F,M	22
O: Cabbage-like			28
O: Meaty		F,S	33
O: Extremely powerful and diffusive, penetrating, only in proper dilution agreeable, coffee-like, caramellic--burnt, sweet F: Dilution < 5 ppm: Pronounced caramellic-coffee-like		F,M,S	17
O: Characteristic, unpleasant			23
O: Strong like roasted coffee			22
F: Coffee			34
	F 0.04 ppb/water F 0.04 ppb/2% protein-hydrolysate solution F 0.003 ppb/skim milk		32
O: Coffee-like		F	22
O: Very strong sulfurated		F,M,S	23
O: Fresh white bread crust	0.04 ppb/water	F	6, 7
O: Roasted bread crust			35
O: Freshly baked bread			36
O: Caramellic		F,M	37
O: Toasted		F,M	37

Continued on next page

4,5-DIHYDROFURANS

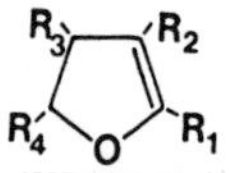

NAME / (SYNONYM)	R_1	R_2	R_3	R_4
2-Methyl-4,5-dihydrofuran-3-thiol	CH_3	SH		
2-Methyltetrahydrofuran-3-ol	CH_3	OH		

FURANONES 2 (5H)

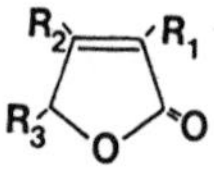

3-Hydroxy-4,5-dimethyl-2(5<u>H</u>)- -furanone	CH	CH_3	CH_3	
3-Hydroxy-4-ethyl-5-methyl- -2(5<u>H</u>)-furanone	OH	C_2H_5	CH_3	
3-Hydroxy-4-methyl-5-ethyl- -2(5<u>H</u>)-furanone	OH	CH_3	C_2H_5	
4-Hydroxy-5-methyl-3(2<u>H</u>)-furanone		OH	CH_3	
4-Mercapto-5-methyl-3(2<u>H</u>)-furanone		SH	CH_3	
2,5-Dimethyl-3(2<u>H</u>)-furanone	CH_3		CH_3	
2,5-Dimethyl-4-methoxy-3(2<u>H</u>)- -furanone	CH_3	OCH_3	CH_3	
4-Hydroxy-2,5-dimethyl-3(2<u>H</u>)- -furanone (Furaneol^{R})	CH_3	OH	CH_3	

TETRAHYDROFURANS 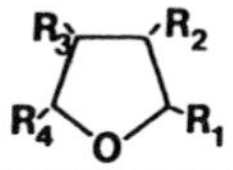

DESCRIPTION	THRESHOLD VALUE	REM.	REF
O: Roasted meat		S	31
O: Fatty		S	31

3 (2H) 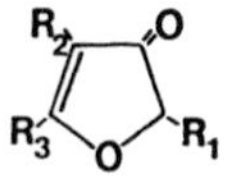

F: Burnt		F	38
O: Typically caramel			29
O: Typically caramel		F	29
	T 0.005-0.01 ppm/ water O 0.5-1 ppm/water		39
O: Maple-like, curry-like		F	38
O: Intense caramel and curry-like			7
O: Caramel-like		F,M	40
O: Roasted chicory root			38
O: Reminiscent of roasted chicory root with an unmistakable undertone of maple sirup			6
O: Sweet, meat-like		S	31
O: Strong odor of freshly baked bread but not very reminiscent of bread aroma		F	7
O: Reminiscent of sherry		F	7
	O 0.03 ppb/water		41
O: Fragrant, fruity, caramel, burnt pineapple F: Burning, sweet		F,M	18
O: Caramel, burnt sugar-like			30
O: Sweet, cotton-candy			42
O: Caramel-like, burnt pineapple which turns into a strawberry-like note as dilution increases			6
	O 0.04 ppb/water		41
	F 1 ppm/water		27
	O 0.1-0.2 ppm/water T 0.03 ppm/water		39

Continued on next page

FURANONES (cont.)

NAME / (SYNONYM)	R_1	R_2	R_3	R_4
2-Hydroxymethyl-4-hydroxy-5-methyl-3(2H)-furanone	CH_2OH	OH	CH_3	
2-Ethyl-4-hydroxy-5-methyl-3(2H)-furanone	C_2H_5	OH	CH_3	

DIHYDROFURAN-3-ONES

NAME / (SYNONYM)	R_1	R_2	R_3	R_4
4-Hydroxy-5-methyl-2,3-dihydrofuran-3-one		OH	CH_3	

TETRAHYDROFURAN-3-ONES

NAME / (SYNONYM)	R_1	R_2	R_3	R_4
2-Methyltetrahydrofuran-3-one	CH_3			
4-Mercaptotetrahydrofuran-3-one (4-Mercapto-3-oxotetrahydrofuran)		SH		
4-Mercapto-5-methyltetrahydrofuran-3-one (4-Mercapto-5-methyl-3-oxotetrahydrofuran)		SH	CH_3	

DESCRIPTION	THRESHOLD VALUE	REM.	REF
O: Unpleasant odor of charred paper		M	42
O: Sweet, reminiscent of shortcake		F	7
F: Intensely sweet			38
O: Caramel-like			40

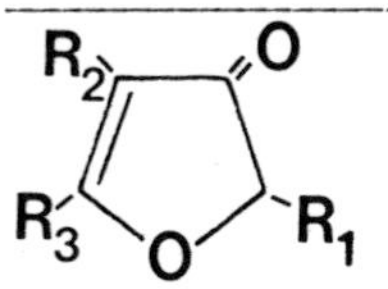

DESCRIPTION	THRESHOLD VALUE	REM.	REF
F: Pleasant, maltol-like		M	43

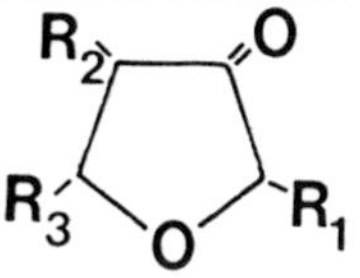

DESCRIPTION	THRESHOLD VALUE	REM.	REF
O: Sweet, roasted		F,M	28
O: Green, meaty, "Maggi-like"		S	31
O: Meaty, "Maggi-like"		S	31

Continued on next page

4H-PYRAN-4-ONES

NAME / (SYNONYM)	R_1	R_2	R_3	R_4
2-Methyl-3-hydroxy-4H-pyran-4-one (Maltol)	CH_3	OH		
2-Ethyl-3-hydroxy-4H-pyran-4-one (Ethylmaltol)	C_2H_5	OH		
2-Methyl-3,5-dihydroxy-4H-pyran--4-one (5-Hydroxymaltol)	CH_3	OH	OH	
2-Ethyl-3-hydroxy-6-methyl-4H--pyran-4-one	C_2H_5	OH		CH_3

2,3-Dihydro-3,5-dihydroxy-6--methyl-4H-pyran-4-one
(5-Hydroxy-5,6-dihydromaltol)

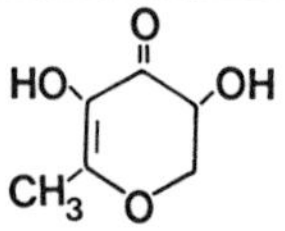

3-Hydroxy-2H-pyran-2-one
(3-Hydroxy-2-pyrone)

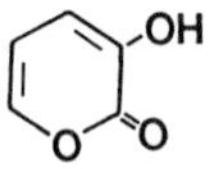

Methyl 4,5-dimethyl-3,6-dihydro-2H-pyran-2-carboxylate

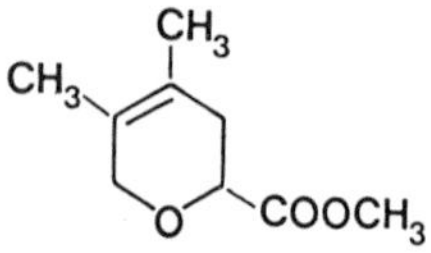

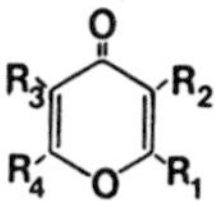

DESCRIPTION	THRESHOLD VALUE	REM.	REFE
O: Warm-fruity, caramellic-sweet with emphasis on the caramellic note in the dry state, while solutions show a pronounced fruity, jam-like odor of pineapple, strawberry type		F,M,S	17
O: Fragrant, caramel F: Bitter, sweetish			18
O: Warm, sweet, fruity; jam-like in solution			23
O: Coffee, malt, caramel			34
	F 20 ppm/water		27
	F 7.1 ppm/water		44
	O 35 ppm/water T 13 ppm/water		39
O: Intensely sweet, fruity-bread-like, pleasant, of immense tenacity F: Sweet, fruity-jam-like, reminiscent of pineapple, strawberry, vanilla and heavy fruit-preserve or syrup, depending upon concentration		S	17
O: Very sweet, caramel-like, of immense tenacity F: Sweet, fruity with initial bitter-tart flavor			23
O: Caramel-like		F	40
O: Preserved tang boiled down in soy sauce		F	7
O: Not reminiscent of maltol			29
O: Odorless		F,S	18
	T > 200 ppm/orange juice		26
	T 30 ppm/orange juice	F	26
O: Interesting green note		F,S	45

Continued on next page

PYRROLES

NAME / (SYNONYM)	R_1	R_2	R_3	R_4
Pyrrole	H			
1-Methylpyrrole (N-Methylpyrrole)	CH_3			
2-Pyrrolecarbaldehyde	H	CHO		
1-Acetonylpyrrole (1-Pyrrolyl-2-propanone)	CH_2COCH_3			
2-Acetylpyrrole (Methyl 2-pyrrolyl ketone)	H	$COCH_3$		
2-Acetyl-5-chloropyrrole	H	$COCH_3$	Cl	
2-Acetyl-5-bromopyrrole	H	$COCH_3$	Br	
2-Formyl-5-methylpyrrole (5-Methylpyrrole-2-carbaldehyde)	H	CHO	CH_3	
1-Ethyl-2-formylpyrrole (N-Ethylpyrrole-2-carbaldehyde)	C_2H_5	CHO		
1-(2-Furfuryl)pyrrole (N-Furfurylpyrrole)	furfuryl			

PYRROLIDINES (TETRAHYDROPYRROLES)

Pyrrolidine	H
1-Acetylpyrrolidine	$COCH_3$

1-Pyrroline

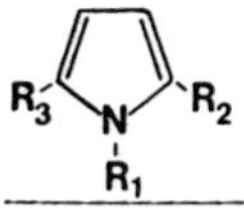

DESCRIPTION	THRESHOLD VALUE	REM.	REF
O: Sweet and warm-ethereal, slightly burnt-nauseating, resembling that of chloroform		F,S	17
O: Powerful and penetrating smoky--tarry, in extreme dilution sweet, woody-herbaceous, slightly animal.		F,S	17
O: Corny, pungent		F,M	37
O: Bready green	O 10 ppb/water	M	46
O: Cookie- or mushroom-like			7
	T 200 ppm/orange juice	F	26
O: Strong almond aroma		F,S	47
O: Strong almond aroma		F,S	48
	T 110 ppm/orange juice	F,M,S	26
O: Pungent			37
	T 2 ppm/orange juice	F,M	26
O: Green, hay-like		F	29

DESCRIPTION	THRESHOLD VALUE	REM.	REF
O: Penetrating amine-type reminiscent of ammonia and piperidine, nauseating and diffusive, of very poor tenacity, repulsive		F,S	17
O: Bread-like		M	46
O: Corn-like		M	49

Continued on next page

2,3-DIHYDRO-1H-PYRROLIZINES

NAME / (SYNONYM)	R_1	R_2	R_3	R_4
5-Acetyl-2,3-dihydro-1H-pyrrolizine	$COCH_3$			
5-Acetyl-6-methyl-2,3-dihydro-1H-pyrrolizine	$COCH_3$	CH_3		
5-Formyl-6-methyl-2,3-dihydro-1H-pyrrolizine	CHO	CH_3		

TETRAHYDROINDOLIZIN-8-ONES

5,6,7,8-Tetrahydroindolizin-8-one	H			
2-Methyl-5,6,7,8-tetrahydroindolizin-8-one	CH_3			

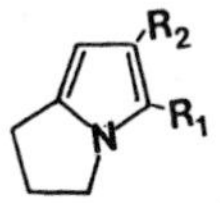

DESCRIPTION	THRESHOLD VALUE	REM.	REF
O: Smoky, amine-like F: Sweet		F,M	30
O: Smoky, medicine-like F: Bitter		M	30
O: Smoky F: Sweet, cinnamic		F,M	30

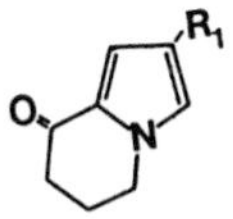

O: Mild smoky, weak amine-like		M	30
O: Mild smoky, weak amine-like		M	30

Continued on next page

PYRIDINES

NAME / (SYNONYM)	R_1	R_2	R_3	R_4
Pyridine				
3-Methylpyridine (β-Picoline)		CH_3		
2-Ethylpyridine	C_2H_5			
3-Ethylpyridine		C_2H_5		
3-Propylpyridine		$\underline{n}$-C_3H_7		
2-Isopropylpyridine	i-C_3H_7			
2-Butylpyridine	$\underline{n}$-C_4H_9			
2-Isobutylpyridine	$\underline{i}$-C_4H_9			
2-Pentylpyridine (2-Amylpyridine)	$\underline{n}$-C_5H_{11}			
2-Methoxypyridine	OCH_3			
2-Ethoxypyridine	OC_2H_5			
2-Pyridinecarbaldehyde (2-Pyridylmethanal)	CHO			
2-Acetylpyridine (Methyl 2-pyridyl ketone)	$COCH_3$			
2-Acetylpyridin-3-ol	$COCH_3$	OH		
3,4-Dimethylpyridine		CH_3	CH_3	

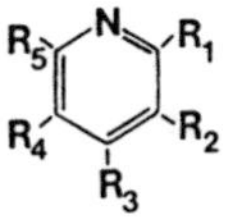

DESCRIPTION	THRESHOLD VALUE	REM.	REF
O: Burnt, pungent, diamine O: Pungent, penetrating and diffusive, generally described as nauseating, but in extreme dilution, warm, "burnt", smoky, of very poor tenacity F: (~1 ppm): Rather sharp, burnt	O 21 ppb/air	F	50 17
	O 0.82 ppm/water		51
F: Green		F,S	52
F: Green		F,S	52
F: Green, smoky			53
F: Tobacco		F	23
F; Sweet, musty, beany			53
F: Green, vegetable		S	52
F: Sweet, green		F,S	52
F: Green pepper		S	52
O: Fatty or tallowy-like	O 0.6 ppb/water	F	54
F: Phenolic		S	52
F: Phenolic		S	52
O: Pungent bitter almond-like, poor tenacity. In extreme dilution quite pleasant, sweet, bitter almond and nut-like			17
O: Popcorn		F	6
O: Bready			46
O: Tobacco-like			23
O: Cracker-like	O 19 ppb/water		55
F: Cereal-like			53
F: Green		F,S	52

Continued on next page

PYRIDINES (cont.)

NAME / (SYNONYM)	R_1	R_2	R_3	R_4
4-Ethyl-3-methylpyridine		CH_3	C_2H_5	
2-Isobutyl-3-methoxypyridine	i-C_4H_9	OCH_3		
3-Isobutyl-2-methoxypyridine	OCH_3	i-C_4H_9		

1,4,5,6-TETRAHYDROPYRIDINES

2-Acetyl-1,4,5,6-tetrahydropyridine	$COCH_3$			

PYRIMIDINES

4,6-Dimethylpyrimidine		CH_3		CH_3
4-Acetyl-2-methylpyrimidine	CH_3	$COCH_3$		

DESCRIPTION	THRESHOLD VALUE	REM.	REF
F: Sweet, nutty		S	52
F: Green pepper		S	52
O: Not very characteristic, but perhaps somewhat camphoraceous	O 11 ppb/water	S	56

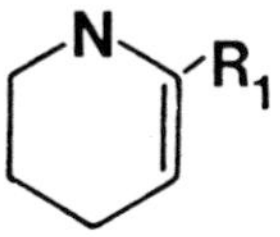

O: Distinct odor of crackers		F,S	57
O: Cracker-like	O 1.6 ppb/water		55

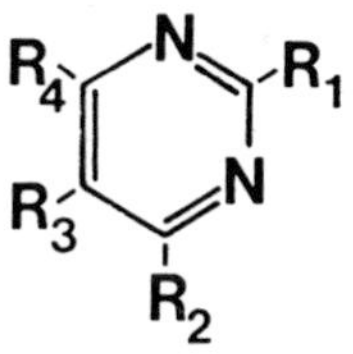

O: Similar to alkyl-substituted pyrazines (i.e. roasted, nut-like)		F	6
Very interesting grilled note		F	6

Continued on next page

PYRAZINES

NAME / (SYNONYM)	R_1	R_2	R_3	R_4
Pyrazine				
2-Methylpyrazine	CH_3			
2-Ethylpyrazine	C_2H_5			
2-Propylpyrazine	$\underline{n}$-C_3H_7			
2-Isopropylpyrazine	i-C_3H_7			
2-Isobutylpyrazine	i-C_4H_9			
2-Pentylpyrazine	$\underline{n}$-C_5H_{11}			
2,3-Dimethylpyrazine	CH_3	CH_3		

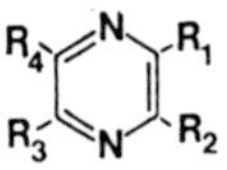

DESCRIPTION	THRESHOLD VALUE	REM.	REF
O: Pungent, sweet, in dilution floral with remote resemblance to heliotrope. Very diffusive, poor tenacity		F,S	17
O: Strong, sweet (slightly ammoniacal)			58
O: Cornlike with bitter note			70
	O 50×10^4 ppb/water		59
F: Roasted hazelnuts	O 17.5×10^4 ppb/water		35
	F $>$100 ppm/beer		60
F: Chocolate, roasted peanuts, grilled chicken		F,S	35
O: Nutty, roasted			52
O: Nutty, green			46
O: Strong basic role; in dilution chocolate character			58
O: Grassy			70
	O 10×10^4 ppb/water		59
	O 105 ppm/water; 27 ppm/oil		61
	F 100 ppm/beer		60
	O 6×10^4 ppb/water		62
O: Nutty, roasted		F,S	52
O: Buttery, rum			65
	O 22 ppm/water; 17 ppm/oil		61
	O 6000 ppb/water		62
	F 10 ppm/beer		60
O: Green, vegetable		F,S	52
O: Green, nutty		F,S	52
O: Green, fruity		F,S	52
	O 400 ppb/water		63
	O 1 ppm/water; 9 ppm/oil	S	61
O: Pungent; in dilution chocolate type		F,S	58
O: Nutty, green			52
O: New leather, linseed oil			65
	O 400 ppm/water		59
	O 2500 ppb/water		62
	O 35 ppm/water		64
	F 50 ppm/beer		60

Continued on next page

PYRAZINES (cont.)

NAME / (SYNONYM)	R_1	R_2	R_3	R_4
2,5-Dimethylpyrazine	CH_3		CH_3	
2,6-Dimethylpyrazine	CH_3			CH_3
2,3-Diethylpyrazine	C_2H_5	C_2H_5		
2,5-Diethylpyrazine	C_2H_5		C_2H_5	
2,6-Diethylpyrazine	C_2H_5			C_2H_5
2-Ethyl-3-methylpyrazine (3-Ethyl-2-methylpyrazine)	C_2H_5	CH_3		
2-Ethyl-5-methylpyrazine (5-Ethyl-2-methylpyrazine)	C_2H_5		CH_3	
2-Ethyl-6-vinylpyrazine	C_2H_5			$CH=CH_2$
2-Methyl-6-propylpyrazine	CH_3			$\underline{n}$-C_3H_7

DESCRIPTION	THRESHOLD VALUE	REM.	REF
F: Characteristic, reminiscent of potato chips		F,S	23
F: Chocolate, grilled chicken, roasted peanuts			35
O: Grassy, "cornnuts"			70
O: Roasted	O 800 ppb/water		66
F in oil: Earthy raw potato	O 1 ppm/water } 2 ppm/oil		67
	O 1800 ppb/water		63
	O 2600 ppb/oil		62
	O 35 ppm/water } 17 ppm/oil		61
	F 25 ppm/beer		60
O: Sweet, "fried", resembling fried potatoes, but not as typical odor as the 2,5-isomer		F,S	17
O: Nutty, roasted	O 200 ppb/water		46
F: Chocolate			35
O: Ether-like with corn note			70
	O 54 ppm/water } 8 ppm/oil		61
	O 9 ppm/water		64
	O 1500 ppb/water		62
	F 3 ppm/beer		60
F: Roasted hazelnuts		F,S	35
	O 20 ppb/water 270 ppb/oil	F,S	62
	O 6 ppb/water	F,S	62
O: Raw potato		F,S	68
O: Nutty, roasted			52
O: Butterscotch, nutty			65
O: Pleasant earthy, nutty			69
	O 130 ppb/water		62
	F 2 ppm/beer		60
O: Nutty, roasted	O 50 ppb/oil	F,S	46
O: Grassy			70
	O 100 ppb/water } 320 ppb/oil		62
	F 1 ppm/beer		60
O: Buttery, baked, potatolike		F	69
O: Burnt, butterscotch	O $<$0.1 ppm/water	F	65

Continued on next page

PYRAZINES (cont.)

NAME / (SYNONYM)	R_1	R_2	R_3	R_4
2-Isobutyl-3-methylpyrazine (3-Isobutyl-2-methylpyrazine)	i-C_4H_9	CH_3		
2,3,5-Trimethylpyrazine	CH_3	CH_3	CH_3	
2,3-Dimethyl-5-ethylpyrazine (5,6-Dimethyl-2-ethylpyrazine)	CH_3	CH_3	C_2H_5	
2,5-Dimethyl-3-ethylpyrazine (3,6-Dimethyl-2-ethylpyrazine)	CH_3	C_2H_5	CH_3	
2,6-Dimethyl-3-ethylpyrazine (3,5-Dimethyl-2-ethylpyrazine)	CH_3	C_2H_5		CH_3
2,3-Diethyl-5-methylpyrazine	C_2H_5	C_2H_5	CH_3	
2,6-Diethyl-5-methylpyrazine	C_2H_5		CH_3	C_2H_5
2,3-Dimethyl-5-butylpyrazine	CH_3	CH_3	n-C_4H_9	
2,3-Dimethyl-5-pentylpyrazine	CH_3	CH_3	n-C_5H_{11}	
2,3-Dimethyl-5-isopentylpyrazine	CH_3	CH_3	i-C_5H_{11}	
2,3-Dimethyl-5-(1-methylbutyl)-pyrazine	CH_3	CH_3	n-$C_3H_7CH(CH_3)$-	

DESCRIPTION	THRESHOLD VALUE	REM.	REF
F: Roasted hazelnuts	O 400 ppb/water	F,S	35
O: Green (bell pepper like), dry and sweet notes	O 130 ppb/water		64
	O 35 ppb/water		62
O: Nutty, roasted		F,S	52
O: Baked potato or roasted peanuts			23
F: Roasted hazelnuts	O 1800 ppb/water		35
O: Very similar to 2,3-dimethyl-pyrazine, but slightly heavier			58
O: Grassy			70
	O 400 ppb/water		59
	O 9 ppm/water, 27 ppm/oil		61
	F 1 ppm/beer		60
F: Chocolate, sweet		F,S	53
O: Nutty, roasted	O 1 ppb/water		46
O: Roasted	O 5 ppb/water	F,S	66
O: Nutty, roasted			46
F: Lard			35
O: Earthy, baked potato-like			69
	O 43 ppm/water, 24 ppm/oil		61
	O 0.4 ppb/water, 24 ppb/oil		62
	F 25 ppb/beer		60
O: Nutty, roasted		F,S	52
F: Walnut, steak			34
O: Nutty, roasted	O 1 ppb/water		46
	O 15 ppm/water, 24 ppm/oil		61
	F 5 ppb/beer		60
O: Nutty, roasted		F,S	52
O: Grassy		F	70
O: Sweet, earthy		F	69
O: Sweet, smoked, caramel-like		S	71
O: Caramel-like, coffee, sweet		S	71
O: Honey-like, sweet		S	71

Continued on next page

PYRAZINES (cont.)

NAME / (SYNONYM)	R_1	R_2	R_3	R_4
2,3-Dimethyl-5-(2-methylbutyl)-pyrazine	CH_3	CH_3	$C_2H_5CH(CH_3)CH_2$-	
2,3-Dimethyl-5-(1,2-dimethylpropyl)-pyrazine	CH_3	CH_3	$(CH_3)_2CHCH(CH_3)$-	
2,3-Dimethyl-5-(2,2-dimethylpropyl)-pyrazine (2,3-Dimethyl-5-neopentylpyrazine)	CH_3	CH_3	$(CH_3)_3CCH_2$-	
2,3-Dimethyl-5-(1-ethylpropyl)-pyrazine	CH_3	CH_3	$(C_2H_5)_2CH$-	
2,3-Dimethyl-5-(1,5-dimethyl-4-hexenyl)pyrazine	CH_3	CH_3	R_3 $(CH_3)_2C{=}CHCH_2CH_2CH(CH_3)$-	
2,3-Dimethyl-5-(3,7-dimethyl-6-heptenyl)pyrazine	CH_3	CH_3	R_3 $CH(CH_3){=}CHCH_2CH_2CH(CH_3)CH_2CH_2$-	
2,3,5,6-Tetramethylpyrazine	CH_3	CH_3	CH_3	CH_3
3,6-Diethyl-2,5-dimethylpyrazine (2,5-Diethyl-3,6-dimethylpyrazine)	CH_3	C_2H_5	CH_3	C_2H_5
2-Isobutyl-3,5,6-trimethylpyrazine	i-C_4H_9	CH_3	CH_3	CH_3
2-(2-Methylbutyl)-3,5,6-trimethylpyrazine	$C_2H_5CH(CH_3)CH_2$	CH_3	CH_3	CH_3
2-Acetylpyrazine	$COCH_3$			
2-Acetyl-3-methylpyrazine	$COCH_3$	CH_3		
2-Acetyl-5-methylpyrazine	$COCH_3$		CH_3	
6-Acetyl-2-methylpyrazine (2-Acetyl-6-methylpyrazine)	CH_3			$COCH_3$
2-Acetyl-3-ethylpyrazine	$COCH_3$	C_2H_5		

DESCRIPTION	THRESHOLD VALUE	REM.	REF
O: Sweet, smoked, caramel-like		S	71
O Sweet, caramel-like		S	71
O: Brown sugar-like, roasted		S	71
O: Honey-like, sweet		S	71
O: Roasted nut		S	71
O: Roasted nut		S	71
O: Fermented soybeans		F,S	4
F: Lard			35
O: Similar to trimethylpyrazine but without the odor intensity	O 1000 ppb/water		58
	O 10 ppm/water 38 ppm/oil }		61
	F > 100 ppm/beer		60
	F 100 ppm/beer	F,S	60
O: Roasted nut, sweet		S	71
O: Sweet, roasted		F,S	71
F: Popcorn-like		F,S	72
O: Popcorn, nutty			52
F: Chocolate, popcorn			35
O: Breadcrust-like, nutty, no green notes, reminiscent of acetamide			64
	O 62 ppb/water		55
F: Cereal, roast grain		F,S	53
O: Nutty, vegetable	O 4 ppb/water		46
F: Popcorn-like		S	72
O: Popcorn		F,S	52
F: Popcorn-like			72
F: Potatoes		F,S	35
F: Reminiscent of slightly roasted potatoes			7

Continued on next page

PYRAZINES (cont.)

NAME / (SYNONYM)	R_1	R_2	R_3	R_4
2-Methoxypyrazine	OCH_3			
2-Ethoxypyrazine	OC_2H_5			
2-Butoxypyrazine	OC_4H_9-n			
2-(Methoxymethyl)pyrazine	CH_2OCH_3			
2-Methoxy-3-methylpyrazine (3-Methoxy-2-methylpyrazine)	OCH_3	CH_3		
2-Methoxy-5-methylpyrazine (5-Methoxy-2-methylpyrazine)	OCH_3		CH_3	
2-Ethyl-3-methoxypyrazine (3-Ethyl-2-methoxypyrazine)	C_2H_5	OCH_3		
2-Ethyl-5-methoxypyrazine (5-Ethyl-2-methoxypyrazine)	C_2H_5		OCH_3	
2-Propyl-3-methoxypyrazine (3-Propyl-2-methoxypyrazine)	n-C_3H_7	OCH_3		
3-Isopropyl-2-methoxypyrazine (2-Isopropyl-3-methoxypyrazine)	OCH_3	i-C_3H_7		
2-Isopropyl-5-methoxypyrazine (5-Isopropyl-2-methoxypyrazine)	i-C_3H_7		OCH_3	
2-Butyl-3-methoxypyrazine	n-C_4H_9	OCH_3		
2-Isobutyl-3-methoxypyrazine (3-Isobutyl-2-methoxypyrazine)	i-C_4H_9	OCH_3		

DESCRIPTION	THRESHOLD VALUE	REM.	REF
O: Sweet, nutty		S	52
O: Not very characteristic	O 700 ppb/water		63
O: Sweet, nutty		S	52
O: Green, vegetable		S	52
F: Nutty, earthy		S	73
O: Ethereal character			58
	O 150 ppb/water		59
O: Reminiscent of hazelnut, almond and peanut		F,S	23
O: Roasted peanuts, nutty, earthy			4
O: in concentration vegetably; in dilution popcorn/potato			58
	O 4 ppb/water		63
	O 3 ppb/water		59
F: Hazelnuts, almonds, peanuts		S	35
O: Green vegetably character			58
	O 15 ppb/water		59
O: Raw potato, earthy, bell pepper		F,S	4
F: Earthy, bell pepper			73
	O 0.4 ppb/water		63
O: Bread-like/mousy		F,S	66
F: Bell pepper		S	73
O: Very similar to bell pepper	O 0.006 ppb/water		63
F: Bell pepper, earthy		F,S	73
O: Potato-like			7
O: Earthy, bell pepper, raw potato			4
O: Strong galbanum note			58
	O 10 ppb/water		59
	O 0.002 ppb/water		63
	O 0.001 ppb/water		74
O: Strong galbanum note		S	58
	O 10 ppb/water		59
F: Bell pepper		S	73
O: Strongly green (bell pepper-like)	O 0.016 ppb/water	F,S	64
O: (Green) bell pepper	O 0.002 ppb/water		24
	O 10 ppb/water		59

Continued on next page

PYRAZINES (cont.)

NAME / (SYNONYM)	R_1	R_2	R_3	R_4
2-Isobutyl-5-methoxypyrazine	i-C_4H_9		OCH_3	
2-Isobutyl-6-methoxypyrazine	i-C_4H_9			OCH_3
3-sec.Butyl-2-methoxypyrazine (2-sec.Butyl-3-methoxypyrazine, 2-Methoxy-3-(1-methylpropyl)-pyrazine)	OCH_3	s-C_4H_9		
2-Methoxy-3-(1-methylbutyl)pyrazine	OCH_3	$CH(CH_3)C_3H_7$-n		
2-Isopentyl-3-methoxypyrazine	i-C_5H_{11}	OCH_3		
2-Hexyl-3-methoxypyrazine (3-Hexyl-2-methoxypyrazine)	n-C_6H_{13}	OCH_3		
2-Methoxy-3-(2-methyloctyl)pyrazine	OCH_3	$CH_2CH(CH_3)C_6H_{13}$		
2-Methoxy-3-isononylpyrazine	OCH_3	i-C_9H_{19}		
2-Methoxy-3-acetylpyrazine	OCH_3	$COCH_3$		
2-Methoxy-3,5-dimethylpyrazine	OCH_3	CH_3	CH_3	
2-Methoxy-3,6-dimethylpyrazine	OCH_3	CH_3		CH_3
2-Methoxy-3-isopropyl-5-methyl-pyrazine	OCH_3	i-C_3H_7	CH_3	
2-Methoxy-3-isopropyl-6-methyl-pyrazine	OCH_3	i-C_3H_7		CH_3
2-Methoxy-3-(α-hydroxyisopropyl)--5-methylpyrazine	OCH_3	$C(OH)(CH_3)_2$	CH_3	
2-Methoxy-3-(α-hydroxyisopropyl)--6-methylpyrazine	OCH_3	$C(OH)(CH_3)_2$		CH_3

DESCRIPTION	THRESHOLD VALUE	REM.	REF
O: Strong green bell pepper note F: Coffee	O 10 ppb/water	S	58 59 35
F: Bell pepper		S	73
F: Bell pepper	O 0.001 ppb/water	S	73 74
F: Bell pepper		S	73
F: Bell pepper		S	73
F: Bell pepper O: Very similar to bell pepper	O 0.001 ppb/water	S	73 63
F: Earthy, bell pepper		S	73
F: Bell pepper		S	73
O: Weak breadcrust-like, green and nutty notes		S	64
O: Bready/mousy	O 4 ppb/water	F,S	46
F: Medicinal, earthy		S	73
O: Strongly green (green bean-like), floral and ethereal undertone, no nutty notes	O 0.05 ppb/water	S	64
O: Strongly green (green bean-like), no nutty or floral notes	O 0.045 ppb/water	S	64
O: Weak, green (bell pepper-like), earthy undertone, chimney soot		F,S	64
O: Weak, green (bell pepper-like), earthy undertone, chimney soot, nutty note		S	64

Continued on next page

PYRAZINES (cont.)

NAME / (SYNONYM)	R_1	R_2	R_3	R_4
2-Isobutyl-3-methoxy-5-methylpyrazine	i-C_4H_9	OCH_3	CH_3	
2-Isobutyl-3-methoxy-6-methylpyrazine	i-C_4H_9	OCH_3		CH_3
2-Methoxy-3,6-diisobutylpyrazine	OCH_3	i-C_4H_9		i-C_4H_9
2-Methoxy-3-acetyl-5-methyl-pyrazine	OCH_3	$COCH_3$	CH_3	
2-Methoxy-3-acetyl-6-methyl-pyrazine	OCH_3	$COCH_3$		CH_3
5-Isobutyl-2-methoxy-3-methyl-pyrazine	OCH_3	CH_3	i-C_4H_9	
5-(2-Methylbutyl)-2-methoxy--3-methylpyrazine	OCH_3	CH_3	$CH_2CH(CH_3)C_2H_5$	
5-Isopentyl-2-methoxy--3-methylpyrazine	OCH_3	CH_3	i-C_5H_{11}	
5-(2-Methylpentyl)-2-methoxy--3-methylpyrazine	OCH_3	CH_3	$CH_2CH(CH_3)C_3H_7$	
2-Ethoxy-5-isobutyl-3-methyl-pyrazine	OC_2H_5	CH_3	i-C_4H_9	
5-Isobutyl-2-isopropoxy-3-methyl-pyrazine	OC_3H_7-i	CH_3	i-C_4H_9	
5-Isobutyl-3-methyl-2-phenoxy-pyrazine	OC_6H_5	CH_3	i-C_4H_9	
5-Isobutyl-3-methyl-2-(methylthio)-pyrazine	SCH_3	CH_3	i-C_4H_9	
5-(2-Methylpentyl)-3-methyl--2-(methylthio)pyrazine	SCH_3	CH_3	$CH_2CH(CH_3)C_3H_7$	
5-Isobutyl-3-methyl-2-(phenylthio)-pyrazine	SC_6H_5	CH_3	i-C_4H_9	
2-Isobutyl-3-methoxy-5,6-dimethyl-pyrazine	i-C_4H_9	OCH_3	CH_3	CH_3
2,5-Dimethoxy-3,6-dimethylpyrazine	OCH_3	CH_3	OCH_3	CH_3

DESCRIPTION	THRESHOLD VALUE	REM.	REF
O: Similar to bell pepper but with some minty notes	O 0.3 ppb/water	S	56
O: Minty camphoraceous in character with slight bell pepper undertone	O 2.6 ppb/water	S	56
F: Earthy, low bell pepper		S	73
O: Weak, breadcrust-like, green musky note, chimney soot		F,S	64
O: Weakly green, unpleasant, no nutty notes, chimney soot		S	64
O: Licorice-woody	O 0.7 ppm/water	S	75
O: Licorice-woody, slightly green	O 1 ppm/water	S	75
O: Licorice-woody, walnut-like	O 1 ppm/water	S	75
O: Burdock, bell pepper	O 0.5 ppm/water	S	75
O: Burdock-like brownish, slightly bell pepper, galbanum green	O 0.2 ppm/water	S	75
O: Licorice-woody, walnut-like	O 1 ppm/water	S	75
O: Bell pepper, cacao bean	O 1 ppm/water	S	75
O: Licorice-woody, walnut-like	O 0.8 ppm/water	S	75
O: Burdock-like, slightly earthy	O 1 ppm/water	S	75
O: Nutty, macadamia-like	O 0.7 ppm/water	S	75
O: Minty-camphoraceous	O 315 ppb/water	S	56
O: Green (bell pepper-like), nutty notes, floral and ethereal undertone	O 180 ppb/water	S	64

Continued on next page

PYRAZINES (cont.)

NAME / (SYNONYM)	R_1	R_2	R_3	R_4
2,5-Dimethoxy-3,6-diisopropyl-pyrazine	OCH_3	$i\text{-}C_3H_7$	OCH_3	$i\text{-}C_3H_7$
2,5-Dimethoxy-3-isopropyl-6-methylpyrazine	OCH_3	$i\text{-}C_3H_7$	OCH_3	CH_3
2,5-Dimethoxy-3-acetyl-6-methylpyrazine	OCH_3	$COCH_3$	OCH_3	CH_3
2,6-Dimethoxy-3-isopropyl-5-methylpyrazine	OCH_3	$i\text{-}C_3H_7$	CH_3	OCH_3
2-Ethoxy-3-methylpyrazine	OC_2H_5	CH_3		
2-Ethoxy-6-methylpyrazine (6-Ethoxy-2-methylpyrazine)	OC_2H_5			CH_3
2-Ethoxy-3-ethylpyrazine	OC_2H_5	C_2H_5		
2-Ethoxy-3-isobutylpyrazine	OC_2H_5	$i\text{-}C_4H_9$		
2-Butoxy-3-methylpyrazine	OC_4H_9	CH_3		
2-Butoxy-3-propylpyrazine	OC_4H_9	$\underline{n}\text{-}C_3H_7$		
2-Methylamino-3-methylpyrazine	$NHCH_3$	CH_3		
2-Dimethylamino-3-methylpyrazine	$N(CH_3)_2$	CH_3		
2-Dimethylamino-6-methylpyrazine	$N(CH_3)_2$			CH_3
2-Dimethylamino-3-isobutylpyrazine	$N(CH_3)_2$	$i\text{-}C_4H_9$		
2-Dimethylamino-6-isobutylpyrazine	$N(CH_3)_2$			$i\text{-}C_4H_9$
2-(Methylthiomethyl)pyrazine	CH_2SCH_3			
2-Methylthio-3-methylpyrazine (2-Methyl-3-methylthiopyrazine)	SCH_3	CH_3		
5-Methylthio-2-methylpyrazine	CH_3		SCH_3	

DESCRIPTION	THRESHOLD VALUE	REM.	REF
O: Green (bell pepper-like), nutty notes, weakly sweet	O 670 ppb/water	S	64
O: Green (bell pepper-like), nutty notes	O 250 ppb/water	S	64
O: Unpleasant, weakly green, earthy notes, chimney soot		S	64
O: Nutty, green (bell pepper-like) woody bynote	O 70 ppb/water	F,S	64
F: Earthy, nutty		S	73
F: Hazelnuts, almonds, peanuts			35
F: Pineapple		S	35
O: Raw potato	O 11 ppb/water	S	56
F: Earthy, bell pepper		S	73
F: Floral, medicinal		S	73
F: Medicinal, earthy		S	73
O: Weak, reminds of roasted peanuts, green and cocoa notes		S	64
O: Like 2-methylamino-3-methylpyrazine but more green and peanut-like, less cocoa notes	O 180 ppb/water	S	64
O: Reminds of unburnt coffee, peanut-like and sweet notes, no cocoa notes	O 1200 ppb/water	S	64
O: Strong cocoa note, reminds of fusel oil, green and burnt undertone	O 1600 ppb/water	S	64
O: Weakly green, cocoa note, no peanut note	O 5000 ppb/water	S	64
O: Weak sulfide note	O 20 ppb/water	S	58
F: Nutty, cracker		S	73
O: Cooked meat, vegetable			58
	O 1 ppb/water		59
O: Meaty, vegetably	O 4 ppb/water	S	58

Continued on next page

PYRAZINES (cont.)

NAME / (SYNONYM)	R_1	R_2	R_3	R_4
2-Methylthio-3-isobutylpyrazine	SCH_3	i-C_4H_9		
2-Methylthio-6-isobutylpyrazine	SCH_3			i-C_4H_9
2-(Furfurylthiomethyl)pyrazine	CH_2-S-furfuryl			
2-Methyl-3-(furfurylthio)pyrazine	CH_3	S-furfuryl		
2-Methyl-5-(furfurylthio)pyrazine	CH_3		S-furfuryl	
5-Methylcyclopenta[b]pyrazine	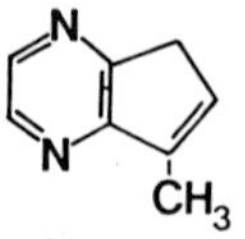			
6,7-Dihydro-5H-cyclopenta[b]pyrazine	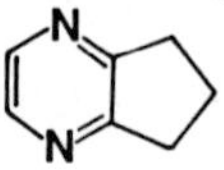			
2-Methyl-6,7-dihydro-5H-cyclopenta-[b]pyrazine	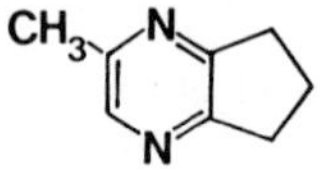			
5-Methyl-6,7-dihydro-5H-cyclopenta-[b]pyrazine	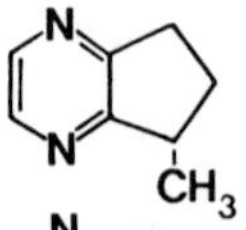			
3,5-Dimethyl-6,7-dihydro-5H-cyclo-penta[b]pyrazine	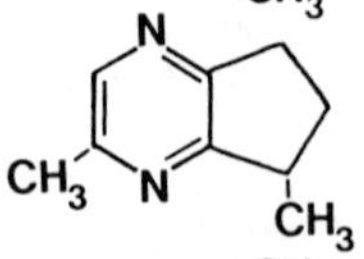			
5,7-Dihydro-5,7-dimethylfuro[3,4-b-]-pyrazine	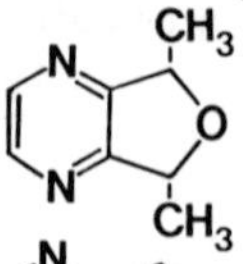			
5,7-Dihydrothieno[3,4-b]pyrazine	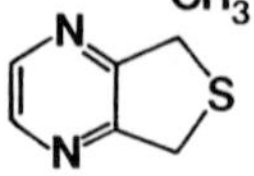			

DESCRIPTION	THRESHOLD VALUE	REM.	REF
F: Bell pepper O: Weakly green, roasted peanuts, cheese-like (Camembert) note	O 0.33 ppb/water	S	73 64
F: Bell pepper		S	73
O: Powerful coffee, cooked meat note, also strong chocolate character	O 1 ppb/water	S	58
F: Roasted coffee-like O: Powerful coffee, cooked meat note	O < 1 ppb/water	S	23 58
F: Roasted, coffee-like O: Powerful coffee, cooked meat note	O < 1 ppb/water	S	23 58
O: Grilled meat		F	66
F: Green, phenolic		F,S	53
O: Earthy, baked, potato-like F: Chocolaty		F,S	69 53
O: Earthy, baked, potato-like F: Peanut		F,S	69 53
O: Earthy, baked, potato-like		F	69
O: Roasted aroma character		S	76
F: Roasted nut		S	76

Continued on next page

PYRAZINES (cont.)

NAME / (SYNONYM)

2-Methyl-5,7-dihydrothieno [3,4-*b*]-pyrazine

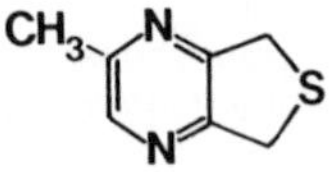

5-Methyl-3,4,5,7-tetrahydro-2*H*-cyclopenta[*b*] pyrazine

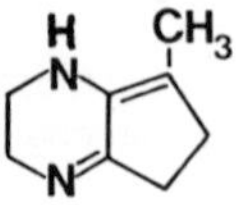

4a,5,6,7,8,8a-Hexahydroquinoxaline

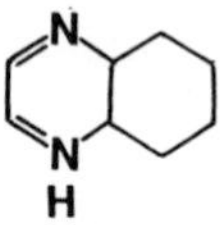

5,7-Dimethyl-1,2,3,4,7,8-hexahydro-quinoxaline

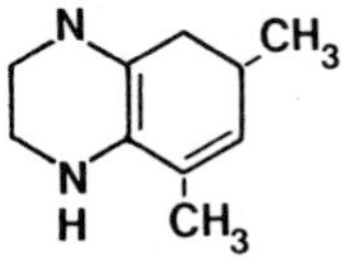

2,3-Dimethyl-4a,5,6,7,8,8a-hexahydro-quinoxaline

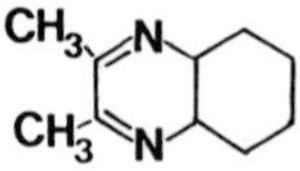

5,7,7-Trimethyl-2,3,4,6,7,8-hexahydro-quinoxaline

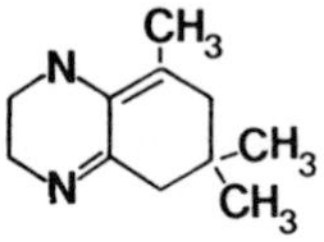

2,3,5,6,7,8,9,10,11,12-Decahydrocyclodeca[*b*]-pyrazine

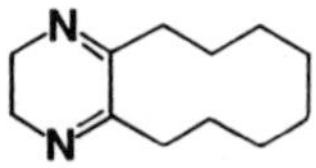

1,5(or 7)-Dimethyl-1,2,3,5,6,7-hexahydrodicyclo-penta[*b*,*e*]pyrazine

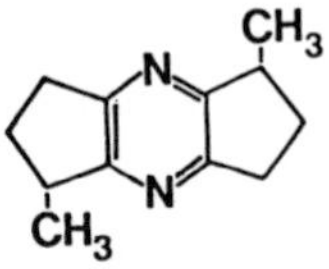

DESCRIPTION	THRESHOLD VALUE	REM.	REF
O: Roasted nut-like		S	76
O: Bready, nut-like		S	77
O: Intense popcorn		S	77
O: Earthy, baked, potato-like		F	69
O: Cedar wood, tobacco, buttery		S	77
O: Sweet, tobacco-like fragrance note		S	77
F: Nut-like, fatty		S	77
O: Roasted bean-like		M	78

Continued on next page

THIOPHENES

NAME / (SYNONYM)	R_1	R_2	R_3	R_4
Thiophene				
2-Methylthiophene	CH_3			
3-Methylthiophene		CH_3		
2-Ethylthiophene	C_2H_5			
2,4-Dimethylthiophene	CH_3		CH_3	
2,5-Dimethylthiophene	CH_3			CH_3
3,4-Dimethylthiophene		CH_3	CH_3	
2-Thienyl alcohol (2-Hydroxythiophene)	OH			
2-Acetylthiophene (Methyl 2-thienyl ketone)	$COCH_3$			
2-Acetyl-3-methylthiophene (Methyl 3-methyl-2-thienyl ketone)	$COCH_3$	CH_3		
2-Acetyl-5-methylthiophene (Methyl 5-methyl-2-thienyl ketone)	$COCH_3$			CH_3
2-Thiophenecarbaldehyde (2-Formylthiophene)	CHO			

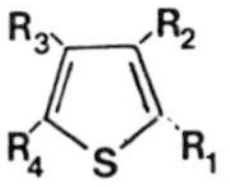

DESCRIPTION	THRESHOLD VALUE	REM.	REF
O of GC eluates: Sickly, pungent		F,M	19
	F 8 ppb/water F 300 ppb/skim milk F 4 ppb/protein-hydrolysate solution 2%		32
O of GC eluates: Green, sweet		F,M,S	19
O at GC exhaust: Onion-like, gasoline-like, paraffinic			31
O: Heated onion or sulfury			79
O: Fatty, winey		F,M	28
O: Styrene-like		F,M	28
O: Fried onion		F,M	28
O: Fried onions	O 1.3 ppb/water	F,M	80
O: Greenish			28
F: Fresh onion		F,M	34
O: Fried onions	O 1.3 ppb/water		80
O: Burnt		F,M	28
O: Mustard-like		F,M	28
O in syrup: 1 g/100 ml: Onion-like; O in coffee: Malty, roasted			6
	F 0.08 ppb/water F 1 ppm/skim milk F 0.3 ppb/protein-hydrolysate solution 2%		32
F: Mustard		F,M	34
O: Nutty			28
O: Sweet, flowery		F,M	28
O: Nutty		F,M	81
O: Reminiscent of benzaldehyde			6
O: Spicy meat			33
O: Coconut-like			28
O: Sharp, sweet-nutty, somewhat roasted grain-like			82
	F 200 ppb/water F 8 ppm/skim milk F 600 ppb/protein-hydrolysate solution 2%		32

Continued on next page

THIOPHENES (cont.)

NAME / (SYNONYM)	R_1	R_2	R_3	R_4
5-Methyl-2-thiophenecarbaldehyde (5-Methyl-2-formylthiophene)	CHO			CH_3
3-Methyl-2-thiophenecarbaldehyde (3-Methyl-2-formylthiophene)	CHO	CH_3		
1-(2-Thienyl)-1-propanone (2-Propionylthiophene)	COC_2H_5			
1-(2- or 3-Thienyl)-propane-1,2-dione	$COCOCH_3$			
2-Methylthiophene-3-thiol	CH_3	SH		
2-Methylthiophene-4-thiol	CH_3		SH	

DIHYDROTHIOPHENES 2,3-(4,5)

2-Methyl-2,3(or 2,5)-dihydrothiophene	CH_3			
2-Methyl-4,5-dihydrothiophene	CH_3			
2-Methyl-2,3-dihydrothiophene-3-thiol	CH_3	SH		
2-Methyl-2,3-dihydrothiophene-4-thiol	CH_3		SH	
3-Mercapto-2-methyl-2,3-dihydro-thiophen-4-ol	CH_3	SH	OH	
2-Methyl-4,5-dihydrothiophene--3-thiol	CH_3	SH		
2-Methyl-4,5-dihydrothiophene--4-thiol	CH_3		SH	
2-Methyltetrahydrothiophene--3-thiol	CH_3	SH		
2-Methyltetrahydrothiophene--4-thiol	CH_3		SH	

DESCRIPTION	THRESHOLD VALUE	REM.	REF
O: Cherry-like		F,M	6
F: Cherry, bitter almond			34
	F 1 ppb/water		1
O: Reminiscent of saffron		F,M	6
O in syrup 1 g/100 ml: Creamy, caramel-like		F,M	6
Imparts a parline-like and woody note to coffee		F,M	6
O: Woody			28
O at GC exhaust: Roasted meat		S	31
O at GC exhaust: Rubbery		S	31

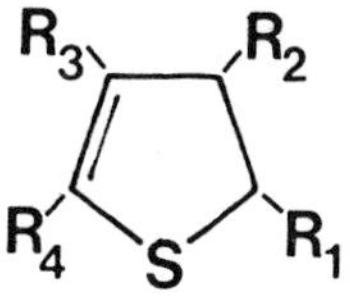

TETRAHYDROTHIOPHENES

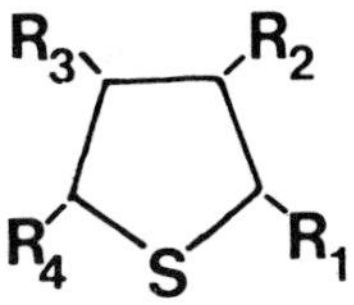

DESCRIPTION	THRESHOLD VALUE	REM.	REF
O at GC exhaust: Cabbage-like		S	31
O: Heated onion or sulfury		M	79
O at GC exhaust: Sweet, roasted meat		S	31
O at GC exhaust: Rubbery, meaty		S	31
O at GC exhaust: Meaty, savory, phenolic		S	31
O at GC exhaust: Meaty		S	31
O at GC exhaust: Roasted meat		S	31
O at GC exhaust: Cis-form, meaty; trans-form, meaty, savory		S	31
O at GC exhaust: Meat-like		S	31

Continued on next page

3-(2<u>H</u>-)-THIOPHENONES

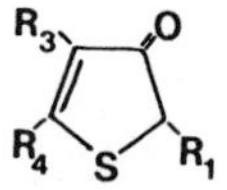

NAME / (SYNONYM)	R_1	R_2	R_3	R_4
4-Hydroxy-5-methyl-3(2H)-thiophenone		=O	OH	CH_3
4-Mercapto-5-methyl-3(2H)--thiophenone		=O	SH	CH_3
4,5-Dihydro-3(2H)-thiophenone		=O		
2-Methyl-4,5-dihydro-3(2H)--thiophenone	CH_3	=O		

3,4-Dimethyl-2,5-dioxo--2,5-dihydrothiophene

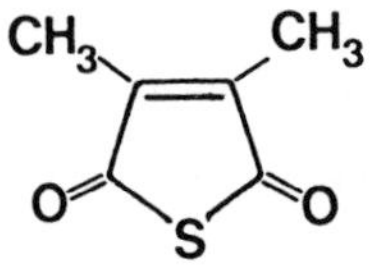

4,5-DIHYDRO-3(2H)-THIOPHENONES 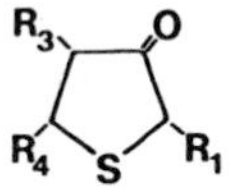

DESCRIPTION	THRESHOLD VALUE	REM.	REF
O at GC exhaust: Popcorn-like		S	31
O at GC exhaust: Nutty		S	31
F: Garlic		F,M	7
F: Green, burnt, coffee		F,M,S	7
O at GC exhaust: Acetylenic			31
O: Onion, a little H_2S		F	83

Continued on next page

AROMATIC SULFUR COMPOUNDS

NAME / (SYNONYM)	
α-Toluenethiol (Benzyl mercaptan, Phenylmethanethiol)	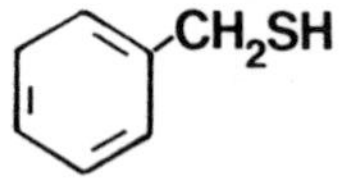
Benzyl methyl sulfide	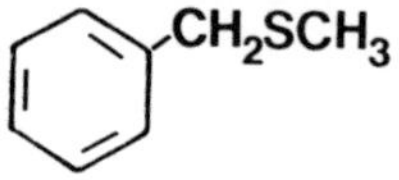
2-Naphthalenethiol (2-Naphthyl mercaptan)	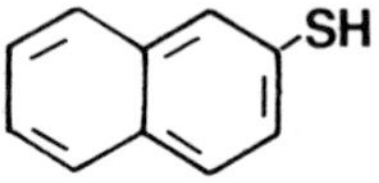

DESCRIPTION	THRESHOLD VALUE	REM.	REF
O "Leek-like" (a more exact description than the older still common "onion-like"). Has the sharpness of leek odor, yet not lachrymatory		F,S	17
O: Repulsive, garlic-like			23
F: Reminiscent of the natural landcress off-flavor	F 1 ppb/milk		84
	F 10 ppb/milk F 100 ppb/butter oil	F,M	84
O: Disagreeable, mercaptan-like		F,S	23

Continued on next page

CYCLIC NON-AROMATIC SULFUR COMPOUNDS

NAME / (SYNONYM)	
Ethylene sulfide (Thiirane)	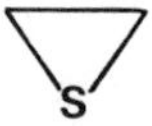
1,2-Dithiacyclopentene (3H-1,2-Dithiole)	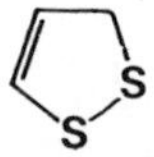
3-Methyl-1,2-dithiolane	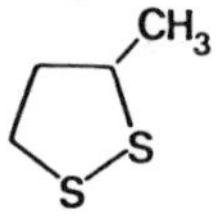
3,5-Dimethyl-1,2-dithiolane	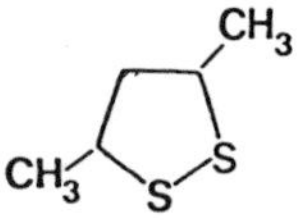
1,3-Dithiolane	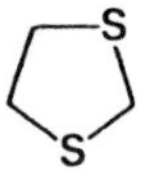
2-Methyl-1,3-dithiolane	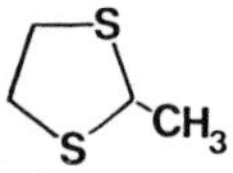
4-Methyl-1,3-dithiolane	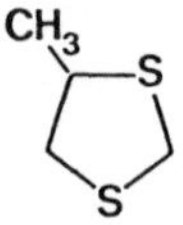
2,4-Dimethyl-1,3-dithiolane	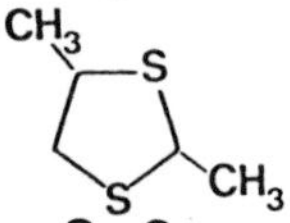
1,2,4-Trithiolane	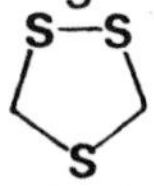
3,5-Dimethyl-1,2,4-trithiolane	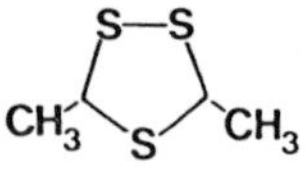
3-Ethyl-5-methyl-1,2,4-trithiolane	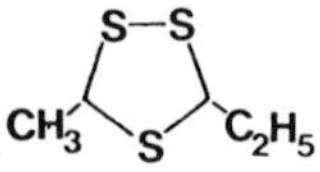
3,5-Diethyl-1,2,4-trithiolane	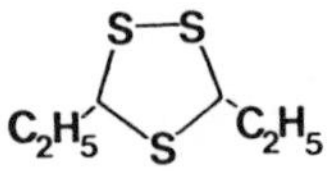

DESCRIPTION	THRESHOLD VALUE	REM.	REF
O: Cooked cabbage		F	6
O of GC eluates: Sickly, cooked cabbage, pungent			19
O: Pleasant smell like cooked asparagus		F	85
T: Oniony metallic with cooked beef or braised and vegetable nuances	T 2 ppb	S	86
Cooked onion, cooked vegetable nutty note	T 10 ppb	S	86
Sweet, sulfury character, in high levels: reminiscent of degrading onions	T 2 ppb	F,S	86
F: Onion and garlic roasted or boiled	20 ppb/water	F,S	34
T: Onion, root-vegetable-like	T 50 ppb	S	86
T: Onion-like with slight metallic background notes	T 20 ppb	S	86
O: Roast beef		M	78
O: Sulfurous			28
O: Characteristic of boiled beef		F,M,S	87
O: Onion-like			88
	F 10 ppb/water		1
O: Garlicky		F,S	88
O: Garlicky		F,S	88

Continued on next page

CYCLIC NON-AROMATIC SULFUR COMPOUNDS (cont.)

NAME / (SYNONYM)	
3,5-Di-(2-methylpropyl)-1,2,4-trithiolane	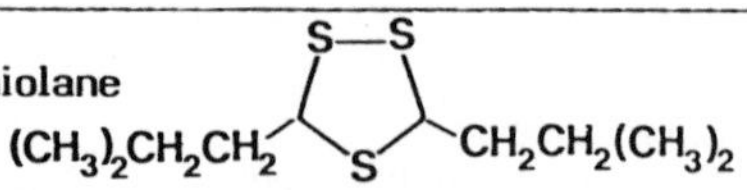
1,2-Dithiane	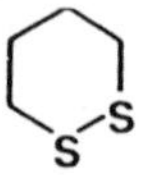
3-Methyl-1,2-dithiane	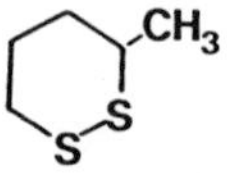
3,5-Dimethyl-1,2-dithiane	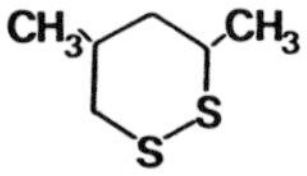
2,6-Dimethyldithiin (2,6-Dimethyl-4H-dithiin)	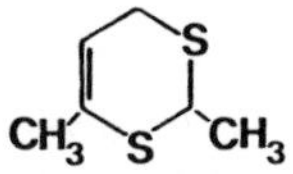
1,3-Dithiane	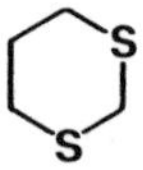
2-Methyl-1,3-dithiane	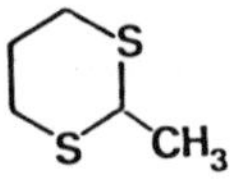
4-Methyl-1,3-dithiane	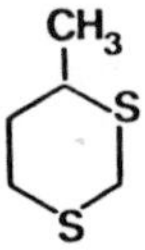
2,4-Dimethyl-1,3-dithiane	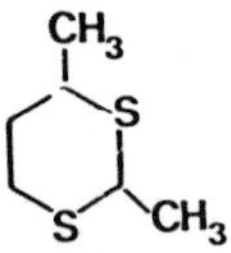
2,4,6-Trimethyl-1,3-dithiane	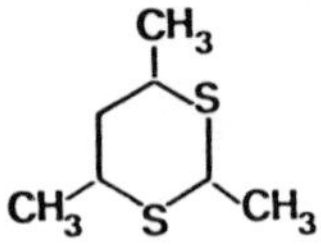
1,4-Dithiane	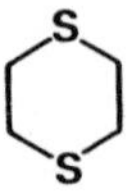

DESCRIPTION	THRESHOLD VALUE	REM.	REF
F at 1 ppm: Roasted, crisp bacon-like, pork rind-like		S	86
Garlic character with slight metallic nuance	T 2 ppb	M,S	86
F: Onion and garlic, roasted or boiled	2 ppb/water	S	34
F: Onion and garlic, roasted or boiled	10 ppb/water	S	34
O: Fuel gas-like		F,S	88
F: Onion and garlic, roasted or boiled	2 ppb/water	F,S	34
T: Onion, garlic-like with a metallic by-note	T 50 ppb		86
Onion-like character with metallic notes	T 20 ppb	S	86
Onion, garlic and tomato like character	T 50 ppb	S	86
Allium-onion like character with slight metallic notes	T 5 ppb	S	86
Root-vegetable-like character	T 10 ppb	S	86
Onion-like or garlic-like character	T 50 ppb	F,S	86

Continued on next page

CYCLIC NON-AROMATIC SULFUR COMPOUNDS (cont.)

NAME / (SYNONYM)	
1,3,5-Trithiane (*s*-Trithiane)	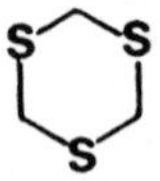
2,4,6-Trimethyl-1,3,5-trithiane (2,4,6-Trimethyl-*s*-trithiane)	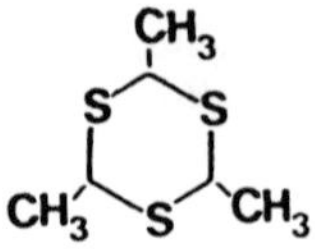
2,4,6-Triethyl-1,3,5-trithiane	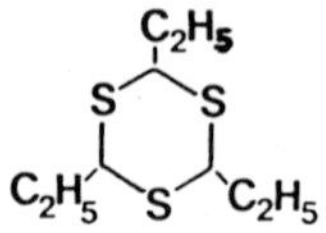
6-Methyl-2,3-dihydrothieno[2,3-*c*]furan (Kahweofuran)	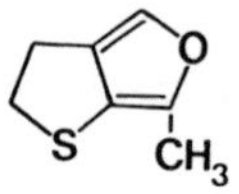

DESCRIPTION	THRESHOLD VALUE	REM.	REF
O: Sulfurous		F,M	78
	F 0.04 ppb/water F 60 ppb/skim milk F 0.08 ppb/protein-hydrolysate solution 2%		32
F dissolved in water: Dusty, earthy, nutty		F,M	89
F dissolved in water: Green, allium (onion, garden cress)		M	89
O in pure state: Violent, sulfury, but in high dilution develops a pleasant roasted and smoky note		F,S	7

Continued on next page

OXAZOLES

NAME / (SYNONYM)	R_1	R_2	R_3
2-Acetyloxazole	$COCH_3$		
4-Butyloxazole		n-C_4H_9	
5-Butyloxazole			n-C_4H_9
4-Heptyloxazole		n-C_7H_{15}	
2,4-Dimethyloxazole	CH_3	CH_3	
4,5-Dimethyloxazole		CH_3	CH_3
4,5-Diethyloxazole		C_2H_5	C_2H_5
4-Methyl-5-propyloxazole		CH_3	n-C_3H_7
4-Methyl-5-butyloxazole		CH_3	n-C_4H_9
4-Methyl-5-hexyloxazole		CH_3	n-C_6H_{13}
4-Ethyl-5-propyloxazole		C_2H_5	n-C_3H_7
4-Ethyl-5-butyloxazole		C_2H_5	n-C_4H_9
4-Ethyl-5-pentyloxazole		C_2H_5	n-C_5H_{11}
4-Propyl-5-methyloxazole		n-C_3H_7	CH_3
4-Propyl-5-ethyloxazole		n-C_3H_7	C_2H_5
4-Butyl-5-methyloxazole		n-C_4H_9	CH_3
4-Butyl-5-propyloxazole		n-C_4H_9	C_3H_7
4-Pentyl-5-methyloxazole		n-C_5H_{11}	CH_3
4-Hexyl-5-methyloxazole		n-C_6H_{13}	CH_3
2,4-Diethyl-5-propyloxazole	CH_3	CH_3	n-C_3H_7
2,4,5-Trimethyloxazole	CH_3	CH_3	CH_3

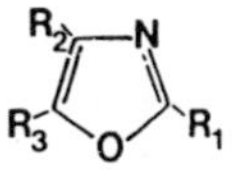

DESCRIPTION	THRESHOLD VALUE	REM.	REF
O: Nutty, popcorn-like		F	90
O: Vegetable-like, green		S	91
O: Green, sweet, vegetable		F	90
O: Vegetable-like		S	91
O: Nutty, sweet		M	37
O: Green, sweet, vegetable		F	90
O: Green, vegetable-like		S	91
O: Vegetable-like, green		S	91
O: Vegetable-like, green		S	91
O: Vegetable-like, green		S	91
O: Vegetable-like, green		S	91
O: Vegetable-like, green		S	91
O: Vegetable-like, green		S	91
O: Green, weak bell pepper-like		S	91
O: Celery-like, green		S	91
O: Bell pepper-like		S	91
O: Bell pepper-like		S	91
O: Bell pepper-like		S	91
O: Vegetable-like, green		S	91
O: Nutty, sweet, green		F	90
F: Like boiled beef		F,S	6
F: Nutty, sweet, green	F 5 ppb		92

Continued on next page

OXAZOLINES

NAME / (SYNONYM)	R_1	R_2	R_3
2-Ethyl-5-methyl-2-oxazoline	C_2H_5		CH_3
5-Methyl-2-propyl-2-oxazoline	n-C_3H_7		CH_3
5-Methyl-2-pentyl-2-oxazoline	n-C_5H_{11}		CH_3
2-Methyl-3-oxazoline	CH_3		
2,4-Dimethyl-3-oxazoline	CH_3	CH_3	
2,4,5-Trimethyl-3-oxazoline	CH_3	CH_3	CH_3
2,4-Dimethyl-5-ethyl-3-oxazoline	CH_3	CH_3	C_2H_5
2,5-Dimethyl-4-ethyl-3-oxazoline	CH_3	C_2H_5	CH_3
2,4,5-Triethyl-3-oxazoline	C_2H_5	C_2H_5	C_2H_5
4,5-Diethyl-2-isopropyl-3-oxazoline	i-C_3H_7	C_2H_5	C_2H_5
4,5-Dipropyl-2-isopropyl-3-oxazoline	i-C_3H_7	n-C_3H_7	n-C_3H_7
2-s-Butyl-4,5-diethyl-3-oxazoline	s-C_4H_9	C_2H_5	C_2H_5
2-Isopropyl-4,5,5-trimethyl-3-oxazoline	i-C_3H_7	CH_3	2xCH_3
2-Isobutyl-4,5,5-trimethyl-3-oxazoline	i-C_4H_9	CH_3	2xCH_3
2-t-Butyl-4,5,5-trimethyl-3-oxazoline	t-C_4H_9	CH_3	2xCH_3

BENZOXAZOLES

2-Methylbenzoxazole	CH_3		

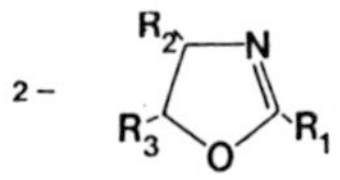

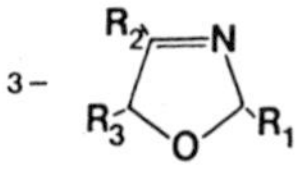

DESCRIPTION	THRESHOLD VALUE	REM.	REF
O: Melon, fruity			35
F: Fresh mint			35
F: Mild, fruity			35
O: Nutty, sweet		F	90
F: Nutty, vegetable, not roasted	F 1.0 ppm/water	F,S	92
O: Nutty, sweet			90
F: Woody, musty, green	F 1.0 ppm/water	F,S	92
O: Characteristic boiled beef			87
F: Nutty, sweet, green, woody	F 0.5 ppm/water	F,S	92
F: Nutty, sweet	F 1.0 ppm/water	F,S	92
O: Very reminiscent of fresh carrots		S	7
O: Green			93
O: Cocoa			7
O: Banana		S	35
O: Cool, banana-like			93
O: Cocoa		S	93
Rum-like note		S	7
O: Cool, menthol-like, unroasted cocoa			93
O: Earthy (fungal) with a butter-like green leaf-like note		S	7
O: Earthy, diacetyl			93
O: Fresh mint and, depending on the concentration, fresh banana or buttery			35
O: Cool, mint-like			93

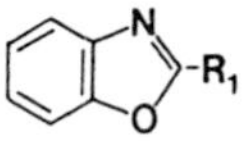

DESCRIPTION	THRESHOLD VALUE	REM.	REF
O: Very sweet, of rather gassy-pungent character when undiluted, becoming floral-sweet, heavy, when diluted		F,S	17

Continued on next page

THIAZOLES

NAME / (SYNONYM)	R_1	R_2	R_3
Thiazole			
2-Methylthiazole	CH_3		
4-Methylthiazole			CH_3
2-Ethylthiazole	C_2H_5		
2-Propylthiazole	$\underline{n}$-C_3H_7		
2-Isopropylthiazole	i-C_3H_7		
2-Butylthiazole	$\underline{n}$-C_4H_9		
2-Isobutylthiazole	i-C_4H_9		
2-Acetylthiazole (Methyl 2-thiazolyl ketone)	$COCH_3$		
4-Acetylthiazole (Methyl 4-thiazolyl ketone)		$COCH_3$	
2-Methoxythiazole	OCH_3		
5-Methoxythiazole			OCH_3
2-Ethoxythiazole	OC_2H_5		

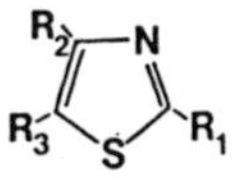

DESCRIPTION	THRESHOLD VALUE	REM.	REF
F in beer: Sulfury, grainy, CS_2	T 23 ppm/beer	F,M	25
O: Pyridine-like			23
O: Nutty, meaty			22
F: Green, vegetable		F,M,S	52
O: Green, nutty		F,M	23
F: Green, nutty		F,M,S	52
F: Green, herby, nutty		M,S	52
F: Green, vegetable		S	52
F: Raw, green, herby		S	52
F in water: Spoiled wine-like, slightly horseradish F in canned tomato juice or tomato paste: More intense, fresh tomato-like	O 2 ppb/water	F,S	94
O: Strong green, resembling that of tomato leaf			7
	O 3.5 ppb/water		95
	F 3 ppb/water		1
F: Nutty, cereal, popcorn		F,M,S	52
O: Cereal bready			46
O: Taco, grassy			70
O: Popcorn-like, strong, nutty-roasted character			82
	F 10 ppb/water		1
F: Nutty, cereal		M,S	52
O: Hazelnut, earthy F: Nutty, steak with bitter aftertaste, baked bread, meat			35
O: Cracker-like	O 170 ppb/water		55
F: Sweet, roasted, phenolic		M,S	52
F: Oatmeal, bread and caramel			35
F: Roasted, meaty, onion		M,S	52
F: Braised vegetables, roast meat, fresh onions			35
F: Phenolic, burnt, nutty		M,S	52
F: Roasted peanuts, roast meat			35

Continued on next page

THIAZOLES (cont.)

NAME / (SYNONYM)	R_1	R_2	R_3
5-Ethoxythiazole			OC_2H_5
2-Butoxythiazole	OC_4H_9-n		
2,4-Dimethylthiazole	CH_3	CH_3	
4,5-Dimethylthiazole		CH_3	CH_3
2,4-Diethylthiazole	C_2H_5	C_2H_5	
2,5-Diethylthiazole	C_2H_5		C_2H_5
4,5-Diethylthiazole		C_2H_5	C_2H_5
2,4-Dipropylthiazole	n-C_3H_7	n-C_3H_7	
4,5-Dibutylthiazole		n-C_4H_9	n-C_4H_9
5-Ethyl-4-methylthiazole		CH_3	C_2H_5
2-Acetyl-4-methylthiazole (Methyl 4-methyl-2-thiazolyl ketone)	$COCH_3$	CH_3	
5-Acetyl-4-methylthiazole (Methyl 4-methyl-5-thiazolyl ketone)		CH_3	$COCH_3$
4-Ethyl-5-propylthiazole		C_2H_5	n-C_3H_7
4-Ethyl-5-butylthiazole		C_2H_5	n-C_4H_9
4-Butyl-5-methylthiazole		n-C_4H_9	CH_3
4-Butyl-5-ethylthiazole		n-C_4H_9	C_2H_5
4-Butyl-5-propylthiazole		n-C_4H_9	n-C_3H_7
2-Methyl-5-methoxythiazole	CH_3		OCH_3

DESCRIPTION	THRESHOLD VALUE	REM.	REF
F: Cooked onion		M,S	52
F: Cooked vegetables (sharp resembling nuts)			35
F: Green vegetable		S	52
O: Skunky-oily	O 18 ppb/water	F,M	55
O: Meat, cocoa	O 0.1 ppb/water		34, 35
F: Roasted, nutty, green		F,M,S	52
F: Braised meat, hazelnut O: Meaty, boiled poultry	O 0.5 ppm		35
	O 470 ppb/water		96
O: Ethereal, musty, earthy	O 6.5 ppb/water	F	55
O: Slightly skunky and green peppery	O 0.09 ppb/water	F	55
	O 5.1 ppb/water	S	96
O: Sweet, fruity, minty	O 26 ppb/water		55
	O 0.19 ppb/water	S	96
F: Nutty, green, earthy		F,M,S	52
O: Green vegetables, unroasted hazelnuts F: Very sharp, dry and earthy parsley	O 0.02 ppm		35
F: Anthranilic, burnt		F,M	7
	F 300 ppb/water		1
F: Roasted, nutty, sulfury		M,S	52
O: Peanut, earthy F: Roasted nuts, sulfury, bitter metallic and earthy	O 0.05 ppm		35
O: Bell pepper	O 0.06 ppb/water	S	96
O: Bell pepper	O 0.12 ppb/water	S	96
O: Bell pepper	O 0.01 ppb/water	S	96
O: Bell pepper	O 0.02 ppb/water	S	96
O: Bell pepper	O 0.003 ppb/water	S	96
F: Cabbagy, sulfury, vegetable		S	52

Continued on next page

THIAZOLES (cont.)

NAME / (SYNONYM)	R_1	R_2	R_3
4-Isobutyl-5-methoxythiazole		$i\text{-}C_4H_9$	OCH_3
4-Isobutyl-5-ethoxythiazole		$i\text{-}C_4H_9$	OC_2H_5
2,4,5-Trimethylthiazole	CH_3	CH_3	CH_3
2,4-Dimethyl-5-ethylthiazole	CH_3	CH_3	C_2H_5
4-Acetyl-2,5-dimethylthiazole	CH_3	$COCH_3$	CH_3
5-Acetyl-2,4-dimethylthiazole	CH_3	CH_3	$COCH_3$
2,4-Dimethyl-5-vinylthiazole	CH_3	CH_3	$CH{=}CH_2$
2,5-Dimethyl-4-butylthiazole	CH_3	$\underline{n}\text{-}C_4H_9$	CH_3
2,5-Diethyl-4-methylthiazole	C_2H_5	CH_3	C_2H_5
2-Isopropyl-4,5-dimethylthiazole	$i\text{-}C_3H_7$	CH_3	CH_3
2-Propyl-4,5-diethylthiazole	$\underline{n}\text{-}C_3H_7$	C_2H_5	C_2H_5
4-Isobutyl-5-methoxy-2- -methylthiazole	CH_3	$i\text{-}C_4H_9$	OCH_3
4-Isobutyl-5-ethoxy-2-methyl- thiazole	CH_3	$i\text{-}C_4H_9$	OC_2H_5
2-Butyl-4-isobutyl-5- -methoxythiazole	$\underline{n}\text{-}C_4H_9$	$i\text{-}C_4H_9$	OCH_3

DESCRIPTION	THRESHOLD VALUE	REM.	REF
F: Green pepper		M,S	52
O: Vegetable soup (minestrone) F: Pepper, onion, celery and green vegetables (peas and haricot beans)			35
F: Cucumber, green potato		M,S	52
F: Cucumber, green pepper, onions	O 0.02 ppm		35
F: Cocoa, nutty		F,M,S	52
O: Cocoa, hazelnut, dark chocolate, green vegetables (haricot beans)	O 0.05 ppm/water		34, 35
F: Nutty, roasted, meaty			52
O: Hazelnut, steak, meat, liver, green vegetables, "Buccu" leaf and black currant	O 0.02 ppm	F,M,S	35
F: Nuts, meat, steak	O 0.002 ppm		34
F: Roasted, meaty, sulfury		S	52
F: Roasted, nutty, meaty		M,S	52
O: Sulfury as in meat F: Boiled beef, chicken and turkey			35
O: Strong, characteristic, nut-like		F	7
O: Sweet, earthy		F	69
F: Green, nutty		S	52
O: Pleasant, nutty		F	90
O: Pleasant, nutty		F	90
F: Green, vegetable		M,S	52
O: Onion and pepper F: Pepper, vegetables (green)			34, 35
F: Green, vegetable, onion		M,S	52
F: Vegetables, sharp and green			35
O: Vegetable soup with touch of barley, potato, green pepper and onion	O 0.0002 ppm	M	35

Continued on next page

THIAZOLINES (DIHYDROTHIAZOLES)

NAME / (SYNONYM)	R_1	R_2	R_3
2-Acetyl-2-thiazoline	$COCH_3$		
2,4-Dimethyl-3-thiazoline		CH_3	CH_3
2,4,5-Trimethyl-3-thiazoline	CH_3	CH_3	CH_3
2-Propyl-3-thiazoline	*n*-C_3H_7		
2-(2-Methylpropyl)-4,5-dimethyl-3-thiazoline	$CH_2CH\ (CH_3)_2$	CH_3	CH_3
2-(1-Methylpropyl)-4,5-dimethyl-3-thiazoline	$CH(CH_3)C_2H_5$	CH_3	CH_3
2-Propyl-2,4,5-trimethyl-3-thiazoline	*n*-C_3H_7 CH_3	CH_3	CH_3

2- 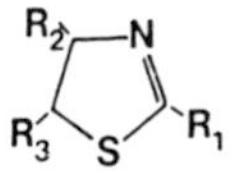 3- 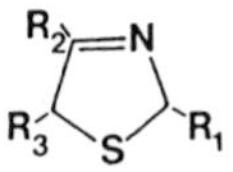

DESCRIPTION	THRESHOLD VALUE	REM.	REF
O: Cracker-like	O 1.3 ppb/water	F	55
O: Freshly baked bread crust			22
	F 3 ppb/water		1
F: Nutty, roasted, vegetable	F 0.02 ppm	F,M,S	92
F: Meaty, nutty, onion-like	F 0.5 ppm	F,M,S	92
F at 0.2-1 ppm: Fresh chopped meat-like with light sour effect T at 2 ppm: Chemical			97
O at 0.2 ppm: Green, vegetable nutty and roasted T at 0.2 ppm: Sweet, green, vegetable, roasted, cocoa powderlike		S	97
O at 2 ppm: Sweet, roasted meat-like, roasted nut-like, dark chocolate-like, baked goods-like and vegetable green-like T at 2 ppm: Sweet, roasted meat-like, roasted nut-like, chocolate-like, vegetable green-like, hydrolyzed vegetable protein-like. Having a hydrolyzed vegetable protein aftertaste and astringent and chocolate-like notes		S	97
O at 2 ppm: Sweet, herbaceous, spicy, chocolate-like, nutty, vegetable-like, hydrolyzed vegetable protein-like, roasted F at 2 ppm: Herbaceous, vegetable, green-like, nutty, roasted, chocolate-like, astringent		S	97
O at 1 ppm: Sweet, green bean-like, cucumber-like, geranium-like and spicy F at 1 ppm: Cucumber-like, green bean-like, spicy, watermelon-like, black pepper-like, astringent		S	97

Continued on next page

BENZOTHIAZOLES

NAME / (SYNONYM)	R_1	R_2	R_3
Benzothiazole	H		
2-(Methylthio)benzothiazole	SCH_3		

1,3,5-DITHIAZINES

NAME / (SYNONYM)	
2,4,6-Trimethyldihydro-1,3,5-dithiazine	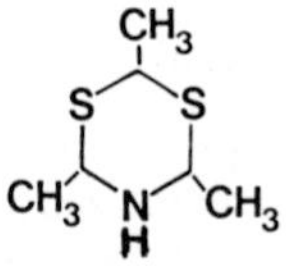
2-Ethyl-4,6-dimethyldihydro-1,3,5-dithiazine	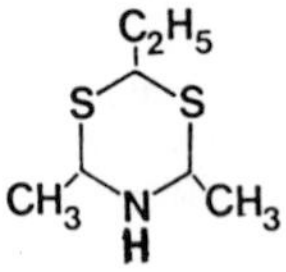
4-Ethyl-2,6-dimethyldihydro-1,3,5-dithiazine	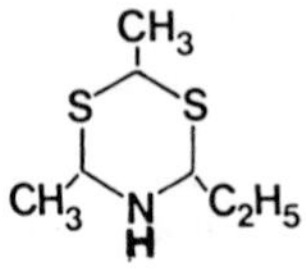

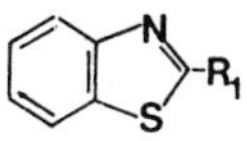

DESCRIPTION	THRESHOLD VALUE	REM.	REF
O: Quinoline-like, rubbery		F	7
O: Similar to pyridine			34
O: Heated rubber-like, heavy and dirty			82
O: Imparts characteristic fatty and smoky odor to meat aroma	O 5 ppb/water	F	6, 7

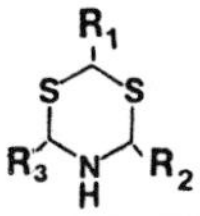

DESCRIPTION	THRESHOLD VALUE	REM.	REF
O: Roasted beef-like		F,M,S	88
Typical note of heated meat			6
O: Cooked beef-like		F,S	88
O: Onion-, cooked beef-like		F,S	88

Continued on next page

ALDEHYDES AND KETONES

NAME / (SYNONYM)	
Ethanal (Acetaldehyde)	CH_3CHO
Butanal (Butyraldehyde)	$\underline{n}$-C_3H_7CHO

DESCRIPTION	THRESHOLD VALUE	REM.	REF
O: Pungent ethereal-nauseating, in high dilution reminiscent of coffee or wine		F,M,S	17
O: Characteristic pungent and penetrating			23
O of GC eluates: Pungent			19
O: Sour, greenhouse			98
O 100°C: Caramel, sweet; 180°C: Burnt sugar			18
O: Green, sweet	O 0.21 ppm/air (recognition)		50
	O 0.12 mg/m^3/air		99
	O 0.041 mg/m^3/air		55
	O 0.005 mg/m^3/air		100
	*		23, 141, 142
	O 15 ppb/water		101
	O 109 ppb/water		102
	O 120 ppb/water		20
F: Green leaves, fruity	F 25 ppb/beer		25
	F 50 ppm/beer		103
O: Very diffusive, penetrating pungent-irritating Only in extreme dilution truly fruity, banana-like, green-fresh odor		F,S	17
O: Characteristic pungent			23
O of GC eluates: Burnt, green, nasty			19
	O <0.013 mg/m^3/air		104, 105
	O 0.042 mg/m^3/air		55
	*		23, 141, 142
	O 9 ppb/water		106
	O 37.3 ppb/water		102
F in beer: Melon, green leaves, varnish	F 1.0 ppm/beer		25
	F 1.0 ppm/beer		103

Continued on next page

ALDEHYDES AND KETONES (cont.)

NAME / (SYNONYM)	
2-Methylpropanal (Isobutyraldehyde)	$(CH_3)_2CHCHO$
2-Methylbutanal (α-Methylbutyraldehyde)	$CH_3CH_2CH(CH_3)CHO$
3-Methylbutanal (Isovaleraldehyde)	$(CH_3)_2CHCH_2CHO$

DESCRIPTION	THRESHOLD VALUE	REM.	REF
O: Extremely diffusive, penetrating pungent and undiluted-unpleasant sour, repulsive. In extreme dilution almost pleasant, fruity banana-like, "overripe-fruit-like"		F,M,S	17
O 100°C: Rye bread, fruity, aromatic; 180°C: Penetrating chocolate			18
O of GC eluates: Green, pungent, sweet			19
O: Harsh, grain, baked poatoes			98
	O 0.14 mg/m³/air		104, 105
	O 0.015 mg/m³/air		107
	O 0.1 ppb/water; 43 ppb/oil		62
	O 0.01 ppm/water		20
	O 2.3 ppb/water		102
F in beer: Banans, melon, varnish, green leaves, bitter	F 1 ppm/beer		25
O: Powerful, choking when undiluted, but in extreme dilution tolerable almost pleasant fruity-"fermented" with a peculiar note resembling that of roasted cocoa or coffee. F: Sweet, slightly fruity-chocolate-like when diluted to < 20 ppm. Pungent at higher concentrations		F,M	17
O of GC eluates: Burnt, sickly			19
O 100°C: Musty, fruity aromatic; 180°C: Burnt cheese			18
	O 1 ppb/water		102
	O 140 ppb/oil		62
F in beer: Green grass, fruity, sour/ /medicinal	F 1.25 ppm/beer		25
O: Very powerful, penetrating, acrid-pungent, causes cough-reflexes unless highly diluted. In extreme dilution fruity, rather pleasant. F: Peach-like, heavy fruity at < 10 ppm			17
O 100°C: Sweet, chocolate, toasted bready; O 180°C: Burnt cheese			18
O of GC eluates: Burnt, green, sickly			19
	O 0.2 ppb/water; 13 ppb/oil		62
	O 7 ppb/water		20
	O 2 ppb/water		102
	F 0.17 ppm/water		44
F in beer: Unripe banana, apple, cherry, cheese	F 0.6 ppm/beer		25

Continued on next page

ALDEHYDES AND KETONES (cont.)

NAME / (SYNONYM)	
2,3-Butanedione (Diacetyl, dimethylglyoxal)	$CH_3COCOCH_3$
2,3-Pentanedione (Acetylpropionyl)	$CH_3COCOC_2H_5$
4-Hydroxy-2,3,5-hexanetrione (Acetylformoin)	$CH_3COCOCH(OH)COCH_3$

DESCRIPTION	THRESHOLD VALUE	REM.	REFE
O: Very powerful and diffusive, pungent, buttery, chlorine--quinone-like in high concentration, oily-buttery in extreme dilution		F,M,S	17
O: Quinonic, buttery, chlorine-like F: Sharp			18
O of GC eluates: Buttery, sickly			19
	O 0.0025 mg/m^3/air		108
	O 0.0026 mg/m^3/air		109
	*		23, 141, 142
	O 2.6 ppb/water		110
	O 6.5 ppb/water		20
	O 2.3 ppm/water		111
F in beer: Diacetyl, butterscotch	F 0.15 ppm/beer		25
O: Buttery, penetrating	F 0.1 ppm/beer		98
O: Oily-buttery, pungent and somewhat "quinone-like", less sharp and volatile than diacetyl. Weak aqueous solutions practically odorless F aqueous solution: Sweet		F,S	17
O of GC eluates: Butter, sickly			19
O: Burnt, grain, malty, burnt butter	F 1 ppm/beer		98
F in beer: Diacetyl, fruity	F 0.9 ppm/beer		25
	T 18 ppm/orange juice	F	26

Continued on next page

CYCLOPENTANONES

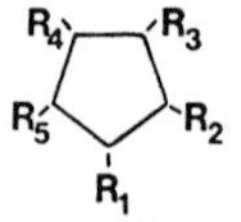

NAME / (SYNONYM)	R_1	R_2	R_3	R_4	R_5
2-Methylcyclopentan-1-one	=O	CH_3			
3-Methylcyclopentan-1-one	=O		CH_3		
2-Mercapto-3-methyl-2- -cyclopenten-1-one	=O	SH	CH_3		
5-Imino-2-methyl-1- -cyclopenten-1-ol	OH	CH_3			=NH
5-Imino-4-methyl-1- -cyclopenten-1-ol	OH			CH_3	=NH
2-Hydroxy-3-methyl-2- -cyclopenten-1-one (Methylcyclopentenolone, Cyclotene)	=O	OH	CH_3		
3-Ethyl-2-hydroxy-2- -cyclopenten-1-one (Ethylcyclopentenolone)	=O	OH	C_2H_5		
3,4-Dimethyl-2-hydroxy-2- -cyclopenten-1-one	=O	OH	CH_3	CH_3	
3,5-Dimethyl-2-hydroxy-2- -cyclopenten-1-one	=O	OH	CH_3		CH_3

CYCLOPENTENONES

1– 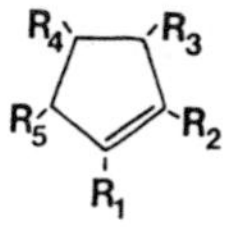

2– 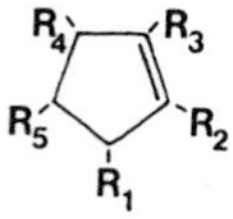

	DESCRIPTION	THRESHOLD VALUE	REM.	REF
O:	Roasted beef		M	78
O:	Roasted beef		M	78
O:	Seaweed-like		M	78
O	Cooked rice		M	78
O:	Rice cracker		M	78
O:	Fragrant, burnt, licorice		F,M,S	18
F:	Sweet			
O:	Strong caramel-like			42
O:	Caramel, maple-like			30
		T 5 ppm/orange juice		26
F:	Sweet, somewhat similar to licorice			23
O:	Strong caramel-like		F,M,S	42
F:	Caramel-like, exhibits flavor-enhancing characteristics			23
O:	Strong caramel-like		F,M	42
O:	Strong caramel-like		F,M	42

Continued on next page

NON-CYCLIC SULFUR COMPOUNDS

NAME / (SYNONYM)	
Hydrogen sulfide	H_2S
Carbon disulfide	CS_2
Dimethyl sulfide (Methyl sulfide, (Methylthio)methane)	CH_3SCH_3

DESCRIPTION	THRESHOLD VALUE	REM.	REF
O: Very unpleasant, chokingly repulsive often described as reminiscent of "rotten eggs" or "decaying seaweed" etc.		F,M,S	17
O: Eggy, sulfidy	O 4.7 ppb/air (H_2S from Na_2S) (recogn)		50
	O 0.47 ppb/air (H_2S gas)(recogn)		
	O 0.0001 mg/m^3/air		112
	O 0.18 mg/m^3/air		99
	O 0.00021-0.0016 mg/m^3/air		113
	O 0.012 mg/m^3/air		114
	O 0.0047-0.009 mg/m^3/air		115
	*		23, 141, 142
	O 10 ppb/water		116
	F 5 ppb/water		1
	F 5 ppb/beer		98
O: Diffusive, chokingly repulsive ethereal-sulfuraceous GC exhaust: Sweet, cabbage-like, almost herbaceous, slightly green		F,S	17
O: Vegetable, sulfidy	O 0.21 ppm/air (recogn)		50
	O 0.07 mg/m^3/air		117
	O 0.05 mg/m^3/air		118
	O 0.08-0.5 mg/m^3/air		119
	*		23, 141, 142
O: Extremely diffusive, repulsive, reminiscent of wild radish, sharp green, cabbage-like. Only in high dilution bearable and most acceptable, pleasant, vegetable-like		F,M,S	17
O of GC eluates: Sulfurous, sickly			19
O: Vegetable, sulfidy	O 1 ppb/air (recogn)		50
	O 0.0094 mg/m^3/air		99
	O 0.002-0.03 mg/m^3/air		113
	O 0.014 mg/m^3/air		120, 121, 122
	O 0.003 mg/m^3/air		123
	*		23, 141, 142
	O 10 ppb/water		124
	O 0.33 ppb/water		106
	O 1 ppb/water		20
	F 12 ppb/water		125
	F 0.03 ppb/water		32
F in beer: Cooked vegetable (corn, onion), garlic, H_2S	F 50 ppb/beer		25
	F 60 ppb/beer		103
	F 6 ppb/skim milk		32
	F 0.02 ppb/protein-hydrolysate solution 2%		32

Continued on next page

NON-CYCLIC SULFUR COMPOUNDS (cont.)

NAME / (SYNONYM)	
Diethyl sulfide ((Ethylthio)ethane)	$C_2H_5SC_2H_5$
Dibutyl sulfide ((Butylthio)butane)	$\underline{n}-C_4H_9SC_4H_9-\underline{n}$
Diallyl sulfide (3-(2-Propenylthio)propene)	$CH_2=CHCH_2SCH_2CH=CH_2$
Dimethyl disulfide ((Methyldithio)methane, Methyl disulfide)	CH_3SSCH_3
Diethyl disulfide ((Ethyldithio)ethane)	$C_2H_5SSC_2H_5$
Dipropyl disulfide ((Propyldithio)propane)	$\underline{n}-C_3H_7SSC_3H_7-\underline{n}$

DESCRIPTION	THRESHOLD VALUE	REM.	REF
	O 0.010 mg/m^3/air	F,M	99
	O 0.0045 mg/m^3/air		117
	O 0.31 mg/m^3/air		126
	*		23, 141, 142
	O 0.11 ppm/water		127
F in beer: Cooked vegetable (onion, garlic), H_2S	F 1.2 ppb/beer		25
	F 3 ppb/beer		103
	O 0.09 mg/m^3/air	F	99
F in beer: Rubber, rotten onion, sulfury	F 0.002 ppb/beer		25
O: Powerful, penetrating garlic-radish-like of poor tenacity Not a lachrymator, but produces some irritation of eyes and mucous membrane F: Green and sharper than diallyl disulfide		F,S	17
	O 0.00065 mg/m^3/air		99
O: Intensely onion-like, very diffusive, nonlachrymatory		F,M,S	17
O of GC eluates: Sulfurous, sickly, cooked cabbage			19
	O 0.003-0.014 mg/m^3/air		113
	*		23, 141, 142
	O 1.2 ppb/water		128
	O 0.16 ppb/water		20
	O 12 ppb/water		129
	F 30 ppb/water		1
	F 21 ppb/water		130
	F 0.06 ppb/water		32
	F 3 ppb/beer		103
	F 0.6 ppb/skim milk		32
	F 0.03 ppb/protein-hydrolysate solution 2%		32
	O 0.0003 mg/m^3/air	F,M	131
	*		23, 141, 142
	O 0.02 ppb/water		124
	F 30 ppb/water		1
F in beer: Garlic, burnt rubber, H_2S	F 0.4 ppb/beer		25
O: Very penetrating and powerful, diffusive, of garlic/onion type but with no lachrymatory effect. Poor tenacity. In extreme dilution more herbaceous-green than truly sulfuraceous.		F,S	17
F: Cooked onions	F 3.2 ppb/water		80

Continued on next page

NON-CYCLIC SULFUR COMPOUNDS (cont.)

NAME / (SYNONYM)	
Diallyl disulfide (3-(2-Propenyldithio)propene)	$CH_2=CHCH_2SSCH_2CH=CH_2$
Methyl propyl disulfide (1-(Methyldithio)propane)	$CH_3SSC_3H_7$-n
Methyl propenyl disulfide (1-(Methyldithio)propene)	$CH_3SSCH=CHCH_3$
Propyl propenyl disulfide (1-(Propyldithio)propene)	n-$C_3H_7SSCH=CHCH_3$
Dimethyl trisulfide (Methyl trisulfide, (Methyltrithio)methane)	CH_3SSSCH_3
Dipropyl trisulfide (1-(Propyltrithio)propane)	n-$C_3H_7SSSC_3H_7$-n
Methyl propyl trisulfide	$CH_3SSSC_3H_7$-n

DESCRIPTION	THRESHOLD VALUE	REM.	REF
O: Pungent, not truly reminiscent of garlic, but heavier, more sulfide-like, obnoxious. Taste - Only in extreme dilutions resembling that of garlic		F,S	17
O: Characteristic garlic			23
	O 0.0072 mg/m^3/air		99
O: Powerful penetrating sulfuraceous herbaceous of onion-type, but non-lachrymatory and not very tenacious. In extreme dilution sweet and more pleasant "natural" odor. F: Almost identical to its odor		F,S	17
F: Cooked onions			23
F: Cooked onions	F 6.3 ppb/water	F	80
F: Cooked onions	F 2.2 ppb/water	F	80
O: Powerful, diffusive, penetrating, reminiscent of fresh onions		F,M,S	23
O of GC eluates: Sulfurous, burnt, cooked cabbage			19
	O 0.0073 mg/m^3/air (recogn)		132
	O 0.01 ppb/water		129
	F 3 ppb/water		1
O: Very powerful and diffusive, garlic-like, penetrating and repulsive when undiluted, but in extreme dilution rather pleasant, sweet-herbaceous, distinctly garlic-like		F,S	17
O: Onion			83
O: Very powerful and penetrating, warm-herbaceous-oily of onion-like character, particularly upon dilution. F in conc. $<$ 5 ppm: Sweet, onion-like and warm-oily		F,S	17

Continued on next page

NON-CYCLIC SULFUR COMPOUNDS

NAME / (SYNONYM)	
Methanethiol (Methyl mercaptan)	CH_3SH
Ethanethiol (Ethyl mercaptan)	C_2H_5SH
1-Propanethiol (Propyl mercaptan)	$\underline{n}$-C_3H_7SH

DESCRIPTION	THRESHOLD VALUE	REM.	REF
O: Rotten cabbage, very diffusive and objectionable		F,M,S	17
O of GC eluates: Sickly, sulfurous, cooked cabbage			19
O: Sulfidy, pungent	O 2.1 ppb/air (recogn)		50
	O 0.081 mg/m^3/air		99
	O 0.0005 mg/m^3/air		133
	O 0.0002 mg/m^3/air		123
	O 0.003 mg/m^3/air		134
	*		23, 141, 142
	O 0.02 ppb/water		106
F in beer: Putrefaction (of egg, cabbage), drains, estery	F 2 ppb/water		25
O: Earthy, sulfidy	O 1.0 ppb/air (recogn)	F,M	50
	O 0.00066-0.0076 mg/ $/m^3$/air		99
	O 0.0001 mg/m^3/air		135
	O 0.00065 mg/m^3/air		136
	*		23, 141, 142
	O 0.016 mg/m^3/air		134
	F 0.008 ppb/water		32
F in beer: Putrefaction (of leek, onion, garlic, egg)	F 1.7 ppb/beer		25
	F 0.01 ppm/beer		98
	F 2 ppb/skim milk		32
	F 0.004 ppb/protein-hydrolysate solution 2%		32
O: Very powerful, penetrating and diffusive, cabbage-like, sulfuraceous, unpleasant unless extremely diluted F in conc. < 1 ppm: Sweet onion-cabbage-like, not unpleasant		F,M,S	17
	O 0.005 mg/m^3/air		99
	O 0.07-0.03 mg/m^3/air		137
	O 0.0016 mg/m^3/air		138
	O 0.018 mg/m^3/air		134
	*		23, 141, 142
	F 0.06 ppb/water		32
F in beer: Putrefaction (of onion, garlic, egg)	F 0.15 ppb/beer		25
	F 2 ppb/skim milk		32
	F 0.3 ppb/protein-hydrolysate solution 2%		32

Continued on next page

NON-CYCLIC SULFUR COMPOUNDS (cont)

NAME / (SYNONYM)	
1-Butanethiol (Butyl mercaptan)	$\underline{n}$-C_4H_9SH
2-Methyl-1-propanethiol (Isobutyl mercaptan)	i-C_4H_9SH
1-Hexanethiol (Hexyl mercaptan)	$\underline{n}$-$C_6H_{13}SH$
3-Methyl-1-butanethiol (Isopentyl mercaptan)	$CH_3CH(CH_3)CH_2CH_2SH$
3-Methyl-2-butene-1-thiol	$(CH_3)_2C{=}CHCH_2SH$
1-(Methylthio)-1-ethanethiol	$CH_3CH(SH)(SCH_3)$
1-(Methylthio)-1-propanethiol	$CH_3CH_2CH(SH)(SCH_3)$
1-(Methylthio)-1-hexanethiol	$\underline{n}$-$C_5H_{11}CH(SH)(SCH_3)$
1-(Ethylthio)-1-ethanethiol	$CH_3CH(SH)(SC_2H_5)$

DESCRIPTION		THRESHOLD VALUE	REM.	REF
O: Powerful, diffusive, reminiscent of cabbage. Often described as skunk-like or just sulfuraceous. In extreme dilution intensely sweet and non-descript. Taste slightly bitter			F,S	17
	O	0.0037 mg/m^3/air		99
	O	0.007-0.04 mg/m^3/air		137
	O	0.003 mg/m^3/air		138
	O	0.003 mg/m^3/air		139
	*			23, 141, 142
	O	6 ppb/water		51
	F	0.004 ppb/water		32
F in beer: Putrefaction (of onion, garlic, egg)	F	0.7 ppb/beer		25
	F	0.8 ppb/skim milk		32
	F	0.001 ppb/protein-hydrolysate solution 2%		32
	O	0.000002-0.0002 mg/ /m^3/air	F	135
	O	0.007 mg/m^3/air		134
	*			23, 141, 142
F in beer: Putrefaction (of onion, garlic, egg)	F	2.5 ppb/beer		25
O: Extremely powerful and diffusive			F,S	17
	O	0.00002 mg/m^3/air		139
	*			23, 141, 142
O: Repulsive, characteristic mercaptan-like			F,S	23
	O	0.0018 mg/m^3/air		99
Intense leek or onion-like- This compound is responsible for the skunky odor of "sunstruck beer"			M	7
O: Strong, meaty, onion-like			F,M	7
O in very aqueous solutions 1-5 ppb: Meaty rather than onion-like				140
	F	5 ppb/water		1
F dissolved in water: Allium (onion), meaty			M	89
F dissolved in water: Allium (onion) black currant, rhubarb and green bell pepper			M	89
F dissolved in water: Allium (onion, leek), black currant			M	89

Continued on next page

NON-CYCLIC SULFUR COMPOUNDS (cont.)

NAME / (SYNONYM)	
1-(Propylthio)-1-propanethiol	$C_2H_5CH(SH)(SC_3H_7)$
Bis(1-mercaptoethyl) sulfide	$CH_3CH(SH)SCH(SH)CH_3$
Bis(1-mercaptopropyl) sulfide	$C_2H_5CH(SH)SCH(SH)C_2H_5$
Bis(1-mercaptoisobutyl) sulfide	$i\text{-}C_3H_7CH(SH)SCH(SH)i\text{-}C_3H_7$
Bis(1-mercaptohexyl) sulfide	$\underline{n}\text{-}C_5H_{11}CH(SH)SCH(SH)C_5H_{11}\text{-}\underline{n}$
3-(Methylthio)-1-propanol (Methionol, methyl 3- -hydroxypropyl sulfide)	$CH_3S(CH_2)_3OH$
3-Methylthio-1-hexanol	$CH_3(CH_2)_2CH(SCH_3)(CH_2)_2OH$
3-(Methylthio)propanal [Methional, 3-(Methylthio)-propionaldehyde]	$CH_3SCH_2CH_2CHO$
3-(Methylthio)hexanal	$C_3H_7CH(SCH_3)CH_2CHO$
1-Methylthio-2-butanone	$C_2H_5COCH_2SCH_3$

DESCRIPTION	THRESHOLD VALUE	REM.	REF
F dissolved in water: Allium (onion), black currant		M	89
F dissolved in water: Allium (onion, chives), meaty		M	89
F dissolved in water: Green, fruity, Allium (onion, garlic)		M	89
F dissolved in water: Allium (onion), mushroom and soup		M	89
F dissolved in water: Allium (onion), green, fatty		M	89
O and F: Powerful, sweet, soup- or meat--like O: In undiluted form displays a rather repulsive odor with perceptible notes of sulfuraceous character. Extreme dilution causes a marked improvement to agreeable and pleasant, food-like note.		F,S	17
Green, fatty and sulfury note		F	7
O: Powerful and diffusive onion-meat--like, in dilution more pleasant, less onion-like, reminiscent of bouillon F 5 ppm: Pleasant, warm meat- or soup--like, at higher conc. with a slight "bite" or pungency		F,M,S	17
O 100°C or 180°C; Potato			18
O: Cooked-cabbage F: "Sunlight" flavor	F 0.05 ppm/water		130
	F 10 ppb/water		1
	O 0.2 ppb/water 0.2 ppb/oil		62
F in beer: Mashed potato, warm, soup-like	F 250 ppb/beer		25
F dissolved in water: Strongly green, privet		M	89
O: Reminiscent of mushroom, with a characteristic garlic undertone O: Mushroom-like		F,S	23 31

Continued on next page

NON-CYCLIC SULFUR COMPOUNDS (cont.)

NAME / (SYNONYM)	
4-Mercapto-4-methyl-2-pentanone	$CH_3C(SH)(CH_3)CH_2COCH_3$
Methyl 3-(methylthio)propionate	$CH_3SCH_2CH_2COOCH_3$
S-Methyl thiohexanoate	$CH_3SCO(CH_2)_4CH_3$
S-Methyl thioheptanoate	$CH_3SCO(CH_2)_5CH_3$

DESCRIPTION	THRESHOLD VALUE	REM.	REF
O: Unpleasant odor of cat urine		F	7
O: Extremely powerful, diffusive penetrating sweet, sulfuraceous, only in dilution becoming endurable and fruity-sweet. Its onion-like character in high conc. changes to "cooked"-fruity in minute conc. F: Fruity-sweet, pineapple-like in conc. < 5 ppm, most pleasant < 1 ppm. The flavor is clearly perceptible and agreeable well below 0.1 ppm		F,M,S	17
	O 0.3 ppb/water	F	123
	O 2 ppb/water	F	123

Literature Cited

1. Schutte, L. CRC Crit. Rev. Food Technol. 1974, 4, 457-505.
2. Nursten, H.E. Food Chem. 1980-81, 6, 263-277.
3. Vernin, G. Parfums, Cosmétiques, Arômes, 1980, (32, April-May), 77-89.
4. Maga, J.; Sizer, C. CRC Crit. Rev. Food Technol. 1973, 39-115.
5. Maga, J. J. Agric. Food Chem. 1981, 29, 691-694.
6. Ohloff, G; Flament, I. Heterocycles 2, 1978, 663-695.
7. Ohloff, G.; Flament, I. Fortschr. Chem Org. Naturst. 1978, 36, 231-283.
8. Shibamoto, T. J. Agric. Food Chem. 1980, 28, 237-243.
9. Suyama, K.; Adachi, S. J. Agric. Food Chem. 1980, 28, 546-549.
10. Maga, J.; Sizer, C. J. Agric. Food Chem. 1973, 21, 22-30.
11. Maga, J. CRC Crit. Rev. Food Sci. Nutr. 1982, 13, 1-48.
12. Maga, J. CRC Crit. Rev. Food Sci. Nutr. 1976, 7, 147-192.
13. Maga, J. CRC Crit. Rev. Food Sci. Nutr. 1975, 6, 241-270.
14. Maga, J. J. Agric. Food Chem. 1978, 26, 1049-1050.
15. Maga, J. CRC Crit. Rev. Food Sci. Nutr. 1981, 12. 295-307.
16. Maga, J. CRC Crit. Rev. Food Sci. Nutr. 1975, 6, 153-176.
17. Arctander, S. "Perfume and Flavor Chemicals", Vol. I, II; published by the author, Montclair, NJ, USA, 1969.
18. Hodge, J. "Chemistry and Physiology of Flavors"; Schultz, H.W.; Day, E.A.; Libbey, L.M., Eds.; AVI Publishing Co., Westport, Conn., 1967; pp. 465-491.
19. Persson, T.; von Sydow, E. J. Food Sci. 1973, 38, 377-385.
20. Mulders, J. Z. Lebensm. Unters.-Forsch. 1973, 151, 310-317.
21. Evans, C.D.; Moser, H.A.; List, G.R. J. Am. Oil Chem. Soc. 1971, 48, 495-498.
22. Baltes, W. Deut. Lebensm. Rundschau 1979, 75, 2-7.
23. "Fenaroli's Handbook of Flavor Ingredients", 2nd ed., Vol. I; Furia, T.E.; Bellanca, N., Eds. CRC Press, Inc., Cleveland, Ohio, 1975.
24. Buttery, R.; Seifert, R.; Guadagni, D.; Ling, L. J. Agric. Food Chem. 1969, 17, 1322-1327.
25. Meilgaard, M. Master Brew. Ass. Am. Techn. Quart. 1975, 12, (3), 151-168.
26. Shaw, P.; Tatum J.; Kew, T.; Wagner, C.Jr.; Berry, R. J. Agric. Food Chem. 1970, 18, 343-345.
27. Brulé, G.; Dubois, P.; Obretenov, T. Ann. Technol. Agric. 1971, 20, 305-312.
28. Sakaguchi, M.; Shibamoto, T. J. Agric. Food Chem. 1978, 26, 1260-1262.
29. Baltes, W. Lebensmittelchem. gerichtl. Chem. 1980, 34, 39-47.
30. Shigematsu, H.; Shibata, S.; Kurata, T.; Kato, H.; Fujimaki, M. J. Agric. Food Chem. 1975, 23, 233-237.
31. Ouweland, G.; Peer, H. J. Agric. Food Chem. 1975, 23, 501-505.
32. Golovnja, R.V.; Rothe, M. Nahrung 1980, 24, 141-154.
33. Rödel, W.; Kruse, H-P. Nahrung 1980, 24, 129-139.
34. Vernin, G. Ind. Aliment. Agric. 1980, 97, 433-449.

35. Vernin, G. Parfums, Cosmétiques, Arômes, 1979, (29, June-July), 77-87.
36. Mulders, E.; Kleipool, R.; ten Noever de Brauw, M.C. Chem. Ind. (London) 1976, 613-614.
37. Shibamoto, T. J. Agric. Food Chem. 1977, 25, 206-208.
38. Manley, C.; Wittack, M.; Fagerson, I. J. Food Sci. 1980, 45, 1096, 1098.
39. Pittet, A.; Rittersbacher, P.; Muralidhara, R. J. Agric. Food Chem. 1970, 18, 929-933.
40. Nunomura, N.; Sasaki, M.; Yokotsuka, T. Agric. Biol. Chem. 1980, 44, 339-351.
41. Pyysalo, T. "Studies on the Volatile Compounds of Some Berries in Genus Rubus, Especially Cloudberry (Rubus chamaemorus L.) and Hybrids between Raspberry (Rubus ideaeus L.) and Artic bramble (Rubus arcticus L.)". Materials and Processing Technology, Technical Research Centre of Finland, Publication No. 14, Helsinki, 1976.
42. Shaw, P.; Tatum, J.; Berry, R. J. Agric. Food Chem. 1968, 16. 979-982.
43. Shimazaki, H.; Tsukamoto, S.; Saito, T.; Eguchi, S.; Komata, Y. U.S. 4,013,800 (1977).
44. Keith, E.; Powers, J. J. Food Sci. 1968, 33, 213-218.
45. Lee, K.; Ho, C-T.; Giorlando, C.; Peterson, R.; Chang, S. J. Agric. Food Chem. 1981, 29, 834-836.
46. Tressl, R. Monatsschr. Brau. 1979, 32, (5), 240-246, 248.
47. Ho, C-T.; Jin, Q.Z.; Lee, K.; Carlin, J. J. Agric. Food Chem. 1982, 30, 362-364.
48. Ho, C-T.; Jin, Q.Z.; Lee, K.; Carlin, J. Lebensm.-Wiss.u.-Technol. 1982, 15, 169-170.
49. Yoshikawa, K.; Libbey, L.M.; Cobb, W.Y.; Day, E.A. J. Food Sci. 1965, 30, 991-994.
50. Leonardos, G.; Kendall, D.; Barnard, N. J. Air Pollut. Control Assoc. 1969, 19, 91-95.
51. Baker, R. J. Am. Water Works, Assoc. 1963, 55, 913-916.
52. Pittet, A.; Hruza, D. J. Agric. Food Chem. 1974, 22, 264-269.
53. Polak's Frutal Works N.V., Douwe Egberts Koninklijke Tobaksfabrik - Koffiebranderijen - Theehandel. Brit. 1,248,380 (1971).
54. Buttery, R.; Ling, L.; Teranishi, R.; Mon, T. J. Agric. Food Chem. 1977, 25, 1227-1229.
55. Teranishi, R.; Buttery, R.; Guadagni, D. "Geruch- und Geschmackstoffe", Int. Symp.; Bad Pyrmont, 1974; Drawert, F., Ed.; Verlag Hans Carl, Nürnberg, 1975; pp. 177-186.
56. Seifert, R.; Buttery, R.; Guadagni, D.; Black, D.; Harris, J. J. Agric. Food Chem. 1972, 20, 135-137.
57. Hunter, I.; Walden, M.; Scherer, J.; Lundin, R. Cereal Chem. 1969, 46, 189-195.
58. Calabretta, P. Perfumer Flavorist 1978, 3, June/July, 33-42.
59. Calabretta, P. Cosmetics Perfumery 1975, 90, June, 74, 76, 79-80.
60. Collins, E. J. Agric. Food Chem. 1971, 19, 533-535.
61. Koehler, P.E.; Mason, M.E.; Odell, G.V. J. Food Sci. 1971, 36, 816-818.

62. Guadagni, D.; Buttery, R.; Turnbaugh, J. J. Sci. Food Agric. 1972, 23, 1435-1444.
63. Seifert, R.; Buttery, R.; Guadagni, D.; Black, D.; Harris, J. J. Agric. Food Chem. 1970, 18, 246-249.
64. Takken, H.; van der Linde, L.; Boelens, M.; van Dort, J. J. Agric. Food Chem. 1975, 23, 638-642.
65. Sizer, C.E.; Maga, J.A.; Lorenz, K. Lebensm.-Wiss.u. -Technol. 1975, 8, 267-269.
66. Tressl, R.; Renner, R.; Kossa, T.; Köppler, H. European Brewery Convention, 16th Congress Proc. 1977, 693-707.
67. Deck, R.E.; Chang, S.S. Chem. Ind. (London) 1965, 1343-1344.
68. Fenaroli, G. Riv. Ital. E.P.P.O.S 1978, 60, 1-16.
69. Coleman, E.; Ho, C-T. J. Agric. Food Chem. 1980, 28, 66-68.
70. Boyko, A.; Morgan, M.; Libbey, L. "Analysis of Food and Beverages; Headspace Techniques," Charalambous, G. (Ed.), Academic Press: New York, 1978; pp. 57-79.
71. Kitamura, K.; Shibamoto, T. J. Agric. Food Chem. 1981, 29, 188-192.
72. Roberts, D. (R.J. Reynolds Tobacco Company), U.S. 3,402,051 (1968).
73. Parliment, T.; Epstein, M. J. Agric. Food Chem. 1973, 21, 714-716.
74. Murray, K.; Shipton, J.; Whitfield, F. Chem. Ind. (London) 1970, 897-898.
75. Masuda, H.; Yoshida, M.; Shibamoto, T. J. Agric. Food Chem. 1981, 29, 944-949.
76. Evers, W.; Katz, I.; Wilson, R.; Thelmer, E. (International Flavors and Fragrances, Inc.) U.S. 3,647,792 (1972).
77. Pittet, A.; Muralidhara, M.; Thelmer, E. (International Flavors and Fragrances, Inc.) U.S. 3,705,158 (1972).
78. Nishimura, O.; Mihara, S.; Shibamoto, T. J. Agric. Food Chem. 1980, 28, 39-43.
79. Arnold, R.; Libbey, L.; Lindsay, R. J. Agric. Food Chem. 1969, 17, 390-392.
80. Boelens, M.; de Valois, P.; Wobben, H.; van der Gen, A. J. Agric. Food Chem. 1971, 19, 984-991.
81. Ho, C.; Coleman, E. J. Food Sci. 1980, 45, 1094-1095.
82. Peterson, R.; Izzo, H.J.; Jungermann, E.; Chang, S. J. Food Sci. 1975, 40, 948-954.
83. Tokarska, B.; Karwowska, K. Nahrung 1981, 25, 565-571.
84. Walker, N.; Gray, I. J. Agric. Food Chem. 1970, 18, 346-352.
85. Tressl, R.; Bahri, D.; Holzer, M.; Kossa, T. J. Agric. Food Chem. 1977, 25, 459-463.
86. Shu, C-K.; Mookherjee, B.; Vock, M. (International Flavors and Fragrances, Inc.) U.S. 4,259,507 (1981).
87. Chang, S.S.; Hiriai, C.; Reddy, B.R.; Herz, K.O.; Kato, A. Chem. Ind. (London) 1968, 1639-1641.
88. Kubota, K.; Kobayashi, A.; Yamanishi, T. Agric. Biol. Chem. 1980, 44, 2677-2682.
89. Boelens, M.; van der Linde, L.; de Valois, P.; van Dort, H.; Takken, H. J. Agric. Food Chem. 1974, 22, 1071-1076.

90. Lee, M-H.; Ho, C-T.; Chang, S. J. Agric. Food Chem. 1981, 29, 686-688.
91. Ho, C-T.; Tuorto, R. J. Agric. Food Chem. 1981, 29, 1306-1308.
92. Mussinan, C.J.; Wilson, R.A.; Katz, I.; Hruza, A.; Vock, M.H. ACS Symp. Series 1976, 26, 133-143.
93. Hetzel, D.S. (Pfizer Inc.) Ger. Offen. 2,120,577 (1971); Chem. Abstr. 76 (1972), 72505.
94. Kazeniac, S.; Hall, R. J. Food Sci. 1970, 35, 519-530.
95. Buttery, R.; Seifert, R.; Guadagni, D.; Ling, L. J. Agric. Food Chem. 1971, 19, 524-529.
96. Buttery, R.; Guadagni, D.; Lundin, R. J. Agric. Food Chem. 1976, 24, 1-3.
97. Vock, M.; Giacino, C.; Hruza, A.; Withycombe, D.; Mookherjee, B.; Mussinan, C. (International Flavors and Fragrances, Inc.) U.S. 4,243,688 (1981).
98. Palamand, S.R.; Hardwick, W.A. Master Brew. Ass. Am. Techn. Quart. 1968, 6, 117-128.
99. Katz, S.H.; Talbert, E.J. U.S. Bureau of Mines, Technical Report No. 480 (1930).
100. Hartung, L.D.; Hammond, E.G.; Miner, J.R. Livest. Waste Manage. Pollut. Abatement, Proc. Int. Symp., 1971, 105-106.
101. Flath, R.A.; Black, D.R.; Guadagni, D.G.; Mc Fadden, W.H.; Schultz, T.H. J. Agric. Food Chem. 1967, 15, 29-35.
102. Amoore, J.E.; Forrester, L.J.; Pelosi, P. Chem. Senses Flavor, 1976, 2, 17-25.
103. Harrison, G. J. Inst. Brew. 1970, 76, 486-495.
104. Hellman, T.M.; Small, F.H. Chem. Eng. Progr. 1973, 69, 75-77.
105. Hellman, T.M.; Small, F.H. J. Air Pollut. Contr. Assoc. 1974, 24, 979-982.
106. Guadagni, D.; Buttery, R.; Okano, S. J. Sci. Food Agric. 1963, 14, 761-765.
107. Amoore, J.E. Chem. Senses Flavor 1977, 2, 267-281.
108. van Anrooij, A. "Desodorisatie langs photochemischen weg", Thesis, Utrecht, 1931.
109. Appel, L. Am. Perfum. Cosmet. 1969, 84, 45-50.
110. Sega, G.M.; Lewis, M.J.; Woskow, M.H. Proc. Am. Soc. Brew. Chem. 1967, 156-164.
111. Stahl, W.H. "Compilation of Odor and Taste Threshold Values Data". ASTM Data Series DS 48, 1973.
112. Henning, H. "Der Geruch", 2nd Ed. Barth, Leipzig, 1924.
113. Lindvall, T. Nord. Hyg. Tidskr. 1970, 51, Suppl. 2, 1-181.
114. Randebrock, E.M. "Gustation and Olfaction"; Ohloff, G.; Thomas, A.F., Eds.; Academic Press: London/New York, 1971; pp. 111-125.
115. Adams, D.F.; Young, F.A.; Luhr, R.A. Tappi 1968, 51, 62A - 67A.
116. Pippen, E.L.; Mecchi, E.P. J. Food Sci. 1969, 34, 443-446.
117. Deadman, K.A.; Prigg, J.A. Research Communication GC 59, The Gas Council, London, 1959.

118. Hildenskiold, R.S. Gig. Sanit. 1959, 24, (6), 3-8 (in Russ.).
119. Baikov, B.K. Gig. Sanit. 1963, 28, (3), 3-8 (in Russ.).
120. Laffort, P. "Theories of Odor and Odor Measurement; Tanyolac, N.N., Ed; Robert College: Bebek, Istanbul, 1968; pp. 247-270.
121. Laffort, P. Olfactologica 1968, 1, 95-104.
122. Laffort, P.; Dravnieks, A. J. Theor. Biol. 1973, 38, 335-345.
123. Guadagni, D.; Buttery, R.; Harris, J. J. Sci. Food Agric. 1966, 17, 142-144.
124. de Grunt, F.E. National Institute for Water Supply. Voorburg, Netherlands, 1975, (unpublished data).
125. Patton, S.; Forss, D.A.; Day, E.A. J. Dairy Sci. 1956, 39, 1469-1470.
126. Ifeadi, N.G. "Quantitative measurement and sensory evaluation of dairy waste odor". Thesis, Ohio State Univ., Columbus, 1972.
127. Kauffmann, M. Z. Sinnesphysiol. 1907, 42, 271-280.
128. Amoore, J.E.; Venstrom, D. J. Food Sci. 1966, 31, 118-128.
129. Buttery, R.; Guadagni, D.; Ling, L.; Seifert, R.; Lipton, W. J. Agric. Food Chem. 1976, 24, 829-832.
130. Patton, S. J. Dairy Sci. 1954, 37, 446-452.
131. Zwaardemaker, H. "Handbuch der physiologischen Metodik III;" Vol. 1, Tigerstedt, R. Ed.; S. Hirzel, Leipzig, 1914, pp. 46-108.
132. Wilby, F.V. J. Air Pollut. Control Assoc. 1969, 19, 96-100.
133. Freudenberg, K.; Reichert, M. Tappi, 1955, 38, 165A-166A.
134. Blanchard, J. Gas (Rome) 1976, 26, 287-309.
135. Stuiver, M. "Biophysics of the sense of smell." Thesis, Groningen, 1958.
136. Endo, R.; Kohgo, T.; Oyaki, T. "10. Research on odor nuisance in Hokkaido (part 2). Chemical analysis of odors." Proc. 8th Annual Meeting Japan Society of Air Pollution, Symposium (1): Hazard nuisance pollution problems due to odors (English translation for APTIC by SCITRAN), 1967.
137. Blinova, A.E. Hyg. Sanit. USSR. 1965, 30, 18-22.
138. Kniebes, D.V.; Chisholm, J.A.; Stubbs, R.C. Gas (Rome) 1969, 19, 211-217.
139. Patte, F. "Le Codage Moléculaire de la Stimulation Olfactive - Clarifications et mises au point." Thesis, Lyon, 1978.
140. Brinkman, H.; Copier, H.; de Leuw, J.M.; Sing, Boen Tjan. J. Agric. Food Chem. 1972, 20, 177-181.
141. van Gemert, L.J.; Nettenbreijer, A.H., Eds. "Compilation of Odor Threshold Values in Air and Water". National Institute for Water Supply, Voorburg, Netherlands. Central Institute for Nutrition and Food Research, TNO, Zeist, Netherlands, 1977.
142. van Gemert, L.J.; Belz, R.; van Straten, S.; Jetten, J. "Compilation of Odor Threshold Values in Air." Supplement III. Report No. R. 6633. Central Institute for Nutrition and Food Research, TNO, Zeist, Netherlands, 1980.

RECEIVED December 14, 1982

13

Mechanism Responsible for Warmed-Over Flavor in Cooked Meat

A. M. PEARSON and J. I. GRAY

Michigan State University, Department of Food Science and Human Nutrition, East Lansing, MI 48824

Studies using a model meat system have demonstrated that phospholipids are the primary contributors to warmed-over flavor (WOF) in cooked meat. Although phosphatidylcholine has little effect upon WOF, phosphatidylethanolamine plays an important role which is related to the propensity of its constituent polyunsaturated fatty acids to autoxidation. Both nitrite and EDTA inhibit WOF as does removal of heme pigments, which suggests that myoglobin may catalyze WOF development. However, purified myoglobin in the model system does not cause autoxidation. It has been demonstrated that WOF is catalyzed by the non-heme iron released from meat pigments during cooking. Evidence has indicated that overheating of meat protects against WOF by producing Maillard reaction products possessing antioxidant activity.

The term "warmed-over flavor" (WOF) was coined by Tims and Watts (1) to describe the rapid development of oxidized flavor in cooked meat upon subsequent holding. The rancid or stale flavor becomes readily apparent within 48 h in contrast to the more slowly developing rancidity that becomes evident only after freezer storage for a period of months. Although WOF was first recognized as occurring in cooked meat, hence the name WOF, it also develops in raw meat that is ground and exposed to air (2, 3). With increased consumption of precooked convenience meat entrees, such as quick-frozen meat dishes and TV dinners, preventing the development of WOF has become of great economic importance (4).

The present review will cover evidence on the mechanism involved in WOF development, with description of a model system that has been utilized to clarify some of the pathways. The role of meat pigments and various lipid components will be reviewed. Finally some procedures for inhibiting WOF in cooked

0097-6156/83/0215-0287$06.00/0

meats will be discussed. Although an earlier review covers some of these points (5), considerable new information is now available and will be presented herein.

Description of Model Meat System

Love and Pearson (6) have described a model meat system in which bovine muscle was ground and extracted with distilled, deionized water at 4°C until it was devoid of color, indicating the removal of all meat pigments, i.e., myoglobin and hemoglobin. Other water-soluble components would also be partially or completely extracted by this procedure. The remaining extracted muscle was then used as a model system to which purified myoglobin, ferrous iron and ferric iron were added back to ascertain their role in WOF.

Essentially the same system has since been used in our laboratory to study the role of phospholipids and of triglycerides in development of WOF using beef and chicken dark and white meat (7). Figure 1 shows not only how the system was prepared for studying the effect of adding back myoglobin but also how chloroform-methanol was utilized to extract the total lipids (8) and then how triglycerides and phospholipids could be separated by silicic acid column chromatography with elution by chloroform and methanol (9).

Mechanism of WOF Development

Role of Myoglobin and Non-Heme Iron. The model meat system containing the indigenous lipids was used to obtain the data presented in Table I. Oxidation was followed by 2-thiobarbituric acid (TBA) numbers obtained using the distillation procedure of Tarladgis *et al.* (10). The data show the effects of different concentrations of metmyoglobin (MetMb) and ferrous iron (Fe^{2+}) upon TBA numbers in the model beef muscle system (6). MetMb concentration had no effect upon autoxidation, whereas Fe^{2+} increased oxidation of the model system with the extent being directly related to the concentration. Thus, it was shown that MetMb does not catalyze oxidation of the model system, but Fe^{2+} causes rapid autoxidation of the residual lipids. These results are in agreement with an earlier report (3), in which both Fe^{2+} and ascorbate were shown to catalyze development of WOF. Thus, it has been clearly shown that myoglobin is not directly responsible for development of WOF, although this concept has been widely accepted (11-14).

A subsequent study (15) using the model meat system showed that removal of the heme pigments or addition of 156 ppm of nitrite significantly inhibited lipid oxidation in cooked meat, which supports the earlier concept that myoglobin (Mb) is involved in WOF. However, it should be borne in mind that removal of heme pigments by leaching will also remove a number

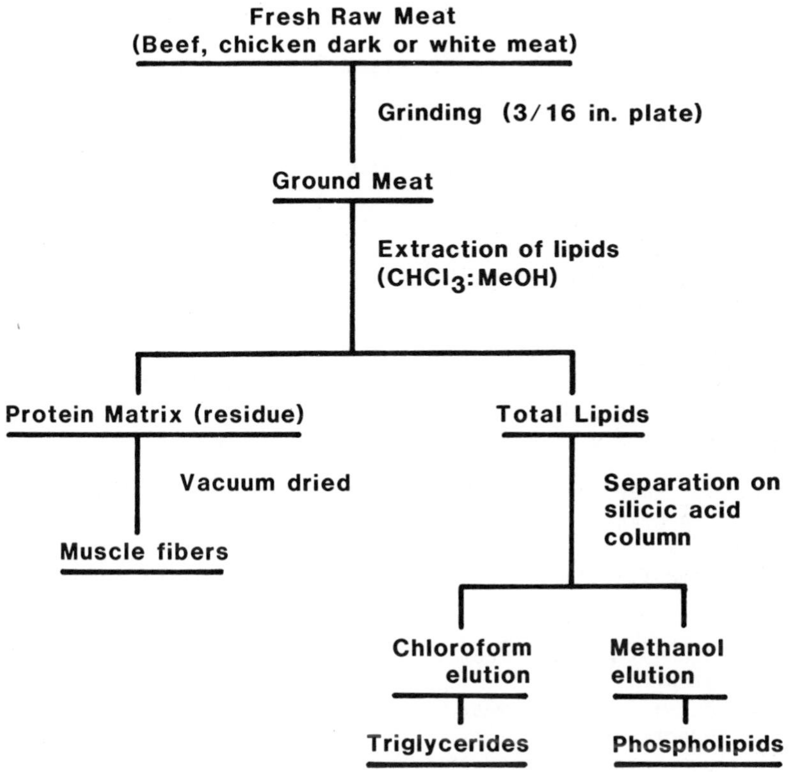

Figure 1. Preparation of model meat system. (Reproduced from Ref. 7. Copyright 1979, American Chemical Society.)

of other compounds including Fe^{2+}. By using the scheme shown in Figure 2, Igene *et al.* (15) demonstrated that the level of free Fe^{2+} greatly increased during cooking, and accelerated

Table I. Effects of Various Concentrations of MetMb and Fe^{2+} on TBA Numbers of Beef Muscle Residue[a]

MetMb		Fe^{2+}	
Conc., mg/g	TBA no.[b]	Conc., ppm	TBA no.[b]
0	1.4	0	1.0
1.0	1.4	1.0	1.5
2.5	1.4	2.0	3.6
5.0	1.4	3.0	4.1
10.0	1.4	4.0	5.4

[a]The reactants were mixed, heated, and stored for 48 h at 4°C before measuring the TBA values. [b]TBA numbers = mg of malonaldehyde produced per 1000 g of meat. Taken from Love and Pearson (6).

lipid oxidation in cooked meat. This indicates that Mb serves as a source of Fe^{2+}, being readily broken down during the cooking process and catalyzing autoxidation. The data showing these effects are summarized in Table II.

Table II. Role of Heme and Non-heme Iron on Development of TBA Numbers in Cooked Beef.[a,b]

Experimental treatment	Mean TBA no.
residue + total raw meat pigment	5.00
residue + total raw meat pigment (chelated)	1.55
residue + total cooked free meat pigment	4.35
residue + total cooked free meat pigment (chelated)	1.46
residue + H_2O_2-treated total meat pigment	6.02
residue + H_2O_2-treated meat pigment (chelated)	1.54

[a]Each experimental treatment used 100 g of beef residue in addition to 50 mL of the concentrated extract. [b]EDTA was used to chelate the inorganic or free iron at a concentration of 2%. Taken from Igene *et al.* (15).

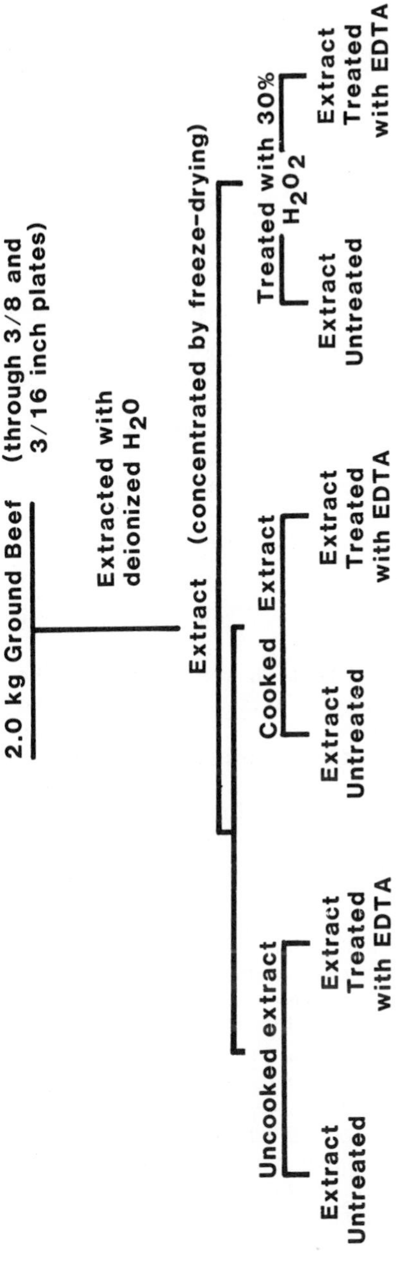

Figure 2. Preparation and design of experiment to compare the effect of heme and non-heme iron on the development of warmed-over flavor. (Reproduced with permission from Ref. 16.)

It is interesting to also note that treatment with H_2O_2 destroys even more of the heme pigments than heating, resulting in still greater oxidation of the meat system (Table III). On the other hand, addition of 2% EDTA markedly reduced autoxidation as shown by much lower TBA numbers.

Table III. Concentrations of Total Iron, Heme Iron, and Free Non-heme Iron in Treated and Untreated Meat Pigment Extract

Experimental treatment	Amount of free Fe^{2+}, μg/g of meat
total iron in fresh meat pigment extract	20.64
non-heme iron in fresh meat pigment extract	1.80
heme iron in fresh meat pigment extract	18.84
total iron in cooked meat pigment filtrate	5.51
free non-heme iron in cooked pigment filtrate	4.18
total iron in H_2O_2-treated meat pigment extract	13.59
free iron in H_2O_2-treated meat pigment extract (chelated[a])	12.33

[a]Chelated by adding 2% EDTA. Taken from Igene *et al.* (15).

Results demonstrate that Mb *per se* is not the catalyst of lipid oxidation in cooked meat. However, cooking destroys part of the Mb, releasing Fe^{2+} which then catalyzes the development of WOF. Although the role of grinding in development of WOF was not studied, it seems likely that it also releases Fe^{2+}. It has recently been shown that solubilization of iron from grinding equipment can increase the free iron content of fish meal, which could also be a factor in autoxidation of fresh ground meat (16).

Role of Phospholipids and Triglycerides Circumstantial evidence has suggested that phospholipids are the major contributors to WOF, with the correlation coefficients between TBA numbers and phospholipid levels being higher than those for TBA numbers and total lipids (17). On adding phospholipids, triglycerides, and total lipids back to a model meat system, it was verified that phospholipids were the major contributors to WOF (7). This conclusion is supported by the data given in Table IV, which demonstrate that the same principles are involved in development of WOF in beef and in both chicken dark and light meat. The data in the same table also show a relationship between sensory scores for WOF and TBA numbers. Although the addition of triglycerides resulted in some increase in TBA numbers, the effect was always considerably lower than

Table IV. TBA Numbers and Sensory Scores for Cooked Beef, Chicken Dark Meat and White Meat Model Systems[a,b]

Meat type	Composition of treatments	Mean TBA no.	Mean sensory scores[c]
beef	0.8% phospholipids	5.76[b]	2.69[a]
	9.2% triglycerides	1.88[a]	3.84[b]
	10% total lipids	6.81[b]	2.64[a]
	control	1.16[a]	3.75[b]
chicken dark meat	0.74% phospholipids	8.48[b]	2.97[a]
	4.3% triglycerides	6.30[a,b]	3.65[a,b]
	5% total lipids	11.74[c]	3.25[a]
	control	5.78[a]	3.88[a,b]
chicken white meat	0.7% phospholipids	5.03[b]	2.80[a]
	4.3% triglycerides	2.99[a]	3.35[a]
	5% total lipids	5.53[b]	3.09[a]
	control	2.18[a]	4.02[b]

[a]There were four replicates for each treatment.
[b]Taste panel score were from 1-5, with 1 very pronounced WOF and 5 no WOF.
[c]All numbers in same column within a meat type followed by the same superscript are not significant at the 5% level. Taken from Igene and Pearson (7).

that of the phospholipids. The additive effect of triglycerides was, however, greater in chicken dark meat than for chicken white meat or beef. This is probably due to the greater degree of unsaturation in the triglycerides from chicken dark meat.

Table V presents information on the relative effects of adding back phosphatidylcholine (PC), phosphatidylethanolamine (PE), and total blood serum phospholipids (TP) to the model meat system. Nitrite was also added but will be discussed later herein. The addition of both PE and TP increased TBA numbers, with the effect of PE being the greatest. Sensory scores also indicated that oxidation of PE, TP, and PC contributed to WOF, with PE again having the greatest effect.

Table V. TBA Numbers and Sensory Scores for Cooked Meat Model Systems Containing Added PC, PE, Serum Phospholipids, or Nitrite[a,b]

Experimental treatment	Mean TBA no.	Mean sensory score
model system only	0.36 $\pm$ 0.07[b]	3.33 $\pm$ 0.42[a,b]
model system + nitrite	0.26 $\mp$ 0.06[a]	4.11 $\mp$ 0.44[b,c]
model system + PC	0.34 $\mp$ 0.03[a,b]	3.36 $\mp$ 0.51[a,b]
model system + PC + nitrite	0.29 $\mp$ 0.07[a]	4.52 $\mp$ 0.27[c]
model system + PE	0.81 $\mp$ 0.13[d]	2.64 $\mp$ 0.58[a]
model system + PE + nitrite	0.36 $\mp$ 0.07[b]	3.61 $\mp$ 0.23[b]
model system + TP	0.62 $\mp$ 0.12[c]	3.29 $\mp$ 0.49[a,b]
model system + TP + nitrite	0.34 $\mp$ 0.05[a,b]	4.04 $\mp$ 0.43[b,c]

[a]Each treatment was replicated four times.
[b]A significant P<0.01) "r" value of -0.62 was found between TBA numbers and sensory score. Numbers in the same column bearing same letter are not significant at 5% level. Take from Igene and Pearson (7).

In summary, it has been shown that phospholipids are the major contributors to development of WOF, with triglycerides playing only a minor role. Studies on the individual phospholipids have demonstrated that PE is the main phospholipid involved in development of WOF.

Comparison of WOF and Normal Oxidation

It is well known that meat lipids are stable during freezer storage of fresh meats for periods of up to 12 months or more (5). Recent work has shown that α-tocopherol levels declined steadily during storage and had largely disappeared by the end of 6 months in freezer storage (18). In contrast to the involvement of phospholipids in development of WOF in cooked meats, Igene et al. (19) have recently shown that the major changes occurring during frozen storage were due to losses in the triglyceride fraction. The phospholipids remained relatively constant during frozen storage for up to 18 months. The extent of deterioration in the triglycerides has been shown to be related to the degree of unsaturation as well as to the length of storage (20).

Although there is some decline in the constituent phospholipids (PC and PE) during frozen storage of uncooked meat, the decline is much greater in cooked meat (21). Drippings collected upon cooking contain primarily triglycerides, while PE is essentially absent suggesting that it is membrane-bound (21). Since there is an increase in the proportion of PE in cooked meat, this along with its propensity to oxidize may account for the faster oxidation in cooked meat, even during freezer storage. Thus, the faster deterioration of cooked meat is evident and even occurs during freezer storage. These studies help in explaining the relative roles of triglycerides and phospholipids in autoxidation of cooked meat and have bearing on our understanding of WOF.

Inhibition of WOF

Phosphate, ascorbate, NaCl, nitrite, Maillard reaction products, and other antioxidants or prooxidants have been reported to influence development of WOF. Their roles have been reviewed (5), but newer evidence is now available and will be covered herein.

Role of Phosphates Tims and Watts (1) showed that addition of phosphates to cooked meat protects against autoxidation. This was found to be the case for pyro-, tripoly-, and hexametaphosphate, but orthophosphate had no antioxidant effect. Sato and Hegarty (3) verified the effects of phosphates in retarding the development of WOF by showing that pyro-, tripoly-, and hexametaphosphate protected cooked ground beef against WOF during storage at 2°C.

The phosphates appear to prevent autoxidation by chelating metal ions, although the exact mechanism is not known. Other work, however, has shown that metal chelators do protect meat against autoxidation (1, 22). EDTA has been shown to inhibit development of WOF by chelating the non-heme iron in cooked meat (15). Thus, evidence suggests that the phosphates probably chelate Fe^{2+} and thus inhibit WOF.

Role of Ascorbates At low levels (< 100 ppm) ascorbic acid has been shown to catalyze development of WOF as shown by increased TBA values (1, 3). At higher levels (>1,000 ppm), however, ascorbic acid retards oxidation (3). Although the mechanism of retardation is not known, it has been suggested that high levels of ascorbates may upset the balance between

Fe^{2+} and Fe^{3+} or else could have its effect by acting as an oxygen scavenger (5).

Tims and Watts (1) showed that a combination of ascorbates and phosphates acted synergistically to retard development of rancidity. Sato and Hegarty (3) verified the antioxidant activity of the combination and suggested that ascorbic acid acts by keeping part of the iron in the Fe^{2+} state. It has been shown that ascorbic acid and phosphates act synergistically in preventing oxidation of cured meat (23), and probably help in explaining the virtual absence of WOF in cured meats.

Role of NaCl The activity of NaCl in initiating color and flavor changes in meat is well known but poorly understood. The effect is further complicated by the fact that NaCl is both a prooxidant and an antioxidant, depending upon the concentration dissolved (24). Lea (25) has discussed the fact that salt may be a prooxidant in some cases and either have no effect or be an antioxidant in other foods. Watts (26) concluded that salt may have antioxidant activity in dilute solutions but on crystallization has a prooxidant effect. On the other hand, Mabrouk and Dugan (24) found that salt inhibited autoxidation in aqueous emulsions of methyl linoleate, its effectiveness as an antioxidant being directly associated with increasing concentrations. They suggested that dissolved oxygen may be eliminated from the system as the NaCl concentration is increased.

Some of the studies on salt are, no doubt, complicated by the fact that salt may contain metal contaminants, which could serve as catalysts of lipid oxidation. Nevertheless, rancidity may still develop in the fat of dry cured hams, even though salt with a low metal content is used (27). The use of an antioxidant in combination with such salt, however, did inhibit rancidity and improve flavor scores. Further work to clarify the role of salt in lipid oxidation is needed before its mechanism is fully understood.

Role of Nitrite Sato and Hegarty (3) reported that 2000 ppm of nitrite completely eliminated WOF, while as little as 50 ppm greatly inhibited its development. Bailey and Swain (28) further confirmed the effectiveness of nitrite in preventing oxidation of fresh meat stored under refrigeration and verified its role in preventing WOF. A concentration of 156 ppm of nitrite has been shown to inhibit WOF development in cooked meat, with a twofold reduction of TBA values for beef and chicken and a fivefold reduction for pork (29). Table V also demonstrates that nitrite inhibits WOF development. Undoubtedly these results explain the higher flavor scores for nitrite as compared to NaCl-containing cured pork (30).

The mechanism by which nitrite prevents or inhibits WOF has been suggested to be related to stabilization of the lipids in the membrane (3, 5), which are normally disrupted and exposed to oxygen by cooking or grinding. Zipser et al. (31) proposed that nitrite forms a stable complex with the iron porphyrins of heat-denatured meat, thus inhibiting WOF. Kanner (32) has demonstrated that S-nitrosocysteine is a potent antioxidant and has suggested that it may be generated in the nitrite curing of meat. Thus, S-nitrosocysteine could also serve as an inhibitor of WOF in cured meat. Since non-heme iron has been shown to be the major lipid prooxidant in uncured, heated meat systems (3, 6, 15), it seems more probable that nitrite stabilizes the heme pigments so that they do not release Fe^{2+} and thus catalyze development of WOF. The inhibition of WOF by EDTA lends further credence to this theory, which as yet is unproven.

Role of Maillard Reaction Products Certain products of the Maillard reaction are known to have antioxidant properties and are produced during retorting of meat. Zipser and Watts (33) first described the development of an antioxidative effect in overcooked meat and suggested that diluted slurries could be used to protect normally cooked meats from oxidation. Sato et al. (34) demonstrated that retorted meat, indeed, possessed strong antioxidant activity against development of WOF. They then demonstrated that the material retained by dialysis of extracts of the retorted meat had no antioxidant activity but that the diffusate possessed strong inhibitory activity against WOF. This suggested that the substances responsible for inhibiting WOF in retorted meat are water soluble and of low molecular weight. Huang and Greene (35) confirmed these findings and suggested that a temperature of about 90°C is required to produce the browning compounds responsible for the antioxidant activity in cooked meat. They further showed that there was a relationship between antioxidant activity and development of the brown color, which was accompanied by an increase in fluorescence.

Porter (36) has reviewed the role of Maillard reaction products as antioxidants in food systems, pointing out the importance of high temperatures (~100°C) to their development in contrast to lower temperatures (~70°C), which accelerate development of WOF. Eichner (37) has shown that browning intermediates -- primarily Amadori rearrangement products -- have strong antioxidant activity even though they are colorless. The mechanism by which these reductone and reductone-like compounds inhibit autoxidation seems to by decomposing hydroperoxides and inactivation of free radicals (37). Sato and Herring (38) have reported that reductic acid and maltol, which are browning products, possess strong antioxidant activity against WOF. Ascorbic acid, a related reductone, has been reported to accelerate WOF at concentrations of 100 ppm or less but to inhibit the reaction at levels of 1,000 ppm or over (3). Lingnert and Eriksson (39) have demonstrated that Maillard reaction products inhibit oxidation in emulsion-type sausages.

The studies reviewed demonstrate that browning products produced on retorting of meat inhibit development of WOF, so that canned meat products are not subject to this flavor defect. The flavor of canned meat is less desirable, however, than that of freshly cooked meat. Nevertheless, the strong inhibitory action of the Maillard reaction products against WOF suggests that they could be useful in preventing development of WOF, so further research in this area could be fruitful.

Role of Other Antioxidants A great deal has been written about the primary antioxidants, which may be used to inhibit development of WOF. Porter (36) has reviewed these compounds, their structures, and probable modes of action so they will not be discussed. The discussion will focus on the other natural antioxidants that may be useful in controlling WOF development.

Sato *et al.* (34) demonstrated that a variety of common meat additives, including cottonseed flour, nonfat dry milk, spray-dried whey, wheat germ, and textured soy flour, inhibited WOF in the meat system. These products may have exerted their inhibitory effect on WOF through the Maillard reaction, since most of them contain some reducing sugars. Pratt (40) reported soybeans and soy protein concentrate had an inhibitory effect upon development of WOF and was able to demonstrate that the active components are water soluble. Fractionation and analysis of the water-soluble fraction showed the antioxidant activity was due to the presence of isoflavones and hydroxylated cinnamic acids (40). This confirms earlier work showing that the flavonoids present in plant extracts inhibit oxidation in sliced roast beef (41).

The addition of smoke to meat also imparts antioxidant properties (42). This has also been shown on application of smoke to fatty fishes (43). Porter (36) has reviewed the antioxidant properties of smoke and concluded that a number of phenolic compounds are responsible. Several of these phenolic components, including phenol, guaiacol and catechol, are not only potent antioxidants but also have antibiotic activity. These components would also serve as inhibitors of WOF, but some of the components of wood smoke are known carcinogens (36, 42).

Summary

The use of model meat systems has been shown to be helpful in elucidating the role of different compounds in development of WOF. Results of these studies have shown that the major catalyst of WOF is Fe^{2+}, which is released from the heme pigments by heating, and presumably by grinding. The phospholipids are the major lipids involved in development of WOF, in contrast to oxidation during frozen storage, where the triglycerides are mainly involved. The role of phosphates, nitrite, salt, ascorbate, other additives, and of Maillard reaction products are discussed in light of their role as possible inhibitors of WOF. These active inhibitors of WOF could play an important role in protecting pre-cooked meat entrees against WOF.

Literature Cited

1. Tims, M.J.; Watts, B.M. Food Technol. 1958, 12, 240-243.
2. Greene, B.E. J. Food Sci. 1969, 34, 110-113.
3. Sato, K.; Hegarty, G.R. J. Food Sci. 1971, 36, 1098-1102.
4. Love, J.D.; Pearson, A.M. J. Am. Oil Chem. Soc. 1971, 48, 547-549.
5. Pearson, A.M.; Love, J.D.; Shorland, F.B. Adv. Food Res. 1977, 34, 1-74.
6. Love, J.D.; Pearson, A.M. J. Food Sci. 1974, 22, 1032-1034.
7. Igene, J.O.; Pearson, A.M. J. Food Sci. 1979, 44, 1285-1290.
8. Folch, J.; Lees, M.; Stanley, G.H.S. J. Biol. Chem. 1957, 226, 497-509.
9. Choudhury, R.B.R.; Arnold, L.K. J. Am. Oil Chem. Soc. 1960, 37, 87-88.
10. Tarladgis, B.G.; Watts, B.M.; Younathan, M.T.; Dugan, L.R. J. Am. Oil Chem. Soc. 1960, 37, 44-48.
11. Tappel, A.L. "Symposium on Foods: Lipids and Their Oxidation," Schultz, H.W.; Day, E.A.; Sinnhuber, R.O., Eds.; Academic Press; New York, 1962; pp. 122-138.
12. Watts, B.M. "Symposium on Foods: Lipids and Their Oxidation," Schultz, H.W.; Day, E.A.; Sinnhuber, R.O., Eds.; Academic Press; New York, 1962; pp. 202-214.
13. Wills, E.D. Biochem. J. 1966, 99, 667-676.
14. Liu, H.P.; Watts, B.M. J. Food Sci. 1970, 35, 596-598.
15. Igene, J.O.; King, J.A.; Pearson, A.M.; Gray, J.I. J. Agric. Food Chem. 1979, 27, 838-842.
16. Soevik, T.; Opstvedt, J.; Braekan, O.R. J. Sci. Food Agric. 1981, 32, 1063-1068.
17. Wilson, B.R.; Pearson, A.M.; Shorland, F.B. J. Agric. Food Chem. 1976, 24, 7-11.
18. Shorland, F.B.; Igene, J.O.; Pearson, A.M.; Thomas, J.W.; McGuffey, R.K.; Aldridge, A.E. J. Agric. Food Chem. 1981, 28, 863-871.
19. Igene, J.O.; Pearson, A.M.; Merkel, R.A.; Coleman, T.H. J. Anim. Sci. 1979, 49, 701-707.
20. Igene, J.O.; Pearson, A.M.; Dugan, L.R. Jr.; Price, J.F. Food Chem. 1980, 5, 263-276.
21. Igene, J.O.; Pearson, A.M.; Gray, J.I. Food Chem. 1981, 7, 289-303.
22. Watts, B.M. J. Am. Oil Chem. Soc. 1950, 27, 48-51.
23. Chang, I.; Watts, B.M. Food Technol. 1949, 3, 332-336.
24. Mabrouk, A.F.; Dugan, L.R., Jr. J. Am. Oil Chem. Soc. 1960, 37, 486-490.

25. Lea, C.H. "Rancidity in Edible Fats;" Chemical Publishing Co.: New York, 1939.
26. Watts, B.M. Adv. Food Res. 1954, 5, 1-52.
27. Olson, D.G.; Rust, R.E. J. Food Sci. 1973, 38, 251-253.
28. Bailey, M.E.; Swain, J.W. Proc. Meat Ind. Res. Conf. 1973, pp. 29-46.
29. Fooladi, M.H.; Pearson, A.M.; Coleman, T.H.; Merkel, R.A. Food Chem. 1979, 3, 283-292.
30. Cho, I.C.; Bratzler, L.J. J. Food Sci. 1970, 35, 668-670.
31. Zipser, M.W.; Kwon, T.W.; Watts, B.M. J. Agric. Food Chem. 1964, 12, 105-109.
32. Kanner, J. J. Am. Oil Chem. Soc. 1979, 56, 74-76.
33. Zipser, M.W.; Watts, B.M. J. Food Sci. 1961, 15, 445-447.
34. Sato, K.; Hegarty, G.R.; Herring, H.K. J. Food Sci. 1973, 38, 398-403.
35. Huang, W.H.; Greene, B.E. J. Food Sci. 1978, 43, 1201-1203.
36. Porter, W.L. "Autoxidation in Food and Biological Systems," Simic, M.G.; Karel, M., Eds.; Plenum Press: New York, 1980, pp. 295-365.
37. Eichner, K. "Autoxidation in Food and Biological Systems," Simic, M.G.; Karel, M., Eds.; Plenum Press: New York, 1980, pp. 367-385.
38. Sato, K.; Herring, H.K. Proc. Recip. Meat Conf. 1973, 26, 64-75.
39. Lingnert, H.; Eriksson, C.E. "Maillard Reactions in Food," Ericksson, C., Ed.; Pergamon Press: Oxford, 1981, pp. 453-466.
40. Pratt, D.E. J. Food Sci. 1972, 37, 322-323.
41. Pratt, D.E.; Watts, B.M. J. Food Sci. 1964, 29, 27-33.
42. Kramlich, W.E.; Pearson, A.M.; Tauber, F.W. "Processed Meats," Avi: Westport, CT; 1972; Chapt. 4.
43. Erdman, A.M.; Watts, B.M.; Ellias, L.C. Food Technol. 1954, 8, 320-323.

RECEIVED October 13, 1982

FOOD TECHNOLOGICAL ASPECTS

14

Maillard Technology: Manufacturing Applications in Food Products

JAMES P. DANEHY and BERNARD WOLNAK
Bernard Wolnak and Associates, Chicago, IL 60601

Just about forty years ago the United States Army Quartermaster Corps was having serious problems with the deterioration of certain stored foods, attributable to non-enzymatic browning reactions. The Corps approached the Corn Products Refining Company (now CPC International) and requested them to undertake research directed toward alleviating their problems. The late Ward Pigman, a carbohydrate chemist, and myself, in charge of the research on proteins, had never heard of the Maillard reaction. Nevertheless, we accepted their invitation, and the result was the first English language review of Maillard chemistry (1), published thirty-one years ago in Advances in Food Research.

Maillard reactions can be involved in the manufacture of foods in at least three quite different ways. First, there is the unconscious role played in the development of flavor* in such traditional processes as the roasting of coffee and cacao beans, the baking of breads and cakes, and the cooking of meats. Second, there is the deliberate use of Maillard technology in the production of artificial (or engineered) foods and flavors. Third, there are the efforts to inhibit undesirable results of Maillard reactions in food processing today.

The Role of Browning in Specific Food Systems

Many bland, or even downright unpleasant-tasting, substances are transformed into some of the most desirable flavors and popular foods by roasting. Thus, those foods, representing such different tastes and aromas as chocolate, bread, roast beef, coffee, and toasted nuts have in common the fact that they are products of the Maillard browning

*Aromas (or odors) are the signals perceived by the olfactory organs, tastes are the signals perceived by the lingual organs, and flavors are the simultaneous perceptions of the other two.

0097-6156/83/0215-0303$06.00/0

reaction. The enormous variety in flavor is due almost entirely to the large number of permutations from the interactions of a relatively few primary reactants, and to the importance of balance between the components finally present. The reproducibility obtained in these seemingly chaotic, and certainly random, systems is as remarkable as the sensitive discrimination of the mammalian olfactory-gustatory system.

During the last forty years a great deal has been learned, both by model studies and by the application of sophisticated analytical methods to foods, about the chemistry which accounts for these flavors. To date, however, there is little or no evidence that the in-depth chemical information has had any impact on the processes by which authentic, "natural" foods are manufactured.

However, during these same four decades about a hundred patents have been granted worldwide for processes and products based on Maillard browning technology. Since Chemical Abstracts has the policy of abstracting only the first issued of several equivalent patents, and listing the later ones in a patent concordance, we have counted only those patents abstracted in Chemical Abstracts. We have, however, used a variety of search and retrieval methods and believe that our bibliography on Maillard patents is substantively complete. Givaudan (2) has stated that over the last twenty years, about a thousand patents in this field have been granted to various flavor companies. Even allowing for the inclusion of all equivalent patents in this sum, it is difficult to believe that this number could be reached.

We have found 45 "standard" patents, i.e., patents which specify mixing one or more amino acids with one or more carbonyl compounds and heating, with some or all of the operating conditions given: temperature, time, water content, pH, innumerable additives. It was pointed out previously that slight changes in initial composition and reaction conditions produce significant changes in the flavor and aroma of the reaction products. But not one of these patents gives a rigid, controlled specification. The wide ranges of operating conditions and the numerous alternatives offered produce such a complete overlap between these patents that not even a knowledgeable chemist and a wily lawyer could distinguish one from the other.

What value might these patents have? Probably very little, both from the standpoint of the holders of the patents, and from the standpoint of those who might hope to learn by studying them.

We have already emphasized the redundancy of the "standard" patents. Not only do they all say substantially the same thing, but they appear to contain little, if anything, that was not fully disclosed in the model studies which have been discussed earlier in this symposium. We believe that it is doubtful if the validity of any one of them could be upheld in court in view of the prior art published in journals. Further, we do not believe that a holder of one of these existing patents could sue successfully for infringement, for two quite different reasons. First, in view of the extreme complexity of the composition of the reaction products, it would be impossible to determine, by examining them, how they were made. Second, in view of the redundancy of the patents themselves, there would be no way to determine whose patent was being infringed.

In order to get reproducible results, it is essential to exercise the most precise control at every stage of the reaction process. The basis for this is not provided by any of the patents, which characteristically give broad ranges.

Maillard reaction products are being manufactured commercially today by detailed, proprietary processes which are not given by any patent.

It is worthwhile to review what has been learned about two well-studied food systems, for upon this scientific base certain useful, artificial flavors and food products have been developed.

Chocolate and Cocoa

One of the world's most popular flavors is determined by a physical-chemical composition which starts with the seeds of the plant, *Theobroma cacao*, and continues with an empirical process discovered and perfected by the Aztecs, or by an earlier society from whom the Aztecs received it.

Two entirely separate stages are essential for the development of this flavor: The fermentation of the beans (seeds) in their mucilaginous pulp enclosure when the pod is opened, and the roasting of the dried, fermented beans. It has long been known that neither aroma nor aroma precursors are present in unfermented cacao beans which, when roasted, develop an odor reminiscent of broad beans (3). Substantially the only sugar present in unfermented beans is sucrose, but a mixture of fructose and glucose accounts for most of the sugar in fermented beans (4,5). Also, in going from unfermented to fermented beans, the concentration of free amino acids increases between three- and four-fold (6).

That summary is based on the reports of a well-conceived and carefully executed research program carried out by Rohan. Mohr et al. (7) extended these studies and was able to draw additional conclusions. First, without exception, free amino acids are much more sensitive to destruction in this system than the peptide-bound amino acids. Second, differences in the stability of amino acids under these conditions are not great—from 25% loss for isoleucine to 68.5% for lysine, over a relatively short period of time. In this system the reducing sugars must be the limiting factor, since the glucose and fructose are completely destroyed or removed. Third, neither cystine nor cysteine are reported to be present, and the only other sulfur-containing amino acid, methionine, is present at a much lower concentration than any other amino acid. Clearly, as we shall see later, cocoa would probably have a considerably different flavor if cysteine or cystine were present in the fermented beans.

Rohan had suggested that the operative reaction in the development of chocolate aroma might be a Strecker degradation of the amino acid fraction. Bailey et al. (8) demonstrated quantitatively that three aldehydes, which could be related to leucine, valine, and alanine, were prominent in the volatiles from a typical sample of roasted, ground cacao beans.

Obviously, however, while a synthetic mixture corresponding to the above would be fragrant, it would certainly not suggest the aroma of cocoa, based on the work reported by many others. What, then is the chemical basis for the aroma and flavor which are characteristic of cocoa and chocolate products? During the period of 1964-1976 more than a dozen reports addressed themselves to this problem. They ranged from the herculean labors of the Firmenich group, which carried out a classical fractionation of 750 kilograms of Arriba (Venezuelan) cocoa, which confirmed the presence of 43 compounds previously reported by others and identified 29 compounds not previously reported (9), to the powerfully instrumented investigations which, to date, have claimed the identification of more than three hundred constituents of cocoa volatiles. The cumulative result of all this effort is the reasonably sure establishment that at least 350 different kinds of organic molecules are present in cocoa volatiles in at least detectable amounts and that a goodly, though indeterminable, number of them are final products of the Maillard reactions. Few, if any, of these molecules would be odorless. But how they combine, in intensity and specificity, to produce the instantly recognizable aroma and flavor of chocolate is still unknown.

In view of the fact that holding large numbers of organic compounds of diverse functionality in a homogeneous system at room temperature, much less at about 100°, is conducive to chemical reaction, we should consider at least two other reasons for the lack of success in attempts to reconstitute the aroma of cocoa. First, some of the compounds which contribute to the aroma of cocoa may have decomposed and are no longer present in the extract analyzed. Second, some of the compounds identified in the extract may be artifacts, not actually present in the cocoa, but synthesized during the extraction and working-up process.

The cocoa and chocolate consumed by the American market are produced by a relatively small number of American and Dutch manufacturers who start with the dried, fermented beans. Current manufacturing practice uses improved machine design and extended automatic control in accord with modern principles of chemical engineering, but it is based solidly on the traditional process, little modified during the last century.

There is no evidence that any of the manufacturers of cocoa and chocolate have adapted any part of the Maillard technology to their manufacturing processes. There are at least two reasons for this. First, the standard processes, as applied to beans of good quality, produce excellent products. Second, while the work just reviewed has given us a rather clear outline as to how chocolate aroma is developed in the roasting of fermented beans, the research work has not yet been done, or reported, that would serve as a basis for improving the industrial processing of cacao beans.

However, there are two related classes of products which do depend, more or less, on Maillard technology: chocolate flavors, and cocoa substitutes (or extenders, as they are more realistically called).

For those foods and beverages which require a chocolate flavor, without the added bulk of cocoa or the fat of chocolate, the flavor houses provide synthetic approximations, extracts of cocoa, or mixtures thereof. The flavorist is an artist, skillful in the blending of essential oils and pure organic compounds. More recently, deliberate use has been made of the products of Maillard reactions between selected ingredients, but operative details are retained as highly confidential information. While some art has been recorded in patents, these data usually list specific organic compounds, and frequently methods for their synthesis. Generally, these patents are of very doubtful value.

Cocoa extenders or substitutes, as products which purport to be serious contenders for a fraction of the cocoa market, are a relatively new phenomenon. So long as cocoa was plentiful, cheap, and of superior quality a cocoa substitute made no sense. But the steady increasing of prices and tightening of supplies of cocoa in view of rising world-wide demand provided the incentive for some companies to undertake limited development of cocoa substitutes. It was no coincidence that their appearance on the market in early 1977 matched the peaking prices of both cacao beans and cocoa.

These cocoa substitutes are of two kinds. First, they consist of otherwise unprocessed bulking agents with added flavor and color. The bulking agents employed are soybean flour, modified food starches, dextrins, or mixtures thereof. They are definitely offered as extenders; none of the manufacturers recommend that they be used as a total replacement for cocoa. Manufacturers include Cargill, Inc., Minneapolis, Minnesota (Cocoa-Max), McCormick & Company, Inc., Hunt Valley, Maryland, (McCormick Cocoa Extenders), and National Starch & Chemical Corp., Bridgewater, New Jersey (N-Liven Cocoa). A.E. Staley Mfg. Co., Decatur, Illinois, entered the field but quickly dropped out.

The second kind of commercial cocoa substitute consists of roasted food products. Clearly, whether intentionally or not, they employ Maillard technology in the same way in which it is employed in producing cocoa and coffee. The ingredients disclosures suggest that no effort has been made to modify or to enhance the flavors by the addition of amino acids or of special sugars (xylose, for example). At least three of these products are worth mentioning.

The carob, or locust bean, tree, Ceratonia siliqua L., is indigenous to Mediterranean shores. It produces flat pods, known as St. John's bread, eight to twelve inches long, from which the seeds are removed and processed to yield locust bean gum. The deseeded pods are dried, broken, and sold as "kibble", which can be roasted and ground to provide a nutrient flavoring agent.

The color range and aromas attainable by varying the times and temperature of roasting, the powdery bulk, complete compatibility with cocoa, its GRAS status, and its price ($\sim$ \$0.60/lb.) give it an initial advantage as a cocoa extender. But it certainly doesn't taste like cocoa, although it is sweet and has a rather pleasant taste.

It has been available for a long time commercially. Indeed, its human use probably antedates that of cocoa, and certainly does in Europe. Its status as a cocoa extender arises simply from temporary economic conditions.

VCR (Viobin Cocoa Replacer) is "....pressure-roasted defatted wheat germ, ground to a fine powder which is similar in color and texture to processed cocoas...The toasting process develops aroma and taste characteristics that complement...cocoa..." Such aroma as it has, and it would not be mistaken for cocoa, can be attributed to Maillard reactions taking place during toasting. It is manufactured and sold by the Viobin Corporation, Monticello, Illinois.

For almost four years Coors Food Products Company, Golden, Colorado, has tried to perfect a product (Cocomost) consisting of "100% brewers' yeast", though processed, of course. Samples were released on a very limited basis for a short period of time in mid-1978 to food manufacturers, and some favorable, but premature, publicity appeared in national magazines which cater to the food industry. The process (proprietary) which produces the alleged cocoa-like flavor is complex and difficult to control. Cocomost has not yet appeared on the market, nor are samples again available.

The fate of cocoa substitutes is entirely dependent on two quite independent factors: quality, and the price of cocoa. It is at least conceivable that a substitute could be developed that would be equal to cocoa. If its price were right we would have a real cocoa replacer. But to date the products are, at best, acceptable adulterants, based on limited development, and no real research work. Nor is the price situation favorable to cocoa substitutes. Swiftly falling prices and accumulating stocks for cocoa from mid-1977 to early 1980 are more of an overreaction than anyone predicted. Based on the information presently available, the outlook for cocoa substitutes is not favorable.

Meat Flavors, Natural and Artificial

Meat is the muscular tissue of the common domestic animals, considered and used as food for human consumption. Meat is at least a three-phase system consisting of (1) a hydrophilic, but water-insoluble, fibrous protein network, (2) a hydrophobic fat deposit held together by membranes, and (3) an aqueous solution containing many soluble, low molecular weight compounds.

Raw meat has very little aroma at room temperature although it is usually possible to distinguish beef, pork, lamb, and chicken by sniffing. The taste of raw meat, which is not at all palatable to most human beings, can be described as somewhat salty and metallic. Raw meat must be cooked in some fashion in order to develop any organoleptically acceptable odor and flavor. Clearly, then, the unheated tissues must contain precursors which undergo thermally induced chemical reaction; these reactions produce both volatile compounds with desirable aromas and nonvolatile compounds which influence the taste; the combination of these two categories determines the flavor. As we proceed, striking analogies between the development of flavors in the cooking of meat and in the food system we have already discussed (cocoa and chocolate) will become apparent.

In a series of investigations (10-19) it was definitely established that the extraction of minced, lean meat (beef, pork, and lamb) with cold water gave solutions which contained salt, lactic acid, glycoproteins, inosinic acid, taurine, glutamine, asparagine, glucose, and some amino acids. Aging, particularly in the case of beef, is important in the development of the various flavor precursors (17). Thus, glycogen undergoes glycolysis to lactic acid almost completely within 24 hours after slaughter. Partial autolysis of the proteins and nucleic acids gives an assortment of peptides and amino acids from the former, and a mixture of inosinic acid and its fragments (inosine, hypoxanthine, ribose-5-phosphate, and ribose) from the latter.

When the filtered, aqueous extract of ground beef is heated, a sequence of aromas is developed, beginning with a faint, blood-like aroma in the barely warm solution, passing through a phase in the boiling solution which gives off the aroma of boiled beef, and terminating in the hot, dried, brown residue (100-150°) with an aroma resembling broiled steak. When this original, filtered, aqueous extract was subjected to dialysis against water in cellulosic sausage casings, and the low-molecular-weight fraction that passed through the membrane was lyophilized to a white powder and then pyrolyzed, it underwent the typical Maillard browning to produce a strong aroma of broiled meat. This aroma was substantially the same, whether the original lean meat was beef or pork. This may result from the fact that the amino acid contents of beef, pork, and lamb are semiquantitatively rather similar.

It appears, then, that there is a general, meaty aroma, common to cooked beef, pork, and lamb (and probably poultry), attributable to the pyrolysis of the mixture of low molecular weight nitrogenous and carbonyl compounds extracted from the lean meat by cold water. But the aromas of roast beef, roast pork, roast lamb, and roast chicken are unmistakably different. The chemical composition of the muscular fat deposits of these animals differ appreciably, and it is to these lipid components that we must look to account for the specific flavor differences. Heating the carefully separated fat alone does not give a meaty aroma at all, much less an animal-specific one. It is the subsequent reactions of pyrolysis products of nonlipid components that give the characteristic aromas and flavors of roasted meats (20).

Since it is precisely at the surface of roasting meat that water concentrations are lowest and temperatures are highest, it is at the meat surface that the flavor and color generating activity during roasting is most prominent. This situation is analogous to the formation of crust and aroma in bread and other baked cereal products. The same facts also account for the significant difference between roasted and boiled meats.

Paralleling the studies of the volatile products of roasted cacao beans and of baked cereal products, and using the same techniques, a great deal of effort has gone into the determination of the compounds present in the volatile fractions of cooked meat. Most of these have been concerned only with beef, either roasted or boiled, but chicken has also received appreciable attention (21). Several lists of compounds isolated from the volatiles of cooked beef have been published (22-24), both cumulative and newly isolated ones. The totals for chicken (as of 1972) and for beef (as of 1977) are more than two hundred each. It must be emphasized again that these are qualitative identifications, not quantitative accountings.

These cumulative tables for cooked meat volatiles are very difficult to distinguish from those published somewhat earlier for cocoa volatiles. Indeed, the larger cumulative tables resemble somewhat abridged versions of the Aldrich and Eastman Kodak catalogs of organic chemicals. Meaningful comparisons are hindered by two quite different facts. First, there is usually no hint as to the fraction of the meat, or even the fraction of the volatiles, that is comprised by a given compound. Second, we are probably getting a great deal of noise with the signals: that is, there must be many compounds which, even though they have odors, would not be missed, if they were absent. We simply cannot believe that more than 200 compounds are required to produce any one of the distinctive roasted food aromas.

But the human nose has no difficulty in distinguishing chocolate from roast beef, and the flavor chemist is trying to catch up with this degree of discrimination.

Chang and Peterson have suggested that lactones, acyclic sulfur compounds, nonaromatic heterocyclic compounds containing ring S, N, or O atoms, and aromatic heterocyclic compounds containing ring S, N or O atoms may be important contributors to meat flavor, even though none of them, alone, smell anything like cooked meat. Wilson and Katz identified 46 sulfur-containing compounds from the volatiles of lean beef pressure-cooked with water at 163° and 182°. Most of these compounds were thiophene or thiazole derivatives, but acyclic thiols, methyl sulfide, and four disulfides were also present. Also, as we shall see later, sulfur compounds (especially cysteine) play a key role in manufacturing artificial meat flavors. Tonsbeek et al. (25,26) have isolated 4-hydroxy-5-methyl-2,3-dihydrofuran-3-one and 4-hydroxy-2,5-dimethyl-2,3-dihydrofuran-3-one, particularly pungent compounds, from cooked beef. Pyrazines are a particularly important class of flavor compounds, but it was not until 1971 that their presence in beef volatiles was reported (27), and by 1973 a total of 33 of them had been identified (28). Recently, Flament et al. (28) identified several pyrrolo[1,2-a]pyrazines.

Despite the large amount of qualitative, if not quantitative, data on the chemical composition of the volatiles from cooked meat, no one has yet claimed anything like a duplication of a meat aroma by the combination of the pure chemicals identified in meat aromas. Once more, the parallel with cocoa and baked products is striking. Just one example points up the elusive relationship between chemical compounds and food flavors. Hydrogen sulfide has the odor which does characterize rotten eggs, yet it appears to be a necessary component of meat aromas. Its odor threshold is 10 parts per billion (ppb), but its concentration in freshly cooked chicken is 20 to 100 times greater. It is generally agreed that the aroma of a food is the sensed perception of an extremely complex interaction of many components, but one reads between the lines the disappointment of some who report new compounds and note that they do not have a meatlike aroma. Chang and Peterson (24) suggest the justifiable fear that some components may have been decomposed or missed, but their hope that a unique component may still be found which alone or in combination will have a characteristic beef aroma, is less justifiable.

The key involvement of organic sulfur compounds in the development of meatlike flavors was announced simultaneously in 1960 by several investigators. In what was the earliest paper to describe deliberate attempts to produce aromas useful in foods via Maillard reactions, Kiely et al. (30) noted that both cysteine and cystine gave meaty odors when heated with reducing sugars. May et al. received several equivalent patents (31-35) in which they claimed that heating cysteine or cystine with furan, or substituted furans, pentoses, or glyceraldehyde gave a meatlike flavor.

It is precisely to the production of meatlike flavors that the great majority of patents based on the Maillard reaction have been directed. Mos of them indicate cysteine or cystine as the essential sulfur-containing compound. Other patents claim alternative sources for sulfur, e.g., derivatives of mercaptoacetaldehyde (36), mercaptoalkylamines (37), S-acetylmercaptosuccinic acid (38), 2-thienyltetrasulfide (39), "a sulfide" (40), and hydrogen sulfide (heated with aqueous xylose without any amino acid) (41).

Two patents (42-43) claim the contribution to meatlike flavors made by thiamine when it is present in the standard pyrolytic mixture. Arnold et al. (44) have reported on the volatile flavor compounds produced by the thermal degradation of thiamine alone. It is generally agreed that the presence of methionine, the other sulfur-containing amino acid in the flavor-developing mixture, produces negative and/or undesirable results. However, one patent (45) specifies methionine in a standard Maillard procedure, and no cysteine.

Shibamoto and Russell (46) heated an aqueous glucose-ammonia-hydrogen sulfide solution at 100° for two hours. Of the 34 major components identified, 2-methylthiophene accounted for 24.9% of the area of the chromatographic peaks ethyl sulfide, thiophene, furfural, and 2-acetylfuran each accounted for 10-11%, and methyl sulfide and 2,5-dimethylthiophene, 7% each. The reaction mixture as a whole was deemed by sensory panel evaluation to have a cooked beef odor.

Once more, although the distributions are expressed quantitatively, there is no information on the yields of these interesting compounds made from glucose, ammonia, and hydrogen sulfide. Earlier, however, in their studies of carbohydrates-ammonia systems (47-49), Shibamoto and Russell found that the amounts of total pyrazines produced, based on the sugars, were in the range of 1-2%.

Wilson's review of thermally produced imitation meat flavors is well worth consulting (50).

To what extent has Maillard technology been exploited in the production of artificial meat flavors? It is difficult to obtain a quantitative answer to that question since either the flavors are produced as proprietary products and sold to food manufacturers, with no way of accounting for the size of the market without access to private records, or else the flavors are produced in-house by the food manufacturers and put in their shelf items, with even less of an opportunity to determine the amounts produced.

Givaudan, in the January, 1979, issue of their house organ (2), gives an appreciative nontechnical introduction to the subject which concludes with this passage:

> "Much of our work in this field has been devoted to the search for more suitable natural intermediates as starting materials. These had to be pretreated prior to their use in browning reactions. The latest development in this field is the utilization of fully natural starting materials which permit to reproduce traditionally known food flavors of greatest perfection.
>
> "...nonenzymatic browning will continue to be an important process for the manufacture of flavors...Especially the combination of enzymatic and nonenzymatic reactions will play an important role in reproducing the exquisite flavors of roasted foods".

At least since 1969, Pfizer Chemicals Division has been manufacturing and marketing imitation beef and chicken flavors under the trademarked name, CORRAL. Their brochure (data sheet No. 638) leaves no doubt as to what the CORRAL products are. The ingredients used to make CORRAL Beef paste are: hydrolyzed vegetable protein, monosodium glutamate, sucrose, vegetable fat, L-arabinose, disodium inosinate, cysteine hydrochloride, β-alanine, and glycine hydrochloride. It is understood that these ingredients have been heated in aqueous solution, and then the mixture largely dehydrated. The ingredients in CORRAL chicken paste are: Hydrolyzed vegetable protein, monosodium glutamate, sucrose, vegetable fat, cysteine hydrochloride, dextrose, L-arabinose, glycine hydrochloride, disodium inosinate, and disodium guanylate. It will be noted that the somewhat different sequences in the two formulations, as well as two additional ingredients in the second, account, at least in part, for the different flavors. Conditions of time, temperature, and water content of the reaction mixture must also influence the final results.

Representative of FIDCO's aggressively advertised SPECTRA imitation meat flavors is FIDCO SPECTRA BN-10 (Nestlé Co.), whose specifications show that it contains 34% sodium chloride and 54% organic solids, including 33% protein. Realistically, that "protein" is almost certainly hydrolyzed vegetable protein (HVP). This product also contains "natural and artificial flavors, monosodium glutamate, beef fat, sodium inosinate, sodium guanylate, and caramel color". It is recommended that it make up 20% by weight of a dehydrated beef soup base. In all likelihood Maillard reaction products are included under "artificial flavors", and may also be present as part of the processing of the HVP with reducing sugars. Considering the size and marketing resources of the company, SPECTRA flavors probably make up an important part of this market. MAGGI's dehydrated soup mixes, more important in Europe than in the United States, are closely related products of another division of Nestlé.

Knox Ingredients Technology (KIT), successor to Knox Gelatine, a wholly owned subsidiary of Thomas J. Lipton, Inc., is a major producer of HVP's, and of a line of "Tastemaker Natural and Artificial Flavorings" based on them. Recent press releases by KIT are quite explicit about the role of Maillard reaction products in the more sophisticated versions of these flavoring agents.

The Edlong Corporation, Elk Grove Village, Illinois, is definitely active in this field. They are particularly interested in the production of precursors by fermentation which may undergo Maillard reactions in the customer's product during cooking. They have just brought out a "Natural Butter Flavor #1-4006" for use in Granola-type items. In addition to the flavor which it develops, it produces a pleasing yellow-brown rather than the gray-brown color often characteristic of Granola products.

Wm. M. Bell Co., Melrose Park, Illinois, is one of the companies active in this technology. They incorporate Maillard reaction products into a variety of flavors, including chocolate.

Borden Industrial Food Products, Northbrook, Illinois, manufacture Wyler Soups and Wyler Brand CB-M flavor concentrates. One of the latter, for example, #78-62 Beef Flavor, contains: hydrolyzed vegetable protein, dextrose, sucrose, vegetable oil, salt, monosodium glutamate, disodium inosinate, disodium guanylate, onion powder, and garlic powder. They are similar to, but not identical with, Pfizer's CORRAL, which also contains arabinose, cysteine, β-alanine, and glycine. Wyler Brand #78-50 Chicken Flavor also contains some chicken.

Chemical Marketing Reporter for June 25, 1979, carried an advertisement of Hydrocal, S.A., Geneva, Switzerland, which announced the availability of "unique new Maillard flavors", manufactured by Frutarom, Ltd., Haifa, Israel. The owner of this operation is Robert Aries, the American chemical engineer who became expatriate in about 1955, and who returned to the United States in 1979.

Alex Fries & Bro., Inc. Cincinnati, Ohio, make meat, cheese, and chocolate flavors based on Maillard technology. They have no information to release on these products other than that in their general brochure.

Fritzsche-Dodge & Olcott, Inc., New York City, incorporates Maillard reaction roducts on a rather broad scale into their meat, chocolate, bread, and malt flavors.

While the advertisements of Haarmann & Reimer, Springfield, New Jersey, emphasize the natural approach, a spokeman agreed that they use Maillard technology, especially in the production of meat flavors.

Literature Cited

1. Danehy, J.P.; Pigman, W.W. Adv. Food Res. 1951, 3, 241-290.
2. Anonymous. The Givaudan Flavourist, International Edition. January, 1979, 3, one page.
3. Rohan, T.A. J. Sci. Food Agric. 1963, 14, 799-805.
4. Rohan, T.A. J. Food Sci. 1964, 29, 456-459.
5. Reineccius, G.A.; Andersen, D.A.; Kavanaugh, T.D.; Keeney, P.G. J. Agric. Food Chem. 1972, 20, 199-202.
6. Rohan, T.A.; Stewart, T. J. Food Sci. 1965, 30, 416-419.
7. Mohr, W.; Roehrle, M.; Severin, Th. Fette, Seifen, Anstrichm. 1971, 515-521.
8. Bailey, S.D.; Mitchell, D.G.: Bazinet, M.L.; Weurman, C. J. Food Sci. 1962, 27, 165-70.
9. Dietrich, P.; Lederer, E.; Winter, M.; Stoll, M. Helv. Chim. Acta 1964, 47, 1581-90.
10. Hornstein, I.; Crowe, P.F. J. Agric. Food Chem. 1960, 8, 494-498.
11. Wood, T. J. Sci. Food Agric. 1961, 12, 61-9.
12. Macy, Jr., R.L.; Naumann, H.D.; Bailey, M.E. J. Food Sci. 1964, 29, 136-41.
13. Macy, Jr., R.L.; Naumann, H.D.; Bailey, M.E. J. Food Sci. 1964, 29, 142-148.
14. Wasserman, A.E.; Gray, N. J. Food Sci. 1965, 30, 801-807.
15. Landmann, W.A.; Batzer, O.F. J. Agric. Food Chem. 1966, 14, 210-214.
16. Zaika, L.L.; Wasserman, A.E.; Monk, Jr., C.A.; Salay, J. J. Food Sci. 1968, 33, 53-58.
17. Wasserman, A.E. J. Agric. Food Chem. 1972, 20, 737-741.
18. Wasserman, A.E.; Spinelli, A.M. J. Food Sci. 1970, 35, 328-332.
19. Jarboe, J.K.; Malbrouk, A.F. J. Agric. Food Chem. 1974, 22, 787-791.
20. Wasserman, A.E.; Spinelli, A.M. J. Agric. Food Chem. 1972, 20, 171-174.
21. Wilson, R.A.; Katz, I. J. Agric. Food Chem. 1972, 20, 741-747.
22. Herz, K.O.; Chang, S.S. Adv. Food Res. 1970, 18, 1-83.
23. MacLeod, G.; Coppock, B.M. J. Agric. Food Chem. 1976, 24, 835-843.
24. Chang, S.S.; Peterson, R.J. J. Food Sci. 1977, 42, 298-305.
25. Tonsbeek, C.H.T.; Plancken, A.J.; Weerdof, T. J. Agric. Food Chem. 1968, 16, 1016-1021.
26. Tonsbeek, C.H.T.; Koenders, E.B.; van der Zijden, A.S.M.; Losekoot, J.A. J. Agric. Food Chem. 1969, 17, 397-400.
27. Flament, I.; Ohloff, G. Helv. Chim. Acta 1971, 54, 1911-1913.
28. Mussinan, C.J.; Wilson, R.A.; Katz, I. J. Agric. Food Chem. 1973, 21, 871-872.
29. Flament, I.; Sonnay P.; Ohloff, G. Helv. Chim. Acta 1977, 60, 1872-1873.
30. Kiely, P.J.; Nowlin, A.C.; Moriarty, J.H. Cereal Sci. Today 1960, 5, 273-274.

31. May, C.G. (Lever Bros. Company) U.S. Pat. 2,934,435 (April 26, 1960).
32. May, C.G.; Morton, I.D. (Lever Bros. Company) U.S. Pat. 2,934, 436 (April 26, 1960).
33. Morton, I.D.; Akroyd, P.; May, C.G. (Lever Bros. Company) U.S. Pat. 2,934,437 (April 26, 1960).
34. May, C.G.; Akroyd, P. (Unilever N.V.) German Pat. 1,058,824 (June 4, 1959.
35. May, C.G.; Akroyd, P. (Lever Bros. Company) U.S. Pat. 2,918,376 (1959).
36. Broderick, J.J.; Linteris, L.L. (Lever Bros. Company) U.S. Pat. 2, 955,041 (October 4, 1960).
37. Ohwa, T. (Lever Bros. Company) Japanese Pat. 72 50387 (December 18, 1972).
38. Mosher, A.J. (Maggi, AG) French Pat. 2,178,924 (December 21, 1973).
39. Katz, I.; Evers, W.J.; Giacino, C. (International Flavors & Fragrances) (December 19, 1972). U.S. Pat. 3,706,577.
40. Heyland, S.; Cerise, L. (Nestle, S. A.) German Pat. 2,837,513 (April 5, 1979).
41. Gunther, R. (Fritzsche, Dodge & Olcott, Inc.) U.S. Pat. 3,642,497 (February 15, 1972).
42. International Flavors & Fragrances. Neth. Appl. 6,603,722 (September 23, 1966).
43. International Flavors & Fragrances. Neth. Appl. 6,609,520 (January 9, 1967).
44. Arnold, R.G.; Libbey, L.M.; Lindsay, R.C. J. Agric. Food Chem. 1969, 17, 390-392.
45. van Pottelsberghe de la Potterie, P.J. (Maggi, AG) German Pat. 2,149,682 (April 13, 1972).
46. Shibamoto, T.; Russell, G.F. J. Agric. Food Chem. 1976, 24, 843-846.
47. Shibamoto, T.; Bernhard, R.A. J. Agric. Food Chem. 1976, 24, 847-852.
48. Shibamoto, T.; Bernhard, R.A. Agric. Biol. Chem. 1977, 41, 143-153.
49. Shibamoto, T.; Bernhard, R.A. J. Agric. Food Chem. 1977, 25, 609-614.
50. Wilson, R.A. J. Agric. Food Chem. 1975, 23, 1032-1037.

RECEIVED November 2, 1982

15

Maillard Reaction Products as Indicator Compounds for Optimizing Drying and Storage Conditions

K. EICHNER

Institut für Lebensmittelchemie der Universität Münster, Piusallee 7, D-4400 Münster, Federal Republic of Germany

W. WOLF

Bundesforschungsanstalt für Ernährung, Engesser Strasse 20, D-7500 Karlsruhe 1, Federal Republic of Germany

Chemical analysis of Maillard reaction intermediates (Amadori compounds) renders possible an early detection of this reaction. On this basis an analytical control of drying processes can be exerted, thus avoiding temperature and moisture ranges which are critical in respect of product quality. With different varieties of carrots it was shown that quality retention can be improved by lowering air temperature during an initial high-temperature drying step and final drying at lower temperatures to a water content providing optimum storage stability.

Physical preservation methods for foods, such as sterilization and drying, are associated with the application of heat. In these cases, because of its high temperature coefficient, the Maillard reaction becomes the dominant deteriorative reaction (1,2,3). It is well known that the Maillard reaction in foods is initiated by the formation of colorless and tasteless intermediates, which preferentially are formed in low-moisture systems (4,5). In this way by reaction of glucose with amino acids fructose-amino acids are formed via Amadori rearrangement of the primary glucosyl-amino acids (1). Fructose-amino acids e.g. have been isolated from freeze-dried apricots and peaches (6,7,8). Amadori compounds arising from aldoses and amino acids are formed during drying of foods of plant origin and can be easily detected by amino acid analysis (5).

During further progress of the Maillard reaction brown discoloration occurs and a great variety of different compounds are formed which partly cause undesi-

0097-6156/83/0215-0317$06.00/0

rable sensory changes (9). Several authors (10-13) tried to correlate analytical with sensory data by using some typical Maillard products such as Strecker degradation products as chemical indicator substances.

The Maillard reaction rate is greatly influenced by temperature and water content; in a certain low-moisture range it reaches a maximum. Hendel and coworkers (14) established a temperature-moisture profile for browning of white potato during drying, from which they were able to calculate the extent of browning at different drying periods. It turned out that the last drying period maximized browning, whereas the interval of the browning maximum made only minor contribution to overall browning response because of the small inherent time intervals. They concluded that the temperature should be lowered during the last drying period, particularly because the activation energy of the Maillard reaction is increased by decreasing the water content.

Using a glucose-glycine browning model, Kluge and Heiss (15) evaluated the permissible reaction time for a given permissible extent of browning for each temperature-water content combination that could occur during drying. Knowing the temperature-water content profiles during drying, they added the reciprocals of the permissible browning times for each time interval, in this way getting portions of the permissible browning extent for any drying time.

From these experiments general conclusions can be drawn with respect to an improvement of the quality of dried products. However, these investigations are based on the measurement of browning being the last step of a multistage deteriorative reaction which normally is already accompanied by the formation of off-flavors. It becomes clear that analytical methods based on the evaluation of the end products of deteriorative reactions will not be satisfactory. Therefore in our own experiments amino acid analysis of Amadori compounds and gas chromatography of volatile Strecker aldehydes were applied to detect the onset of the Maillard reaction well before detrimental sensory changes occurred.

Drying of foods must be looked at in connection with storage conditions, because -- especially at higher water contents -- the Maillard reaction may continue. It may occur during storage at a greater rate, particularly if it has been initiated during drying by formation of Amadori compounds (5). If shelf life is limited by the Maillard reaction, such life may be increased on the other hand by lowering the water content of the product (16).

In former experiments (5) we have shown that chemical analysis for Amadori compounds (mainly consisting of fructose-glutamic acid) and isovaleraldehyde, formed by Strecker degradation of the amino acids leucine and isoleucine, can be used for an early detection of undesirable quality changes caused by the Maillard reaction. In order to demonstrate the usefulness of these compounds as indicator substances for quality improvement of dried products, we performed drying experiments with carrots as an example of plant products.

Procedure

Preparation of the material and drying. Fresh carrots were cut into cubes (edge length: 1 cm), blanched with boiling water for about 2 min, placed on wire screens (single product layer) and dried with upstream circulating hot air (about 3 m/s). After various time intervals, the screens were removed from the dryer and the loss of water was determined by weighing. In samples assigned for chemical analysis the water content was determined by vacuum drying at 70 °C for 4 h. Samples having higher water contents were stabilized by freeze-drying prior to analysis.

For the constant-temperature heating experiments different product water contents were obtained by storage of freeze-dried carrots over saturated salt solutions (17).

Determination of Amadori compounds and browning. Dried carrots (1.5 g) were homogenized in a mixer with 25 ml deionized water ($7x10^3$ rpm) and centrifuged for 30 min at 5 °C ($20x10^3$ rpm). For evaluation of the extent of browning the extinction values of the extracts were measured after diluting them with water (1:10). The Amadori compounds were determined using an amino acid analyzer (Biotronic LC-2000). To 2 ml of the carrot extract 2 ml 0.1 N HCl and 1 ml of buffer concentrate A were added; the mixture was diluted with water to a total volume of 10 ml and 0.1 ml of the resulting solution was injected onto the amino acid analyzer.

Analytical conditions. The analytical column contained a strongly acid cation exchange resin, Dionex DC-6A; column height: 20 cm; column diameter: 0.6 cm; column temperature: 36 °C. The washing column contained Dionex DC-3Li^+; column height: 11 cm. Buffer concentrate A: Li citrate, 1.8 M Li^+; pH 2.38. Buffer A: Li citrate, 0.18 M Li^+, pH 2.38. Buffer B:

Li citrate, 0.23 M Li^+, pH 3.10. Buffer program (min): A: 0-17; B: 17-60; LiOH (0.4 M): 60-85; A: 85-140. Buffer flow: 30 ml/h; ninhydrin flow: 20 ml/h.
For calculation of the molar ratio of Amadori compounds of peak C in the amino acid chromatogram (Figure 1) the following formula was used:

$$C\ (\text{mol\%}) = \frac{1{,}7 \times [C] \times 100}{1{,}7 \times [C] + [\text{Thre}] + [\text{Ser}] + [\text{Asp-NH}_2] + [\text{Glu}] + [\text{Glu-NH}_2]}$$

The concentrations are expressed as integration units of the peak areas (counts), the factor 1.7 being the ninhydrin color-correction factor.

Head-space gas chromatographic determination of isovaleraldehyde Dried carrot samples (0.5 g) were placed in septum vessels (24 ml) and 5 ml of water containing isobutyl alcohol as an internal standard (1 : 20.000 v/v) were added. After sealing with a septum the head-space vessels were kept in a thermostat at 85 °C. After 30 min 1 ml of the head-space gas was withdrawn with a syringe heated at 85 °C and injected into a gas chromatograph. The area of the isovaleraldehyde peak (5) was determined by using an integrator. For determination of the absolute isovaleraldehyde concentrations known amounts of solutions containing known concentrations of isovaleraldehyde were added to the samples and the resulting peak areas in the gas chromatogram were measured. In this way the isovaleraldehyde peak areas could be correlated with the isovaleraldehyde concentrations present in the standard solutions and the sample.

Analytical conditions: 2-m glass column (0.2 mm), filled with Chromosorb 101. Column temperatures: 6 min 120 °C, heated to 140 °C (temperature gradient: 4 °C/min). Carrier gas flow: 25 ml N_2/min; detector: FID. Retention times for isobutyl alcohol (standard): 667 s and for isovaleraldehyde: 869 s.

Results

Optimization of drying by the use of Amadori compounds Figure 1 shows an amino acid chromatogram of dried carrots. Peak C in the chromatogram represents the Amadori compounds fructose-threonine, -serine, -asparagine, -glutamic acid and -glutamine formed by reaction of glucose with the respective amino acids. These compounds are formed without induction period at all experimental conditions (5). The molar ratio of peak C (mol %) can be used as a measure for the latent heat impact during drying, which is causing a reduction of shelf life. In order to find out, which period of time produces Amadori compounds, we performed drying experiments on carrots. Figure 2 represents an air drying experiment with carrot cubes applying an air temperature of 110 °C. As the product temperature rises, the formation of Amadori compounds starts and parallels temperature increase until a maximum is reached, beyond which decomposition of these browning intermediates predominates, whereas browning does not become significant till this point.

Further investigations aimed at elucidation of temperature and moisture ranges critical in respect of food quality impairment were done. Figure 3 shows the drying course of the carrot variety "Bauer's Kieler Rote" at an air temperature of 110 °C, projecting the decrease in moisture content (weight % related to wet matter), the rise of product temperature, and the increase in concentration of Amadori compounds (C, mol %). Figures 4 and 5 give results of the corresponding drying experiments at air temperatures of 90 ° and 60 °C, respectively. From these experiments the strong influence of temperature on formation of Amadori compounds becomes evident, related to the fact that the Maillard reaction has a much higher temperature coefficient than the water vaporization rate. Figure 6 summarizes the described results by plotting the concentrations of Amadori compounds (C, mol %) versus water content (related to wet matter). The strong increase in formation of these reaction intermediates at lower moisture contents may be attributed on the one hand to the longer retention times in this moisture interval, on the other to an increase in the rate of formation of Amadori compounds below a water content of about 20 %, approaching a maximum, as will be shown later. The dashed lines in Figure 6 mark the limit of perceptibility of undesirable sensory changes (line 1) and the tolerable upper limit of quality decrease (line 2) caused by the Maillard reaction. During drying to a final water content of 7 %, this upper limit is exceeded at an air temperature of 110 °C, whereas at an air temperature of 90 °C only the limit of sensory perceptibility is crossed; an air temperature of 60 °C does not lead to any change in sensory quality owing to Maillard reaction products. Figure 7 in the same way presents the formation of brown products during air drying of the carrot variety mentioned. Browning was determined by measuring the extinction values of water extracts of carrot samples at 420 nm. By comparing Figure 7 with Figure 6, it again becomes clear that visible browning starts much later than the formation of Maillard reaction intermediates.

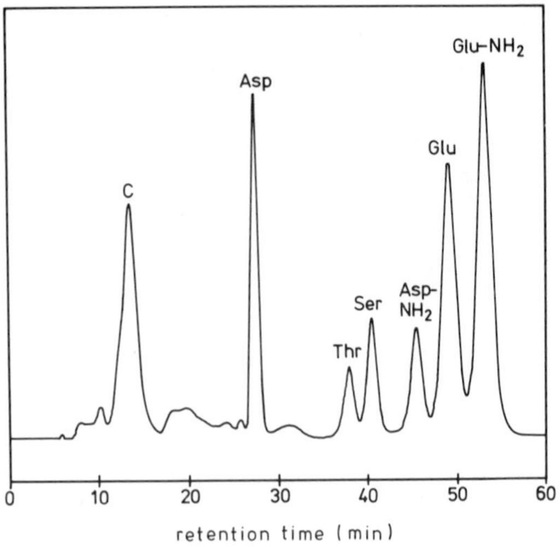

Figure 1. Shortened chromatogram of amino acids in air-dried carrots. Peak C represents Amadori compounds formed by reaction between glucose and the amino acids threonine, serine, asparagine, glutamic acid, and glutamine.

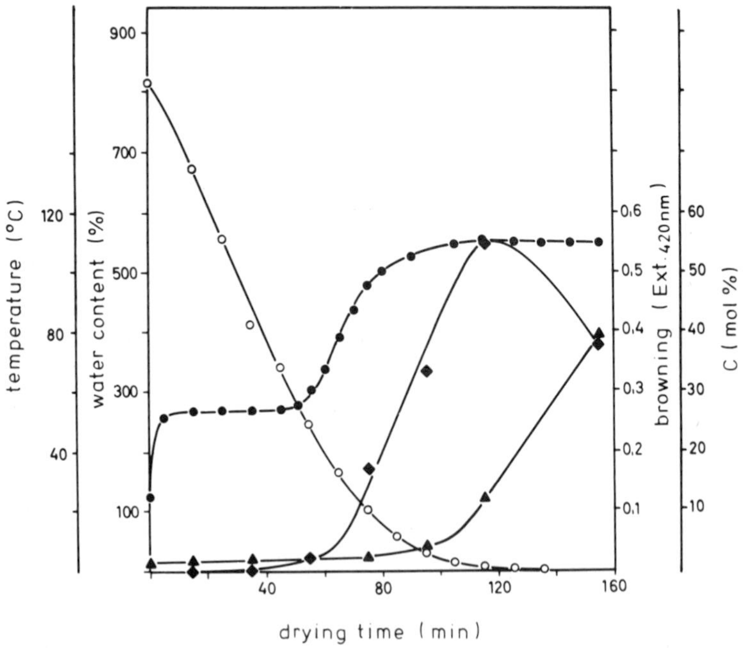

Figure 2. Formation of browning intermediates (Amadori compounds corresponding to peak C in Figure 1) and browning during air drying (110 °C) of carrot cubes. Key: ○, water content related to dry matter; ●, product temperature; ♦, formation of Amadori compounds (mol %); and ▲, browning (excitation wavelength 420 nm).

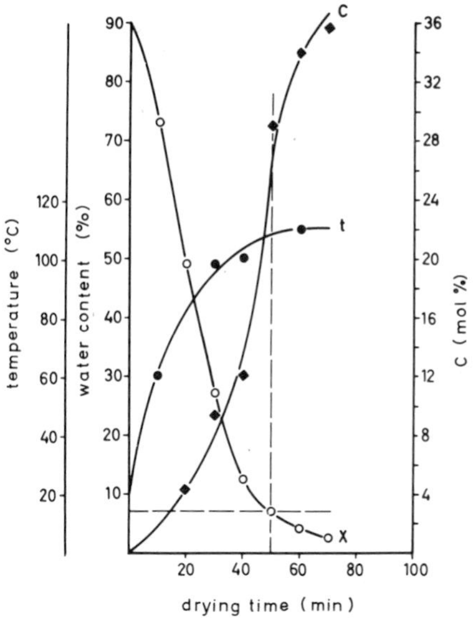

Figure 3. One-step air drying of the carrot variety "Bauer's Kieler Rote" at an air temperature of 110 °C. Key: ◆, mol % of Amadori compounds corresponding to peak C in Figure 1; ●, product temperature (°C); and ○, water content (%, related to wet matter). The dashed lines are associated with drying to a final water content of 7% guaranteeing a sufficient shelf life.

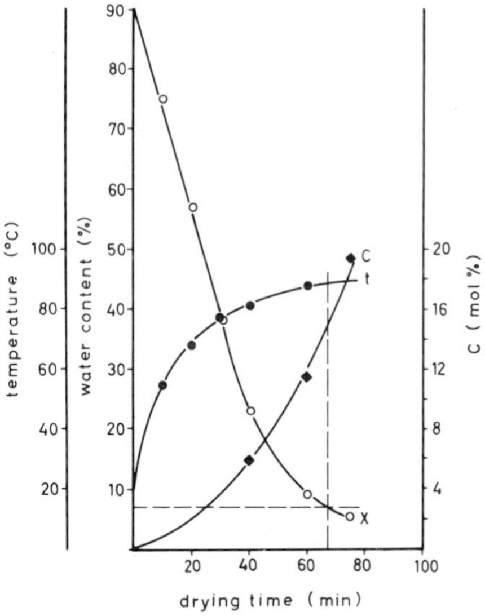

Figure 4. One-step air drying of the carrot variety "Bauer's Kieler Rote" at an air temperature of 90 °C. Key: same as Figure 3.

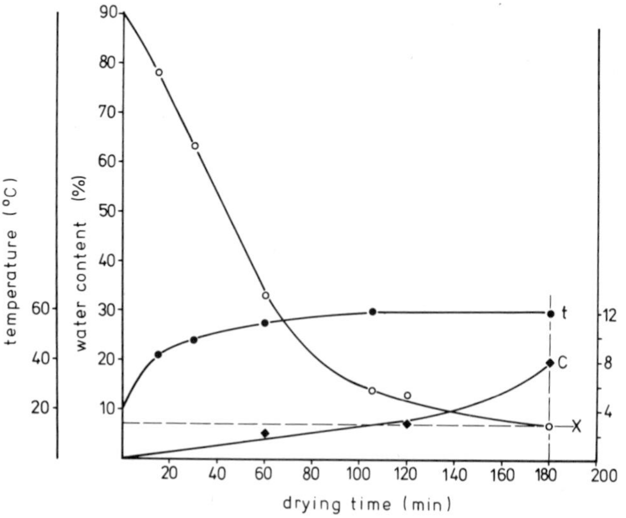

Figure 5. One-step air drying of the carrot variety "Bauer's Kieler Rote" at an air temperature of 60 °C. Key: same as Figure 3.

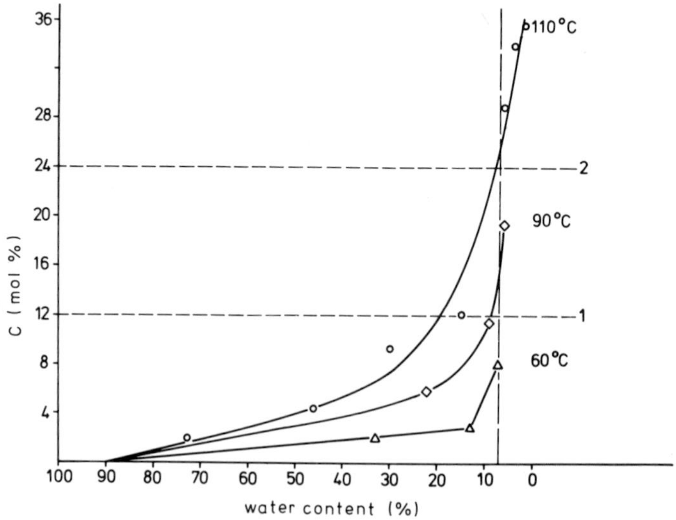

Figure 6. Increase of the concentrations of Amadori compounds during air drying of carrot variety "Bauer's Kieler Rote" dependent on water content and air temperature. Key: △, 60 °C; ◇, 90 °C; ○, 110 °C; dashed line 1, sensory perceptibility limit; and dashed line 2, sensory quality limit.

Furthermore Figure 7 shows that by carrot drying with an air temperature of 110 °C the upper limit of brown discoloration of the product is not reached at a final water content of 7 %, whereas at 90 °C no visible color changes can be detected. Therefore, during carrot drying the flavor quality limit is reached much earlier than the corresponding limit of brown discoloration.

Furthermore it should be elucidated, how far good quality retention during drying is possible without requiring too long drying times, which occur at low air temperatures. This should be possible by applying a two-step drying process, where a great deal of water is removed quickly at higher temperatures, whereas in the critical lower moisture interval the temperature is lowered to minimize the Maillard reaction. Figure 8 shows the results of such experiments. The drying process first was performed along the dashed line at 110°C for 10 min, 20 min, and 30 min; subsequently the air temperature was lowered to 60 °C for final drying. Figure 8 illustrates that Amadori compounds at this temperature increase but not before a water content of about 10 % is reached; drying to a final water content of 7 % results in a good product quality. Extending the final drying process for obtaining lower water contents could cause a certain loss of quality even at a lower air temperature.

Tables I, II, and III summarize the results of the described drying experiments.

Table I

Air drying of the carrot variety "Bauer's Kieler Rote" (Air temperature: 110 °C)

Drying time (min)	Water content (%)	C (mol %)	Isovaleraldehyde (ppm)	Sensory evaluation	
				Color	Flavor
10	73	2.0	1.1	typical	typical
20	46	4.3	1.4	"	"
30	27	9.4	2.0	"	"
40	15	12.1	3.5	"	"
50	6	29.0	9.4	sl.brown	burnt
60	4	33.9	9.6	brown	"
70	2.3	35.6	18.0	"	str.burnt
90	1.4	37.4	26.2	str.brown	scorched

(sl. = slightly; str. = strongly)

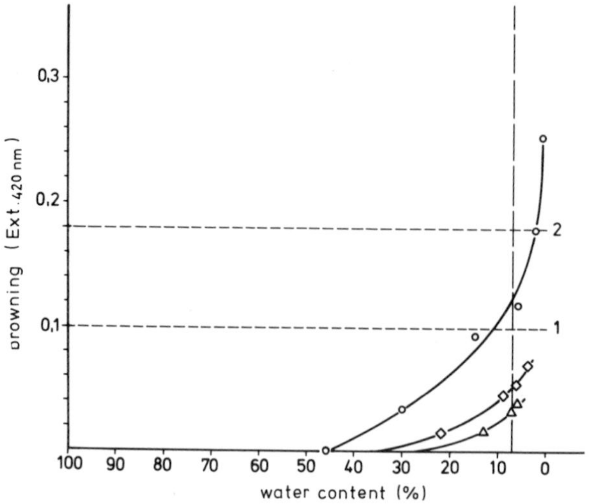

Figure 7. Increase of browning during air drying of the carrot variety "Bauer's Kieler Rote" dependent on water content and air temperature. Key: △, 60 °C; ◇, 90 °C; ○, 110 °C; dashed line 1, limit of detectable visible browning; and dashed line 2, quality limit relative to browning.

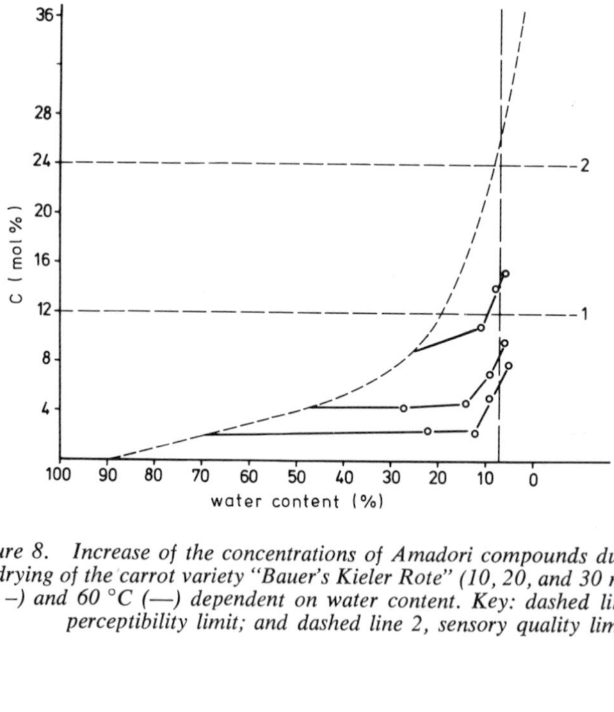

Figure 8. Increase of the concentrations of Amadori compounds during two-step air drying of the carrot variety "Bauer's Kieler Rote" (10, 20, and 30 min at 110 °C (– – –) and 60 °C (—) dependent on water content. Key: dashed line 1, sensory perceptibility limit; and dashed line 2, sensory quality limit.

Table I shows that an increase of Amadori compounds occurs parallel with an increase of isovaleraldehyde formed by Strecker degradation of the amino acid leucine (18). It becomes evident from Table I that the flavor impression "burnt" arises if certain concentrations of isovaleraldehyde are exceeded; this flavor change is increased by increasing isovaleraldehyde concentrations. By this means an analytical control of undesirable sensory changes caused by the Maillard reaction in carrots is available.

Table II

Air drying of the carrot variety "Bauer's Kieler Rote"

Air temp. (°C)	Drying time (min)	Water content (%)	C (mol %)	Isovaleraldehyde (ppm)	Sensory evaluation	
					Color	Flavor
90	40	22	5.9	0.9	typical	typical
	60	9	11.5	1.4	"	"
	75	6	19.4	2.0	"	sl.changed
	90	4	20.4	2.7	"	" "
60	60	33	2.1	0.8	typical	typical
	120	13	2.9	0.3	"	"
	180	7	8.0	0.5	"	"
	210	6	5.1	1.1	"	"

From Table II it can be seen that at air temperatures of 90 °C and 60 °C there is only a minor increase in isovaleraldehyde concentrations, correlating with very little or no sensory changes.

Table III shows the results of the described two-step air drying experiments; in accordance with Figure 8 only a small increase of Amadori compounds was observed during the second drying step.

The influence of the carrot variety on Maillard reaction Samples of six freeze-dried carrot varieties were equilibrated at room temperature to a water activity of 0.33 (17), corresponding to an average water content of 6.3 % (related to wet matter). Then the samples were heated to 55 °C for 30 h and the concentrations of Amadori compounds as well as the corresponding sensory changes were determined. The results are listed in Table IV. The amount of Amadori compounds formed by the heating process seems to be correlated

Table III

Air drying of the carrot variety "Bauer's Kieler Rote"

Air temp. (°C)	Drying time (min)	Water content (%)	C (mol %)	Sensory evaluation	
				Color	Flavor
10 min 110°/ 60°	60	22	2.5	typical	typical
	90	12	2.3	"	"
	120	9	5.2	"	"
	240	5	7.9	"	"
30 min 110°/ 60°	40	21	10.3	typical	typical
	75	11	10.9	"	"
	105	8	14.1	"	"
	160	6	15.3	sl. brown	sl. burnt

Table IV

Reducing sugar and amino acid content/browning activity of different carrot varieties (heated at 55 °C for 30 h) (water content: 6.3 %)

Variety	Glucose + Fructose (mmol/g)	Amino acids (mmol/g)	C (mol %)	Sensory evaluation	
				Color	Flavor
Pariser Markt	1.61	0.13	65.6	typical	sl. burnt
Rubin	1.34	0.11	64.8	sl. brown	burnt
Nantaise	1.21	0.11	61.3	typical	sl. changed
Kundulus	0.86	0.11	48.5	"	sl. burnt
Rubika	0.27	0.20	22.6	sl. brown	" "
Bauer's K. Rote	0.16	0.23	13.6	typical	sl. changed

with the concentrations of reducing sugars present. Furthermore Table IV shows that the ratio of Amadori compounds is not a general measure for sensory changes to be expected; the tolerable concentration limits must be determined separately for each carrot variety.

In Figure 9 the formation of Amadori compounds dependent on water content is presented for three different carrot varieties (heating time: 20 h at 55 °C). It is remarkable that the varieties "Pariser Markt" and

"Kundulus" show a very distinct maximum in reaction rate, similar to the varieties "Rubin" and "Nantaise", whereas the varieties "Bauer's Kieler Rote" as well as "Rubika" show only a very flat characteristic.

From these findings it becomes evident that the lower moisture interval is very critical in respect to quality impairment caused by the Maillard reaction; they explain why final drying for too long times has a detrimental effect. Moreover, as already mentioned, the activation energy of the Maillard reaction is increased by decreasing the water content (14). For these reasons application of low air temperatures during final drying becomes imperative.

Drying experiments using different carrot varieties (Figure 10) show differences in formation of Amadori compounds dependent on the variety as already shown in Table IV. The analytical results of these drying experiments are presented in Table V, which indicates that one-step air drying at 90 °C does not provide the same good sensory quality for all varieties; "Nantaise", "Rubika", and "Bauer's Kieler Rote" show the lowest tendency to undergo undesirable sensory changes induced by the thermal treatment.

Table V

Air drying of different carrot varieties (air temp. 90 °C)

Variety	Water content (%)	C (mol %)	Isovaleraldehyde (ppm)	Sensory evaluation	
				Color	Flavor
Pariser Markt	5.5	57.4	1.48	sl. brown	sl. burnt
Rubin	5.5	42.8	1.88	brown	burnt
Nantaise	6.0	41.8	0.59	sl. brown	typical
Kundulus	6.0	21.9	0.69	" "	caramellic
Rubika	7.0	16.8	1.84	" "	sl. changed
Bauer's K. Rote	7.5	13.6	0.79	" "	" "

According to Table V the tolerable limits of isovaleraldehyde concentration seem to be variety dependent: while "Rubika" containing 1.8 ppm isovaleraldehyde does not exhibit distinct undesirable flavor changes, "Rubin" having the same isovaleraldehyde concentration and "Pariser Markt" with an even lower amount of isovaleraldehyde already show a burnt flavor character. These findings may be attributed to the fact that

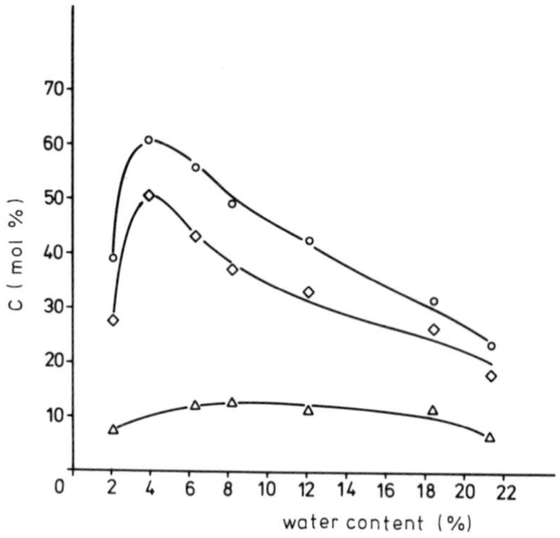

Figure 9. Formation of Amadori compounds in different varieties of carrots by heating for 20 h at 55 °C, dependent on water content. Key: ○, *Pariser Markt;* ◇, *Kundulus; and* △, *Bauer's Kieler Rote.*

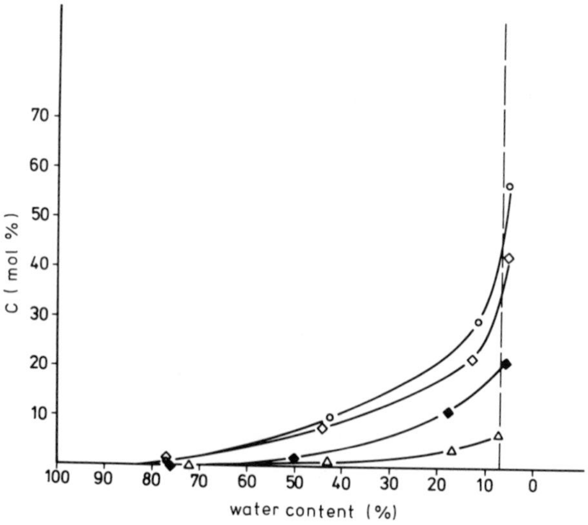

Figure 10. Increase of the concentrations of Amadori compounds during air drying for 90 min of different carrot varieties at an air temperature of 90 °C dependent on water content. Key: ○, *Pariser Markt;* ◇, *Rubin;* ◆, *Kundulus; and* △, *Bauer's Kieler Rote.*

"Rubika" has a higher concentration of the amino acid leucine; therefore in this case more isovaleraldehyde may be generated compared to other volatile products contributing to off-flavor.

The general conclusion that a two-step drying process will improve product quality was verified for different carrot varieties by drying at an air temperature of 90 °C in the first step and subsequent final drying at 70 °C (Figure 11). As shown in Table VI, in this case the concentrations of Amadori compounds and isovaleraldehyde are greatly lowered, resulting in a good product quality.

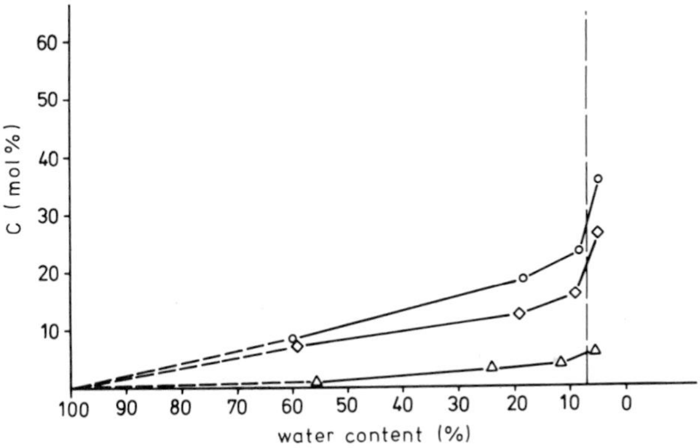

Figure 11. Increase of the concentrations of Amadori compounds during the two-step air drying of different carrot varieties, dependent on water content. Key: – – –, 25 min at 90 °C; —, 155 min at 70 °C; ○, Pariser Markt; ◇, Rubin; and △, Bauer's Kieler Rote.

Table VI

Air drying of different carrot varieties (air temp.: 25 min 90 °C/155 min 70 °C)

Variety	Water content (%)	C (mol %)	Isovaleraldehyde (ppm)	Sensory evaluation	
				Color	Flavor
Pariser Markt	5.0	35.6	0.66	typical	typical
Rubin	5.0	26.2	0.71	sl. brown	sl. changed
Nantaise	4.5	20.3	0.41	typical	typical
Kundulus	5.0	14.6	0.56	"	"
Rubika	6.5	9.8	1.12	"	"
Bauer's K. Rote	5.5	5.8	0.94	"	sl. changed

Conclusion

The drying experiments described provide starting points for optimization of drying processes. Since temperature and water content change simultaneously during drying, further experiments should be performed for investigating the influence of temperature and water content on Maillard reaction rate independently. From the reaction rate constants and the apparent activation energies dependent on water content, the course of Maillard reaction during any drying process could be calculated.

Acknowledgements

We are obliged to the German Association "Arbeitsgemeinschaft Industrieller Forschungsvereinigungen" (AIF) for financial support for this research work.

Literature Cited

1. Hodge, J. E. J. Agr. Food Chem. 1953, 1, 928.
2. Reynolds, T. M. Adv. Food Res. 1963, 12, 1.
3. Reynolds, T. M. Adv. Food Res. 1965, 14, 168.
4. Eichner, K. "Water Relations of Foods"; Duckworth, R. B., Ed.; Academic Press, New York, 1975; Section 5, p. 417.
5. Eichner, K.; Ciner-Doruk, M. "Maillard Reactions in Food"; Eriksson, C., Ed.; Progress in Food and Nutrition Science, Pergamon Press, Oxford,1981, p. 115.
6. Anet, E. F. L. J.; Reynolds, T. M. Nature 1956, 177, 1082.
7. Anet, E. F. L. J.; Reynolds, T. M. Aust. J. Chem. 1957, 10, 182.
8. Huang, I-Yih; Draudt, H. N. Food Technol. 1964, 18, 124.
9. Reynolds, T.M. Food Technol. Aust. 1970, 22, 610.
10. Buttery, R. G.; Teranishi, R. J. Agr. Food Chem. 1963, 11, 504.
11. Sapers, G. M. J. Food Sci. 1970, 35, 731.
12. Ferretti, A.; Flanagan, V. P. J. Agr. Food Chem. 1972, 20, 695.
13. Persson, T.; v. Sydow, E. J. Food Sci. 1974, 39, 537.
14. Hendel, C. E.; Silveira, V. G.; Harrington, W. O. Food Technol. 1955, 9, 433.
15. Kluge, G.; Heiss, R. Verfahrenstechnik 1967, 1, 251.

16. Görling, P. Ind. Obst- und Gemüseverwertung 1962, 47, 673.
17. Rockland, L. B. Anal. Chem. 1960, 32, 1375.
18. Schönberg, R.; Moubacher, R. Chem. Rev. 1952, 50, 261.

RECEIVED January 18, 1983

16

Characterization of Antioxidative Maillard Reaction Products from Histidine and Glucose

H. LINGNERT and C. E. ERIKSSON
SIK–The Swedish Food Institute, Box 5401, S-402 29 Göteborg, Sweden

G. R. WALLER
Oklahoma State University, Department of Biochemistry, Stillwater, OK 74078

Maillard reaction products (MRP) from histidine and glucose were fractionated by various methods to isolate the antioxidative products. Upon dialysis through membranes with a nominal molecular weight cut-off of 1000 daltons, antioxidative components were concentrated in the retentate. Further purification of the antioxidative material was obtained by isoelectric precipitation of the retentate at pH 5.0. The precipitate was considerably more antioxidative than the supernatant. The precipitate was fractionated by preparative electrophoresis. The most antioxidative electrophoresis fraction is now being analyzed by spectrometric methods. Its C, H, N, O content was determined and EPR studies showed it to contain stable free radicals. EPR studies of several fractions of the histidine-glucose reaction mixture showed good agreement between intensity of EPR signal and antioxidative effect. The free radicals are suggested to be of importance for the antioxidative mechanism of the MRP.

Unsaturated fatty acids in foods are very susceptible to oxidation by oxygen in the air during processing and storage. The oxidation results initially in the formation of fatty acid hydroperoxides by a free radical chain mechanism. The hydroperoxides are subject to several further reactions forming secondary products such as aldehydes, ketones, and other volatile compounds, many of which are odorous and cause rancid flavor in the food. This development of rancid flavor limits the storage stability of a large number of food products.

The foods can be protected against lipid oxidation either by the addition of antioxidants or by packaging in vacuum or inert gases to exclude oxygen. The antioxidants can be of various types. They can work as "chain-breakers" that interfere with the free radical chain reaction, as "metal inactivators", that bind otherwise pro-oxidative metals, or as "peroxide destroyers", which react with hydroperoxides to give stable products by nonradical processes (1).

0097-6156/83/0215-0335$06.00/0

The most common antioxidants are phenols, such as butylated hydroxyanisole (BHA) or butylated hydroxytoluene (BHT). Increasing interest has, however, been directed towards the utilization of normal food constituents with antioxidative properties (2). Among those, the Maillard reaction products (MRP) might be of special importance, since they are so widespread in foods.

The first report on antioxidative effect of MRP was made by Franzke and Iwainsky (3). Shortly afterward Griffith and Johnson (4) reported that the addition of glucose to cookie dough resulted in a better stability against oxidative rancidity during storage of the cookies. Research on antioxidative MRP was then mainly performed by groups in Japan. A symposium on Maillard Reactions in Food held in Uddevalla, Sweden, 1979 included also the aspect of antioxidative properties. The contributions on this subject contained also brief reviews (5, 6, 7). Most of the work has been done on model systems. Some applications in food systems have, however, also been reported (8 - 11).

Identification of antioxidative MRP

Knowledge about the chemical structure of the antioxidative MRP is very limited. Only a few attempts have been made to characterize them. Evans, et al. (12) demonstrated that pure reductones produced by the reaction between hexoses and secondary amines were effective in inhibiting oxidation of vegetable oils. The importance of reductones formed from amino acids and reducing sugars is, however, still obscure. Eichner (6) suggested that reductone-like compounds, 1,2-enaminols, formed from Amadori rearrangement products could be responsible for the antioxidative effect of MRP. The mechanism was claimed to involve inactivation of lipid hydroperoxides.

On the other hand, Yamaguchi, et al. (5) found most of the antioxidative effect in the melanoidin fraction. By various chromatographic methods they purified an antioxidative product from glycine and xylose having a molecular weight of approximately 4500.

Possibly several different compounds formed by the Maillard reaction can exhibit antioxidative properties. Their formation might be dependent on what reactants are used and on the reaction conditions (temperature, time, water content etc.)

The aim of the present work was to characterize the antioxidants formed in the reaction between histidine and glucose in order to elucidate the mechanism of their antioxidative action. The combination histidine-glucose was chosen since it previously was found to be one of the most effective combinations in model systems (13).

Synthesis of MRP

Maillard reaction products were obtained by refluxing 100 ml of distilled water containing 0.1 mol L-histidine monohydrochloride monohydrate and 0.05 mol D-glucose for 20 h. The pH of the reaction mixture was adjusted to 7.0 with potassium hydroxide before starting the reaction.

Measurement of antioxidative effect

The antioxidative effect of the various fractions of MRP was evaluated by a previously described polarographic method (14).

Separation of antioxidative MRP

In the course of trying to isolate the antioxidative MRP, we found that part of the antioxidative effect was lost during the isolation processes. This could be explained in two ways. Either the antioxidants were unstable or there were different antioxidants in the reaction mixture acting synergistically, resulting in a lower effect when such antioxidative components had been separated. Since recombination of the various fractions failed to restore the original antioxidative effect, we studied the stability of the antioxidants in more detail. The antioxidants were then found to be sensitive to oxygen (15). With the knowledge of the instability of the antioxidants, special care was used to avoid contact with oxygen in every single separation step. All solvents were degassed and bubbled with nitrogen prior to use and the various fractions were frozen as soon as possible.

Dialysis Previously we reported that the antioxidants from histidine and glucose could be concentrated by ultrafiltration (7). The results indicated that the antioxidative compounds had a molecular weight of more than 1000. Therefore, we further studied the possibility of separating the antioxidants with respect to their molecular size by dialysis. Dialysis tubing with a molecular weight cut-off of 1000 daltons was used (Spectrapor 6, Spectrum Medical Industries Inc., Los Angeles, CA). A 15-ml portion of the Maillard reaction mixture was transferred to each of five dialysis tubes. The contents of the five tubes were dialyzed together against 6.5 l of degassed, nitrogen-bubbled, distilled water at $5^{o}C$. After 12 h one of the tubes was withdrawn and the other four were transferred to a new dialysis tank containing 6.5 l of water and dialysis continued for another 12-h period. Again one of the tubes was withdrawn while dialysis was renewed for the other three for another 12-h period and so on. Figure 1 shows the antioxidative effect and the total amount of material in each of the five retentates and in each of the five dialysates. The total amounts of material (open bars) are given as percentage of the original material in one dialysis tube. The antioxidative effect (filled bars) is compared on an equal-weight basis.

After the third dialysis no further improvement of the antioxidative effect of the retentate was obtained, even though small amounts of less antioxidative material still passed the membrane and were found in the dialysate. After dialysis for 3 x 12 h only 3.5 % of the original material remained in the retentate.

Precipitation After the fourth dialysis some precipitate could be observed in the retentate in the dialysis tube. The precipitation was found to be pH dependent; the precipitate was reversibly dissolved when pH was either increased or decreased. The influence of pH on the

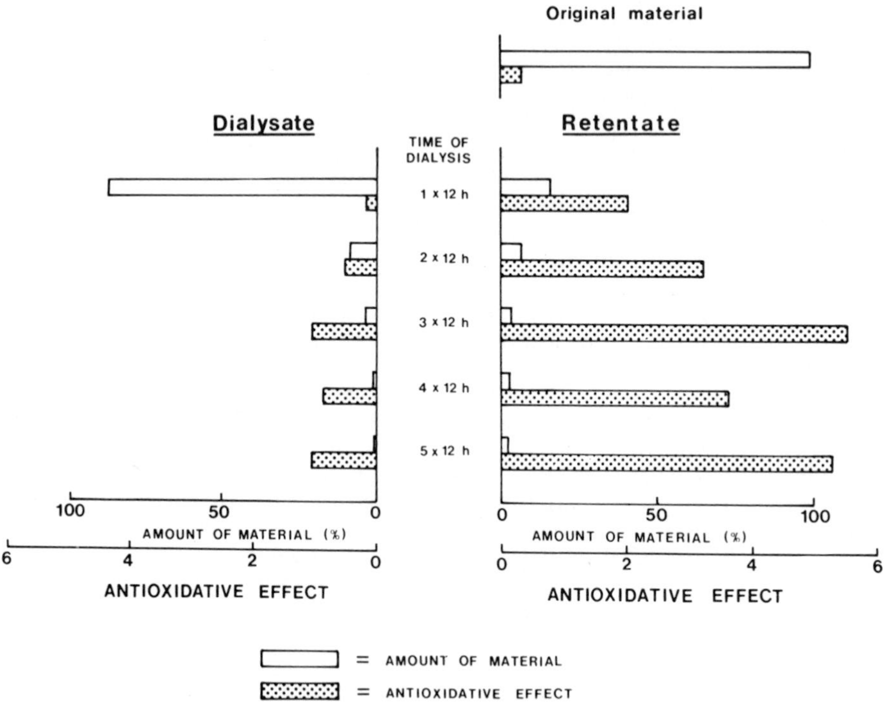

Figure 1. Antioxidative effect and amount of material in retentates and dialysates after dialysis of histidine–glucose reaction mixture up to five times 12 h through dialysis tubing with a molecular weight cut-off of 1000 daltons.

precipitation is shown in Figure 2, where the absorption at 450 nm of the solution is given as a function of pH. Since the precipitate was brown colored, the color intensity of the supernatant could in this way be used as a measure of the extent of precipitation. It can be seen that maximum precipitation was obtained at pH 5.0. At pH lower than 3 or higher than 7 no precipitation could be observed.

This precipitation occurred only with the dialyzed material. Small amounts of precipitate were obtained already with the first rententate, when adjusting the pH to 5.0. The crude reaction mixture was, however, fully soluble all over this pH range. Obviously salts or other material of low molecular weight prevent the precipitation. Addition of salts to the dialyzed material was also shown to decrease the extent of the isoelectric precipitation.

Since the precipitate was by far more antioxidative than the supernatant or the original retentate, when compared on a weight basis, the isoelectric precipitation could be used as one further step in the purifiaction sequence.

Electrophoresis A solution of the precipitate was fractionated by preparative paper electrophoresis at pH 1.9 (acetic acid - formic acid buffer). A 30-mg portion of the precipitate was dissolved in 0.45 ml of the buffer and applied as a 20-cm band on the paper (Whatman Chromatography Paper 3 $\frac{M}{M}$). The sample was chromatographed at 3000 V for 30 min.

Table I. Antioxidative effect and amount of material in each fraction after fractionation of precipitate by paper electrophoresis.

Fraction No.	Amount of material (%)	Antioxidative effect
Original precipitate	100	2.0
1	8	0.6
2	5	0.5
3	31	1.1
4	43	1.7
5	22	0.7

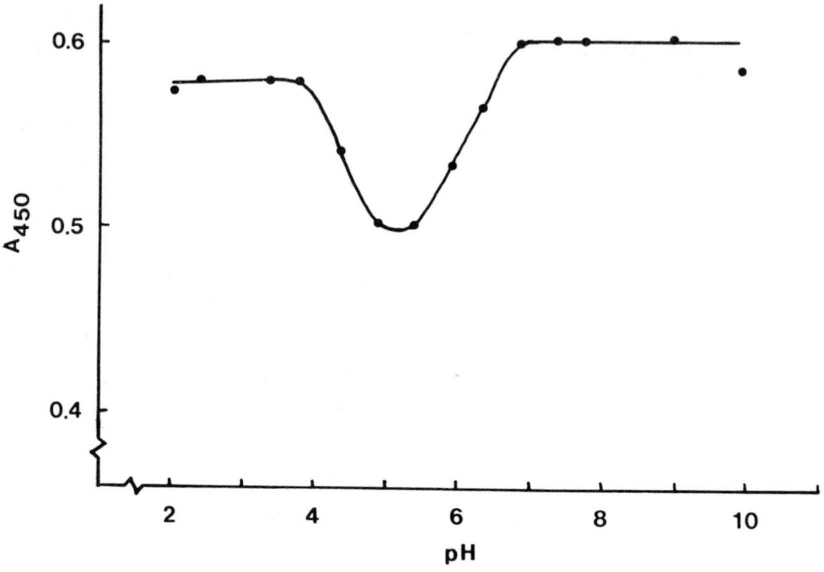

Figure 2. Influence of pH on the formation of precipitate in the retentate of Maillard reaction products from histidine and glucose.

No distinct bands were obtained in the chromatogram, except for an intense fluorescent band of rather low R_f value. The chromatogram was, therefore, divided into five fractions, one being the fluorescent one. The material in each fraction was extracted with acetic acid and its weight and antioxidative effect were determined. The results are given in Table I. The amount of material is expressed as percent of the original precipitate and its antioxidative effect is compared on an equal weight basis. Fraction No. 2 was the fluorescent one. All the fractions were more or less brown colored. The most colored fraction (No. 4) was found to contain most of the material and was also the most antioxidative. The fluorescent fraction was the smallest one and also the least antioxidative.

As can be seen in Table I all the electrophoresis fractions exhibited less antioxidative effect than the precipitate, possibly as a result of oxidation during the separation, in spite of all precautions taken to avoid the contact with oxygen.

Table II summarizes the yield and the antioxidative effect of products obtained in the various steps in the purification of the antioxidants from histidine and glucose. The yield is expressed as percent of the starting material, the crude Maillard reaction mixture. The antioxidative effect of the various fractions is compared to that of the crude reaction mixture on a weight basis, the crude reaction mixture being given the value 1. The relative antioxidative effect of 6 for the retentate means, for example, that the retentate gives the same antioxidative effect as the crude reaction mixture with only one sixth of the amount of material. In the table is also shown the calculated "total antioxidative effect" ("yield " x "relative antioxidative effect").

Table II. Yield and antioxidative effect of material at various steps of the purification of MRP from histidine and glucose.

	Yield (%)	Relative antioxidative effect	Total antioxidative effect
Crude reaction mixture	100	1	100
Retentate	2	6	12
Precipitate	0.2	12	2.4
Electrophoresis fraction No. 4	0.1	6	0.6

Table II shows that as little as 0.2% of the original material remained in the precipitate, the most antioxidative fraction. This fraction contained only 2.4% of the original antioxidative effect. It is obvious that much of the original antioxidative effect was lost during the purification. It should, however, be kept in mind that the main aim in the purification was not to get maximum yield but to get as pure fractions as possible.

Characterization of purified fraction

Electrophoresis fraction No. 4 was chosen for further chemical characterization, in spite of the fact that it was not as antioxidative as the precipitate. It was, however, considered to be more homogeneous. At least it was purified from the fluorescent material. Still, its purity is uncertain. The fact that no well defined bands were obtained in the electrophoresis indicates that the fractions consisted of more than one compound, possibly differing slightly in molecular weight and/or content of charged groups. Another explanation to the lack of distinct bands could be unselective adsorption between the compound and the paper. This is supported by the fact that rechromatography of the five electrophoresis fractions failed to reproduce the same fractions distinctly.

Despite the uncertainty regarding the purity, attempts are in progress to characterize electrophoresis fraction No. 4 by methods such as mass spectrometry, nuclear magnetic resonance spectrometry, and infrared spectrometry. The aim is to at least get an idea of the molecular weight of the compound(s) and its content of specific functional groups that might explain its antioxidative properties.

Elementary composition The C, H, N, and O content of the electrophoresis fraction No. 4 is given in Table III. The table also shows that the fraction had an ash content of 2.9%. The origin of this relatively high content of inorganic matter is not clear.

Table III. Composition of electrophoresis fraction No. 4.

C	54.3%
H	5.4%
N	12.6%
O	24.5%
Ash	2.9%

Electron Paramagnetic Resonance studies Analysis by electron paramagnetic resonance spectrometry (EPR) showed the electrophoresis fraction to contain stable free radicals. A strong EPR signal was obtained at the g factor value of 2.0035 ± 0.0003. The involvement of free radicals in the Maillard reaction has previously been reported (16). Recently also Lessig and Baltes (17) reported the content of extremely stable free radicals in melanoids obtained from the reaction between glucose and 4-chloroaniline.

Examination of various fractions from the purification sequence by EPR showed that, on the whole, there was a good agreement between the intensity of the EPR signal and the antioxidative effect. This is shown in Figure 3. The increasing antioxidative effect in the sequence crude reaction mixture - retentate - precipitate is accompanied by an increased EPR signal. The supernatant, being less antioxidative, also gave a less intense EPR signal. Furthermore, when the retentate was incubated in air for 75 h both the antioxidative effect and the EPR signal were decreased. When it was incubated for the same length of time under nitrogen an increase of both antioxidative effect and EPR signal was noticed, for some reason. There is no direct proportionality between the antioxidative effect and the EPR signal, but the tendency is obvious.

These findings are of interest with respect to the mechanism of the antioxidative action of the MRP. The stable free radicals of the MRP might interact with the free radicals formed in the lipid oxidation and thus lead to an inhibition of the lipid oxidation chain mechanism.

Concluding remarks

A constant observation when the MRP were separated by various methods was that antioxidative effect was found in many different fractions. Both the dialysates and the retentates from dialysis were antioxidative to some extent. All the electrophoresis fractions exhibited some antioxidative effect. Attempts to separate the MRP by column chromatography on Sephadex G-50 have resulted in several fractions with some antioxidative effect, and so on. This indicates that several antioxidative products are formed by the Maillard reaction, possibly differing in molecular size and chemical structure, but perhaps with one single antioxidative functional group in common, such as a free radical function. However, it can not be excluded that the MRP contain a few entirely different antioxidants with different modes of action. Various mechanisms have also been suggested. Eichner (6) claimed MRP to inactivate the hydroperoxides formed by the lipid oxidation. There are also reports on the complex binding of metals by MRP (18, 19).

Another possible explanation of the observation that antioxidative effect is found in several fractions of MRP could be that the antioxidative compound is strongly adsorbed on other MRP and is for that reason difficult to isolate. If that should be the case, the electrophoresis fraction could consist of small amounts of highly effective antioxidants

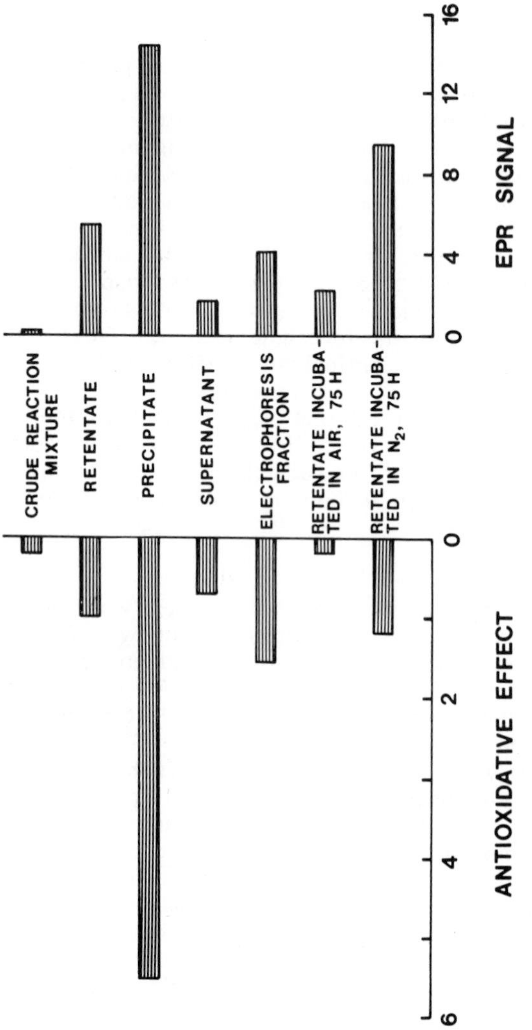

Figure 3. Antioxidative effect and intensity of electron paramagnetic resonance signal for various fractions of Maillard reaction products from histidine and glucose.

of low molecular weight, possibly free radicals, adsorbed on the main component of higher molecular weight. Caution must therefore be used when drawing conclusions about the antioxidative mechanism based on information about the main chemical structures in the fraction .

Literature cited

1. Frankel E.N. Prog. Lipid Res. 1980, 19, 1-22.
2. Dugan, L.R. "Autoxidation in Food and Biological Systems"; Plenum Press: New York, 1980; p. 261.
3. Franzke, C.; Iwansky, H. Deut. Lebensm.-Rundschau 1954, 50, 251-254.
4. Griffith, T.; Johnson, J.A. Cereal Chem. 1957, 34, 159-169.
5. Yamaguchi, N.; Koyama, Y.; Fujimaki, M. Prog. Food Nutr. Sci. 1981, 5, 429-439.
6. Eichner, K. Prog. Food Nutr. Sci. 1981, 5, 441-451
7. Lingnert, H.; Eriksson, C.E. Prog. Food Nutr. Sci. 1981, 5, 453-466.
8. Tomita, Y. Kagoshima Daigaku Nogakubu Gukujutusu Hokoku 1972, 22, 115-121; Chem. Abstr. 1973, 78, 96215h.
9. Lingnert, H. J. Food Process. Preserv. 1980, 4, 219-233.
10. Lingnert, H.; Lundgren, B. J. Food Process. Preserv. 1980, 4, 235-246.
11. Vandewalle, L.; Huyghebaert, A. Med. Fac. Landbouww. Rijksuniv. Gent 1980, 45, 1277-1286.
12. Evans, C.D.; Moser, H.A.; Cooney, P.M.; Hodge, J.E. J. Am. Oil Chem. Soc. 1958, 35, 85-88.
13. Lingnert, H.; Eriksson, C.E. J. Food Process. Preserv. 1980, 4, 161-172.
14. Lingnert, H.; Vallentin, K.; Eriksson, C.E. J. Food Process. Preserv. 1979, 3, 87-103.
15. Lingnert, H.; Waller, G.R. J. Agric. Food Chem. In press.
16. Namiki, M.; Hayashi, T. J. Agric. Food Chem. 1975, 23, 487-491.
17. Lessig, U.; Baltes, W. Z. Lebensm. Unters. Forsch. 1981, 173, 435-444.
18. Kajimoto, G.; Yoshida, H. Yukagaku 1975, 25, 297-300; Food Sci. Techn. Abstr. 1976, 8, 6N222.
19. Gomyo, T.; Horikoshi, M. Agric. Food Chem. 1976, 40, 33-40.

RECEIVED November 4, 1982

NUTRITIONAL ASPECTS

17

The Effect of Browned and Unbrowned Corn Products on Absorption of Zinc, Iron, and Copper in Humans

P. E. JOHNSON, G. LYKKEN, J. MAHALKO, D. MILNE, L. INMAN, and H. H. SANDSTEAD

U.S. Department of Agriculture, Agricultural Research Station, Grand Forks Human Nutrition Research Center, Grand Forks, ND 58202

W. J. GARCIA and G. E. INGLETT

U.S. Department of Agriculture, Northern Regional Research Center, Peoria, IL 61604

Retention and excretion (i.e., balance) of iron, zinc, and copper were determined for men consuming a constant diet which contained either corn flakes or corn grit porridge. After a meal containing ^{65}Zn-labeled corn flakes or grits, subjects were counted in a whole body counter to determine absorption and retention of the radioisotope. Stable isotopes, ^{54}Fe, ^{67}Zn and ^{65}Cu, were mixed with an identical meal. Absorption was calculated from fecal excretion of the stable tracer. Balance, stable isotope, and radioisotope absorption did not differ significantly between diets for any of the three metals. Long-term ^{65}Zn retention was reduced when corn flakes were fed. Urine of subjects consuming corn flakes contained higher-molecular-weight amino-containing substances than that of subjects eating corn grits. Urinary zinc was bound by the heavier compounds.

Toasted and browned foods are rich in products of the Maillard reaction. Much effort has been devoted to investigating the effect of the Maillard reaction on protein quality, flavor, and aroma. Recently the influence of Maillard products on metal metabolism has also been investigated.

In 1975 Freeman et al. (1) reported excessive urinary zinc losses in patients undergoing total parenteral alimentation (TPN). These losses were not observed when the same patients were given the TPN solution via a nasogastric tube, or when they were given a similar solution prepared by autoclaving the sugar and amino acid mixtures separately. Sugar-amine compounds were found in the TPN solutions prepared by autoclaving amino acids and dextrose together, but not in those made from solutions autoclaved separately. It was hypothesized that compounds

Published 1983, American Chemical Society

produced from the reaction of dextrose with amino acids were chelating zinc and increasing its excretion in the urine.

The same investigators subsequently reported increased urinary excretion of copper and iron, as well as zinc, in patients receiving TPN (2). They also detected sugar-amine compounds in the TPN solutions and in the patients' urine and plasma.

When van Rij and co-workers (3) infused patients with various amino acid preparations, they found urinary zinc excretion to be higher with a sugar-containing preparation than when they used the same preparation without sugar. A unique zinc-binding substance was present in the sugar-containing solution and in the urine of patients infused with it.

Although there is evidence that at least some products of the Maillard reaction are absorbed from the digestive tract (4-7), there has been little work on the question of whether or not browning in ordinary diets may affect metal absorption or excretion. In rats, a diet containing 3% Maillard products caused a significant decrease in retention of calcium, phosphorus, magnesium, and of copper in axenic but not holoxenic animals (8). Rats fed a diet of browned egg albumin for 12 months had lower hemoglobins and hematocrits than pair-fed control rats (9). However, anemic rats fed diets containing casein and glucose heated together, absorbed more iron than rats fed diets to which glucose was added after heating (10).

The present study was designed to investigate metal absorption in humans eating conventional diets containing a typical browned food, corn flakes, compared to absorption from diets as free as possible from browning.

Materials and Methods

Adult male subjects age 23-65 in good general health participated in this study while living in the Metabolic Unit of the Center. They gave consent after being informed of the purpose of the research and its potential hazards. This project was approved by the Human Studies Committees of the University of North Dakota School of Medicine and of the USDA Science and Education Administration. Informed consent and experimental procedures were consistent with the Declaration of Helsinki.

Diets The diet was a weighed diet on a three-day rotating-menu cycle, with energy distributed as 45% carbohydrate, 15% protein, and 40% fat. A typical menu is shown in Table I.

The diet was planned with a test breakfast to compare the availabilities of zinc, iron, and copper from browned corn products (corn flakes) with those from unbrowned corn products (hot corn grit cereal). The breakfast served for all 3 days of the menu cycle at all energy levels was identical, with only the

Table I

Labeled Cornflakes Basal Diet
3000 Kcal
Day 2

Breakfast	Lunch
Orange juice	Grapefruit juice
Cornflakes or corn cereal	Beef stew w/green beans
Bread, white, w/o crust	Mashed potatoes
Margarine	Cheese
Milk, NFDM prepared*	Bread, white, w/o crust
Sugar	Margarine
Evening	**Snack**
Chicken a la king	Cheese
Rice pilaf	Sugar cookie
Bread, white, w/o crust	
Butter	
Pears	
Milk, NFDM prepared*	

*nonfat dried milk

corn products changing. A constant dry weight of corn grits (53 g) was served as 56 g corn flakes (5% moisture) or 58 g corn grits (9% moisture). Coffee and tea were not allowed at breakfast.

All ingredients of the diet were purchased in lots large enough to be used for the entire study. Food items were weighed to 0.01 g during preparation. Preparation procedures and serving times were kept as constant as possible. Salt and pepper were weighed and served for an entire day; all other foods were completely consumed at specific meal times. Salt, pepper, coffee, and tea were allowed in constant amounts that were adjusted to the volunteer's preference. The level of browning products in the diet was reduced by serving bread without crust, and by baking cookies and desserts in a microwave oven. Mineral intake was 16.2 ± 0.7 mg Fe/day, 10.4 ± 0.6 mg Zn/day and 1.1 ± 0.04 mg Cu/day.

The corn products were prepared at the USDA Northern Regional Research Center, Peoria, IL. The unlabeled products were made from unenriched, degermed yellow corn grits, ground to 40 mesh. The corn grits were made into hot corn grit cereal, using water and salt, or into toasted corn flakes, using water to make an initial batter. Corn flakes and grits labeled with ^{65}Zn were prepared from intrinsically labeled corn endosperm-hull flour which was made from corn grown and milled as reported previously (11).

Three volunteers received both the corn flakes and the corn grit cereal in a crossover design. A fourth volunteer received the grit diet and then withdrew from the study. Each diet period began with 21 days of equilibration, which was followed by 21 days of balance. The equilibration period allowed time for any metabolic or physiological changes which might have occurred in the subjects due to a change in diet. During the balance period, duplicates of all meals were prepared and all urine and feces were collected for analysis. "Balance" is the difference between intake and excretion of a nutrient. Stable isotopes and radioisotopes were fed in separate test meals three days apart at the beginning of the balance periods on each diet.

Radioisotopes ^{65}Zn-labeled corn products were served once in the form of corn flakes and once in the form of corn grit cereal to each volunteer. At the time of feeding the radioactively labeled cereal, a portion was measured to obtain the desired level of radioactivity and then the corresponding unlabeled product was added to obtain the total desired weight of cereal. Each test meal contained 0.1-0.5 μCi ^{65}Zn.

Radioactivity in each volunteer was determined by counting in a whole body counter at 0.5 and 4 hours after ingestion of the labeled meal and at 1, 2, 3, 4, 6, 13, 20, 27 and 34 days after the meal. The whole body counter consists of a bed placed between two movable 28 x 10 cm NaI(TI) gamma scintillation detectors located in a steel room. Correction for self absorption and geometry was done by the method of Cohn et al. (12). Correction for body potassium-40 and environmental ^{219}Rn and ^{215}Bi was also made. Absorption and retention were determined by computer-fitting of time vs ^{65}Zn retention data.

Stable Isotopes Test meals at the beginning of each diet period were labeled extrinsically with ^{65}Cu, ^{67}Zn or ^{70}Zn, and ^{54}Fe or ^{57}Fe. The stable isotopes were obtained in the oxide form from Oak Ridge National Laboratory. Each volunteer received 2 mg ^{65}Cu (99.69 atom %), 4 mg ^{54}Fe (97.61 atom %), or ^{57}Fe (90.24 atom %) and 4 mg ^{67}Zn (93.11 atom %) or 3.3 mg ^{70}Zn (65.51 atom %, 5 mg total Zn). The isotopes were dissolved in hydrochloric acid, neutralized to pH 4, and mixed with orange juice at breakfast. Duplicate solutions of metal isotopes were added to duplicate diets prepared for analysis. Feces were collected in 3-day pools for 18 days following ingestion of the isotopes. Unabsorbed isotopes appeared in the feces and the levels of enrichment was measured using mass spectrometry. Metals in fecal ash were separated by anion exchange chromatography, chelated with tetraphenylporphyrin, and assayed as described elsewhere (13). Fractional absorption was calculated by determining the difference in the amount of stable isotope ingested compared to that recovered in the feces.

Metal Balances Duplicates were prepared of all meals consumed by the volunteers during the 21-day balance periods. All urine and feces were collected in three-day pools. Diets were homogenized and weighed aliquots taken for analysis. Feces were freeze-dried, weighed, pulverized and blended, and weighed aliquots taken for analysis. Total urinary output was recorded and aliquots taken for analysis. Diets and feces were digested with nitric and perchloric acids, and metals were determined using an inductively coupled argon plasma spectrometer. Urinary metals were measured using atomic absorption spectrophotometry.

Metal binding by corn flakes and grits was measured using the method of Camire and Clydesdale (14). Free amino groups in both cereals were measured after a pepsin-pancreatin digestion (5).

Zinc-binding components of urine were examined using modified gel chromatography (15). Urine (3 ml) was chromatographed on Sephadex G-25 columns (2.5 x 40 cm) equilibrated with a buffer containing 10 ppm Zn as $Zn(NO_3)_2$ and 10 mM Tris buffer, pH 7.4. Fractions of 3 ml were analyzed by atomic absorption spectroscopy. Void volume of the column was determined with blue dextran.

Changes in the molecular weight distribution of urinary components were examined by chromatography of urine (3 ml) on Sephadex G-25 columns (2.5 x 40 cm) with 0.1 M phosphate buffer, pH 8 (5). The resulting fractions were hydrolyzed with HCl and reacted with ninhydrin (5). Absorbance was determined at 570 nm.

Results and Discussion

Metal Balances Balance data for iron, zinc, and copper are shown in Table II. These figures represent the average balance in mg/day over a 21-day period. The mean balance for all subjects does not differ significantly between diet periods for any of the three metals. Subject #2055, who showed the greatest difference in balance between diet periods, was 65 years old, while the other three subjects averaged 25 years old. This subject also had the most erratic pattern of defecation, which may have had an influence on his balance data.

Subject #2070 was inadvertently given 20 mg ^{65}Cu rather than 2 mg with the corn grit test meal. This probably caused the slightly positive copper balance he had while on the corn grit diet.

Total urinary metal excretion was not significantly different between diet periods for any of the three metals. Average urinary excretion was 0.50 $\pm$ .08 mg Zn/day, 0.11 $\pm$ .04

Table II

Iron, Zinc, and Copper Balance
(mg/day)

Subject	Iron		Zinc		Copper	
	Flakes	Grits	Flakes	Grits	Flakes	Grits
2055	-2.73	3.32	-0.69	1.73	-0.25	-0.03
2056	3.33	3.80	2.60	1.01	0.00	-0.04
2065	-	2.26	-	0.40	-	-0.03
2070	3.34	1.18	2.20	1.18	0.01	0.12
mean ± SD	1.31 ± 3.50	2.64 ± 1.17	1.37 ± 1.79	1.08 ± 0.54	-0.08 ± 0.15	0.00 ± 0.08

mg Fe/day, and 0.05 ± .01 mg Cu/day. The lack of any change in urinary metal excretion is consistent with the report of Freeman et al. (1), who observed no change in urinary zinc excretion when patients were fed either the TPN solution via nasogastric tube or a typical hospital diet.

Stable Isotope Absorption Absorption values for stable iron, zinc, and copper from the breakfast test meal are given in Table III.

Table III

Absorption of Stable Iron, Zinc, and Copper from Breakfast Meal

Subject	Iron		Zinc		Copper	
	Flakes	Grits	Flakes	Grits	Flakes	Grits
2055	0%	0%	57%	47%	36%	61%
2056	34%	46%	26%	37%	56%	74%
2065	-	0%	-	25%	-	66%
2070	6%	14%	47%	48%	70%	49%
mean ± SD	13 ± 18%	15 ± 22%	43 ± 16%	39 ± 11%	54 ± 17%	65 ± 11%

There were no significant differences in stable tracer absorption between the two diets. All the absorption values appear to be fairly comparable to those reported in the literature. Iron absorption in adult males is generally less than 10% (16). This meal would be classified as a high-availability meal (17), with non-heme iron availability of 8%, due to the ascorbic acid content of the orange juice. Volunteer #2056 absorbed high amounts of iron, 34% and 46% from flakes and grits, respectively. He also had the most positive iron balance (Table II), which averaged 23% of intake, but ranged as high as 61% retention. This volunteer has participated in other studies in our laboratory and has consistently absorbed high amounts of iron (13), although his hematological status was normal at the time of this study. His serum ferritin was 27.7 ng/ml upon admission (normal ≥20 ng/ml) and his hemoglobin was 15.4 g/dl.

There are few data in the literature concerning zinc absorption from meals. However, King, et al. (18), also using stable tracers, reported 46% absorption of zinc from a formula meal by adult women who did not use oral contraceptives. Using ^{65}Zn tracers, Sandström, et al. (19) reported 15.7% absorption from a meal containing white bread, milk, and cheese. Absorption from white bread alone averaged 38%. Zinc absorption from other meals varied from an average of 8% from wholemeal bread to 36% from chicken (19, 20).

Absorption of copper in adult women consuming a formula diet has been reported to average 57.7 ± 17.7% (18). Our absorption figures for copper fall into this range.

Absorption of ^{65}Zn Absorption of zinc from a ^{65}Zn-labeled meal is given in Table IV and compared to measurements of zinc

Table IV

Comparison of ^{65}Zn and Stable Zn Absorption Measurements

		^{67}Zn	^{70}Zn	^{65}Zn
2055	Grits	47%	-	53%
	Flakes	-	57%	55%
2056	Grits	37%	-	35%
	Flakes	-	26%	51%
2065	Grits	25%	-	30%
2070	Grits	-	48%	39%
	Flakes	47%	-	46%

absorption from an identical meal labeled with stable zinc. The ^{65}Zn and stable zinc-labeled meals were separated by three days. The stable zinc-labeled meal contained 6.2-7.1 mg Zn, while the ^{65}Zn labeled meals contained 2.1 mg Zn. It has been found that the total amount of zinc in a meal will influence both total and percent absorption (19), which might explain some of the difference between absorption values between the two methods. However, the differences in absorption measured by the two methods are of the same size as the precision of the methods except for two cases using ^{70}Zn. In only one case is the absorption of the stable isotope significantly less than absorption of the radioisotope, which suggests that the size of the zinc dose given is not the major determinant of the difference between ^{65}Zn and stable zinc measurements.

It appears that ^{67}Zn is useful as tracer for determining zinc absorption, but that ^{70}Zn is not as suitable. This is not surprising, since ^{67}Zn determinations are more accurate and precise than ^{70}Zn determinations (13).

Retention of ^{65}Zn could be described as

$$R(t) = A_1 \exp(-L_1 t) + (1-A_1)\exp(-L_2 t)$$

where R(t) is the normalized retained ^{65}Zn activity and t is the time in days after ingestion. The parameters for this equation for the two diets are given in Table V.

Table V

^{65}Zn Retention Parameters

	A_1	L_1 (day^{-1})	L_2 (day^{-1})
Grits	0.14 ± .08	0.09 ± .01	0.0018 ± .0005
Flakes	0.25 ± .08	0.10 ± .03	0.0018 ± .0005

The larger average value of A_1 for the cornflakes breakfast indicates that a larger fraction of the absorbed zinc had a short term retention ($t_{1/2}$ = 7 day) and that a smaller fraction ($1-A_1$) had a long term retention ($t_{1/2}$ = 383 day). Thus ^{65}Zn retention was reduced when toasted flakes were fed.

Metal Binding in Corn Products Binding of iron, zinc and copper by corn flakes and grits was determined at pH 5, 6, and 7. Corn flakes bound more copper and less iron than did corn grits, as shown in Table VI.

Table VI

Ratio of Metal Bound
Flakes/Grits

	Fe	Zn	Cu
pH 5	0.76 ± 0.07*	1.76 ± 0.27	1.55 ± 0.18
pH 6	0.63 ± 0.07	1.03 ± 0.08	1.26 ± 0.04
pH 7	0.47 ± 0.01	1.02 ± 0.01	1.54 ± 0.05

*Mean ± SD of 3 determinations

Camire and Clydesdale (14) found an effect of toasting on binding of zinc, magnesium and ferrous iron by wheat bran. For zinc the effect was seen only at pH 5. We also see an effect on zinc binding only at pH 5.

The free amino content of both cereals was also measured after a pepsin-pancreatin digestion. The corn flakes contained 3.1 ± 0.4 mg leucine equivalents/g and the corn grits, 13.5 ± 2.1 mg leucine equivalents/g.

Zinc Binding by Urine Urine from subjects during both dietary treatments was chromatographed to examine patterns of zinc binding. When rats were fed toasted diets, they excreted more amino-containing compounds of molecular weight 3,000-5,000 than when they consumed untoasted diets (5). We observed the same phenomenon in our subjects (Figures 1 and 2). Typically, subjects excreted higher-molecular-weight materials in the urine while on the corn flake diet than when on the corn grits diet.

Insulin (molecular weight 5733) eluted at a Ve/Vo of 1.0-1.1. L-Phenylalanylglyclglycine (MW 279) eluted at Ve/Vo = 1.95 and L-glycylphenylalanylphenylalanine (MW 369) at Ve/Vo = 2.18. Molecular weight markers between these sizes were not available.

When urine was chromatographed with a zinc-saturated buffer, the higher-molecular-weight zinc-binding substances predominated on the corn flake diet, while the lower-molecular-weight substances zinc-binding peak was largest when grits were eaten. The amino-containing substances and the higher-molecular-weight zinc-binding peak coincided (Figure 2) on the corn flake diet. Apparently Maillard products are being

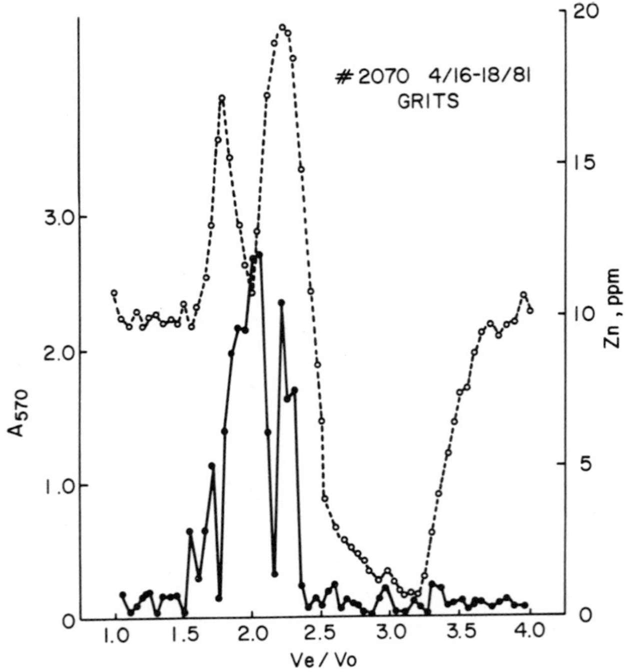

Figure 1. Typical chromatogram of urine from a subject eating corn grits. Key: ●, ninhydrin-positive substances after hydrolysis (absorbance at 570 nm); ○, zinc concentration in eluate.

Chromatography was performed on 2.5 × 40 cm column of Sephadex G-15, with either 0.1 M phosphate buffer, pH 8 (—) or 10 ppm as $Zn(NO_3)_2$ and 10 mL Tris buffer pH 7.4 (---).

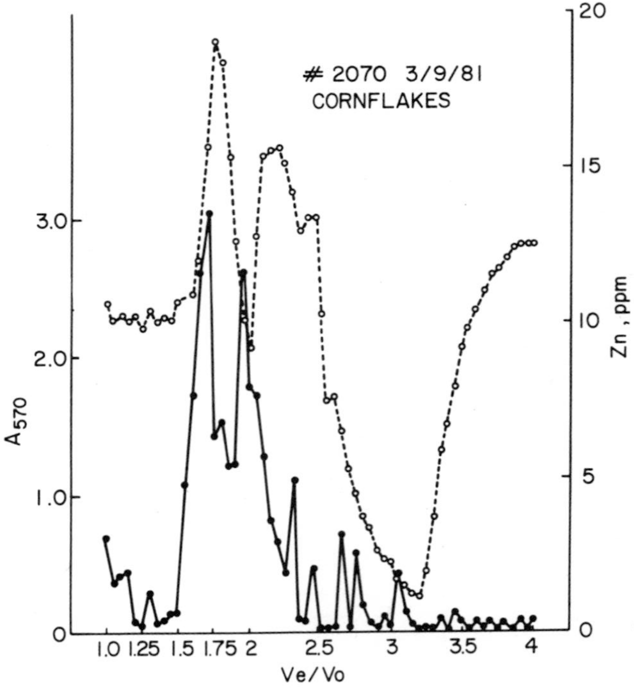

Figure 2. Typical chromatogram of urine from a subject eating corn flakes. Key: ●, ninhydrin-positive substances after hydrolysis (absorbance at 570 nm); ○, zinc concentration in eluate. (chromatography conditions same as Figure 1.)

excreted in the urine and are capable of binding urinary zinc. However, no difference in total urinary excretion of zinc, iron, or copper was observed between the diets.

Although quantitative measures of browning in total diets have not, to our knowledge, been made, most conventional diets contain substantially more browned foods than the diet used in this study. Nevertheless, differences in ^{65}Zn retention and in the molecular-weight distribution of urinary zinc-binding substances were seen. It is likely that these differences would be less subtle if the experimental diet were more highly browned. Future experiments in this laboratory will examine this question.

Literature Cited

1. Freeman, J.B.; Stegink, L.D.; Meyer, P.D.; Fry, L.K.; Denbesten, L. J. Surg. Res. 1975, 18, 463.
2. Stegink, L.D.; Freeman, J.B.; DenBesten, L.; Filer, L.J., Jr. Prog. Food Nutr. Sci. 1981, 5, 265.
3. van Rij, A.M.; Godfrey, P.J.; McKenzie, J.M. J. Surg. Res. 1979, 26, 293.
4. Erbersdobler, H. Z. Tierphysiol., Tiernährg. u. Futtermittelkde. 1971, 28, 171.
5. Pronczuk, A.; Pawlowska, D.; Bartnik, J. Nutr. Metabol. 1973, 15, 171.
6. Sgarbieri, V.G.; Amaya, J.; Tanaka, M.; Chichester, C.O. J. Nutr. 1972, 103, 657.
7. Valle-Riestra, J.; Barnes, R.H. J. Nutr. 1970, 100, 873.
8. Andrieux, C.; Sacqnet, E.; Guégnen, L. Reprod. Nutr. Develop., 1980, 20, 1061.
9. Kimiagar, M.; Lee, T.-C.; Chichester, C.O. J. Agr. Food Chem. 1980, 28, 150.
10. Miller, J. Nutr. Rept. Intl. 1979, 19, 679.
11. Garcia, W.J.; Hodgson, R.H.; Blessin, C.W.; Inglett, G.E. J. Agr. Food Chem. 1977, 25, 1290.
12. Cohn, S.H.; Dombrowski, C.S.; Pate, H.R. Phys. Biol. Med. 1967, 14, 645.
13. Johnson, P.E. J. Nutr. 1982 (in press).
14. Camire, A.L.; Clydesdale, F.M. J. Food Sci. 1981, 46, 548.
15. Evans, G.W.; Johnson, P.E.; Brushmiller, J.G.; Ames, R.W. Anal. Chem. 1979, 51, 839.
16. Kuhn, I.N.; Monsen, E.R.; Cook, J.D.; Finch, C.A. J. Lab. Clin. Med. 1967, 71, 715.
17. Monsen, E.R.; Halberg, L.; Layrisse, M.; Hegsted, D.M.; Cook, J.D.; Mertz, W.; Finch, C.A. Am. J. Clin. Nutr. 1978, 31, 134.
18. King, J.C.; Raynolds, W.L.; Margen, S. Am. J. Clin. Nutr. 1978, 31, 1198.

19. Sandström, B.; Arvidsson, B.; Cederblad, Å.; Björn-Rasmussen, E. Am. J. Clin. Nutr. 1980, 33, 739.
20. Sandström, B.; Cederblad, Å. Am. J. Clin. Nutr. 1980, 33, 1778.

RECEIVED October 5, 1982

Nutritional Value of Foods and Feeds of Plant Origin: Relationship to Composition and Processing

J. E. KNIPFEL, J. G. McLEOD, and T. N. McCAIG
Agriculture Canada, Research Station, Research Branch, Swift Current, Saskatchewan, S9H 3X2 Canada

Historically, alterations in amino acid availability and protein digestibility have been considered as the primary effects of processing reactions leading to reduced nutritional values of food and feeds. Predominant among these reactions are Maillard and crosslinking types. In plant materials used as foods and feeds, including both grains and roughages, reactions among nitrogenous compounds and complex polysaccharides such as hemicellulose appear to be of considerable importance in determining nutritive value responses to processing. Heating of alfalfa increased the content of Neutral Detergent Fibre (NDF) and the nitrogen contents of both NDF and Acid Detergent Fibre (ADF). Associated with increased NDF and increased N content of the fibre fractions was a reduction in organic matter digestibility by rumen bacteria, and in N digestibility by both rumen bacteria and pepsin digestion in vitro. In lower quality roughages, a large proportion of total N was associated with NDF and ADF. Ammonia treatment resulted in decreases in NDF and increases in digestibility and intake of wheat straw by sheep. Following ammoniation, increased levels of N in both NDF and ADF occurred. Investigations with rye grain suggested that the pentosan fraction is of prime importance in the adverse effects observed when rye is fed to animals. Increased chick performance observed following supplementation of rye with amino acids or high-quality protein suggests that pentosans adversely affected digestibility of rye protein through unknown mechanisms. Arabinose was easily released from rye hemicellulose by acid hydrolysis; following

0097-6156/83/0215-0361$06.00/0

heating of arabinose and xylose at 40°C with lysine for 1 hr, browning reactions were observed, raising the possibility that Maillard-type reactions occur at relatively low temperatures in plant materials.

The hemicellulose fraction of plant food and feed materials appears to be of considerably greater importance in determining nutritional value of such foods and feeds than has been generally recognized.

Processing of foods and feeds by man has been carried out from prehistoric times, as a method of preservation or for the purpose of increasing either palatability or nutritive value of many foodstuffs. During the past quarter-century, however, there has been a tremendous and ever-increasing use of foods and feeds which have been subjected to various types of processing. This dramatic increase in processing of foods and feeds occurred as a result of factors such as increased populations and has resulted in considerable pressure to develop new processing technologies and new sources of food and feed.

In conjunction with this increase in processing of foods and feeds there has been increased research into the effects of processing on the nutritive value of these materials, and a number of symposia have been held to discuss these effects during the past decade. From these discussions it has become apparent that there have been significant advances in knowledge of certain aspects of nutritional consequences of processing of foods and feeds. In many instances there is little known of the interactions among food or feed constituents that occur naturally or following processing. This is of particular concern when dealing with the wide variety of potential feed- and foodstocks which may be derived from nonconventional sources such as forages.

In this presentation the interrelationships of carbohydrates and proteins in cereal grains and fibrous feeds will be discussed and related to selected aspects of processing of these materials.

Investigations of Nutritive Value of Rye

Nutritive value of foods and feedstuffs depends to a large degree on protein level and quality, i.e., the relative amounts of the component amino acids compared to the requirements of the animal for various metabolic functions. The cereal grains are notoriously low in certain essential amino acids. Usually lysine is the first or second limiting amino acid. The grain of rye (*Secale cereale* L.) exhibits an amino acid profile superior to that of other cereal grains, especially wheat (1,2,3,4,5). Despite this fact, lysine is still the first limiting amino acid in rye in most instances (6,7).

In recent years factors such as alkylresorcinals (8), trypsin inhibitors, (9,10) and water-soluble polysaccharides have been implicated in the poor feeding quality of rye. Investigations have shown that levels of 15 to 20% rye grain in the diet depress the appetite and growth of chicks (8, 11, 12, 13). Higher levels of rye in the diet decrease utilization. Although the literature contains many reports on feeding rye to swine and poultry, there are relatively few on the effects of rye on ruminants.

The implication of alkylresorcinals in the low feeding quality of rye (8) has been largely disproved (14,15). Protein supplementation studies (16) revealed that rye diets supplemented with low levels of low-quality protein accentuated the detrimental effects of feeding rye while supplementation with high levels of high-quality protein greatly reduced the effects of feeding rye grain. Addition of various milling fractions of rye to chick diets showed that the growth-depressing factor was distributed throughout all such fractions of the grain (14). In addition, the growth depression could be alleviated by water extraction of the rye (17). Water extraction of the grain also eliminated the wet-feces problem associated with feeding rye to chicks.

Reduced nitrogen retention was reported when chicks were fed rye diets (18). Supplementation of those diets with amino acids increased the retention of only the supplemented amino acids and not those contributed by the rye. These observations are compatible with the trypsin inhibitor hypothesis of other researchers (9,10). On the other hand, examination of the differences between endosperm and embryo and trypsin inhibitors of barley, wheat, and rye has revealed that, in contrast to certain trypsin inhibitors from leguminous seeds, those from the cereal grains appeared to be relatively weak, nonstoichiometric inhibitors of trypsin (19).

Although protein efficiency indices (PEI) calculated from animal feeding trials (20) do give an estimate of the bioavailability of proteins, they represent an integration of many factors working individually or in combination. It is possible that other factors such as crosslinking of proteins, amino acid balance or protein-carbohydrate interactions are much more important in producing the adverse effects of rye grain than is trypsin inhibitor.

Rye-type growth depression in chicks fed pectin has been reported (21). Significant levels of pectin were reported to be present in rye grain (22) (as high as 8%). In our laboratory no pectin (as polygalacturonic acid) could be detected in rye grain.

Antibiotics have been shown to alleviate the effects of feeding rye to some extent (11,12,17,18,23). These observations led to the postulation that rye in the diet stimulated the proliferation of an adverse microflora in the gut. This is further substantiated by an adaptive response in chicks fed rye (24).

Rye does contain a significant amount of a polysaccharide (Table I) composed mostly of two pentose sugars, arabinose and xylose. This polysaccharide constitutes about 6 to 9% of the grain by weight and can be separated into a water-soluble and a water-insoluble fraction. These polysaccharide levels are of the same order as those previously determined as pectin (22). Both fractions have been isolated and substituted into diets of a wheat base (25). The detrimental effects of feeding rye were observed when either of the solubility fractions were fed. Both fractions of rye polysaccharide were more powerful inhibitors of chick growth than were alternative polysaccharides used by other researchers (26). The polysaccharides of rye, often referred to as pentosans, were similar to those of wheat flour in composition and have been studied mainly with respect to effects in the breadmaking process. The polysaccharides of rye are highly hygroscopic (27), produce viscous solutions, and should be studied further to determine their mode of action, composition, and nutrient-binding properties.

Table I. Pentosan^{+} levels of rye, wheat, and triticale

Crop	Cultivar	% Pent	Crop	Cultivar	% Pent	Crop	Cultivar	% Pent
Rye	Frontier	6.0	Wheat	Neepawa	5.9	Triti-cale	Welsh	6.5
	Cougar	6.4		Leader	5.9		Carman	5.8
	Puma	6.6		Norstar	4.8		Mapache	5.6
	Kodiak	9.6		Lemhi	4.8		Mean	6.0
	Musketeer	5.7		Pitic	5.9			
	Mean	6.9		Glenlea	5.3			
				NB 320	5.7			
				Norquay	5.5			
				Mean	5.5			

$^{+}$Pentosan was determined by the method of Dische and Borenfreund (28) on grains grown at Swift Current, 1981

Generally, there are higher levels of pentosans in rye than in wheat as shown in Table II. Triticale, a hybrid of wheat and rye, has pentosan levels intermediate to the parental species. The arabino-xylan fractions of the water-soluble polysaccharides of wheat and rye have been examined (27). In general, the molecular weight of rye pentosans was considerably higher than that of wheat pentosans but component sugar ratios were similar. The component sugars of the pentosans were determined (27) by the formation of alditol acetates and gas chromatography (29). The sample was first hydrolyzed in 1N H_2SO_4 for 4 hr at 100°C, which is relatively strong acid hydrolysis. In forage, it has been shown that hydrolysis with dilute acid can release arabinose from the hemicellulose fraction while leaving xylose polymerized. An experiment was set up to determine if arabinose could be released from rye pentosans by gentle hydrolysis and to test the possibility of nonenzymatic browning at physiologic temperatures using sugar-lysine solutions.

Rye pentosans were extracted by the method of Medcalf _et_

al. (27). Samples of the pentosans were hydrolyzed in 2 ml of .05 N H_2SO_4 for 1 hr at 99°C. After hydrolysis the solution was neutralized with Na_2HPO_4 and prepared for gas chromatography. Chromatography indicated that arabinose had been released from the polysaccharide but almost no xylose was detected. This is similar to results reported by Burdick and Sullivan (30), that arabinofuranoside linkages are much more easily cleaved than the pyranoside linkage of the main xylan chain.

Many studies of nonenzymatic browning have been carried out using model systems of monosaccharides and amino acids. Given the ease of release of arabinose from the pentosan, there is the possibility that Maillard reactions could take place if the temperature is appropriate.

Four sugar-lysine solutions were prepared and heated in a water bath. In addition, a pentosan-lysine solution and the pentosan, lysine, and sugars alone were heated (Table II).

Table II. Rating of color intensity (browning) when sugar-lysine solutions were heated for 1 hr

Temp. (°C)	Lysine mixed with					Single compounds					
	Glu	Ara	Xyl	Gal	Pent	Glu	Ara	Xyl	Gal	Pent	Lys
92	++++	++++	++++	++++	+	-	-	-	-	+	-
78	+++	++++	++++	+++	+	-	-	-	-	+	-
60	+	++	++	+	+	-	-	-	-	+	-
40	-	+	+	-	-	-	-	-	-	-	-

The pentose sugars gave more intense browning and exhibited browning reactions with lysine at lower temperatures than did the hexoses (Table II). Maillard (31) observed browning reaction between glycine and xylose at temperatures of 40 and 34°C when the aqueous mixtures were incubated for 20 hr, while glycine-glucose solutions required more extended periods of time to give browning. In the present study, we were able to detect browning when either xylose or arabinose were incubated at 40°C for 1 hr while higher temperatures were necessary to cause noticeable browning with glucose- or galactose-lysine solutions heated for 1 hr. The pentosan color development occurred in the presence or absence of lysine (Table II) only at 60°C or above and may be reflective of reactions other than those of the Maillard type. On the other hand, it is conceivable that there may have been some protein or other nitrogenous compounds present in the pentosan which did undergo Maillard reactions. The demonstrations of Maillard (31) that browning reactions could occur at physiological temperatures, and the results reported in this presentation, suggest that nutritional alterations due to Maillard-type reactions may be much more prevalent than is generally considered, for food- and feedstuffs not subjected to extensive processing.

Nutritive Value of Fibrous Feed Materials in Relation to Processing

In the 1940's Pirie (32) suggested that protein extracted from alfalfa could represent a potential protein source for monogastric animals, including humans. Further development of this concept into a commercial process was undertaken in several countries and it is expected that further development in this area will occur. It is, however, beyond the scope of this presentation to cover this subject; the reader is referred to Bickoff, et al. (33) and Pirie (34) for more detailed treatment.

Feedstuffs containing relatively high levels of fibre, such as forages and cereal crop residues, have traditionally been considered as feeds for ruminant species. It was usually assumed under accepted production schemes that the rumen microbial activity would supply sufficient quantities of microbial amino acids to the lower digestive tract if an adequate quantity of nitrogen was supplied. As production systems intensified and the productivity of the ruminant, particularly the dairy cow, increased, it was realized that a limiting factor in production was the supply of amino acids reaching the lower digestive tract (35,36). This supply of amino acids, however, represents a combination of both microbial protein and feed protein. Since the rumen microbial protein has a biological value of about 55 to 70%, the feeding of a protein with a higher biological value than that of the microbial protein will be of value only if the feed protein is not extensively degraded in the rumen and used to resynthesize microbial protein of lower biological value. Several processing methods have been employed to reduce the degradability of high biological value proteins in the rumen. Descriptions of these processes and the alterations in nutritive value accompanying the treatments have been reviewed in detail earlier (37,38,39,40).

Effects of Heat Treatment on Alfalfa

Previously it was reported (41) that dry heating of alfalfa at 105°C for up to 1440 min resulted in a reduction of in vitro organic matter digestibility (IVOMD) by rumen microbes. Increases in Acid Detergent Insoluble N (ADIN) and Neutral Detergent Fibre (NDF) were also observed. For the convenience of the reader these data have been included with more recent observations in this study.

As shown in Table III, NDF, suggested to represent the hemicellulose-cellulose-lignin fraction of the plant cell (45), increased at 480 and 1440 min of heating, while ADF, representing cellulose + lignin (45), did not. This resulted in an increase in the hemicellulose fraction of alfalfa from 7.1% to 12.2%. No significant changes in either cellulose or lignin occurred. The effect of heat treatment therefore appeared to be primarily an increase in the hemicellulose component of NDF, although it must be realized that the term "hemicellulose" represents a fraction obtained by difference which may contain compounds other than hemicellulose.

Table III. Alterations in alfalfa fraction following heating

Fraction (%)	Time of heating (min) at 105°C							
	0	10	30	60	120	240	480	1440
NDF[1]	35.3	35.3	35.1	35.5	35.8	35.1	37.5	40.6
ADF[2]	28.2	28.7	28.3	28.5	29.1	28.5	29.2	28.4
Hemicellulose[3]	7.1	6.6	6.8	7.0	6.7	6.6	8.3	12.2
Cellulose[4]	27.0	27.1	27.4	27.9	27.7	27.6	27.6	27.4
Lignin[5]	1.0	1.6	0.9	0.6	1.4	0.9	1.6	1.0

[1]Neutral Detergent Fibre (42)
[2]Acid Detergent Fibre (43)
[3]NDF minus ADF
[4]By methods of Crampton and Maynard (44)
[5]ADF minus cellulose

Examination of the partition of N in the fibre fractions (Table IV) shows a slight decrease in total N at the two longest heating times. ADIN showed a gradual increase as the duration of heating increased, as was shown (46,47,48,49) with increasing heat damage to alfalfa. This is more obvious when the ratios of ADIN to total N are calculated and shows about a 50% increase in ADIN at the longest duration of heating. NDIN increased more rapidly than did ADIN following heating and also to a much greater extent, particularly at the longest duration (Table IV) of heating. The ratio of ADIN to NDIN (Table IV) also indicated that before heating almost all of the fibre N (94.1%) was associated with ADF (ligno-cellulose) while after 1440 min of heating, 58.3% of the fibre N was associated with hemicellulose. The amount of insoluble protein increase in the hemicellulose fraction from t = 0 to t = 1440 can be calculated as:

$$NDIN_{1440} - NDIN_0 = \text{Increased NDIN (Kjeldahl conversion factor)} = \text{\% Insoluble Protein Increase, or}$$
$$1.20 - 0.34 = 0.86 \times 6.25 = 5.38\%$$

From Table III the NDF fraction increased from 35.3% to 40.6%, i.e., by 5.3%, which is virtually identical to the calculated increase in insoluble protein. These data suggest that the increase in NDF caused by heating was a result of protein association with the hemicellulose fraction. Several authors (35,45) suggested that Maillard reactions might cause binding of the protein to indigestible carbohydrate fractions of the cell wall while the proposal that a fraction of hemicellulose is associated with protein (47) is also consistent with the observations of Tables III and IV.

In vitro rumen fermentation of the heated alfalfas showed (Table V) a decrease in organic matter digestibility (OMD) which was of the same order of magnitude as the increase in NDIN, suggesting that the NDIN fraction was resistant to microbial attack at rumen pH (6.9). The addition of HCl-pepsin to give a pH of

Table IV. Partition of N in alfalfa subjected to heating

(%)	Time of heating (min) at 105°C							
	0	10	30	60	120	240	480	1440
Total N	2.85	2.89	2.85	2.85	2.85	2.85	2.75	2.75
ADIN[1]	0.32	0.29	0.32	0.37	0.37	0.39	0.44	0.50
ADIN/N_{tot}	11.2	10.0	11.2	13.0	13.0	13.7	16.0	18.2
NDIN[2]	0.34	0.48	0.64	0.64	0.70	0.74	0.82	1.20
NDIN/N_{tot}	11.9	16.6	22.5	22.5	24.6	26.0	29.8	43.6
ADIN/NDIN	94.1	60.2	49.8	57.8	52.8	52.7	53.7	41.7

[1]Acid Detergent Insoluble N
[2]Neutral Detergent Insoluble N

1.3 resulted in an increase in organic matter digestibility (OMDP) over that caused by microbial degradation (Table V). The organic matter solubilized by pepsin-HCl increased as the duration of heating of the alfalfa increased; this suggested that a large proportion of the increase in the NDIN fraction observed earlier (Table IV) constituted material solubilized by the acid-pepsin solution.

In vitro Nitrogen Digestibility (ND) decreased with increased heating of the alfalfa (Table V) when the microbial degradation was considered. This observation is consistent with those of previous workers (46,48,49), who observed decreases in ruminant digestion of alfalfa following heat treatment. Pepsin-HCl digestion removed 62.9% of the N left in the residue following microbial digestion, for unheated alfalfa, and this proportion decreased when the alfalfa was heated for 30 minutes or more (Table V). Total *in vitro* N digestibility therefore decreased as the duration of heating increased, even though the OMDP did not indicate this to be the case. Earlier studies (43,47) demonstrated that the solubility of hemicellulose increased markedly following treatment of plant tissue with proteolytic enzymes. The present data indicate that this solubility in HCl-pepsin may be a factor which masked differences in nutritive value caused by excessive heating when the *in vitro* OMDP was used for assessment.

Table V. *In vitro* rumen fermentation of heated alfalfa

Component (%)	Time of heating (min) at 105°C							
	0	10	30	60	120	240	480	1440
OMD	58.7	58.5	56.6	58.2	57.2	56.2	56.5	54.9
OMDP	65.6	66.8	65.9	67.2	65.8	66.5	66.2	65.8
OMSP	6.9	8.3	9.3	9.0	8.6	10.3	9.7	10.9
ND (% total N)	71.2	71.6	68.4	69.8	68.4	67.7	66.5	62.2
NDP (% res. N)	62.9	62.4	51.1	53.5	51.1	53.9	52.8	49.4
Total ND	89.3	89.3	84.6	86.0	84.6	85.1	84.2	80.9

At the present time the actual mechanisms by which the digestibility of the protein was reduced have not been elucidated. As already noted, a number of workers have suggested that Maillard reactions may cause binding of the protein to indigestible carbohydrates in the cell wall (35,45), while available N may also be reduced substantially through several crosslinking reactions (37,38) or through interactions with other plant components (40). In earlier studies (41,50) cellulose did not appear to react with casein, egg, or soy proteins to cause decreases in nutritive value upon heating, which is consistent with the results of the present study showing no changes in cellulose content with heating (Table III). The increase in ADIN, however, suggests that there was binding of N by components of this fibre fraction, perhaps to a variety of carbohydrate and/or noncarbohydrate compounds. As well, the increase in NDIN indicated that protein interaction with the hemicellulose fraction of the fibre occurred rapidly and to a significant extent.

Chemical Treatment Effects on Fibrous Feeds

Processing of highly fibrous feeds to improve the nutritional value of these materials for ruminants has been practiced since the beginning of the 20th century. The Beckmann (51) procedure has been used in Europe until very recently; this process involved soaking straw with NaOH to improve cellulose digestibility. A number of other procedures to improve the digestibility of the fibre fraction of roughages have been developed more recently. These include ammoniation, refined NaOH procedures, steaming, and other treatments, which have been reviewed by several authors (52,53). Several of these treatments have been developed into commercial processes in Europe and North America. The primary effect of treatment of roughages to improve digestibility for ruminants is to provide a more readily available source of energy for some phase of production. With roughages, a strong positive correlation exists between digestibility and intake, however, so that by increasing digestibility of roughages an increase in intake will also occur and the available energy to the animal (and thus production) is likely to be greater than would be expected from the increase in digestibility. In order to achieve acceptable levels of production, however, other nutrient deficiencies, especially that of N, must be corrected. Ammoniation has been chosen as a method of choice for treatment of crop residues and other low-quality roughage in Western Canada, since it does not result in any increase in cation return to the land in manure, is readily available commercially for on-farm use, and also may supply a large part of the N deficit in low-quality roughages (54).

Very limited information exists on the actual mechanism(s) by which ammonia or NaOH exert their effects on low-quality forages, although the mechanisms are thought to be similar to those proposed by Tarkow and Feist (55) for hardwoods, involving both chemical and physical changes, which will be further discussed in a later section.

Following treatment of barley straw with 5% NaOH, Israelsen, et al. (56) observed a decrease in NDF of 14.6 percentage units, while Braman and Abe (57) found the ADF fraction of wheat straw decreased by 5.7 units following NaOH treatment. Rexen (58) showed a substantial decrease in ADF of barley straw following NaOH treatment, but no effect of NaOH on rice straw. Itoh, et al. (59) observed significant decreases in NDF of ammoniated rice straw, rice hulls, and orchard hay. Treatment with 3.5% ammonia (Table VI) considerably decreased NDF for both chaff and straw, while ADF showed little change. Cellulose and lignin contents similarly changed only to a minor extent following ammoniation, with the major change in the fibre fraction appearing to be a decrease in hemicellulose. A substantial increase in both N content and in vitro OMD also occurred upon ammoniation. In an earlier study, Kernan et al. (60) showed that ammoniation doubled the ADIN content of wheat straw and increased that of sweet clover hay by 25%, and these workers proposed that a Maillard-type reaction between ammonia and some component of the cell wall carbohydrate fraction would account for the increase in ADIN.

Table VI. Effects of 3.5% anhydrous NH_3 on crop residue

%	Straw	Straw + NH_3	Chaff	Chaff + NH_3
N	0.54	1.40	0.62	1.96
NDF	78.8	73.8	70.7	65.6
ADF	47.9	47.7	37.9	38.3
Hemicellulose	30.9	26.1	32.8	27.3
Cellulose	41.1	41.9	34.3	35.5
Lignin	6.8	5.8	3.6	2.8
In vitro OMD	38.8	53.7	47.8	60.0

Following ammoniation of wheat straw or chaff, or crested wheatgrass or Altai wild ryegrass hays (Table VII), increases in NDIN and ADIN were observed. In all of the untreated roughages the NDIN represented 30% or more of total N, which decreased as a percentage of N following ammoniation even though the absolute amount of NDIN increased. Of the N absorbed by the feedstuff after ammoniation the major proportion was present in the non-fibre fraction of the cell. Except for untreated wheat straw, 50% or more of NDIN was present in hemicellulose (Table VII) as indicated by the ratio of ADIN to NDIN. A significant decrease in hemicellulose content of both straw and chaff occurred following ammoniation, which suggests that the amount of N taken up by hemicellulose was greater per unit of weight than is apparent when the results are expressed as a proportion of total sample weight.

It is possible that N associated with both NDF and ADF may

Table VII. Ammoniation effects on roughage N (%) distribution

Roughage	N	NDIN	ADIN	NDIN/N	ADIN/N	ADIN/NDIN x 100
Wheat straw						
Untreated	0.54	0.20	0.12	37.0	22.2	60.0
$+NH_3$	1.28	0.32	0.16	25.0	12.5	50.0
Wheat chaff						
Untreated	0.68	0.24	0.10	35.3	14.7	41.7
$+NH_3$	1.64	0.34	0.16	20.7	9.8	47.1
Crested wheatgrass						
Untreated	0.60	0.20	0.08	33.3	13.3	40.0
$+NH_3$	1.36	0.28	0.12	20.6	8.8	42.9
Altai wild ryegrass						
Untreated	0.76	0.24	0.10	31.6	13.2	35.3
$+NH_3$	1.48	0.42	0.16	28.4	10.8	38.1

have been removed during the extraction procedures for these fractions. In view of the decrease in hemicellulose content of both roughages a substantial proportion of the N determined as soluble N may have, in fact, been associated with hemicellulose; this should be investigated further.

In vivo digestibility data obtained with sheep using 75% straw and 25% oats (Table VIII) show that ammoniation increased the digestibilities of all ration components measured. In the fibre fraction all components were increased in approximately the same order of magnitude in digestibility. In addition, hemicellulose was somewhat higher in digestibility than cellulose both before and after ammoniation. Of considerable interest would be information regarding the forms and availability of N to the animal from both untreated and treated straw. The apparent digestibility of N increased following ammoniation but this increase was not as pronounced as would have been expected from the increase in N content (Table VII), which was presumably in a readily available form since it was not associated with the fibre fraction. Abidin and Kempton (64) recently reported that 65% of the N in ammoniated barley straw was potentially degradable in the rumen, which is of the same order of magnitude as the results of Table VIII.

Intake of digestible organic matter and digestible energy increased about 50% following ammoniation, while digestible N intake increased more than twofold (Table IX), showing that ammoniation increased the nutritive value of straw markedly for the ruminant. The increase in intake of ammoniated roughages, which has been consistently observed in our studies, has been considerably greater than increases in digestibility caused by

Table VIII. Straw component digestibility response to ammoniation

Digestibility of:	Treatment	
	Untreated	$+NH_3$
N	52.8	57.8
Energy	59.8	64.2
OM	62.4	67.2
NDF	57.7	66.0
ADF	53.6	62.2
Cellulose	61.3	68.9
Hemicellulose	64.0	72.9

ammoniation. This observation indicates that a major effect of ammoniation has been to increase susceptibility of the fibre fraction to microbial degradation and thus to increase the rate of digestion to a greater extent than the amount of digestibility. However, a number of fundamental questions remain regarding the basic mechanisms of action of ammonia (or other treatments) on the fibre fraction, and the effects upon the availabilities of the native protein (N) and carbohydrate fractions.

Table IX. Intake of digestible components of straw following ammoniation

Intake[1] of:	Treatment	
	Untreated	$+NH_3$
Digestible N	0.39	0.90
Digestible energy	180	265
Digestible organic matter	40.6	59.9

[1]g of Kcal/kg ^{75}

Proposal for Some Nitrogen-Carbohydrate Interactions in Cereal Grains and Roughages

In view of the extreme complexity of molecular species comprising plant cells, obviously any model or explanation for nitrogen (protein) interactions with carbohydrates must be extremely simplistic in relation to total events occurring in the plant cell, either following any form of processing or in relation to reactions occurring under "natural" conditions. For the purposes of attempting to derive a useful working hypothesis for studies of nutritive value in cereal grains such as rye, for low-quality roughage evaluation, and for developing techniques for improvement of nutritive value, the diagrammatic structure for the interaction of the hemicellulose fraction with nitrogenous materials such as protein or ammonia has been adopted from the original proposal of Tarkow and Feist (54), and is shown in Figure 1.

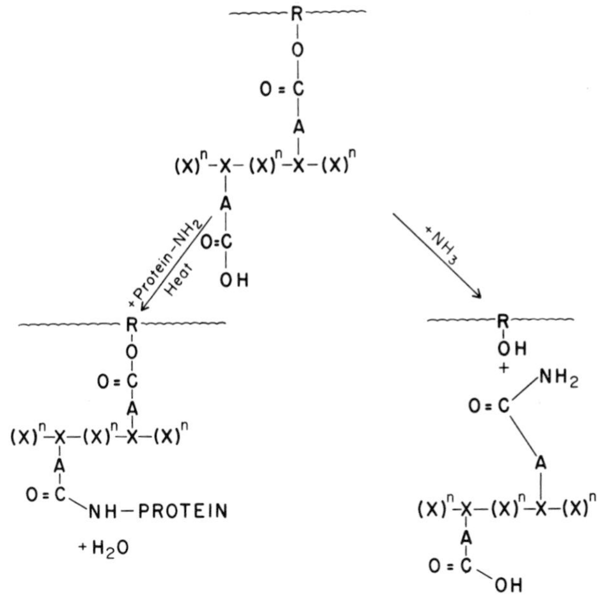

Figure 1. A simplified model for plant fiber interactions.

Hemicellulose may be considered as a polymer of pentosans although a number of hexoses and other materials are usually present also. The simple structure can be represented by a xylan chain, with side chains of several sugars but with arabinose prominent. The arabinose units may be bonded to other compounds (R, e.g., lignin) by ester linkages with these bonds acting as crosslinks to limit swelling or dispersion of the polymer segments in water (54). There may also be free carbonyl groups as shown in Figure 1, which could react with protein in amide or Maillard-type reactions. In Figure 1, for simplicity, the carbonyls shown are carboxyl groups which would undergo reaction to produce amide linkages with protein. However, there are undoubtedly arabinose and other monosaccharide residues present as branches of the xylan chain which would have aldehyde or keto end groups available for classical Maillard reactions. While we have not established that this is indeed the case, the evidence of increased N association with hemicellulose following heat treatment is strongly suggestive. Other authors (46,48,49) have demonstrated the occurrence of the Maillard reaction in forages, although there is considerable controversy over the exact molecular species reacting.

From this type of schematic representation it would be expected that a more highly branched xylan chain would give more potential for reaction of the arabinose residues, either through crosslink formation or by Maillard-type reactions. Indirect evidence supporting this concept can be derived from pentosan analysis of the hemicellulose fraction of grasses and legumes.

Alfalfa hemicellulose has been observed (61) to contain an arabinose:xylose ratio of about 1:5.5, indicating that the hemicellulose is not highly branched. In grass species, the ratio has been observed to be in the order of 1:2.6 (62) for grass-alfalfa mixtures, but closer to unity for grasses (63). These data indicate that the hemicelluloses of grasses are more highly branched, and perhaps crosslinked, than is that of alfalfa, which is consistent with the observation (60) that ammoniation response is much greater with grasses than with alfalfa.

That lignin is intimately involved in the susceptibility of these structural carbohydrates to degradation is well documented. Considerable available evidence (65,66,67) indicates that the ester linkages between lignin and hemicellulose are major factors in this process. The involvement of proteins is less well defined, although earlier studies (46,47) demonstrated the close relationship between lignin and protein in heated forages. In a number of studies it is apparent that the determination of lignin could have included a substantial amount of artifact lignin, which has been suggested (47) to result from Maillard reactions and be unavailable as a source of amino acids to the animal.

The evidence of increases in NDIN and ADIN in heated alfalfa suggests that Maillard-type phenomena and likely other reactions are involved in the decrease in availability of N occurring following heat damage. The effects of ammonia in breaking ester linkages and thus increasing susceptibility of fibrous fractions to bacterial attack is well documented, but whether this, in turn, may render native proteins more available is not known. Furthermore, evidence from studies of rye suggests that the hemicellulose (pentosan) fraction of rye grain is involved in effects upon amino acid availability and in intake and digestibility of the grain by monogastric animals.

While Maillard reactions may represent only a portion of the interactions responsible for alterations in nutritive value of plant materials, the recognition that such reaction may occur at temperatures that can be considered as physiological in plants raises a considerably wider spectrum of possibilities for alterations in nutritive value of feed- and foodstuffs than has generally been assumed. As pressure for increased production efficiency of animals and increased use of nonconventional nutrient sources increases, the effects of processing upon nutritive value of these materials will be studied in greater detail.

Conclusions

1. The hemicellulose (pentosan) fraction of rye grain appears to be intimately involved in poor feeding value of rye, perhaps as a result of naturally occurring Maillard reactions.
2. Heat treatment of alfalfa results in increases in protein content of the fibre fraction of the feed. Associated with this increase is a decrease in in vitro digestibility of OM

by rumen microbes, and reductions in both ruminal and HCl-pepsin digestion of N in vitro.
3. Ammoniation of low-quality roughages generally reduces the NDF content of the roughage while having lesser effects upon ADF or lignin. These changes are associated with increased digestibility and intakes of treated roughages by sheep.
4. The hemicellulose fraction of diverse plant feedstuffs appears to be of considerable importance in determining nutritional value and response to processing.

Literature Cited

1. Jones, D.B.; Caldwell; Widness, K.D. J. Nutr. 1948, 35, 639-650.
2. Stromnaes, A.-S.; Kennedy, Barbara M. Cereal Chem. 1957, 34, 96-200.
3. Kihlberg, R.; Ericson, L.-E. J. Nutr. 1964, 82, 385-394.
4. Knipfel, J.E. Cereal Chem. 1969, 46, 313-317.
5. Riley, R.; Ewart, J.A.D. Genet. Res. (Camb.) 1970, 15, 209-219.
6. Kubiczek, R.; Rakowska, M. Acta Agrobot. 1974, 27, 105-113.
7. Kubiczek, R.; Molski, B.; Rakowska, M. Hodowla Rosl. Aklimat Nascien. 1975, 19, 617-631.
8. Wiringa, G.W., Wageningen, Holland. Publ. 156, 68 pp.
9. Knoblauch, C.J.; Elliott, F.C.; Penner, D. Crop Sci. 1967, 17, 269-272.
10. Tanner, D.G.; Knoblauch, C.J.; Reinbergs, E.; Su-Wong, H. Cereal Res. Comm. 1980, 8, 461-467.
11. Moran, E.T., Jr.; Lall, S.P.; Summers, J.D. Poultry Sci. 1969, 48, 939-949.
12. MacAuliffe, T.; McGinnis, J. Poultry Sci. 1971, 50, 1130-1134.
13. Misir, R.; Marquardt, R.R. Can. J. Anim. Sci. 1979a, 58, 691-701.
14. Misir, R.; Marquardt, R.R. Can. J. Anim. Sci. 1978c, 58, 717-730.
15. Fernandez, R.; Lucas, E.; McGinnis, J. Poultry Sci. 1973, 52, 2252-2259.
16. Misir, R.; Marquardt, R.R. Can. J. Anim. Sci. 1978b, 58, 703-715.
17. Misir, R.; Marquardt, R.R. Can. J. Anim. Sci. 1978d, 58, 731-742.
18. Marquardt, R.R.; Ward, A.T.; Misir, R. Poultry Sci. 1979, 58, 631-640.
19. Mikola, J.; Kirsi, M. Acta Chem. Scand. 1972, 26, 787-795.
20. Knoblauch, C.J.; Elliott, F.C. Crop Sci. 1977, 17, 267-269.
21. Wagner, D.D.; Thomas, O.P. Poultry Sci. 1977, 56, 615-619.
22. McNab, J.M.; Shannon, D.W.F. Br. Poultry Sci. 1975, 16, 9-15.
23. Wagner, D.D.; Thomas, O.P. Poultry Sci. 1978, 57, 971-975.

24. Wagner, D.D.; Thomas, O.P.; Graber, G. Poultry Sci. 1978, 57, 230-234.
25. Antoniou, T.; Marquardt, R.R. Poultry Sci. 1981, 60, 1898-1904.
26. Vohra, P.; Shariff, G.; Kratzer, F.H. Nutr. Repts. Intl. 1979, 19, 463-467.
27. Medcalf, D.G.; D'Appolonia, B.L.; Gilles, K.A. Cereal Chem. 1968, 45, 539-549.
28. Dische, Z.; Borenfreund, E. Biochem. Biophys. Acta 1957, 23, 639-642.
29. Sawardecker, J.S.; Sloneker, J.H.; Jeanes, Allene. Anal. Chem. 1965, 37, 1602-1604.
30. Burdick D.; Sullivan, J.T. J. Anim. Sci 1963, 22, 444.
31. Maillard, L.-C. Ann Chim. (Paris) 1916, [9]5, 258-317.
32. Pirie, N.W. Nature 1942, 149, 251.
33. Bickoff, E.M.; Booth, A.N.; de Fremergy, D.; Edwards, R.H.; Knuckles, B.E.; Miller, R.E.; Saunders,, R.M.; Kohler, G.O. Nutr. Clin. Nutr. 1975, 1(2), 319.
34. Pirie, N.W. Nutr. Clin. Nutr. 1975, 1(2), 341.
35. Clark, J.H. Nutr. Clin. Nutr. 1975, 1(2), 261.
36. Hogan, J.P. J. Dairy Sci. 1975, 58, 1164.
37. Broderick, G.A. Nutr. Clin. Nutr. 1975, 1(2), 211.
38. Broderick, G.A. Adv. Exp. Med. Biol. 1977, 86B, 531.
39. Friedman, M.; Broderick, G.A. Adv. Exp. Med. Biol. 1977, 86B, 545.
40. Knipfel, J.E. Adv. Exp. Med. Biol. 1977, 86B, 559.
41. Knipfel, J.E. Prog. Food Nutr. Sci. 1981. 5, 177.
42. Van Soest, P.J.; Wine, R.H. J. Assoc. Off. Anal. Chem. 1967, 50, 50.
43. Van Soest, P.J. J. Assoc. Off. Anal. Chem. 1963, 46, 829.
44. Crampton, E.W.; Maynard, L.A. J. Nutr. 1938, 15, 383.
45. Van Soest, P.J. J. Assoc. Off. Anal. Chem. 1965, 48, 785.
46. Van Soest, P.J. J. Dairy Sci. 1962, 45, 664.
47. Van Soest, P.J. J. Anim. Sci. 1964, 23(3), 838.
48. Goering, H.K.; Waldo, D.R. Proc. Cornell Nutr. Conf. 1974, p. 75.
49. Goering, H.K. J. Anim. Sci. 1976, 43, 869.
50. Knipfel, J.E.; Botting, H.G. and McLaughlan, J.M. Nutr. Clin. Nutr. 1975, 1(1), 375.
51. Beckmann, E. Sitzber. Preuss. Akad. 1919, 275, 85.
52. Sundstol, F.; Coxworth, E.; Mowat, D.N. World Anim. Rev. 1978, 26, 13.
53. Jackson, M.G. FAO Anim. Prod. Hlth. Paper, 1978, p. 10.
54. Tarkow, H.; Feist, W.C. Adv. Chem. Ser. 1969, 95, 197.
55. Kernan, J.; Coxworth, E.; Nicholson, H.; Knipfel, J. SRC Tech. Rep. 114, 1981.
56. Israelsen, M.; Rexen, B.; Thomsen, K.V. Anim. Feed Sci. Technol. 1978, 3, 227.
57. Braman, W.L.; Abe, R.K. J. Anim. Sci. 1977, 46, 496.
58. Rexen, F. Feedstuffs 1979, Oct. 15.

59. Itoh, H.; Terashima, Y.; Tohrai, N. Jap. J. Zootech. Sci. 1979, 50(1), 54.
60. Kernan, J.A.; Coxworth, E.W.; Moody, M.J. SRC Rep. C-814-K-1-B-80, 1980.
61. Myrhe, D.V.; Smith, F. Agric. Food Chem. 1960, 8, 359.
62. Yang, M.T.; Milligan, L.P.; Mathison, G.W. Can. J. Anim. Sci. 1981, 61, 1087.
63. Milligan, L.P. Unpublished results, 1981.
64. Abidin, Z.; Kempton, T.J. J. Anim. Feed Sci. Technol. 1981, 6, 145.
65. Morrison, I.M. Biochem. J. 1974, 139, 197.
66. Hartley, R.D.; Jones, E.C. Phytochemistry 1976, 15, 1157.
67. Lau, M.M.; Van Soest, P.J. J. Anim. Feed Sci. Technol. 1981, 6, 123.

RECEIVED October 5, 1982

19

Effect of the Maillard Browning Reaction on the Nutritive Value of Breads and Pizza Crusts

C. C. TSEN, P. R. K. REDDY, and S. K. EL-SAMAHY
Kansas State University, Department of Grain Science and Industry, Manhattan, KA 66506

C. W. GEHRKE
University of Missouri, Experiment Station Chemical Laboratory, Columbia, MO 65211

Bread is an excellent staple supplying key nutrients: carbohydrate, protein, minerals and vitamins. The Maillard browning reaction could significantly reduce the nutritive value of bread when baked or toasted. Conventional baking produced a much darker bread crust and crumb than microwave baking or steaming. Rat-feeding tests demonstrated that protein efficiency ratios (PERs) of breads were significantly improved by substituting steaming or microwave baking for conventional baking. Toasting bread reduces the PER from 0.90 to 0.32. Significant differences in PER were found between fermented doughs before and after baking, and between bread crust and crumb. The high-temperature, short-time baking process used in balady bread and pizza crust reduced their nutritive value significantly. The nutritive loss in bread and pizza crust was largely due to the destruction of lysine in those products; to a lesser extent baking caused the lysine to become unavailable.

Bread, in various forms, is the most popular staple in the world. It is prepared from fermented dough made mainly

0097-6156/83/0215-0379$06.00/0

of wheat flour. Wheat flour, like other cereal flours, is low in lysine. The Maillard browning reaction induced by baking or toasting can aggravate the lysine deficiency and reduce the nutritive value of bread.

Lysine is an essential amino acid in human nutrition. Although people in developed nations most often obtain lysine from foods other than bread, the nutritive value provided by lysine in bread is highly important to the majority of people living in developing nations. This paper reports effects of baking or toasting on the nutritive value of breads and related products such as pizza crusts.

Effects of Conventional Baking, Microwave Baking and Steaming

Browning of breads, as induced by baking, is a complex process consisting mainly of the Maillard reaction between flour starch or other carbohydrates and flour protein or other protein-rich additives, and the caramelization of sugar and other carbohydrates.

A number of early investigators, notably Rosenberg and Rohdenburgh (1, 2), Rosenberg, et al. (3), Clarke and Kennedy (4), Ericson, et al. (5), Hutchinson, et al. (6), Jansen and Ehle (7), and Jansen, et al. (8), have shown that lysine is the limiting amino acid in bread, and that baking can aggravate the lysine deficiency in bread. Sabiston and Kennedy (9) found that the protein efficiency of all the breads was approximately 20% less than that of their unbaked ingredients. Gotthold and Kennedy (10) showed the protein quality of steamed bread was superior to that of conventionally baked bread. All these studies point out the deleterious effect of the browning on the nutritive value of bread.

Tsen, et al. (11) found that the nutritive value, as expressed in protein efficiency ratio (PER), of bread was increased significantly if bread was baked with microwave energy or steaming instead of conventional baking. Microwave baking or steaming did not brown the bread crust; accordingly, less browning takes place by these processes than conventional baking. As shown in Table I, rats fed conventional bread diet gained only 22.2 g compared with 40.5 g or 41.0 g by those fed steamed bread or microwave-baked bread diet. Only 7.6 or 7.3 g, respectively, of steamed bread or microwave-baked bread produced the same gain in weight by rats as 11.5 g of conventionally baked bread. The feed conversion ratios indicate clearly that a great deal of wheat bread or flour can be saved by modifying conventional baking processing or by replacing it with microwave baking or steaming.

Protein efficiency ratios of conventionally baked, microwave-baked, and steamed breads, presented in Table II, were 0.79, 1.20, and 1.25, respectively. The significantly lower PER of conventionally baked breads indicates that more

Table I. Weight gain, feed intake, and feed conversion ratio rats fed indicated bread diets 28 days

Bread[a,b] in diet	Weight gain (mean[c] ±SEM) (g)	Feed intake (mean[c] ±SEM) (g)	Feed conversion ratio
88% WF + 12% SF(S)	111.5±5.9A	469.0±15.8A	4.2
88% WF + 12% SF(M)	109.0±9.1A	435.3±22.1AB	4.0
WF + 0.2% L(S)	102.0±4.8AB	440.2±13.3AB	4.3
WF + 0.2% L(M)	94.2±5.6B	422.5±14.3B	4.5
Casein (control)	79.2±4.2C	288.7±10.7E	3.6
WF + 0.2% L(C)	61.7±4.6D	363.7±13.8C	5.9
88% WF + 12% SF(C)	49.3±1.2E	328.3±12.3D	6.6
WF(M)	41.0±4.0E	297.3±19.5DE	7.3
WF(S)	40.5±2.3E	308.7±20.6DE	7.6
WF(C)	22.2±1.1F	254.3±12.1	11.5

[a]WF, WF + 0.2% L, and 88% WF + 12% SF indicate breads prepared from wheat flour, wheat flour fortified with 0.2% L-lysine monohydrochloride, and wheat flour fortified with 12% soy flour (88% parts wheat flour and 12 parts defatted soy flour), respectively.

[b](C), (M), and (S): Conventional baking, microwave baking, and steaming, respectively.

[c]Duncan's Multiple Range Test (1955): Means without a letter in common differ significantly ($p < 0.05$).

Table II. Protein efficiency ratios of bread-diets fed rats 28 days

Bread in diet	(mean±SEM)	Adjusted PER
Casein (control)	2.74±0.12B	2.50
88% WF + 12% SF(M)	2.47±0.09C	2.25
88% WF + 12% SF(S)	2.36±0.05C	2.15
WF + 0.2% L(S)	2.30±0.04CD	2.09
WF + 0.2% L(M)	2.23±0.12D	2.03
WF + 0.2% L(C)	1.69±0.11E	1.54
88% WF + 12% SF(C)	1.50±0.04F	1.36
WF(M)	1.37±0.09G	1.25
WF(S)	1.32±0.06G	1.20
WF(C)	0.87±0 05H	0.79

[a]WF, WF + 0.2% L, and 88% WF + 12% SF indicate breads prepared from wheat flour, wheat flour fortified with 0.2% L-lysine hydrochloride, and wheat flour fortified with 12% soy flour (88% parts wheat flour and 12 parts defatted soy flour), respectively.

[b](C), (M), and (S): Conventional baking, microwave baking, and steaming, respectively.

[c]Duncan's Multiple Range Test (1955): Means without a letter in common differ significantly ($p < 0.05$).

lysine becomes lost or less available nutritionally with conventional baking than with either microwave baking or steaming.

From PER and net protein ratio (NPR) tests, Palamidis and Markakis (12) later also found that the PER of bread mix (ingredients) or bread mix with 4% nonfat dry milk (NFDM) were 0.75 or 1.17, as compared to 0.46 or 0.75 for their corresponding breads, respectively. The findings also clearly show that baking can damage the nutritive quality of bread protein significantly and that fortification of NFDM can raise the PER of bread mix and wheat (white) bread. Their NPR values ranked the protein quality of samples in the same order as the PERs (Table III).

Lysine, Soy or Milk Fortification

Fortifying wheat flour with 0.2% lysine or 12% soy flour raised the PER to 1.54 or 1.36 respectively, for conventionally baked bread; to 2.09 or 2.15 for steamed bread; and to 2.03 or 2.25 for microwave-baked bread (Table II).

The marked increases in PERs with lysine and soy or milk fortifications (Tables II and III), regardless of baking method, confirm that wheat bread is deficient in lysine. Conventionally baked bread's high response to lysine than soy or milk fortification indicates that conventional baking can aggravate the lysine deficiency of wheat bread much more than does microwave baking or steaming.

Comparing the Nutritive Values Between Fermented Doughs Before and After Baking and Between Bread Crust and Crumb

Although a number of investigators such as Sabiston and Kennedy (9) and Palamidis and Markakis (12) compared the protein quality of bread mix (a mixture of ingredients) and finished bread, the comparison did not take dough fermentation into consideration. During dough fermentation and proof, yeast and a host of enzymes, particularly amylases and proteases, act on dough starch and other carbohydrate, proteins, and lipids to produce more reactants for the Maillard reaction and caramelization. As a result of the fermentation and proof, dough volume increases greatly to facilitate the browning reactions. At the same time, the dough's pH is reduced because of the production of carbon dioxide and organic acids such as lactic anc acetic acids. As Dworschak, et al. found, the fermentation could increase the biological value of bread (13).

Tsen, et al. (14) have recently observed the deleterious effect of baking by feeding rat with diets prepared from fermented and proofed dough before and after baking and from bread crust and crumb. PERs (adjusted) were found to be 1.34

and 0.92 for diets with fermented dough before and after baking, respectively. A significant drop in PER resulted directly from baking.

In another study by Tsen et al.(14), a substantial PER difference was observed between diets with bread crust (0.36) and bread crumb (1.35). Palamidis and Markakis (12) also found that the PERs of their bread and its crumb were 0.46 and 0.91, respectively, while the crust showed a negative PER, -0.23. Hansen, et al.(18) reported that bread crumb and crust had respective PERs of 1.36 and 0.62. The difference in PERs for bread crumb or crust reported by the three groups of investigators is largely due to different raw materials and processing conditions for the breads. Nevertheless, the marked difference in PER between bread crumb and crust indicates clearly that the browning reaction can reduce the nutritive value of bread (Table IV).

Effect of Toasting

Toasting, like baking, can reduce the nutritive value of bread, as found by Tsen and Reddy (15). In one of their feeding studies, rats receiving untoasted crumbs averaged 27.5 g gain compared with 17.4, 11.1, and 7.1 g, respectively, by those getting light-, medium-, and dark-toasted bread slices. It also was evident that toasting significantly reduced the PERs: to 0.64, 0.45, and 0.32, respectively, for light-, medium-, and dark-toasted slice samples compared with 0.90 for the untoasted crumb sample, as shown in Table V.

Palamidis and Markakis (12) conducted a thorough study to evaluate not only the PER of toasted products but also their net protein ratios (NPRs) and digestibilities. They reported that light toasting reduced the PER of bread to 0.40 (NPR 1.11) and dark toasting to 0.16 (NPR 0.95). The protein digestibility decreased as the intensity of toasting increased.

Effects of High-Temperature and Short-Time Baking on the Nutritive Value of Balady Bread and Pizza Crust

In addition to white bread, studies have been done on some exotic breads, such as balady bread, widely consumed in Egypt and other Mid-East countries, by El-Samahy and Tsen (16) and some popular foods such as pizza in the U.S., Italy, and many other countries by Tsen, *et al*. (16). Balady bread and pizza crust, like white bread, are prepared from fermented dough, but they are baked at a much higher temperature for a shorter period.

Chemical characteristics and amino acid contents varied only slightly among balady breads baked at temperatures ranging from 248-343°C for 3.5-7.0 min. However, bread

Table III. PER (adjusted to PER = 2.50 for casein) and NPR values of bread products

	Adjusted PER[1]	NPR[1]
Bread mix	0.75C	1.51C
Bread	0.46B	1.22B
Crumb	0.91D	1.62D
Crust	-0.23A	0.69A
LTB[2]	0.40B	1.11B
DTB[2]	0.16A	0.95A
NFDM-mix[3]	1.17E	1.84E
NFDM-bread[3]	0.75C	1.51C
Casein	2.50	3.23

[1]Values with different letters differ significantly ($p < 0.05$), according to Duncan's multiple range test.

[2]LTB: Light-toasted bread
DTB: Dark-toasted bread

[3]NFDM: Nonfat dry milk

Table IV. Protein efficiency ratios and digestibility of bread diets

Bread in diet	PER (adjusted)	Digestibility
Casein (control)	2.90 (2.50)A	91.7 A
Bread (whole)	1.02 (0.88)C	86.3 B
Bread crust	0.42 (0.36)D	83.4 C
Bread crumb	1.47 (1.35)E	87.1 B

Table V. Feed intake, weight gain, and protein efficiency ratios of bread diets fed rats 28 days

Bread in diet	Feed intake (mean[a] SEM) (g)	Weight gain (mean[a] SEM) (g)	PER (mean[a] SEM)	Adjusted PER
Casein (control)	376.5±18.5A	107.1±8.9A	2.84±0.26A	2.50
Untoasted (crumb)	266.5±32.8B	27.5±6.6B	1.02±0.19B	0.90
Light-toasted	234.5±25.4BC	17.4±4.1C	0.73±0.14C	0.64
Medium-toasted	212.2±31.2CD	11.1±5.3CD	0.51±0.22CD	0.45
Dark-toasted	189.5±19.6D	7.1±3.2D	0.36±0.13D	0.32

[a]Duncan's Multiple Range Test (1955): Means without a letter in common differ significantly ($p < 0.05$).

protein quality deteriorated significantly as a result of raised baking temperature or prolonged baking time as shown in the reduction in PER listed in Table VI. A difference in baking time of only 1.4 min significantly reduced the PER from 1.13 to 0.87 when the breads were baked at 327°C. Increasing the baking temperature merely 16°C from 327° to 343°C reduced the PER from 1.13 to 0.95 when both breads were baked for 5.0 min. The significant reductions in PER point out the importance of controlling temperature and time when baking balady bread or related products such as pizza crust.

As for pizza, the high-temperature and short-time baking reduced the lysine and, to a lesser extent, tyrosine, cystine, and threonine in the crust. As shown in Table 7, the total lysine content loss ranged from 7.1% for whole wheat pizza crust to 19.4% for a commercial pizza crust (17). The difference between total and available lysine for each pizza crust was small, indicating that the nutritive loss in pizza crust should be attributed largely to the destruction of a portion of lysine in such crust.

Lysine Involvement

Wheat flour is the most important ingredient in breadmaking. Accordingly, the changes in flour-protein components by thermal processing should be considered first. From their studies on thermal processing of flour in a closed heat exchanger, Hansen, et al. (18) concluded that high processing temperatures (108, 150, and 174°C) could cleave some flour protein into peptides. With the production of the peptides, lysine, arginine and cystine contents were reduced, as shown in Table 8. Enzymatic rates of pepsin and trypsin on heated flour protein also were decreased.

In addition to flour, some sugar, amylases, protein additives such as non-fat dry milk and soy flour, and shortening, used as minor ingredients for breadmaking, also are involved in the browning reaction during baking. Baking conditions, particularly temperature and time, affect the browning reaction.

Of all amino acids involved in the browning reaction, lysine with its ε-amino group is especially susceptible to side reaction and crosslinking and becomes unavailable. Lysine, like other amino acids, can also be decomposed by heating. The early work of Lea and Hannan showed that the destruction and cross-linking of lysine with sugars took place in the reaction between glucose and casein (19). Since then, many workers have demonstrated the deleterious effect of the browning reaction on the availability of lysine in foods by heating or baking as reviewed by Reynolds (20) and recently by Dworschak (13).

Table VI. Protein efficiency ratios (PERs) of bread diets fed rats for 28 days

Diets Containing Bread Baked at Temperature (°C)	Time (min)	PER[a,b]	Adjusted PER
Casein diet		2.99±0.09A	2.50
248	6.4	1.65±0.06B	1.38
327	6.4	1.03±0.03F	0.87
327	5.0	1.35±0.04D	1.13
343	5.0	1.13±0 08EF	0.95
327	3.6	1.52±0.03BC	1.27
343	3.5	1.26±0.08ED	1.05
288	7.0	1.38±0.07CD	1.16
		0.17[c]	-

[a]Mean ± SEM.

[b]Duncan's multiple range test (1955); Means without a letter in common differ significantly ($p < 0.05$).

[c]Least significant difference.

TABLE VII. Amino acids contents in wheat flours and pizza crusts

Amino acid	Amino acid, g per 100 g protein (corrected to 100% recovery basis)									
	WF[a]	WWF	PCW	PCWB	PCWW	PCWWB	PCC	PCCB	PCCM	PCCMB
Aspartic acid	4.30	5.39	4.19	4.12	5.33	5.36	4.72	4.49	4.79	4.69
Threonine	2.56	2.89	2.66	2.65	2.98	2.91	2.75	2.74	2.94	2.83
Serine	4.93	5.06	4.96	4.88	4.97	5.09	4.83	4.79	4.93	4.94
Glutamic acid	34.44	31.28	34.44	34.46	30.25	30.50	33.35	33.83	33.04	33.32
Proline	10.20	9.55	9.53	9.90	9.05	9.10	9.16	9.23	8.86	9.08
Glycine	3.48	4.09	3.51	3.57	4.27	4.17	3.58	3.53	3.49	3.48
Alanine	3.23	4.04	3.25	3.28	3.50	3.48	3.12	3.15	4.13	4.09
Half cystine	1.72	1.66	1.62	1.56	1.64	1.56	1.55	1.49	1.71	1.61
Valine	3.12	3.21	2.82	2.86	3.12	3.10	2.90	2.70	2.89	2.99
Methionine	1.51	1.50	1.47	1.52	1.50	1.50	1.50	1.41	1.56	1.56
Isoleucine	2.53	2.42	2.31	2.36	2.40	2.39	2.33	2.32	2.40	2.45
Leucine	6.31	6.40	6.10	6.19	6.25	6.30	6.16	6.12	6.29	6.29
Tyrosine	3.01	3.12	3.08	3.00	3.25	3.13	3.18	3.16	3.12	3.05
Phenylalanine	4.38	4.54	4.63	4.66	4.52	4.57	4.73	4.85	4.70	4.69
Histidine	3.19	3.20	3.35	3.51	3.50	3.47	3.48	3.64	3.44	3.41
Arginine	3.50	4.39	4.31	4.12	5.11	5.07	4.50	4.62	4.30	4.17
Lysine	2.51	2.92	3.00	2.76	3.54	3.41	3.20	2.83	3.35	3.05
	Amino acid (g/100g sample)									
Lysine total	--	--	0.31	0.26	0.42	0.39	0.31	0.25	0.35	0.31
Lysine available	--	--	0.29	0.24	0.39	0.37	0.29	0.23	0.33	0.29

[a]WF, WWF, PCW, PCWB, PCWW, PCWWB, PCC, PCCB, PCCM, and PCCMB indicate white flour, whole wheat flour, pizza crusts prepared from white flour, unbaked and baked; from whole wheat flour, unbaked and baked; from commercial frozen pizza, unbaked and baked; and from commercial pizza mix, unbaked and baked, respectively.

TABLE VIII. Effect of thermal processing on the chemical score of hard red wheat flour

Amino acid	Egg (mg amino acid/g N)	Wheat flour (33% moisture)			
		Unheated %	108°C-10 min %	150°C-10 min %	174°C-10 min %
Isoleucine	393	55.2	53.9	56.2	54.7
Leucine	551	77.3	76.2	78.8	77.5
Lysine	436	31.4[a]	30.1[a]	22.0[a]	15.4[a]
Methionine	210	67.6	68.1	59.5	58.6
Cystine	152	108.6	113.2	92.8	47.4
Phenylalanine	358	83.8	82.4	84.9	82.7
Tyrosine	260	71.9	58.9	70.4	68.5
Threonine	320	50.6	50.0	53.1	48.8
Valine	428	59.6	57.9	60.3	59.4

[a]Essential amino acid content is expressed as a percentage of the egg standard. The lowest percentage (underlined) is the chemical score.

Several early workers reported the availability of lysine in breads (4, 10, 21). Palamidis and Markakis (12) reported that the total lysine content decreased with baking or toasting. Available lysine contents significantly decreased as the heat exposure increased. This correlates highly with PER. However, existing techniques, such as fluorodinitrobenzene (FDNB) (21) and trinitrobenzenesulfonic acid (TNBS) (22), for chemically determining available lysine suffer interferences and inaccuracies.

Hwang (24) did his thesis work under Gehrke's direction to develop a rapid and accurate method for measuring available lysine in proteins and foods. He found that cyanoethylation of lysine residue in the protein matrix with acrylonitrile gave two acid-stable products, N-ε-mono-(carboxyethyl)lysine (MCEL) and N,N-ε,ε-di(carboxyethyl)-lysine (DCEL). He made subsequent determination of available lysine values by amino acid analyses (cation exchange chromatography, CIE) or automatic enzymatic determination (AED). He found that the TNBS method gave highly inconsistent data for available lysine in flour and breads. Use of the CIE and AED methods provided data that were in agreement. With these new methods, he measured the total and available lysine in bread, bread crust and crumb samples (Table IX).

As shown in Table IX, the lysine availability (%) showed changes for the three samples. However, the unavailable lysine (total lysine minus available lysine) contents in bread (whole), bread crust and crumb were only 0.04, 0.05, 0.03%, respectively. Table 7 shows that the unavailable lysine contents for all pizza crusts, baked and unbaked, varied only from 0.02 to 0.03%. These data indicate the reduction of lysine caused by baking is mainly shown by the total lysine analysis. It appears then that there is no need to run available lysine determinations for such bakery foods. This finding also suggests that the nutritive loss of bread and pizza crusts was primarily due to the destruction of lysine in those products; to a lesser extent baking caused it to become unavailable.

Discussion

Bread is an excellent staple supplying key nutrients; carbohydrates, proteins, minerals, and vitamins. Its nutritive value affects a great majority of the human population. The marked increase in nutritive value of microwave or steamed breads, observed in the present study, indicates that those processes deserve attention or that conventional baking warrants some modification.

The importance of controlling temperature and time also should be emphasized, particularly when products such as balady bread or pizza crust, baked by a high-temperature and

Table IX. Total and available lysine in bread, bread crust, and bread crumb

Sample	Cation exchange chromatography			Automatic enzymatic determination		
	Lysine (g/100g sample) Total	Available	Availability %	Lysine (g/100g sample) Total	Available	Availabililty %
Bread (whole)	0.26	0.22	84.6	0.27	0.23	85.2
Bread crust	0.20	0.15	75.0	0.20	0.15	75.0
Bread crumb	0.32	0.29	90.6	0.30	0.27	90.0

short-time process, are discussed. Although toasting is the customary way of serving bread for breakfast, it is certainly advisable to toast the slice lightly, not darkly, to preserve the nutritive value of bread.

In addition to the destruction and cross-linking of lysine by baking, protein digestibility is reduced as a result of the browning reaction. In spite of the deleterious effect of the browning reaction on the nutritive value of bread and pizza crust, browning plays an important role in producing the unique golden brown crust and flavor. Baker et al. (25) have pointed out years ago that fermentation followed by the formation of a brown crust is essential to the development of full flavor and aroma of bread. Thus we must compromise or optimize to produce a high-quality bread and related products with a minimum loss of nutritive values during processing.

Acknowledgments

Contribution No 82-576-A, Department of Grain Science and Industry, Kansas Agricultural Experiment Station, Kansas State University, Manhattan, KS 66506. Financial support from NC-132 project is gratefully acknowledged.

Literature Cited

1. Rosenberg, H. R.; Rohdenburgh, E. L. J. Nutr. 1951, 45, 593.
2. Rosenberg, H. R.; Rohdenburgh, E. L. Arch. Biochem. Biophys. 1952, 37, 461.
3. Rosenberg, H. R.; Rohdenburgh, E. L.; Baldini, J. T. Arch. Biochem. Biophys. 1954, 49, 263.
4. Clarke, J. K.; Kennedy, B. M. J. Food Sci. 1962, 27, 609.
5. Ericson, L. E.; Larsson, S.; Lid, G. Acta Physiol. Scand. 1961, 53, 85.
6. Hutchinson, J. B.; Moran, T.; Pace, J. Brit. J. Nutr. 1959, 13, 151.
7. Jansen, G. R.; Ehle, S. R. Food Tech. 1965, 19, 1439.
8. Jansen, G. R.; Ehle, S. R.; Hause, N. L. Food Tech. 1964, 18, 367.
9. Sabiston, A. R.; Kennedy, B. M. Cereal Chem. 1957, 34, 94.
10. Gotthold, M. L. M.; Kennedy, B. M. J. Food Sci. 1964, 29, 227.
11. Tsen, C. C.; Reddy, P. R. K.; Gehrke, C. W. J. Food Sci. 1977, 42, 402.
12. Palamidis, N.; Markakis, P. J. Food Process. Preserv. 1980, 4, 199.
13. Dworschak, E. CRC Crit. Rev., Food Sci. and Nutr. 1980, 13, 1.

14. Tsen, C. C.; unpublished data.
15. Tsen, C. C.; Reddy, P. R. K. J. Food Sci. 1977, 42, 1370.
16. El-Samahy, S. K.; Tsen, C. C. Cereal Chem. 1981, 58, 546.
17. Tsen, C. C.; Bates, L. S.; Wall, L. L. Sr.; Gehrke, C. W. J. Food Sci. 1982, 47, 674.
18. Hansen, L. P.; Johnston, P. H.; Ferrel, R. E. "Protein Nutritional Quality of Foods and Feeds"; Friedman, M. Ed.; Marcel Dekker: New York, NY, 1975; 393.
19. Lea, C. H.; Hannan, R. S. Biochim. Biophys. Acta 1950, 5, 433.
20. Reynolds, T. M. Adv. Food Res. 1965, 14, 167. 20.
21 Calhoun, W. K.; Hepburn, F. N.; Bradley, W. B. J. Nutr. 1960, 70, 337.
22. Carpenter, K. J. Biochem. J. 1960, 77, 604.
23. Kakade, M. L.; Liener, I. E. Anal. Biochem. 1969 27, 273.
24. Hwang, D. H. M.S. Thesis, University of Missouri, Columbia, MO, 1978.
25. Baker, J. C.; Parker, M. K.; Fortmann, K. L. Cereal Chem., 1953, 30, 22.

RECEIVED January 18, 1983

20

Loss of Available Lysine in Protein in a Model Maillard Reaction System

BARBEE W. TUCKER, VICTOR RIDDLE,[1] and JOHN LISTON

University of Washington, Institute for Food Science and Technology, College of Ocean and Fishery Sciences, Seattle, WA 98195

A model Maillard reaction system was developed to measure the loss of available lysine in a pure protein system after reaction with two keto-aldehydes normally present in food systems. Bovine plasma albumin and malonaldehyde or methylglyoxal reactions were measured in an aqueous solution. Conditions of reaction were varied over a range representative of conditions during handling and storage of food. pH was adjusted between 5 and 8. Temperatures in the range 20-60°C were tested. Reactant concentrations of 0.07 to 35.0 mM carbonyl compound and 1 to 10 g/L albumin were used. Loss of available lysine was measured by a trinitrobenzenesulfonic acid method for residual lysine. With malonaldehyde, loss of available lysine decreased as pH increased, but with methylglyoxal, an increased pH increased the lysine loss. Both carbonyl compounds caused significant increasing loss of lysine with increasing temperature and concentration. Results indicated that methylglyoxal reacts with protein to effect a loss of available lysine, but more slowly than does malonaldehyde. Temperature had the greatest effect on lysine loss, but the concentration of the carbonyl compound and the pH of the reaction mixture also influenced the rate.

Of the eight amino acids essential for man and other animals, lysine is the most easily damaged by processing and/or storage of food. This damage or modification results in a reduction in nutritional availability of lysine. Available lysine can be

[1] Deceased

0097-6156/83/0215-0395$06.00/0

utilized for metabolism and is distinguished from total lysine, which includes damaged or bound lysine. It has been shown in experiments with growing chicks that no correlation exists between growth and total lysine while there is a good correlation between growth and unbound lysine, suggesting that lysine is not used biologically unless the ε-amino group is free (1). Plant proteins are often low in lysine content and it is the limiting amino acid in most cereals. Processing damage compounds the problem. Although lysine is relatively abundant in animal protein, heat treatment and/or storage may render it nutritionally unavailable by promoting its irreversible reaction with carbonyl compounds (i.e. Maillard reactions) to form indigestible colorless browning intermediates. Carbonyl compounds are present in foods as reducing sugars and sugar breakdown products, lipid oxidation products, or carbonyl groups of proteins.

Maillard reactions may be divided into three stages as described by Mauron (2): early, advanced, and final. Early Maillard reactions involve a condensation between a carbonyl compound and a free amino group of a protein--in this instance the ε-amino of of lysine--which is rapidly converted via Schiff bases and the Amadori rearrangement to the biologically unavailable deoxyketosyl compound (3). Whatever the source of the carbonyl group, the basic reaction rendering the lysine unavailable as a biochemical component is the same. Although visible browning may not occur, lysine loss is irreversible and this early-forming product is the major form of blocked lysine in foods. The second stage involves a number of reactions producing volatile or soluble substances and the third gives insoluble brown polymers.

Although the early Maillard reaction products are reported to have antioxidant properties and, in fact, can be utilized by processors to inhibit lipid oxidation in animal protein foods such as fish products, there is an accompanying lysine loss (4,5).

Malonaldehyde (MA) is a major end product of oxidizing or rancid lipids and it accumulates in moist foodstuffs (6). Several MA-protein systems have been studied. Chio and Tappel combined RNAase and MA to demonstrate fluorescence attributed to a conjugated imine formed by crosslinking two ε-amino groups with the dialdehyde (7). Shin studied the same reaction and found it to be dependent on pH and reactant concentrations (8). Crawford reported the reaction between MA and bovine plasma albumin (BPA) also to be pH dependent, and of first order kinetics with a maximum rate near pH 4.30. At room temperature 50-60% of the ε-amino groups were modified--40% in the first eight hours, the remainder over a period of days (9).

Another carbonyl compound often found in food products is the keto aldehyde methylglyoxal (MGA). Having the same empirical formula and molecular weight as MA, MGA has been of interest primarily as a glycolytic by-pass intermediate. In foods derived from animal tissue MGA is probably formed from hydrated glyoxalate (dihydroxyacetate, DHA), the latter accumulating as a result of impaired glycolysis following death and subsequent freezing, thawing, and/or storage (10). Riddle and Lorenz clearly demonstrated that MGA can be produced nonenzymatically as the result of a polyvalent-anion-catalyzed reaction (11). It also occurs as a product of autoclaved glucose, unsaturated fatty acid oxidation, and food irradiation (12,13,14). Hodge demonstrated its participation in browning reactions and it or its precursor DHA is currently a component of quick- or instant-tanning lotions (15).

At the pH of most foods and in moist systems MA exists as the enol tautomer and an enamine derivative would be the initial product while MGA would form, more slowly, the more highly colored imine derivative.

This study was undertaken to further investigate the relationships between food system conditions and damage to lysine.

Experimental

A model system demonstrating the nutritional destruction of lysine in bovine plasma albumin (BPA) by reaction with either a dialdehyde (MA) or a keto-aldehyde (MGA) was studied in relation to reaction rates as affected by pH, temperature, reaction time and carbonyl concentration. The BPA was Fraction V obtained from Schwartz/Mann and had a molecular weight of 69×10^3 with sixty lysine residules/mole, an assayed content of 11.4%. It was dissolved in 0.0200 M phosphate-citrate buffer adjusted to the desired pH. Malonaldehyde was prepared by acid hydrolysis of its bis-(dimethyl acetal). An aqueous solution of pyruvic aldehyde was diluted with distilled water and phosphate-citrate buffer to give an MGA solution of the desired pH (16).

Available lysine in the model system was determined by an adaptation of the method of Kadade and Leiner (17). After preliminary rate studies the parameters chosen for measurement were: pH--5,6,7,8; temperature--20, 40 and 60°C in a water bath; time--usually 24 hours with samplings at 10 min and 1,2,4, and 8 or 10 hours (occasionally after 2 or 4 days @ 20°C); BPA concentration--1 g/L, 10 g/L; carbonyl compound concentration--0.07 to 35.0 mM or carbonyl:lysine--1:10 to 50:1.

Experimental procedures employed generally were as follows: The BPA solution was measured into a culture tube, the carbonyl compound at the desired concentration added and the two gently mixed. Zero time samples were withdrawn and the capped tubes were placed in a Beckman thermocirculator at the experimental temperature. Blank (no carbonyl) and control (no BPA) tubes were included with each run. Samples were withdrawn at specified intervals and assayed for available lysine.

In addition to determining protein damage as indicated by per cent loss of available lysine, the amount of carbonyl compound remaining after reaction with BPA was measured. MA was assayed by Buttkus's modification of the method of Sinnhuber and Yu (18,19). MGA was assayed by the method of Vogt as modified by Riddle (20,11).

Results and discussion

Among the variables tested (pH, temperature, and reactant concentration) temperature was the most influential parameter with each carbonyl compound at all levels tested and reactant concentration was significant when carbonyl-lysine ratios were above 1. In contrast, alteration of pH within the pH 5-8 range tested (that commonly found in protein food systems) had a negligible effect. Although preliminary studies indicated optimal pH values for these reactions to be close to 5 for MA and near 8 for MGA, rate changes when the experiments were conducted at pH 6 were slight--especially as compared with rate changes occurring with increases in temperature and reactant concentration. Therefore, the graphs presented are based upon data using a pH of 6.

Fig. 1 shows the loss of available lysine after 12 hours of reaction of BPA with either MA or MGA as a function of temperature and carbonyl concentration. As the reaction temperature was increased the lysine loss was significantly increased with each carbonyl compound although the loss with MA was considerably greater than with MGA. At 60°C, not an unlikely temperature for a variety of processing procedures, as much as 80% of the lysine was lost with MGA and 90% with MA--much of this loss occurring during the first 2-4 hours as shown in Fig. 2. These are significant nutritional losses and illustrate the damage possible to foods either through prolonged heat processing or, perhaps more important, by storage in hot climates. Even after 10 min the loss was more than 20% while it exceeded 50% in 4 hours. Studies of the MA-lysine reaction by other authors, as discussed previously, were conducted at 37°C or lower. There are a few reports of lysine losses measured as a function of increasing temperature to more than 100°C; some of these reports correlate these losses with browning and others with moisture content (21-26). However, most studies of high temperature effects have been on browning rather than nutrient loss.

The effects of carbonyl concentration are illustrated in Figs. 3 and 4. MA exhibited a more obvious relationship between carbonyl concentration increases and lysine loss than did MGA. It is evident that loss occurs to a greater extent with the di- rather than the keto-aldehyde. This is consistent with the previously mentioned theory that the enamine derivative forms more readily than the imine.

The carbonyl:lysine ratio used influenced the lysine loss. MA at a ratio of 50:1 caused a significantly greater lysine loss than equimolar amounts of MGA, but each demonstrated a definite, though not linear, increase in reactivity with increasing

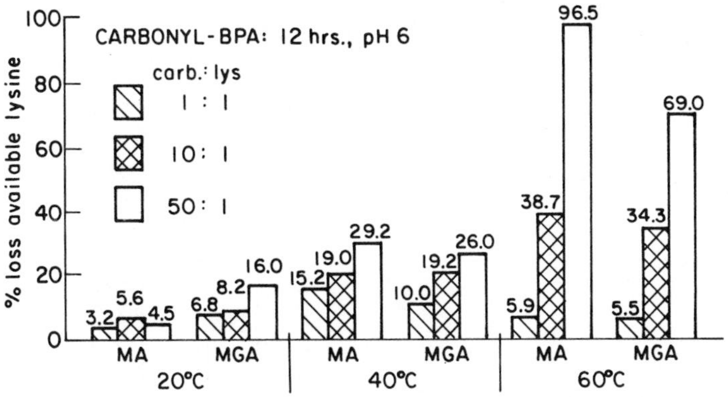

Figure 1. Variation of lysine loss with temperature, nature of carbonyl reagent, and ratio of reagents.

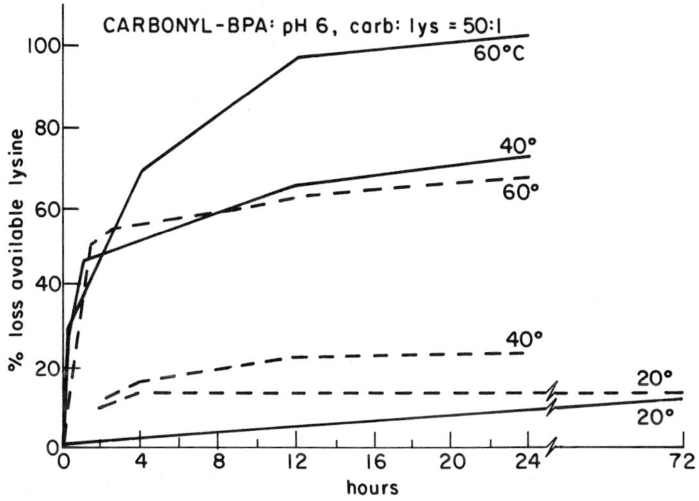

Figure 2. Rate of lysine loss at fixed reactant ratio with change of temperature and nature of carbonyl reagent. Key: —, malonaldehyde (MA); and – – –, methylglyoxal (MGA).

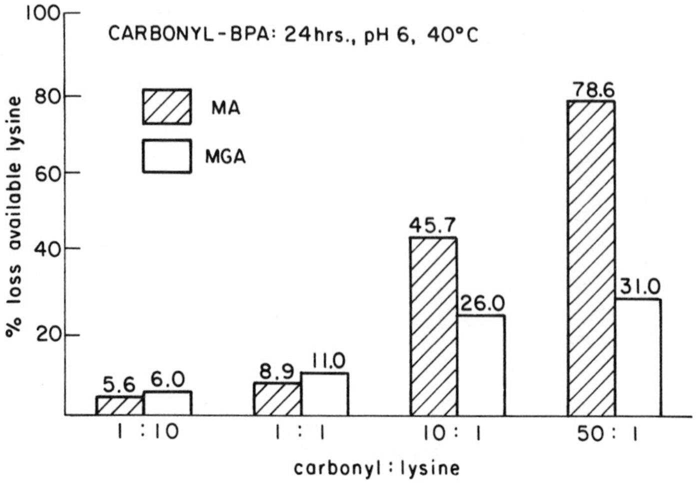

Figure 3. Variation of lysine loss with ratio of reactants.

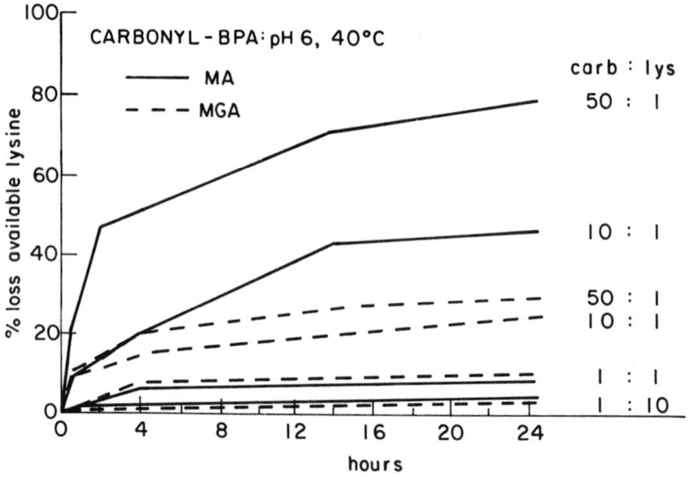

Figure 4. Rate of lysine loss at various ratios of malonaldehyde and methylglyoxal to bovine plasma albumin.

concentrations. At molar ratios of carbonyl:lysine near 1 very little effect was noted with an increase in temperature (Fig. 1)--an indication that foods with a low content of carbonyl compounds might maintain lysine stability even under processing conditions.

Loss of the carbonyl component was measured as an indication of its reaction with BPA. Results are illustrated as functions of the various parameters in Tables I and II. At each concentration level of carbonyl compound, increasing temperatures were accompanied by increases in loss of the carbonyl component. Increasing the lysine:MGA to 100:1, in effect creating conditions for a first order reaction, decreased the influence of temperature, yet even at 20°C more than 75% of the MGA reacted.

Control samples of MGA in which no BPA was present showed a loss of 42.7% of the MGA at 60°C, but no significant loss at 20 or 40°C. These large losses of MGA at the higher temperature are attributed in part to polymerization and intramolecular Cannizzaro-type reactions. Corrected results, taking this phenomenon into account, are shown in Table II. Even after correcting for self-reactions, both carbonyl compounds reflected losses in excess of the lysine loss (mole for mole), which might be due to reactions occurring between the carbonyl compounds and other active sites of the protein such as the α-amino group of the N-terminal aspartate and/or thiol residues (27). Therefore, it is difficult to correlate losses of MGA with decreased lysine availability.

Because this model system employed dilute aqueous solutions, the effect of water activity (A_w) on the loss of available lysine was not a parameter. During storage of dried foods A_w as well as temperature is a consideration. Labuza reported a loss of available lysine in dry milk at A_w = 0.44 to 0.68 (28).

Conclusions

In general both di- and keto-aldehydes react with lysine residues of BPA at temperatures which may occur during storage under tropical conditions or relatively mild heat processing to effectively decrease the available lysine. The di-aldehyde, through formation of an enamine derivative, reacts more rapidly and to a greater extent than the keto-aldehyde at pH values likely to be encountered in foods. The conjugated imine derivative of the keto-aldehyde, however, contained a more highly colored chromophore. Therefore, damage to the lysine by MGA may be accompanied by more "browning" than the greater lysine damage with MA. Temperature as well as reactant concentration significantly increases these losses. Rapid cooling and low storage temperatures after heat processing of protein foodstuffs appear to be required in minimizing nutritional lysine loss.

TABLE 1

Variation of loss of carbonyl compound
(by reaction with BPA at 1:10 ratio)
with temperature, source of carbonyl group, and ratio of reagents

	LOSS CARBONYL (%)		
	20°C	40°C	60°C
MA - pH 5	2.7	8.6	16.0
MGA - pH 7	6.0	47.0	92.5

TABLE 2

Variation of loss of MGA (by reaction with BPA in 24 hr at pH 7)
with ratio of reactants and temperature

MGA:lysine	% LOSS MGA		
	20°C	40°C	60°C
1:10	6	47	93 53[#]
1:100	76	84	100 57[#]

[#]Corrected for self-reaction.

Acknowledgments

We wish to acknowledge support for some publication costs from the Egtvedt Food Research Fund, University of Washington.

Literature cited

1. Roach, A.G.; Sanderson, P.; Williams, D.R. J. Sci. Food Agr. 1967, 18, 274.
2. Eriksson, C., Ed.; "Maillard Reactions in Food"; Pergamon Press: Oxford, 1981.
3. Ibid; p. 159-176.
4. Ibid; p. 421-428.
5. Ibid; p. 441-442.
6. Meyer, L.H. "Food Chemistry"; Reinhold Book Corp.: NY, 1960.
7. Chio, K.S.; Tappel, A.L. Biochemistry 1969, 8, 2827.
8. Shin, B.K.; Huggins, J.W.; Carrasay, K.L. Lipids 1972, 7, 229.
9. Crawford, D.L.; Yu, T.C.; Sinnhuber, R.O. J. Food Sci. 1967, 32, 332.
10. Fujimaki, S. Agr. Biol. Chem. 1968, 32, 46.
11. Riddle, V.; Lorenz, F.W. J. Biol. Chem. 1968, 243, 2718.
12. Englis, D.T.; Hanahan, D.J. J. Am. Chem. Soc. 1945, 76, 51.
13. Cobb, W.Y.; Day, E.A. J. Am. Oil Chem. Soc. 1965, 42, 110.
14. Scherz, H. Radiat. Res. 1970, 43, 12.
15. Hodge, J.E. J. Agr. Food Chem. 1953, 1, 928.
16. Tucker, B.W. M.S. Thesis, Univ. of Washington, 1974.
17. Kakade, M.L.; Leiner, I.E. Analyt. Biochem. 1969, 27, 273.
18. Buttkus, H. J. Food Sci. 1967, 32, 432.
19. Sinnhuber, R.O.; Yu, T.C. Food Res. 1958, 23, 626.
20. Vogt, M. Biochim. Z. 1929, 211, 17.
21. Taira, H.; Sukurai, Y. Jap. J. Nutr. Food 1966, 18, 359.
22. Miller, E.L.; Carpenter, K.T.; Morgan, C.B. Brit. J. Nutr. 1965, 19, 249.
23. Miller, E.L.; Carpenter, K.J.; Milner, C.K. Brit. J. Nutr. 1965, 19, 547.
24. Carpenter, K.J.; Morgan, C.B.; Lea, C.G.; Parr, L.J. Brit. J. Nutr. 1962, 16, 451.
25. deGroot, A.P. Food Tech. 1963, 17, 339.
26. Mauron, J.; Egli, R.H. Arch. Biochem. Biophys. 1955, 59, 433.
27. Buttkus, H. J. Am. Oil Chem. Soc. 1972, 49, 613.
28. Labuza, T.P. CRC Crit. Rev. Fd. Tech. 1972, 3, 217-240.

RECEIVED November 2, 1982

Effect of a Glucose–Lysine Reaction Mixture on Protein and Carbohydrate Digestion and Absorption

RICKARD ÖSTE, INGER BJÖRCK, ARNE DAHLQVIST, MARGARETHA JÄGERSTAD, and PER SJÖDIN

University of Lund, Department of Applied Nutrition, Chemical Center, P.O. Box 740, S-220 07 Lund, Sweden

HANS SJÖSTROM

The Panum Institute, Department of Biochemistry C, Copenhagen, Denmark

A low-molecular-weight fraction from a glucose-lysine reaction mixture reduced the plasma level of lysine originating from dietary protein, when given to rats (1.5% w/w in diet). This fraction inhibited in vitro trypsin, carboxypeptidase A, and carboxypeptidase B as well as aminopeptidase N of the brush border. A high-molecular-weight fraction from the reaction mixture strongly inhibited, competitively, invertase and lactase. When intubated into rats, however, it did not affect the absorption of sucrose (ratio sucrose/reaction fraction 100:1 w/w). The low-molecular-weight fraction had only a slight effect on the carbohydrate -hydrolyzing enzymes.

When proteins are heated together with carbohydrates a decrease in the nutritive value is frequently observed. The degree of this decrease is dependent on a number of factors, including water activity, type and amount of reducing sugar, type of protein, as well as the extent of heat treatment (1). The loss of protein quality may be expressed as reduced biological value (BV) and digestibility (TD), as apparent in a conventional net protein utilization (NPU) assay with rats. When the heat treatmeant is mild, a loss in the BV often corresponds to the loss of lysine caused by the Maillard reaction, provided lysine is the limiting amino acid in the protein. When the heating is more pronounced, the reduction in BV is often found to be greater in the protein and may also cause a reduction of TD. Reviews of the effect of the Maillard reaction in protein nutrition have recently been published by Mauron (2) and Dworschak (3).

There are some reports on possible mechanisms behind these effects of severe heat treatment. Adrian (1) has shown that water-soluble premelanoidins from a glucose-glycine reaction mixture reduce the protein digestibility and affect the utilization of

0097-6156/83/0215-0405$06.00/0

absorbed amino acids. Valle-Riestra (4) suggests that the reduced uptake of a severely heated glucose-egg albumen mixture is the result of a lowered pancreatic secretion of digestive enzymes. The observed decrease in biological value (the result of enhanced urinary nitrogen excretion) may partly be explained by an uptake of indigestible low-molecular residues from the gut (4, 5, 6).

The present study was performed in order to examine whether the compounds in a glucose-lysine reaction mixture *per se* may influence the digestion and uptake of proteins. Since such an effect was noted, the study was extended to include the digestion and uptake of carbohydrates as well.

Glucose-lysine reaction mixture

Equimolar amounts of glucose (4.50 g) and L-lysine hydrochloride (4.55 g) were dissolved in 100 ml of distilled water and the solution was boiled under reflux for 24 h. The pH was kept constant during the reaction by a pH-stat-controlled addition of 5N NaOH. The pH-stat setting was 6.5. After the reaction the pH of the mixture at room temperature was about 5.3. The mixture was dialyzed against 3x2 l distilled water in a Spectrapor sack, which according to the manufacturer (Spectrum Medical Industries, Inc., Los Angeles, U.S.A.) had an exclusion limit of 6000-8000 daltons. The dialysate was evaporated, the residue dissolved in 0.1 M ammonium-acetate buffer, and placed on a ConA-Sepharose column to separate unreacted glucose. The eluate was concentrated and constituted the low-molecular-weight (LMW) fraction used in this study. The yields from repeated preparations were 7.4-7.7 g. The retentate from the dialysis was centrifugated to remove insoluble material, and then concentrated by evaporation. The yield was 0.9-1.2 g and constituted the high-molecular-weight (HMW) fraction.

Effects on protein utilization

Animal assays. When assessing the nature of a protein quality reduction, the conventional methods of protein quality measurement have certain limitations. For example, a reduced TD in an NPU assay is not necessarily the result of a reduced protein digestion. Obviously the same result will be obtained if the increase of fecal nitrogen is caused by an enhanced excretion of endogenous protein or if there is a fixation of metabolic nitrogen by the colonic micro-flora.

To be able to specifically study the digestion of an exogenous protein in rats, a method was elaborated as follows: A solution of [U-^{14}C]-lysine was injected in the wing-vein of a laying hen at the time of maximal egg-albumen synthesis. The eggs were recovered, the egg-white separated and dialyzed to remove glucose and then lyophilized. An acid hydrolysate was separated by thin layer chromatography and subjected to autoradiography. It was thereby shown that the radioactivity present in the egg-white was derived almost exclusively from protein-bound lysine (45000 dpm/mg protein). Only traces of radioactivity could be found in non-lysine residues.

A control diet was prepared as follows: Tha labelled egg-white protein was mixed with a basal diet in the proportion 2:98. The basal diet contained 10% casein, 10% sucrose, about 70% starch and 5% maize oil, as well as vitamins and minerals. A small amount of [^{3}H]-lysine was added to this mixture, giving a ratio of tritium to (^{14}C) dpm of approximately 10:1. The experimental diet was obtained by adding a small amount ot the glucose-lysine reaction mixture to the control diet.

These diets were then fed to male Sprague-Dawley rats, weighing 100-120 g each. The rats were placed in individual cages and were fasted overnight. Early in the morning each rat was given 1.5 g of the control or the experimental diet to be consumed _ad libitum_. We have observed that overnight fasted rats in the _morning immediately_ will consume at least 2 g of food, provided the animal room is kept dark and the food is not too untasty. In these experiments all the rats completely consumed the portion within 15 min. This procedure was accurate enough to obviate intubation as a means for administering the diets. Exactly two hours after finishing the meal each rat was anesthesized with ether, blood was sampled by heart puncture, and the (^{14}C)/(^{3}H)-ratio in the plasma was measured.

In the first experiment the experimental diet contained 1.5% of the LMW fraction. Table I shows the (^{14}C)/(^{3}H)-ratios found in the plasma. The ratio obtained in the control diet was about the same as that in the diet (0.120). In the group fed with the experimental diet a statistically significant 15% decrease was observed. In the second experiment the experimental diet contained 1.7% of the HMW fraction. With this fraction a somewhat unexpected effect was observed. There was a small, but highly significant increase in the (^{14}C/^{3}H)-ratio (Table I). This could be explained by a decrease in the uptake of free lysine from the small intestine, since a considerable part of the originally protein-bound (^{14}C)-labelled lysine might be absorbed in the form of small peptides (8). Table I also shows the absolute values of (^{3}H) and (^{14}C) in the plasma of the rats fed the HMW fraction. The standard deviations of these data are higher than the one calculated from the (^{14}C)/(^{3}H)-ratio and the increased average of (^{14}C) in the experimental group is not statistically significant. The levels of (^{3}H) are about the same in both groups. These result did not indicate a specific affect on the lysine uptake.

Enzyme assays As shown previously the LMW fraction had a repressing effect on the protein digestion in the _in vivo_ experiment. Accordingly, it was of interest to study _in vitro_ the effect of this fraction on the kinetics of reactions catalyzed by proteases and peptidases present in the gastro-intestinal tract.

In Table II are shown the results from kinetic studies with commercially available gastric and pancreatic enzymes. Trypsin was strongly inhibited, at least at a low concentration of casein as substrate. The hydrolysis of benzoylarginine ethyl ester (BAEE) by trypsin was non-competitively inhibited, giving a 30% reduction of V_{max} at 0.5 mg/ml of the LMW fraction. Carboxypeptidase A, and to a lesser extent carboxypeptidase B, were non-competitively inhibited as well. Pepsin and chymotrypsin were not affected by the conditions used in these assays.

The action of the gastric and pancreatic enzymes causes the release of small peptides as well as free amino acids, the peptide fraction being the quantitatively dominant one (9). Thus further hydrolysis is crucial, if the dietary protein is to be completely utilized by the organism. The final stages of hydrolysis is associated with the intestinal mucosal cells. Larger peptides are probably hydrolyzed by enzymes at the brush border membrane. Di- and tripeptides may be absorbed as such and hydrolyzed intracellularly (9, 10).

There are at least three peptidases in the brush border of the small intestine: Aminopeptidase A, which has an affinity for peptide-bound acid amino acids (11), aminopeptidase N, which has a broad specificity (12), and dipeptidylpeptidase IV, which releases dipeptides from the N-terminal end of peptides with a preference for X-PRO terminals (13). In Table III are shown the effect of a low concentration of the LMW fraction on the activity of these enzymes in extracts of hog intestine. Aminopeptidase N was found to be strongly inhibited by 0.25 mg/ml of the fraction. Aminopeptidase A and dipeptidylpeptidase IV were not inhibited.

Around 40% of the LMW fraction is absorbed from the small intestine in the rat (14). A part of the fraction may thus be present in the epithelial cells at the time of intracellular peptide hydrolysis. The effect of the LMW fraction on the activity of two cytosol enzymes, present in a preparation from hog intestine, is shown in Table III. Glycylleucine dipeptidase, which has a broad specificity (15), was inhibited at 0.7 mg/ml of the fraction, while proline dipeptidase, which catalyzes the hydrolysis of X-PRO (16), was not.

Since the glucose-lysine reaction mixture used in this study consisted of a number of different substances, it was of interest to study whether the observed inhibitory effect *in vitro* could be attributed to some specific compound(s). In order to obtain a separation, an aliquot of the LMW fraction, radiolabelled by [U-^{14}C] glucose added to the reactants, was applied on a Sephadex G-50 column and eluted with water. The UV absorbance was recorded and the eluate was collected in fractions. The degree of inhibition effected by small samples of equal volume from each fraction and exerted on carboxypeptidase A and purified aminopeptidase N was determined as well as the radioactivity

TABLE I
Measurements of (^3H) and (^{14}C) in rat plasma

Experiment	(^{14}C)/(^3H)	(^{14}C)***	(^3H)***
LMW fraction			
Control (6)	0.117 ±0.014		
Experimental (7)	0.102 ±0.010*		
HMW fraction			
Control (10)	0.100 ±0.0027	1969 ±235	18251 ±1382
Experimental (10)	0.106 ±0.0036**	1817 ±164	18568 ±1846

All values are mean ± standard deviation. A new control diet was prepared before each experiment. Within parenthesis number of rats. From Öste et al.(7)

* P<0.05
** P<0.01
*** Dpm/ml plasma

TABLE II
Effect of the LMW fraction on the activity of gastric and pancreatic enzymes in vitro.

Enzyme	Substrate	Conc. of LMW fraction*	Results
Pepsin	Hemoglobin	1.0	No inhibition.
Trypsin	Casein	1.0	Inhibition at low substrate conc.
	BAEE	0.5	Non-competitive inhibition. V_{max} reduced 30%.
Chymotrypsin	Casein	1.0	No inhibition.
Carboxypeptidase A	N-Benzoyl-GLYPHE	0.5	Non-competitive inhibition. V_{max} reduced 30%.
Carboxypeptidase B	N-Benzoyl-GLYARG	0.5	Competitive inhibition?

* mg/ml
From Öste et al. (7)

TABLE III
Effect of the LMW fraction on the activity of enzymes of the intestinal mucosa.

Enzyme	Substrate	Conc. of LMW fraction*	Results
<u>Brush border membrane enzymes</u>			
Amino-peptidase A (EC 3.4.11.7)	GLU-p-nitro-anilide	0.25	No inhibition
Amino-peptidase N (EC 3.4.11.2)	ALA-p-nitro-anilide	0.25	Mixed-type inhibition? V_{max} reduced 60%
Dipeptidyl-peptidase IV			No inhibition
<u>Cytosol enzymes</u>			
Glycyl-leucine dipeptidase (EC 3.4.13.11)	GLYLEU	0.7	Inhibition V_{max} reduced ca. 50-60%
Proline dipeptidase (EC 3.4.13.9)	ALAPRO	0.7	No inhibition

* mg/ml
From Öste et al.(7)

of each fraction. The results are shown in Figure 1. The inhibition of carboxypeptidase A seemed to approximately coincide with the degree of UV absorbance. Aminopeptidase N was inhibited to a varying extent by different fractions. It was evident that the inhibition could not be ascribed to any single compound. On the other hand, the inhibitory effect could not be a feature common to the quantitatively dominant fractions in the mixture, since the amount of radioactivity found in each fraction, which should roughly reflect the total amount of Maillard reaction compounds, did not correspond to the degree of inhibition of the enzymes. In addition, the Sephadex fractions affected the activity of the two enzymes to a different degree, indicating that not necessarily the same compounds were responsible for the inhibitory effect of the LMW fraction noted with different enzymes.

Effect on carbohydrate utilization

Since the glucose-lysine reaction mixture inhibited enzymes of protein digestion, an effect on carbohydrate digestion could also be expected. The quantitatively most important dietary carbohydrate is starch, which is hydrolyzed by salivary and pancreatic α-amylase into maltose and oligosaccharides. These are then hydrolyzed into glucose by amyloglucosidase, isomaltase and maltase present in the brush border of the epithelial cells. Sucrose is hydrolyzed by brush border invertase and lactose by brush border lactase. Trehalase splits trehalose present in mushrooms. The monosaccharides formed in these processes, glucose, fructose and galactose, are actively transported (for references, see (17)).

Table IV shows the effect ot fhe LMW fraction on the activity of some of these enzymes in vitro. Maltase, lactase and invertase were competitively inhibited at a concentration of 10 mg/ml. When the effects of a range of concentrations (2.5-20 mg/ml) of the LMW fraction were studied, it was revealed that the inhibition was not of the pure competitive type. Table V shows the effect of the HMW fraction. Low concentrations had to be used in the assays, as the intense brown color of this fraction interfered with the spectrophotometric measurements. In spite of this a strong competitive inhibition of lactase and of invertase was found. Maltase was also inhibited, and, to a lesser extent, even trehalase. α-Amylase from saliva was not affected at the concentration tested.

In order to study whether the pronounced inhibition of invertase might be of significance for the in vivo uptake of sucrose, a sucrose solution containing 1 % (w/w) of the HMW fraction was intubated into fasting rats. The rats were killed at various time intervals after ingestion, the gastro-intestinal tracts were removed, and the contents of total carbohydrates were determined (the detailed procedure has been described earlier, see ref. 19). The results, given in Figure 2, show that no difference in the amount of sugar absorbed could be observed.

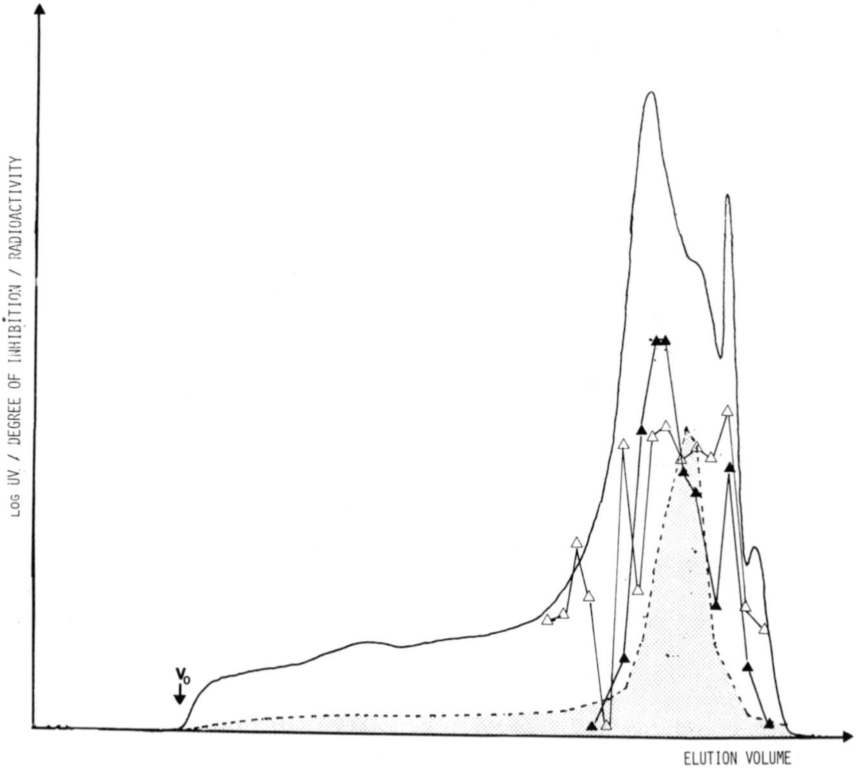

Figure 1. Separation of the low molecular weight (LMW) fraction on Sephadex G-50 superfine (2.6 × 90 cm, eluent water) (7). Key: —, UV absorbance; shaded area, relative distribution of a (^{14}C)-label on reactant glucose; △, inhibition of aminopeptidase N; and ▲, inhibition of carboxypeptidase A.

TABLE IV
Effect of the LMW fraction on carbohydrate-hydrolyzing enzymes

Enzyme	Substrate	Conc. of LMW fraction*	Results
α-Amylase	Soluble starch	4.6	No inhibition
Maltase	Maltose	2.5-20	Mixed-type inhibition $K_{mapp} = 2K_m$ at 10 mg/ml
Lactase	Lactose	2.5-20	Mixed-type inhibition $K_{mapp} = 2.7 K_m$ at 10 mg/ml
Invertase	Sucrose	10	Inhibition $K_{mapp} = 1.2K_m$
Trehalase	Trehalose	10	Weak inhibition

* mg/ml
From Öste et al.(18)

TABLE V
Effect of the HMW fraction on carbohydrate-hydrolyzing enzymes

Enzyme	Substrate	Conc. of HMW fraction*	Results
α-Amylase	Soluble starch	0.8	No inhibition
Maltase	Maltose	1.4	Competitive inhibition? $K_{mapp} = 2.3\ K_m$
Lactase	Lactose	1.4	Competitive inhibition? $K_{mapp} = 13\ K_m$
Invertase	Sucrose	2.4	Competitive inhibition? $K_{mapp} = 8\ K_m$
Trehalase	Trehalose	2.4	Inhibition

* mg/ml
From Öste et al. (18)

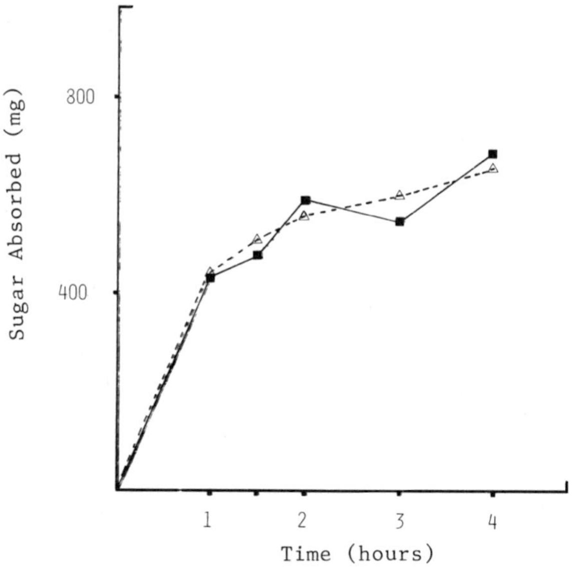

Figure 2. Total sugar absorbed from rat intestine at different times after the ingestion of 760 mg sucrose in solution with (■), and without (△) 8 mg of the high molecular weight (HMW) fraction of the glucose–lysine reaction mixture (18).

Discussion

It is clear that the LMW fraction from the glucose-lysine reaction mixture affected, down to a concentration of at least 0.25 mg/ml, the rates of hydrolysis catalyzed by some, but not all, of the enzymes involved in protein digestion in vitro. Carboxypeptidase A and aminopeptidase N were inhibited by a number of substances in the mixture, some being more effective than others. Efforts to characterize these compounds are in progress.

The question is, whether the inhibitory effect of the LMW fraction might explain the reduced uptake of lysine from orally given egg-white protein as observed in vivo in rats. The total intake per rat of the LMW fractions was 22.5 mg. Assuming a total volume of digestive juices around 5 ml, the concentration in the intestine should most likely be high enough to bring about a notable inhibition, even with a high degree of binding of the compounds to non-specific proteins in the intestine. The inhibitory effects observed in vitro were non-competitive or of the mixed type. This means that even under conditions of excess substrate concentration, which is possibly found, at least initially, in the lumen of the intestine, an effect on the substrate turnover is to be expected. Provided this effect will be rate-limiting for the overall digestion and absorption, a reduced blood uptake, or at least uptake rate, of lysine might be the result, as observed in the in vivo assay.

The nutritional signification of the effect of the LMW fraction is not clear. We do not know if the observed reduced uptake, or uptake rate, will in fact lead to an equivalently reduced total lysine utilization by the rat. The levels of plasma amino acids from the diet two hours after intake is of course not an indisputable measure of amino acid availability. However, the level of total lysine in the plasma of the rat reaches a maximum about two hours after a meal given as described above (unpublished observation). This indicates that a change in dietary amino acid content is probably best reflected in plasma two hours after ingestion. Another question to be answered is if other essential amino acids are affected to the same extent as lysine. This is not necessarily the case, since not all of the enzymes involved in proteolysis or in the final intestinal hydrolysis were affected. The results obtained, however, do support the findings of Adrian (1), who reported that water-soluble products in a glucose-glycine reaction mixture modified an enzymatic proteolysis in vitro performed with pepsin, trypsin and erepsin. He also observed an effect on the protein digestion performed by rats. The HMW fraction gave a significant rise of the dietary protein-derived lysine concentration in plasma. This fraction thus had an effect opposite to the LMW fraction. We have at present no explanation to this observation.

The carbohydrate-hydrolyzing enzymes were inhibited by fairly high concentrations of the LMW fraction. These inhibitions was of the mixed type with a pronounced competitive element. To bring about a substantial inhibition the concentration of the LMW fraction had to be of the same magnitude as the substrate concentration. The effects of the LMW fraction are therefore probably of little practical importance.

The HMW fraction strongly inhibited invertase and lactase at surprisingly low concentrations (1.4-2.4 mg/ml). When given to rats together with sucrose in a ratio of 1:100, no effect on the uptake was observed. A higher concentration of the HMW fraction could of course have had an effect, but we believe such conditions would too much deviate from what is to be expected from consuming food products. Moreover, in rat and in man, the hydrolytic capacity for maltose and sucrose exceeds the transport capacity for the monosaccharides formed (19,20). In the case of sucrose digestion, the effect of the HMW fraction is thus probably of little practical significance. The strong inhibition of lactase, on the other hand, deserves further investigation since the lactase activity in adult man is rate limiting for the utilization of lactose (21).

Literature cited

1. Adrian, J. World Rev. Nutr. Diet. 1974, 19, 71-122.
2. Mauron, J. "Maillard Reactions in Food", Prog. Fd. Nutr. Sci. 5; Eriksson, C., Ed.; Pergamon: Oxford, 1981; pp 5-35.
3. Dworschak, E. CRC Crit. Rev. Food Sci. Nutr. 1980, 13, 1-40.
4. Valle-Riestra, J.; Barneds, R.H. J. Nutr. 1970, 100, 873-882
5. Pronczuk, A.; Pawlowska, D.; Bartnik, J. Nutr. Metabol. 1973, 15, 171-180.
6. Ford, J.E.; Shorrock, C. Br. J. Nutr. 1971, 26, 311.
7. Öste, R.; Sjöström, H. To be published.
8. Adibi, S.A. "Nutrition and Gastroenterology"; Winick, M., Ed.; John Wiley & Sons: New York, 1980; pp 55-75.
9. Adibi, S.A.; Mercer, D.W. J. Clin. Invest. 1973, 52, 1586-1594.
10. Das, M.; Radhakrishnan, A.N. World Rev. Nutr. Diet. 1976, 24, 58-87.
11. Danielsen, E.M.; Norén, O.; Sjöström, H.; Ingram, J.; Kenny, A.J. Biochem. J. 1980, 189, 591-603.
12. Sjöström, H.; Norén, O.; Jeppesen, L.; Staun, M.; Svensson, B.; Christiansen, L. Eur. J. Biochem. 1978, 88, 503-11.
13. Svensson, B.; Danielsen, M.; Staun, M.; Jeppesen, L.; Norén, O.; Sjöström, H. Eur. J. Biochem. 1978, 90, 489-498.
14. Sjödin, P. To be published.
15. Norén, O.; Sjöström, H.; Josefsson, L. Biochim. biophys. acta 1973, 327, 446-456.
16. Sjöström, H.; Norén, O.; Josefsson, L. Biochim. biophys. acta 1973, 327, 457-470.

17. Dahlqvist, A. "Biochemistry of Carbohydrates, II", vol. 16; Manners, D.J., Ed.; University Park: Baltimore, 1978; pp 179-207.
18. Öste, R.; Björck, I.; Dahlqvist, A.; Jägerstad, M.; Sjödin, P. To be published.
19. Dahlqvist, A.; Thomson, D.L. J. Physiol. 1963, 167, 193-209.
20. Dahlqvist, A.; Thomson, D.L. Acta Physiol. Scand. 1963. 59, 111.
21. Dahlqvist, A. J. Clin. Invest. 1962, 41, 463.

RECEIVED October 19, 1982

Determination of Available Lysine by Various Procedures in Maillard-Type Products

H. F. ERBERSDOBLER [1] and T. R. ANDERSON [2]

Institut für Physiologie, Physiologische Chemie und Ernährungs-physiologie der Tierärztlichen Fakultät der Ludwig-Maximilians-Universität München, Veterinärstrasse 13, D 8000 München, Federal Republic of Germany

Three methods of reactive-lysine determination employing the reagents Acid Orange 12, succinic anhydride, and dansyl chloride, as well as methods for total lysine and furosine using ion exchange chromatography, were compared with the method for fluorodinitrobenzene(FDNB)-reactive lysine as reference, in their ability to estimate available lysine in heat-damaged soya protein samples. All methods showed at least some measure of sensitivity to increased severity of lysine damage, with samples heated alone being least damaged followed by those heated with lactose, glucose, then xylose. Under more severe conditions lactose showed reactivity comparable to that of xylose, probably owing to partial hydrolysis of the disaccharide. Total lysine overestimated FDNB lysine at all levels of heat damage. Acid Orange 12 and succinic anhydride values correlated well with FDNB lysine for both non-Maillard ($r = 0.93$) and Maillard ($r = 0.90$) material. The dansyl method gave good relative values but certain problems prevented exact quantitative calculations. The chemical methods correlated quite well with plasma lysine and lysine digestibility methods although the absolute values varied considerably between methods.

The nutritional availability of lysine in foods of plant and animal origin may be significantly decreased by the ready involvement of the ε-NH_2 group of lysine in intra- and intermolecular

[1] Current address: Department of Human Nutrition and Food Science, University of Kiel, Düsternbrooker Weg 17-19, D 2300 Kiel, Federal Republic of Germany

[2] Current address: Biochemistry Department, University of Zululand, Private Bag X1001, Kwa-Dlangezwa 3886, South Africa

0097-6156/83/0215-0419$06.00/0

crosslinking in proteins as well as in reactions with various food constituents. The most common involvement is that together with reducing sugars in the Maillard condensation since this reaction occurs also under mild and moderate conditions such as storage of the food. Some of the intermediate or end products have been reported to be toxic (1, 2, 3). Although the final answer as to the nutritional consequences of processing damage to lysine can obviously be given only by in vivo analysis, recently emphasis has been placed on methods that not only give an acceptable estimate of available lysine but are simple and rapid enough to be automated for use in general laboratory practice (e.g.4, 5).

In this report various chemical and in vivo methods are compared in their ability to assess nutritional damage in various model soya proteins representing mainly the Maillard-type modification of lysine.

Materials and Methods

Model protein samples. Isolated soya protein (Purina Brand Assay Protein RP 100) was employed in these studies and found to contain on a dry matter basis 96% crude protein, 1.4% ash, 0.2% crude fiber, and 2% N-free extract. This material was used to prepare 48 samples, which were heated in airtight metal containers (250-ml) at temperatures of 90, 110, 130° C for 0.5, 1, 2, and 4 h respectively.

The samples were prepared by a mixture of 90 parts of soya protein with 10 parts of tap water or of 80 parts of soya protein with 10 parts of glucose and 10 parts of tap water or instead of glucose an equivalent amount of lactose or xylose according to their molecular mass compared on water-free basis to the glucose. In this way for instance the lactose samples contained 71 g soya protein, 19 g lactose and 10 g tap water.

A further 3 samples were prepared by heating equal masses of isolated soya protein and tap water in open vessels for 24 h at either 95, 138, or 160° C.

Crude protein and total lysine plus furosine Total nitrogen was determined by macro Kjeldahl digestion and protein estimated as N x 6.25. Total lysine values were obtained from conventional amino acid analyses carried out on 500-mg samples following digestion with 800 ml of 6M HCl under reflux by use of a Biotronic LC 6000 or Kontron Liquimat III amino acid analyzer. Furosine was determined in the same way using 300 ml 7.8 M HCl as described in (6) with an amino acid analyzer (7).

Fluorodinitrobenzene (FDNB)-reactive lysine The direct FDNB method of Carpenter (8) was used with some further (9, 10) modifications.

Succinic anhydride (SA)-reactive lysine The method of Anderson (11) was employed. Duplicate 12- to 15- mg portions of each sample (particle size < 80 µm) were accurately weighed into 20-ml glass scintillation vials, 3 ml of 6M guanidine hydrochloride and 0.1 ml of 5M NaOH added, and the whole was magnetically stirred at ca. 50° C for 15 min. After cooling to ca. 25°C, the suspension was treated with solid [1,4-^{14}C] -succinic anhydride with a specific activity of 0.026 µCi per µmol (Radiochemical Centre, Amersham, England). The reagent was added in small portions over a period of 30 min to give an approximately 80-fold molar excess over total lysine content. Vigorous stirring was maintained throughout the addition process during which a further two aliquots (0.1 and 0.05 ml) of 5M NaOH were added after ca. 10 and 20 min respectively.

The mixture was then treated with 5 ml of alkaline hydroxylamine reagent (20 ml hydroxylamine hydrochloride titrated to pH 13 with 3.5M NaOH) for 5 min and the protein precipitated with 5% trichloroacetic acid. After centrifuging (5000 g) for 5 min the supernatant was carefully decanted and the precipitate washed twice with 10-ml aliquots of absolute ethanol. The material was then dissolved in 1 ml, of 0.2M NaOH, mixed with 10 ml of Insta-gel scintillator (Packard Instrument (Pty) Ltd.) , and the radioactivity counted in a Packard Tri Carb Liquid Scintillation Spectrometer.

Dansyl chloride (DAN)-reactive lysine The method of Christoffers (12) was employed. Duplicate 19- to 21-mg portions of each sample (particle size <80 µm) were accurately weighed into 10-ml centrifuge tubes and with good mixing first 1 ml of 0.5M sodium bicarbonate (70 h old) and then 4 ml of absolute ethanol were added. The whole was centrifuged at 5000 g for 0.5 min, the supernatant decanted, and the residue mixed with 8 ml of freshly prepared dansyl chloride solution (1 mg/1 ml 95% ethanol). The tubes were then incubated for 1 h in a water bath at 30° C, the whole centrifuged at 5000 g for 0.5 min, and the supernatant decanted. The residue was washed three times with 4 ml of 94% ethanol and then suspended in 5 ml of deionized water in a cylindrical glass cuvette. Fluorescence intensity was measured using a Biotronic Model BT 1010 amino acid- protein- filter fluorometer, an instrument specially designed to minimize the problems associated with measuring the fluorescence of insoluble suspended material.

Dye-binding (DB) lysine using Acid Orange 12 The dye-binding capacity method as a rapid indicator of lysine of Hurrell and Carpenter (13, see also 5) was used with some modifications. This type of procedure has been widely used for many years.

Plasma lysine The concentration of free lysine in the portal plasma of rats was measured 45, 60, 75, 90, 12o, and 180 min after feeding a test meal of various selected proteins to adult rats which were trained to consume their meals within 15 min. Relative lysine availability was calculated from the difference of concentrations between the fasting level and the levels at different times after the uptake of the test meal. The exact method has been described (14, 15).

Lysine digestibility in rats Certain selected samples were analyzed for their lysine digestibility in rats according to a described method (14, 15, 16).

Results and Discussion

In Figure 1 it is clear that all five procedures showed at least some measure of sensitivity to lysine damage in all heated samples and that generally with increased severity of heat treatment a corresponding decrease in reactive lysine was detected. The SA and DAN methods and to a smaller degree the DB method were more sensitive than the FDNB method, particularly for the most severely damaged samples. The total lysine method was least sensitive.

There was also general agreement between methods in terms of relative values for the different types of heat treatments. The samples heated alone showed least lysine damage followed by those heated with lactose, glucose, and then xylose, particularly in the gently and moderately treated samples. Lewis and Lea (17) found the same order of reactivity of these sugars when heated with casein at 37° C. The present results however indicate that this order alters under more severe heat treatment (110° C, 4 h or 130° C, 2, and 4h) in that lactose is observed to cause damage similar to or even higher than that of xylose. This was probably due to the partial hydrolysis of lactose to glucose and galactose, a step which was reported (18) to considerably increase the susceptibility of milk powder for lactose-intolerant infants, to Maillard reactions.

Although other newer chemical methods are starting to take over as a method of choice for reactive lysine (19), in the present work the classical direct FDNB method was selected as standard reference since many times it has been shown to give results that correspond closely with those from both biological evaluation and in vitro enzymic digestion (19, 20).

Table I shows a series of simple linear regression equations calculated from the data plotted in Figure 1. For the samples heated alone total lysine correlated ($r = 0.55$) badly with FDNB lysine and overestimated it at all levels of heating. One would however expect this overestimation to be more profound since isopeptides are acid-labile and hence all lysine in this form is measured by total amino acid analysis. Also in the Maillard samples total lysine values overestimated FDNB lysine. The overestimation in the case of the less damaged samples can be explained by the fact that e.g. the "early" Maillard product

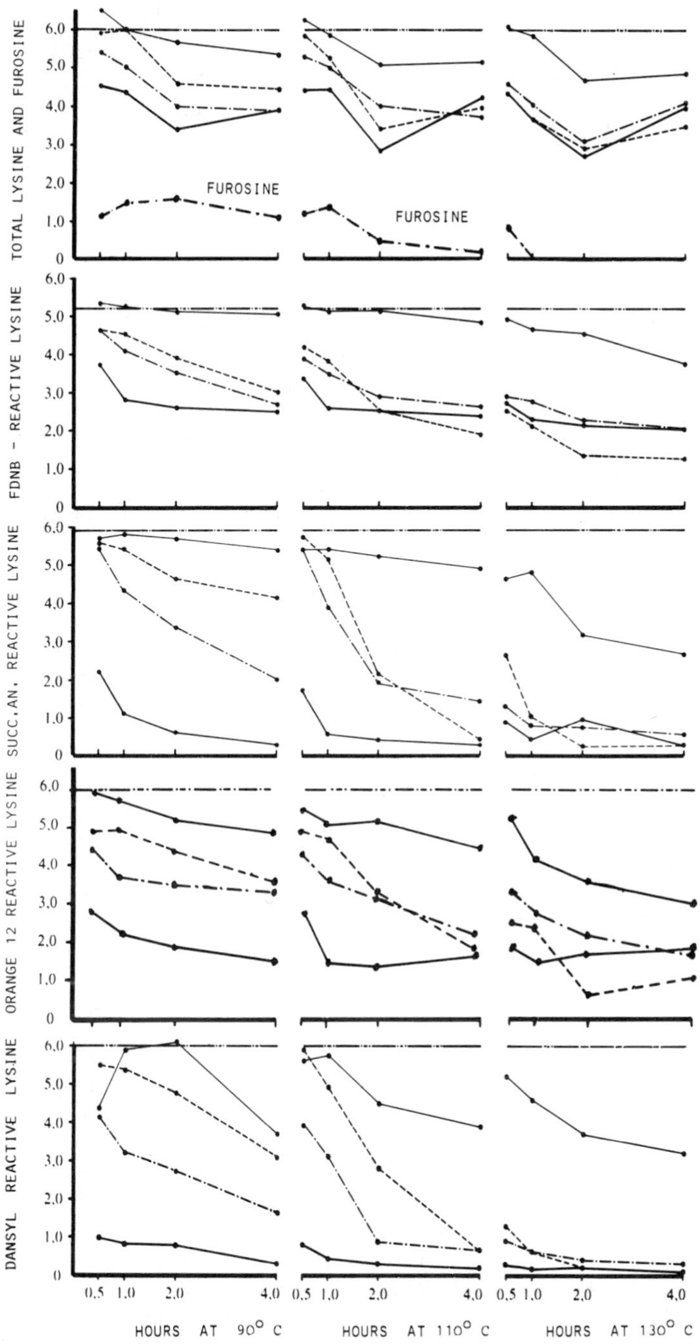

Figure 1. Comparison of the sensitivity of the various chemical procedures to lysine damage in a variety of heat-treated model soya protein samples. (All lysine data are given in g/16 g of N) Key: • — • — , unheated; —, heated alone; — • —, with glucose; —, with xylose; and – – –, with lactose.

fructoselysine, which is both poorly digested and totally unavailable as a source of lysine (20, 21, 22), releases 40-50% of its lysine content during acid hydrolysis (3, 6, 20-22). Although it was found (13) that total lysine values gave a good estimate of lysine availability in more severely damaged "late" Maillard material since under these conditions the majority of lysine is totally destroyed, in the present study FDNB lysine was grossly overestimated in all cases. Fructoselysine, measured as furosine, however was only detectable in the less damaged products (see figure 1,). Nevertheless the results of total lysine determination correlated (r = 0.85) quite well with the FDNB values.

The SA method correlated quite well with the FDNB method, showing a greater sensitivity than total lysine but giving absolute values that were in some cases very different from that of FDNB. In the lower range the SA values were below FDNB values but above about 4 g lysine/16 g N, the SA levels were slightly higher. The difference may arise from the fact that FDNB tends to react with the ε $-NH_2$-linkage of blocked lysine (22). There may be also differences due to the failure of the various reagents to fully penetrate the high crosslinked material. The SA method may therefore better reflect the influence of poor digestibility on lysine availability.

Table I. Simple linear regression equations; correlation between FDNB-reactive lysine and lysine determined by the other methods

Isopeptide samples (n = 15)						
$Lysine_{FDNB}$	=	0.88 $Lysine_{TOTAL}$	-	0.28;		r = 0.55
$Lysine_{FDNB}$	=	0.51 $Lysine_{SA}$	+	2.36;		r = 0.93
$Lysine_{FDNB}$	=	0.08 $Lysine_{DAN}$	+	1.01;		r = 0.85
$Lysine_{FDNB}$	=	0.59 $Lysine_{DB}$	+	2.05;		r = 0.93
Maillard samples (n = 36)						
$Lysine_{FDNB}$	=	0.91 $Lysine_{TOTAL}$	-	0.95;		r = 0.85
$Lysine_{FDNB}$	=	0.41 $Lysine_{SA}$	+	2.02;		r = 0.90
$Lysine_{FDNB}$	=	0.06 $Lysine_{DAN}$	+	1.91;		r = 0.85
$Lysine_{FDNB}$	=	0.65 $Lysine_{DB}$	+	1.10;		r = 0.90

SA = succinic anhydride; DAN = dansyl chloride; DB = dye-binding with Acid Orange 12.

Attempts to determine absolute values for the DAN method were complicated mainly by the problem that fluorescence intensity was found to increase with decreasing particle size (even below 80 µm). For materials of the same particle size, as in the present study, the results correlated (r = 0.85) quite well with the FDNB lysine and showed good reproducibility, attributable to the unique design of the fluorometer used (12).

The dye-binding method with Acid Orange 12 was very similar in results to succinic anhydride, showing regression equations between (near to) SA and DAN. The DB method also underestimated FDNB lysine in the severely heat-damaged samples.

Comparison with in vivo procedures Although the FDNB procedure proved to be a suitable reference method, there is no doubt that all methods should be ultimately compared to in vivo procedures. For this reason selected samples were also analyzed by plasma amino acid and digestibility methods. Preliminary results (Table II) show that plasma lysine results correlated very well with results for lysine digestibility and FDNB lysine (r = 0.95), reasonably well with those for dansyl chloride lysine, succinic anhydride reactive lysine and dye binding lysine, but poorly with total lysine. Although the absolute values were in many cases very different, it is apparent that all methods except total lysine can be used to at least indicate the extent of lysine damage.

Table II. Relative percentage of the lysine availability in selected samples as determined by various methods (unheated total lysine - 6.0 g /16 g N - = 100%).

Sample & Treatment	Plasma Lys	Digest. Lys	Total Lys	FDNB Lys	SA Lys	DAN Lys	DB Lys
90°C, 2 h	85	nd	93	85	95	98	87
110°C, 1 h	76	nd	97	85	90	92	85
95°C, 24 h	113	98	100	85	88	81	81
138°C, 24 h	48	80	95	63	25	48	28
160°C, 24 h	0	13	80	30	8	21	17
90°C, + gluc, 2 h	38	nd	67	58	57	44	58
110°C, + gluc, 1 h	31	nd	83	58	65	49	59
90°C, + lac, 2 h	66	nd	77	65	77	76	73
110°C, + lac, 1 h	39	nd	87	63	87	79	79
Correlations with Plasma lysine (r)	--	0.94	0.61	0.95	0.76	0.86	0.74

FDNB = fluorodinitrobenzene; SA = succinic anhydride; DAN = dansyl chloride; DB = dye-binding with Orange 12; nd = not determined

If the four reactive-lysine methods studied are compared on the basis of simplicity and rapidity, the FDNB method is not preferred since it is fairly complicated, takes ca. 20 h per assay, and reqires special precautions due to the vesicant effects of the

FDNB on the skin. The other methods are on the other hand relatively simple and rapid, their assay procedures taking 1,5 - 2 h to complete. The succinic anhydride method does however have the problem of requiring radioisotope facilities, which could limit its use in the typical quality control laboratory.

In conclusion it can be said that all the tested methods for availability, but not that for total lysine, are most definitely useful as simple and rapid methods for detecting lysine damage in "isopeptide", "early" Maillard and "late" Maillard type material and the results correlate fairly well with FDNB, plasma lysine test, and lysine digestibility. But further work will be needed if we are to become confident about their ability to give absolute measures of nutritionally available lysine. In any case, making the assumption that reactive lysine will always be equivalent to nutritionally available lysine should be done with caution since there are many factors that affect the final utilization of lysine. Under these circumstances the prime aim of such reactive-lysine methods should surely be to furnish a good relative estimate of all categories of lysine damage that can occur in both plant and animal protein and to achieve this by a method that is simple, rapid, and economical.

Acknowledgments

One of us (TRA) is indebted to the Council for Scientific and Industrial Research, Pretoria, and Fedfood, Isando, Republic of South Africa for their generous financial assistance.

Literature cited

1. Masters, P. M.; Friedman, M. In: Whitaker, J. R., Fujimaki, M., Eds.; "Chemical Deterioration of Proteins"; American Chemical Society: Washington, DC, 1980; pp. 165-194.
2. Erbersdobler, H. F.; Alfke, F.; Hänichen, T.; Schlecht, C.; v. Wangenheim, B. Ernähr. Umschau 1977, 24, 375.
3. Erbersdobler, H. F.; Brandt, A.; Scharrer, E.; v. Wangenheim, B. In: Eriksson, C., Ed. "Progress in Food and Nutrition Science (Maillard Reactions in Food)"; Pergamon Press: Oxford, 1981, 5, pp. 257-263.
4. Anderson, T. R.; Quicke, G. V. South Africa Food Rev. 1980, 7, 113-123.
5. Hurrell, R. F.; Carpenter, K. J. In: Eriksson, C., Ed. "Progress in Food and Nutrition Science (Maillard Reactions in Food)"; Pergamon Press: Oxford, 1981, 5, pp. 159-176.
6. Brüggemann, J.; Erbersdobler, H. F. Z. Lebensm. Untersuch. Forsch. 1968, 137, 137-143.
7. Erbersdobler, H. F.; Holstein, A.-B.; Lainer, E. Z. Lebensm. Untersuch. Forsch. 1979, 168, 6-8.
8. Carpenter, K. J. Biochem. J. 1960, 77, 6o4-61o.

9. Erbersdobler, H. F.; Zucker, H. Z. Tierphys. Tierern. Futtermittelk. 1964, 19, 224-255.
1o. Booth, V. H. J. Sci. Food Agric. 1971, 22, 658-664.
11. Anderson, T. R. "Determination of available lysine in foodstuffs"; M. Sc. Thesis, Biochemistry, Univ. of Natal, Pietermaritzburg, 1980.
12. Christoffers, D. "Fluorometrische Lysinbestimmung in Getreidemehlen"; Doctoral Thesis, Technical University of Hannover, 1976.
13. Hurrell, R. F.; Carpenter, K. J. Proc. Nutr. Soc. 1976, 35, 23A.
14. Erbersdobler, H. F.; Dümmer, H.; Zucker, H. Z. Tierphys. Tierern. Futtermittelk. 1968, 24, 136-152.
15. Grimm, F. "Untersuchungen zur Aminosäuren-Verfügbarkeit nach Erhitzung von Proteinen mit verschiedenen Kohlenhydraten"; Doctoral Thesis, Vet. Med. Faculty, University of Munich, 1973.
16. Brüggemann, J.; Erbersdobler, H. F. Z. Tierphys. Tierern. Futtermittelk. 1968, 24, 55-67.
17. Lewis, V. M.; Lea, C. H. Biochim. Biophys. Acta 1959, 4, 532-534.
18. Burval, A.; Asp, N.; Bosson, A.; San Jose, C.; Dahlqvist, A. J. Dairy Res. 1978, 45, 381-389.
19. Mauron, J. In: Birch, G. G.; Parker, K. J., Ed. "Food and Health"; Applied Science Publishers Ltd.: London, 1980, pp. 389-413.
20. Erbersdobler, H. F. In: Cole, D. J. A.; Boorman, K. N.; Buttery, P. J.; Lewis, D.; Neale, R. J.; Swan, H. Eds. "Protein Metabolism and Nutrition"; Butterworths: London-Boston, 1976, pp. 139-157.
21. Erbersdobler, H. F. In: Friedman, M., Ed. "Protein Crosslinking (Nutritional and Medical Consequences)"; Plenum Publishing Corp.: New York, 1977, pp. 367-378.
22. Finot, P. A.; Mauron, J. Helv. Chim. Acta 1972, 55, 1153-1164.

RECEIVED October 5, 1982

MAILLARD REACTIONS IN VIVO

23

Nonenzymatic Glycosylation and Browning of Proteins In Vivo

V. M. MONNIER[1] and A. CERAMI
The Rockefeller University, New York, NY 10021

The discovery, in the past decade, that Amadori products form readily in vivo and that the levels of glycosylated hemoglobins reflect the glycemia over several weeks has greatly stimulated the interest in the Maillard reaction in vivo. The fact that long-lived proteins such as lens crystallins, collagen and elastin become pigmented and highly crosslinked suggests that the Maillard reaction could be involved in the general aging process. In this chapter, we review recent studies on the Maillard reaction with proteins in vivo.

In 1912, L. C. Maillard (1) observed that solutions of amino acids heated in the presence of reducing sugars developed yellow-brown color, and he hypothesized that this reaction could occur in vivo and be of importance in diabetes. The biological importance of this observation, however, was not recognized for a long time by medical scientists. Instead, it was found initially to be of practical relevance to the storage and processing of foods, where the reaction of sugars with amino groups led to a series of unwanted changes in food quality and appearance. Therefore, most of today's knowledge concerning the chemistry of the Maillard reaction has been provided by food and agricultural chemists.

In 1962, Rhabar discovered that red blood cells in diabetics had elevated amounts of a minor component of hemoglobin (2). This hemoglobin, called hemoglobin A_{1c} (HbA_{1c}) (3), was found to contain a 1-amino-2-deoxyfructose molecule attached to the N-terminus of the β chain (4,5). Since various studies indicated that the glycosylation of hemoglobin occurred in the absence of an enzyme (6,7,8), the term "nonenzymatic glycosylation" has been generally accepted to designate the formation of Amadori products in vivo (Figure 1, top). The observation that, in the diabetic, the levels of HbA_{1c} reflect the integrated glycemia over the pre-

[1] Current address: Case Western Reserve University, Institute of Pathology, Cleveland, OH 44106

0097-6156/83/0215-0431$06.00/0

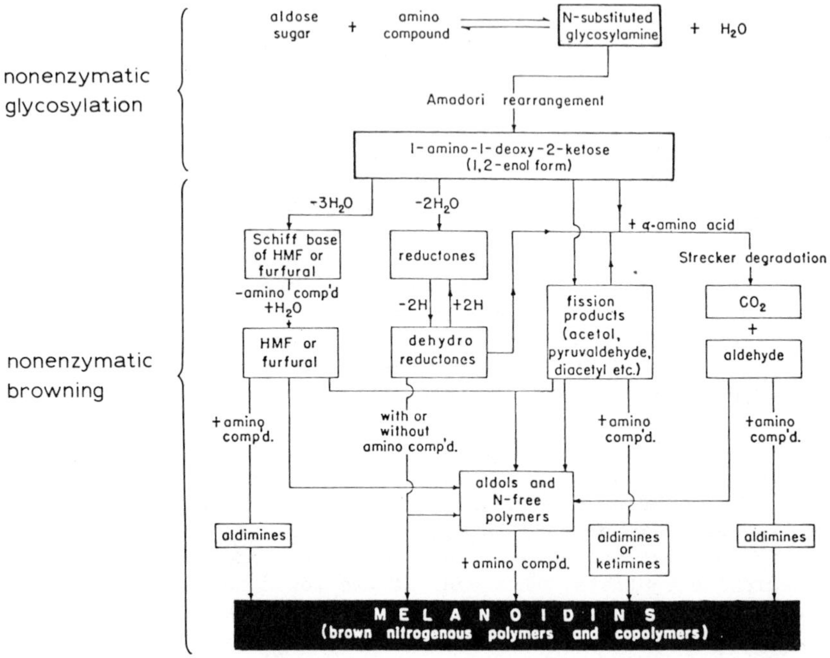

Figure 1. Amadori rearrangement of glycosylated proteins. Integration of known reactions leading to browning in sugar–amine systems. (Reproduced from Ref. 101. Copyright 1953, American Chemical Society.)

ceding 3 to 4 weeks suggested that it could be used to assess overall carbohydrate control (9). Furthermore, since various complications of diabetes occur in tissues not dependent on insulin for the uptake of glucose, it was hypothesized that increased nonenzymatic glycosylation could be involved in some of the long-term sequelae of diabetes (10). The fact that proteins with little or no turnover, such as lens crystallins and collagen, become pigmented and insoluble with age, suggests to us that the later stages of the Maillard reaction could be responsible for these changes. Nonenzymatic glycosylation can affect biological functions of proteins in a very different way from intermediate and late products of the Maillard reaction. We therefore use the term "nonenzymatic browning" to designate specifically protein changes characterized by the presence of fluorescent, pigmented and crosslinked adducts generated during the later stages of the Maillard reaction (11).

In this article, we will review studies on both nonenzymatic glycosylation and nonenzymatic browning *in vivo*. Structural, functional, analytical, and kinetic aspects will be discussed. Clinical aspects have been reviewed recently elsewhere (12,13,14) and will therefore not be considered in this chapter.

Nonenzymatic Glycosylation

The observation that many complications of diabetes are reversible upon tight control of glycemia spurred an interest in the potential role of increased protein glycosylation in the long-term sequelae. Particular attention has been devoted to proteins exposed to high concentrations of glucose such as those in the blood, lens, nerves, vascular and basement membranes, and skin. These are also the tissues that are most severely affected by the disease. The proteins that are susceptible to nonenzymatic glycosylation include cellular, extracellular, and membrane proteins (Table I). Whereas glucose is the main sugar involved in glycosylation of secretory, structural, and membrane proteins, other sugars, in particular the phosphorylated intermediates of glycolysis, can form adducts with intracellular proteins (Figure 2).

Detection of Amadori Products Various methods are now available for the detection and quantification of Amadori products in body proteins (Figure 3). The furosine method, developed by Finot et al. (15), allows the quantification of glycosylated lysine residues. It has been used recently by Dolhover, et al. (16) in studies on lipoproteins. Alternatively, the Amadori products can be reduced with tritiated sodium borohydride to both label the sugar and prevent its partial destruction upon acid hydrolysis. The stable adduct, e.g., ε-*N*-(1-deoxyglucitolyl)-lysine can then be separated by cation exchange chromatography (17,18) or by affinity chromatography using *m*-aminophenylboronic acid coupled to a solid support (19,20). This method has been

Table I: Nonenzymatically Glycosylated Proteins

Localization	Protein	Reference
	Hemoglobin	72
	A_{Ia1}	35
	A_{Ia2}	7,29,35,70
Intracellular	A_{Ib}	36
	A_{Ic}	4,5,6,8,70
	A_o	69
	S	38,73
	Crystallin α,β,γ	18,40,41,42,43,44,55
	Cathepsin B	47
	Plasma Proteins	48,56,57
	Albumin	39,49,50,67
	Ferritin	51
	Insulin	52,53,54
	Lipoproteins	75,76,77,78
	Collagen	
Extracellular	Skin	58,59,71
	Tendon	60
	Aorta	61
	Glomerular basement membrane	62,64,74
	Lens capsule	63
	Urine	
	Peptides and amino acids	19
	Erythrocytes	65,68
Membrane	Endothelium	66
	Myelin	21,46

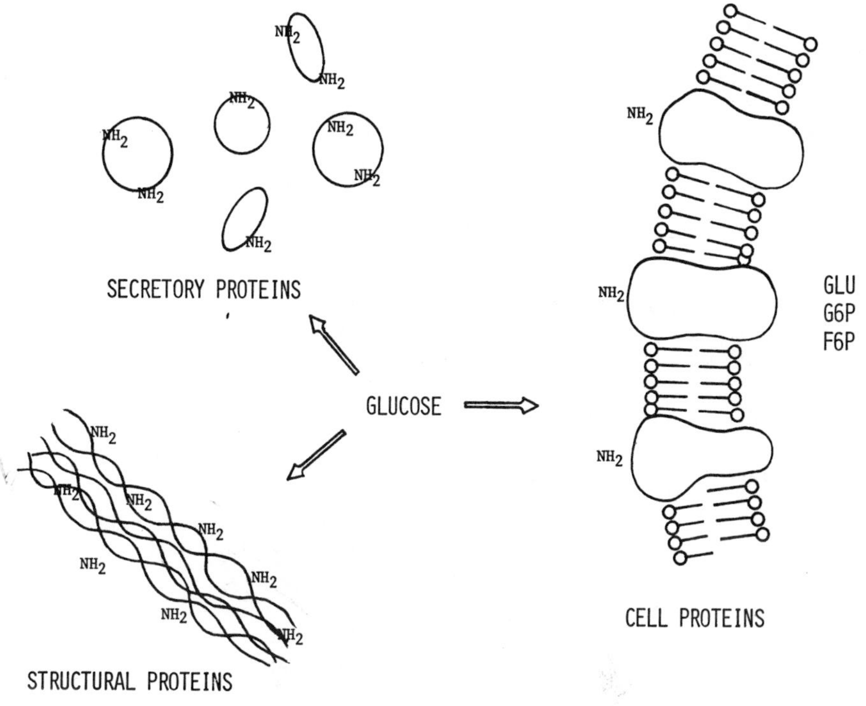

Figure 2. Possible reaction sites of glucose and other reducing sugars with extracellular, membrane, and intracellular proteins.

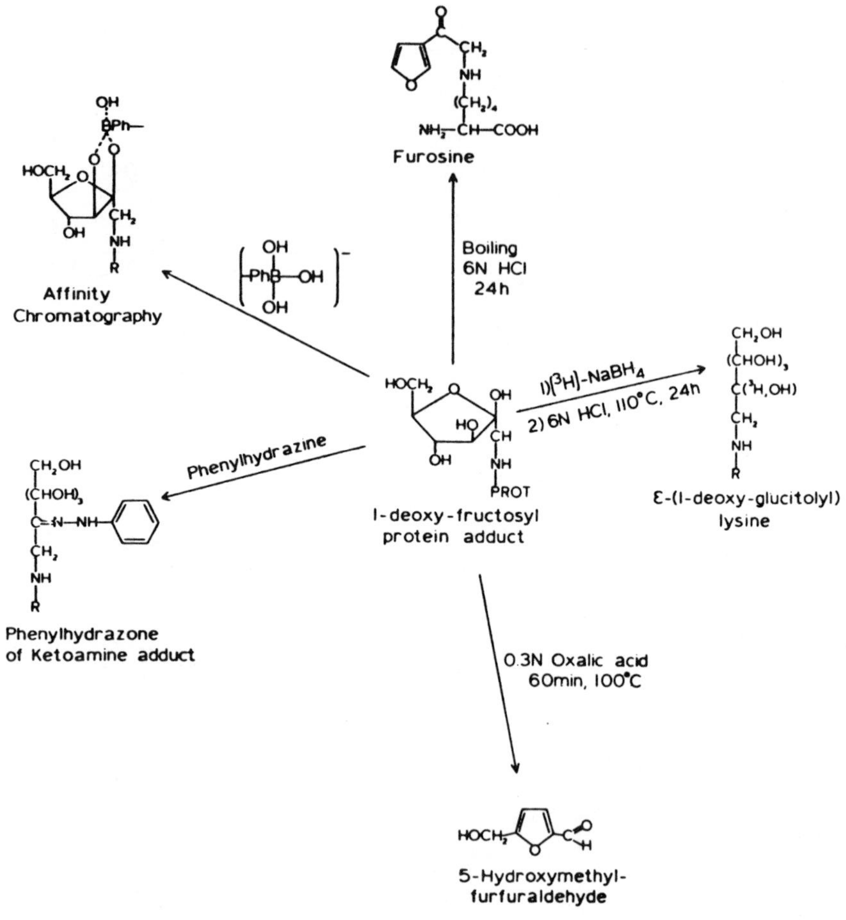

Figure 3. Various methods for the detection and quantification of Amadori products.

successfully used to detect and quantify glycosylated peptides and amino acids in the urine of diabetic patients (19). It also can be used to isolate nonenzymatically glycosylated proteins such as glycohemoglobin (20). Flückiger and Winterhalter (6) have used mildly acidic conditions to release 5-hydroxymethyl-furfuraldehyde from hemoglobin. This is then treated with thiobarbituric acid (TBA) to form a chromophore which can be quantified spectrophotometrically. This method is widely used for routine purposes. However, it requires careful standardization and freedom from contaminants that can react with TBA, such as lipid peroxidation products. Amadori products can also be derivatized with hydrazine. The hydrazone, e.g., phenylhydrazone, is stabilized by reduction with borohydride and quantified spectrophotometrically. This method has been used to detect hemoglobin adducts with glyceraldehyde (26). Finally, some glycoproteins can be separated by cation exchange chromatography (3,22,35), electrophoresis (23,24), or immunoprecipitation (25). Some of these methods have found clinical applications for the estimation of glycohemoglobins in diabetic patients. Hurrel, et al. (27) found that ferricyanide reacts with Amadori products in milk to form a chromophore which can be measured quantitatively.

Kinetics of the Glycosylation Reaction Several factors influence the kinetics of the glycosylation reaction. The rate of condensation of the carbonyl group with the free amine depends on the pK of the protonated amino group and the type and concentration of the sugar. The formation of Amadori products *in vivo* most probably occurs nonenzymatically. Studies on the biosynthesis of hemoglobin A_{1c} in mice have indicated that it was synthesized throughout the life of the red cell at a constant rate of 0.1% per day (28). In diabetic mice, the reaction rate was increased by a factor of 2.7. Haney and Bunn (29) studied the reaction kinetics of glucose and glucose-6-phosphate (G6P) with deoxy- and carboxy-hemoglobin. The adduct formation of deoxy-Hb was twenty times as fast with deoxy-Hb as with carboxy-Hb. The conformation of carboxy-Hb and deoxy-Hb are different, which suggests that conformational changes can strongly influence the glycosylation rate. Similar observations have been made by Stevens, *et al.* (30), who showed that the rate of adduct formation was dependent upon the concentrations of both hemoglobin and G6P.

The kinetics of the nonenzymatic glycosylation of hemoglobin to form hemoglobin A_{1c} have been analyzed in detail by Higgins and Bunn (31). Assuming a bimolecular reaction for the condensation reaction and first-order kinetics for the Amadori rearrangement, they determined the rate constants for the glycosylation of the N-terminal valine of the β-chain:

$$
\begin{array}{ccccc}
\beta\text{-NH}_2 & + &
\begin{array}{c}
\text{HC=O} \\
| \\
\text{HCOH} \\
| \\
\text{HOCH} \\
| \\
\text{HCOH} \\
| \\
\text{HCOH} \\
| \\
\text{CH}_2\text{OH} \\
\\
\text{glucose}
\end{array}
&
\underset{\underset{k_{-1}}{\longleftarrow}}{\overset{k_1}{\longrightarrow}}
&
\begin{array}{c}
\text{HC=N-}\beta\text{A} \\
| \\
\text{HCOH} \\
| \\
\text{HOCH} \\
| \\
\text{HCOH} \\
| \\
\text{HCOH} \\
| \\
\text{CH}_2\text{OH} \\
\\
\text{aldimine} \\
\text{(Schiff base)}
\end{array}
&
\overset{\text{Amadori}}{\underset{k_2}{\longrightarrow}}
&
\begin{array}{c}
\text{CH}_2\text{-NH-}\beta\text{A} \\
| \\
\text{C=O} \\
| \\
\text{HOCH} \\
| \\
\text{HCOH} \\
| \\
\text{HCOH} \\
| \\
\text{CH}_2\text{OH} \\
\\
\text{keto amine}
\end{array}
\end{array}
$$

$$
\text{HbA}_{1c} \underset{\text{rapid}}{\overset{}{\rightleftharpoons}} \text{pre-A}_{1c} \xrightarrow{\text{slow}} \text{HbA}_{1c}
$$

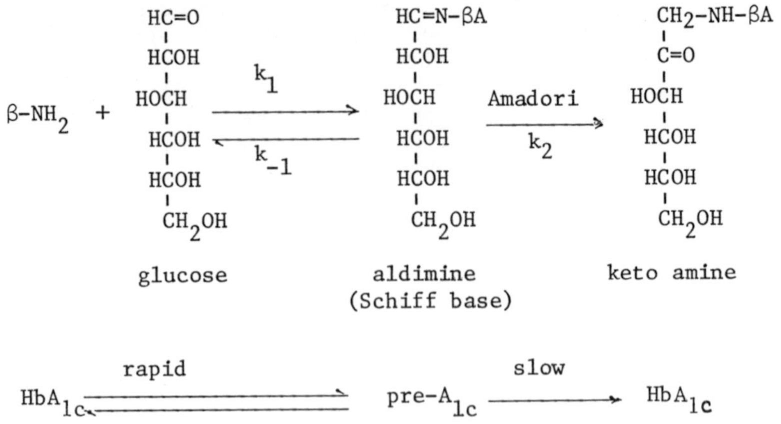

Glucose concentrations were estimated with purified, ^{14}C-labelled glucose. To determine k_1, the overall rate of Schiff base formation at several sites of hemoglobin was first measured by incubating hemoglobin A_O with $NaCNBH_3$. At neutral pH, this reagent selectively reduces Schiff base linkages without affecting carbonyl groups ($\underline{34}$). The overall condensation rate, k_1', was found to be $0.9 \pm 0.2 \times 10^{-3}$ $mM^{-1}h^{-1}$. Since the rate of condensation of D-$[^{14}$C]-glucose was 2/3 that of HbA_O, the rate of condensation of the β-NH$_2$ terminus, k_1, was estimated at 0.3×10^{-3} $mM^{-1}h^{-1}$. As anticipated, k_1' rose with increasing pH. However, it was not significantly affected by increasing the concentration of hemoglobin to about two-thirds the level found in normal red blood cells. The overall first-order rate constant (k_{-1}) for the dissociation of glucose from the aldimine complex was calculated from incubations of $[^{14}$C]-glucose with hemoglobin in the absence of $NaCNBH_3$, assuming $k_{-1} \gg k_2$. The value found for k_{-1} was 0.35 h^{-1}. The rate of the Amadori rearrangement at the β-NH$_2$ terminus was estimated as for k_{-1} and with chromatographic quantification of the keto amine HbA_{1c}. The mean value for k_2 was 0.0055 h^{-1}, which is very close to the one obtained from kinetic data in humans ($k_2 = 0.006$ h^{-1}) ($\underline{37}$). These data indicate that the rate of the Amadori rearrangement is 1/60 that of the dissociation of the Schiff base to hemoglobin and glucose. For the normal red cell with a 5 mM glucose concentration, calculated and estimated data on nonenzymatic glycosylation can be summarized as follows:

Type of Hemoglobin	Glycosylation Site	Sugar	% of Total Hb	Ref.
A_o			90%	
	ε-NH_2	glucose	8% of HbA_o	69,72
	β-NH_2	glucose	4% of HbA_o	69,72
A_{1a_1}	β-NH_2	fructose-1,6-P	0.5%	35
A_{1a_2}	β-NH_2	glucose-6-P	0.5%	7,35
A_{1b}		not glycosylated		36
A_{1c}	β-NH_2	glucose	4. %	4,5,6,8,70
Pre-A_{1c} (Schiff base)	β-NH_2	glucose	0.5%	31

The rate of incorporation of sugars into proteins depends strongly on the type of sugar. The reactivity of reducing sugars generally increases with the rate of mutarotation. With lens crystallins and radiolabelled sugars, we found the order to be fructose < glucose < glucose-6-P ≃ galactose < xylose (11). This observation is of special interest in the formation of experimental sugar cataracts, since lens opacities appear faster in rats fed with xylose than in those fed galactose, and faster in galactosemic than in diabetic rats. Similar observations were made by Bunn and Higgins (33). These authors measured the reactivity of various monosaccharides with hemoglobin and found that glucose was the least reactive of the aldohexoses. Therefore, they propose that glucose probably emerged as the primary metabolic fuel because the high stability of its ring structure limits the potentially deleterious nonenzymatic glycosylation of proteins.

Finally, caution should be exerted when determining kinetic data with radiolabelled sugars. Trueb, et al. (32) found that labelled glucose from several manufacturers contained radioactive impurities of unknown structure which often reacted faster than glucose itself. Purification methods can be found in reference 31.

Structure and Function of Glycosylated Proteins. A wide variety of proteins has been studied for nonenzymatic glycosylation both in vitro and in vivo (Table I). Using NMR spectroscopy, Koenig and Cerami (28) obtained conclusive evidence for the presence of 1-deoxyfructosylvaline in hemoglobin A_{1c}. They showed that fructosylvalylhistidine from HbA_{1c} was identical with the synthetic glycopeptide. Various studies on hemoglobin glycosylation have also come from the laboratories of Bunn, Gabbay, Gallop and Winterhalter, and have already been discussed above. Acharya and Manning (38) studied the reactive sites of hemoglobin S (HbS) with glyceraldehyde, which readily forms Amadori products with hemoglobin (26). About 23% of the glyceraldehyde incorporated into hemoglobin was present at Val-1 (β) and less than 5% at Val-1 (α). The reactive lysine residues of the β-chain were Lys-82, Lys-59 and Lys-120. They incorporated 27%, 12% and 9.6% respectively of the total glyceraldehyde. The most reactive lysine residue of the α-chain was Lys-16, which incorporated 30% of total glyceraldehyde.

Apart from hemoglobin, various other proteins have been shown to be glycosylated by one of the methods described above. With the exception of insulin and albumin, however, the specific glycosylation sites remain to be elucidated.

Nonenzymatic glycosylation, theoretically, can alter protein function in several ways. The isoelectric point (pI) drops, particularly if the sugar is phosphorylated. The presence of an additional sugar on a free amino group may lead to decreased enzymatic activity, decreased ligand binding affinity, changes in conformation, modified protein half-life, blocking of proteolytic sites, or suppression of crosslinking. In addition, it could lead to a selective receptor-mediated uptake of the protein or the formation of anti-hapten antibodies. Several observations support these theoretical considerations which could be of importance in the long-term sequelae of diabetes.

Inhibition of ligand binding has been observed in glycohemoglobins, since 2,3-diphosphoglycerate (2,3-DPG), a regulator of hemoglobin function, cannot bind to the N-terminal valine of the β-chain (22). The oxygen affinity of the two acidic minor hemoglobins, A_{1a} and A_{1a_2}, is lower than that of HbA_o at neutral pH, whereas the affinity of HbA_{1c} is slightly increased. Although diabetic blood has an elevated affinity (about 2 mm Hg), this change is not important enough to account for the clinical symptoms. Decreased binding of, e.g., drugs, may also be found in albumin, since aspirin competes with the glycosylation of lysine residue 199 (48).

Changes in conformation have been postulated in crystallins incubated with hexoses (18). As a result of the incubation, light-scattering disulfide-linked high-molecular-weight protein aggregates were formed which were similar to those of experimental galactose or diabetic cataract. The phenomenon was investigated by Liang and Chakrabarti (55), who found changes in the tertiary structure of α-crystallins that had been incubated with glucose-6-phosphate.

No alteration of protein half-life has been detected with glycosylated albumin (39) and ferritin (51). However, Jones and Peterson (45) found that fibrinogen survival, which was reduced in diabetic patients with elevated glycemia, could be increased to normal value shortly after insulin therapy. This suggests that a labile Schiff base could play a role in the regulation of fibrinogen turnover.

A decrease in enzymatic activity upon glycosylation has been reported for cathepsin B (47). The conversion of proinsulin (PI) to insulin (I) was impaired. The PI/I ratio increased by a factor of 3 when the glycosylated enzyme was tested.

Decreased digestibility of nonenzymatically glycosylated glomerular basement membrane has been demonstrated by Lubec and Pollak (62). They hypothesized that thickening of diabetic glomerular basement membranes could be due to reduced proteolytic degradation. Decreased digestibility has also been shown in skin collagen from diabetic patients (59). It correlated well with the increased amounts of keto amine linkages in the same samples. Overall, the results indicated that collagen of juvenile-onset diabetics had undergone accelerated aging.

Amadori products may act as recognition signals for the selective uptake of nonenzymatically glycosylated proteins by cells. McVerry (67) observed that mice injected repetitively with glycosylated serum proteins developed basement membrane thickening as in diabetes. A possible explanation for these observations may be found in the observation of an increased uptake of glycosylated albumin by endothelial cells *in vitro* (80). Studies on the uptake of glycosylated low-density lipoproteins (LDL) are somewhat controversial. Gonen, *et al.* (76) found that glycosylated LDLs were degraded faster by rat peritoneal macrophages than were nonglycosylated ones. Kim and Kurup (77) observed that extensively glycosylated LDLs had a decreased catabolism in rats. However, Kraemer and Reaven (78) found no difference between LDL binding or degradation by fibroblasts of diabetic patients and fibroblasts of normal individuals.

Finally, membrane proteins are also increasingly glycosylated in diabetes (65). Reid, *et al.* (68) recently identified an anti-M alloagglutinin in juvenile-onset diabetes that would agglutinate M-positive cells that had been pre-incubated in glucose. This suggests that Amadori products could act as haptens and elicit an immune reaction towards glycosylated tissues.

In summary, many, if not all, body proteins can be glycosylated to different degrees, both *in vitro* and *in vivo*. The degree of impaired function, however, may depend upon the total amount of glycosylation.

Nonenzymatic Browning

Although nonenzymatic glycosylation may affect practically every protein *in vivo*, it is likely that nonenzymatic browning will occur only in proteins that have a slow turnover or none at all, such as lens crystallins, collagen, elastin and proteoglycans. In some tissues, these proteins are, in effect, "stored" for a lifetime and undergo some characteristic changes, many of which have been observed in stored and processed foodstuffs (Table II).

In the aging human lens, and particularly during cataract formation, protein aggregation (81), decreased protein solubility

Table II: Some Characteristics of Aging Proteins

Physical Changes	Chemical Changes
Denaturation	Oxidation (Trp,Tyr,His, Cys,Met)
Aggregation	Deamidation
Insolubility	Decreased digestibility
Pigmentation	Racemization
Fluorescence	Cross-linking
	Adduct formation

(82), and accumulation of yellow (82) and fluorescent (83) pigments has been demonstrated. Several lines of evidence support the possibility that nonenzymatic browning may be in part responsible for these changes. 1) Amadori products of lysine residues are present in human lens crystallins (41) and have been found to accumulate in the lens with age (44). 2) A loss of lysine residues has been observed, particularly in advanced cataracts (82). 3) Browning of proteins with glucose can occur not only upon heating but also at physiological pH and temperature (84). 4) Increased amounts of glucose and glucose-6-phosphate have been reported in human lenses from older individuals (85).

To investigate the presence of nonenzymatic browning products in the lens, we have incubated bovine lens crystallins with reducing sugars for 10 months under physiological conditions (30). The spectroscopic properties of these protein solutions, which had developed yellow color, were compared with those of human normal and cataractous lens proteins. Similar fluorescence-excitation spectra with maxima at 360 nm and 470 nm and a shoulder at 400 nm were obtained from bovine lens proteins incubated with glucose-6-phosphate and from proteins isolated from cataracts. Non-disulfide covalent crosslinks were also detected in protein solutions incubated with hexoses, particularly upon incubation with glucose-6-phosphate. Human lenses were then analyzed for the presence of yellow, fluorescent and borohydride-reducible protein adducts, possibly generated during nonenzymatic browning. Such adducts were detected in acid hydrolysates of heavily pigmented proteins from cataractous lenses (86). Typically, a 20% to 40% decrease in lysine and histidine residues was noted in these lenses, suggesting that these amino acids had been modified by the cataractous process. These adducts were purified by gel filtration, cation exchange and reverse phase chromatography. Two borohydride-reducible amino acid derivatives were found to co-chromatograph with browning products synthesized from glucose and α-tert.-butyloxycarbonyllysine prepared at 37°C and pH 7.4. The structure of these compounds has not been elucidated yet. However, Olsson and colleagues (87) have isolated a series of bicyclic products from the reaction of D-glucose and methylamine which, hopefully, will serve as models for the structures of browning products.

Preliminary indications for the occurrence of nonenzymatic browning in connective tissue can be found in the literature. For example, LaBella and associates (88,89) have shown that aortic elastin is associated with a covalently bound fluorescent material that accumulates with age, in parallel to the resistance of the native protein to solubilization and degradation by elastase. The age-related accumulation of yellow pigments and fluorogens with excitation-emission maxima at 350/440 and 370/460 nm similar to those reported for synthetic browning products involving glucose and lysine (30) has also been found in human tendon collagen (90).

The potential for the occurrence of nonenzymatic browning in collagen is supported by the recent demonstration by Schnider and

Kohn (59) that nonenzymatic glycosylation and resistance to pepsin digestion were greatly increased in tendon and skin collagen of diabetic, and particularly juvenile-onset diabetic, patients (Figure 4). Increased glycosylation has also been observed in collagen isolated from the aorta of diabetic rats (61). These observations suggest that accelerated formation of cross-links may occur in diabetes as a result of excessive nonenzymatic browning and glycosylation of lysine or hydroxylysine residues. These studies point to the possible acceleration of the aging process in diabetes. In effect, several complications of diabetes mellitus resemble the general characteristics of aging which occur in collagen-rich tissues. These include arteriosclerosis (91), stiffening of arteries (92), stiffening of lungs (93), peri-articular rigidity (94), and osteoarthritis (95). The thickening of capillary and glomerular basement membrane which is observed in diabetes and aging could be due to nonenzymatic browning of proteolytic digestion sites. In this respect, it is of interest to note that DeBats and Rhodes (96) observed an increase in blue autofluorescence in diabetic kidneys when compared with age-matched controls.

Conclusions

Nonenzymatic glycosylation has been shown to occur *in vivo* with various proteins and to affect protein function in several ways. The amounts of Amadori products are increased, on an average, by a factor of two in diabetes, particularly in proteins from tissues that are not dependent on insulin for the uptake of glucose. A linear accumulation of glyco-adducts was demonstrated in aging collagen. Several reports now indicate that nonenzymatic glycosylation could explain some of the acute or chronic molecular changes observed in diabetes.

Studies on nonenzymatic browning are not as far advanced as those on glycosylation. However, the similarities between changes found in tissues with long-lived proteins and foodstuffs stored in the presence of reducing sugars, as well as preliminary analytical data, indicate that the browning process occurs *in vivo*. It may explain, in part, changes in solubility of aging and diabetic proteins.

The Maillard reaction with reducing sugars which we have described above is one of the classes of carbonyl-amine reactions which occurs *in vivo*. There is now an increasing interest among medical researchers in the role of carbonyl-amine reactions. Cyanate, an antisickling agent, produces neuropathy and cataracts (97). The cyanate adduct of lysine, homocitrulline, has been found in the human lens (98). Acetaldehyde, the first metabolite of ethanol, forms adducts with hemoglobin (99) and could be involved in some of the chronic sequelae of alcoholism. Formaldehyde has long been known to be a health risk factor in workers chronically exposed to it. Finally, steroids, such as

16-α-hydroxyestrone and cortisol, have recently been found readily to form adducts with proteins (100). Such steroid-protein adducts could play a role in diseases such as systemic lupus erythematosus and complications due to chronic therapy with antiinflammatory steroids.

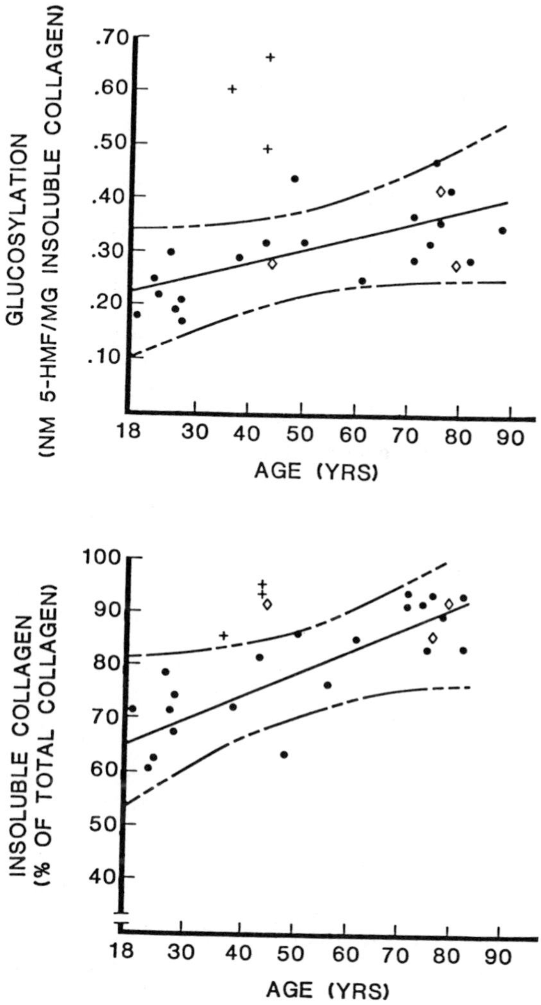

Figure 4. TOP: The amount of keto amine-linked glycosylation of insoluble collagen plotted as a function of subject's age. BOTTOM: The amount of insoluble collagen plotted as a function of subject's age. Key: ○, normal; +, juvenile-onset diabetic; and □, maturity-onset diabetic.

The solid line represents the regression equation. The 99.9% confidence bands are represented by the dashed lines. The regression lines and confidence bands are derived from the data on nondiabetics.

Literature Cited

1. Maillard, L. -C. C. R. Hebd. Séances Acad. Sci. 1912, 72, 599-601.
2. Rahbar, S. Clin. Chem. Acta 1968, 22, 296-98.
3. Allen, D. W.; Schroeder, W. A.; Balog, J. J. Am. Chem. Soc. 1958, 80, 1628-34.
4. Koenig, R. J.; Blobstein, S.; Cerami, A. J. Biol. Chem. 1977, 252, 2992-97.
5. Bookchin, R.M.; Gallop, P. M. Biochem. Biophys. Res. Commun. 1968, 32, 86-93.
6. Flückiger, R.; Winterhalter, K. H. FEBS Lett. 1976, 71, 356-60.
7. Stevens, V.J.; Vlassara, H.; Abati, A.; Cerami, A. J. Biol. Chem. 1977, 252, 2998-3002.
8. Bunn, H. F.; Haney, D. N.; Gabbay, K. H.; Gallop, P. M. Biochem. Biophys. Res. Commun. 1975, 67, 103-09.
9. Koenig, R. J.; Peterson, C. M.; Jones, R. L.; Saudek, C.; Lehrman, M.; Cerami, A. N. Eng. J. Med. 1975, 295, 417-20.
10. Cerami, A.; Stevens, V. J.; Monnier, V. M. Metabolism 1979, 28, 431-37.
11. Hodge, J. E. Adv. Carbohyd. Chem. 1955, 10, 162-205.
12. Bunn, H. F. Am. J. Med. 1981, 70, 325-30.
13. Jovanovic, L.; Peterson, C. M. Am. J. Med. 1981, 70, 331-38.
14. Monnier, V. M.; Cerami, A. Clin. Endocrinol. 1982, in press.
15. Finot, P. A.; Bricout, J.; Viani, R.; Mauron, J. Experientia 1968, 24, 1097-99.
16. Schleicher, E.; Wieland, O. H. J. Clin. Chem. Biochem. 1981, 19, 81-7.
17. Schwartz, B. A; Gray, G. R. Arch. Biochem. Biophys. 1977, 181, 542-49.
18. Stevens, V. J.; Rouzer, C. A.; Monnier, V. M.; Cerami, A. Proc. Natl. Acad. Sci. U.S.A. 1978, 75, 2918-22.
19. Brownlee, M.; Vlassara, H.; Cerami, A. Diabetes 1980, 29, 1044-47.
20. Mallia, A. K.; Hermanson, G. T.; Krohn, R. I.; Fujimoto, E. K.; Smith, P. K. Anal. Lett. 1981, 14, 649.
21. Flückiger, R.; Winterhalter, K. H. "Biochemical and Clinical Aspects of Hemoglobin Abnormalities"; W. S. Coughey, Ed.; Academic Press, Inc., New York, 1978; p 205-14.
22. McDonald, M. J.; Bleichman, M.; Bunn, H. F.; Noble, R. W. J. Biol. Chem. 1979, 254, 702-07.
23. Caspers, M.; Posey, Y.; Brown, R. K. Anal. Biochem. 1977, 21, 166-80.
24. Spicer, K. M.; Allen, R. C.; Buse, M. G. Diabetes 1978, 27, 384-88.
25. Javid, J.; Pettis, P. K.; Koenig, R. J.; Cerami, A. Br. J. Haematol. 1978, 38, 329-37.
26. Acharya, A. S.; Manning, J. M. J. Biol. Chem. 1980, 255, 7218-24.

27. Hurrel, R. F.; Guignard, G.; Finot, P. A. Abstract, 13th Annual Meeting of the Swiss Societies for Experimental Biology. Experientia Suppl. 1981.
28. Koenig, R. J.; Cerami, A. Proc. Natl. Acad. Sci. U.S.A. 1975, 72, 3687-91.
29. Haney, D. N.; Bunn, H. F. Proc. Natl. Acad. Sci. U.S.A. 1976, 73, 3534-38.
30. Monnier, V. M.; Cerami, A. Science 1981, 211, 491-93.
31. Higgins, P. J.; Bunn, H. F. J. Biol. Chem. 1981, 256, 5204-08.
32. Trüeb, B.; Holenstein, C. G.; Fischer, R. W.; Winterhalter, K. H. J. Biol. Chem. 1980, 255, 6717-20.
33. Bunn, H. F.; Higgins, P. J. Science 1981, 213, 224.
34. Borch, R. F.; Bernstein, M. D.; Durst, H. D. J. Am. Chem. Soc. 1971, 93, 2897-904.
35. McDonald, M. J.; Shapiro, R.; Bleichman, M.; Solway, J.; Bunn, H. F. J. Biol. Chem. 1978, 253, 2327-32.
36. Krishnamoorthy, R.; Gacon, G.; Labie, D. FEBS Lett. 1977, 77, 99.
37. Bunn, H. F.; Itaney, D. N.; Kamin, S.; Gabbay, K. H.; Gallop, P. M. J. Clin. Invest. 1976, 57, 1652-59.
38. Acharya, A. S.; Manning, J. M. J. Biol. Chem.1980, 255, 1406-12.
39. Pfoffenbarger, P. L.; Megna, A. T. Biochemistry 1980, 9, 485-88.
40. Monnier, V. M.; Stevens, V. J.; Cerami, A. J. Exp. Med. 1979, 50, 1098-1107.
41. Pande, A.; Garner, W. H.; Spector, A. Biochem. Biophys. Res. Commun. 1979, 89, 1260-66.
42. Ansari, N. H. M.; Awasthi, Y. L.; Srivastava, S. K. Appl. Ophthalmol. 1980, 10, 459-73.
43. Chiou, S. H.; Chylack, L. T.; Bunn, H. F.; Kinoshita, J. H. Biochem. Biophys. Res. Commun. 1980, 95, 894-901.
44. Chiou, S. H.; Chylack, L. T.; Tung, W. H.; Bunn, H. F. J. Biol. Chem. 1981, 256, 5176-80.
45. Jones, R. J.; Peterson, C. M. J. Clin. Invest. 1979, 63, 425-429.
46. Vlassara, H.; Brownlee, M.; Cerami, A. Proc. Natl. Acad. Sci. U.S.A. 1981, 78, 5190-92.
47. Coradello, H.; Pollak, A.; Pugano, M.; Leban, J.; Lubec, G. IRCS Med. Sci. 1981, 9, 766-67.
48. Day, J. F.; Thorpe, R.; Baynes, J. W. J. Biol. Chem. 1979, 254, 595-97.
49. Dolhofer, R.; Wieland, O. H. FEBS Lett. 1979, 103, 282-86.
50. Guthrow, C. E.; Morris, M. A.; Day, J. F.; Thorpe, S. R.; Baynes, J. W. Proc. Natl. Acad. Sci. U.S.A. 1979, 76, 4258-61.
51. Zaman, Z.; Verwilghen, R. L. Biochim. Biophys. Acta 1981, 699, 120-24.
52. Amaya, F. J.; Lee, T. C.; Chichester, C. O. J. Agric. Food Chem. 1976, 24, 465-67.
53. Dolhofer, R.; Renner, R.; Wieland, O. H. Diabetologia 1981, 21, 211-215.

54. Brownlee, M.; Cerami, A. Science 1979, 206, 1190-91.
55. Liang, J. N.; Chakrabarti, B. J. Biol. Chem., in press.
56. Kennedy, A.; Mehl, T. D.; Riley, W. J.; Merimee, T. J. Diabetologia 1981, 21, 94-98.
57. Kennedy, A.; Kandell, T. W.; Merimee, T. J. Diabetes 1979, 78, 1006-10.
58. Tanzer, M. C.; Fairweather, R.; Gallop, P. M. Arch. Biochem. 1972, 151, 137-41.
59. Schnider, S. L.; Kohn, R. R. J. Clin. Invest. 1981, 67, 1630-35.
60. Schnider, S. L.; Kohn, R. R. J. Clin. Invest. 1980, 66, 1179-81.
61. Rosenberg, H.; Modrak, J. B.; Hassing, J. M.; Al-Turk, W. A.; Stohs, S. J. Biochem. Biophys. Res. Commun. 1979, 91, 498-501.
62. Lubec, G.; Pollack, A. "The Glomerular Basement Membrane"; Renal Physiology, Karger Basel, 1981; 3, p 4-8.
63. Heathcote, J. G.; Bailey, A. J.; Grant, M. E. Biochem. J. 1980, 190, 229-237.
64. Cohen, M. P.; Urdanivia, E.; Surma, M.; Wu, V. Y. Biochem. Biophys. Res. Commun. 1980, 95, 765-69.
65. Miller, J. A.; Gravallese, E.; Bunn, H. F. J. Clin. Invest. 1980, 65, 896-901.
66. Williams, S. K.; Devenny, J. J.; Bitensky, M. W.; Proc. Natl. Acad. Sci. U.S.A. 1981, 78, 2392-97.
67. McVerry, B. A.; Hopp, A.; Fisher, G.; Huehns, E. R. Lancet 1980, I, 738-40.
68. Reid, M. E.; Ellison, S. S.; Barker, J. M.; Lewis, T.; Avoy, D. R. Vox Sanguinis 1980, 45, 85-90.
69. Bunn, H. F.; Shapiro, R.; McManus, M.; Garrick, L. M.; McDonald, M. J.; Gallop, P. M.; Gabbay, K. H. J. Biol. Chem. 1979, 254, 3892-98.
70. Fitzgibbons, J. F.; Koler, R. D.; Jones, R. T. J. Clin. Invest. 1976, 58, 820-24.
71. Robins, S. P.; Bailey, A. J. Biochem. Biophys. Res. Commun. 1972, 48, 76-84.
72. Shapiro, R.; McManus, M. J.; Zalut, C.; Bunn, H. F. J. Biol. Chem. 1980, 255, 3120-3127.
73. Sosenko, J. M.; Flückiger, R.; Platt, O. S.; Gabbay, K. H. Diabetes Care 1980, 3, 590-93.
74. Cohen, M. P.; Yu-Wu, V. Biochem. Biophys. Res. Commun. 1981, 100, 1549-54.
75. Schieicher, E.; Deufel, T.; Wieland, O. H. FEBS Lett. 1981, 129, 1-4.
76. Gonen, B.; Jacobson, B.; Farrar, P.; Schonfeld, G. Diabetes Suppl. Abstract, 1981, 30, p 47A.
77. Kim, H. -J.; Kurup, I. V. Diabetes Suppl. 1 Abstract, 1981, 30, p 47A.
78. Kraemer, F. B.; Reaven, G. M. Diabetes Suppl. 1 Abstract, 1981, 30, p 47A.

79. Bailey, A. J.; Robins, S. P.; Tanner, M. J. A. Biochim. Biophys. Acta 1976, 434, 51-57.
80. Williams, S. K.; Devenny, J. J.; Bitensky, M. W. Proc. Natl. Acad. Sci. U.S.A. 1981, 2392-97.
81. Jedziniak, J. A.; Kinoshita, J. H.; Yates, E. M.; Mocker, L. O.; Benedek, G. B. Exp. Eye Res. 1973, 15, 185.
82. Pirie, A. Invest. Ophthalmol. Vis. Sci. 1968, 7, 634-49.
83. Satoh, H.; Bando, M.; Nakajima, A. FEBS Lett. 1973, 129, 1-4.
84. Mohammed, A.; Olcott, H.S.; Fraenkel-Conrat, H. Arch. Biochem. 1949, 24, 157-63.
85. Klethi, D.; Nordmun, J. Arch. d'Ophthalmol. (Paris) 1975, 35, 891-96.
86. Monnier, V. M.; Cerami, A. submitted.
87. Olson, K.; Pernehalm, P. A.; Pepoff, T.; Theander, O. Acta Chem. Scand. 1977, B31, 469-74.
88. LaBella, F. S. J. Gerontol. 1962, 17, 8-13.
89. LaBella, F. S.; Lindsay, W. G. J. Gerontol. 1963, 18, 111-18.
90. LaBella, F. S.; Paul, G. J. Gerontol. 1964, 19, 54-59.
91. Kannel, W. B.; McGee, D. L. J. Am. Med. Assoc. 1979, 241, 2035-38.
92. Pillsbury, H. C.; Hung, W.; Kyle, M. C.; Freis, E. D. Am. Heart J. 1974, 87, 783-90.
93. Schuyler, M. R.; Niewoehner, D. E.; Inkley, S. R.; Kohn, R. R. Am. Rev. Respir. Dis. 1976, 113, 37-41.
94. Grgic, A.; Rosenbloom, A. L.; Weber, F. T.; Giorduno, B. N. Eng. J. Med. 1975, 292, 372.
95. Waine, H.; Nevinny, D.; Rosenthal, J.; Joffe, I. B. Tufts Folia Med. 1961, 7, 13-19.
96. DeBats, A.; Rhodes, E. L. Lancet 1974, 1, 137-38.
97. Tellez-Nagel, I.; Korthals, J. K.; Vlassara, H. K.; J. Neuropath. Exp. Neurol. 1977, 36, 351-63.
98. Harding, J. J.; Rixon, V. C. Exp. Eye Res. 1980, 31, 567-71.
99. Stevens, V. J.; Fantl, W. J.; Newman, C. B.; Sims, R. V.; Cerami, A.; Peterson, C. M. J. Clin. Invest. 1981, 67 361-69.
100. Bucala, R.; Fishman, J.; Cerami, A. Proc. Natl. Acad. Sci. U.S.A. 1982, 79, 3320-24.
101. Hodge, J. E. ,J. Agric. Food Chem. 1953, 1, 928.

RECEIVED October 5, 1982

Maillard Reactions of Therapeutic Interest

L. MESTER, L. SZABADOS, and M. MESTER
Centre National de la Recherche Scientifique, Institut de Chimie des Substances Naturelles, Gif-sur-Yvette, 91190, France

Some Maillard-type compounds formed in vitro or in vivo have well established therapeutic effects, e.g., browning products of D-glucose and glycine promote the growth of lactobacilli in microflora of rats. Deoxyfructoserotonin formed synthetically or in vivo inhibits the multiplication of Mycobacterium leprae and shows anti-stress activity. The Maillard reaction prevents lysine-rich proteins from aggregating platelets. Enzymic cleavage of Maillard-type compounds in vivo conteracts the detrimental effects of the Maillard reaction in diabetes, e.g., cataract formation or thickening of the basal membrane of blood vessels. The enzymic cleavage of Maillard-type compounds depends on the concentrations of cytochrome P-450 and NADPH.

Maillard, who first observed the facile transformation of reducing sugars with amines and proteins, attributed to this reaction some biological significance (1). Paradoxically, this aspect was completely neglected until recently, whereas the role of the reaction in food chemistry generated a paramount interest. Thus, many thousands of papers have been devoted to understanding such processes as nonenzymatic browning, preservation of foodstuffs, and formation of flavor (2).

The occurrence of Maillard reactions in vivo was observed for the first time only a few years ago. In 1975 we reported (3) the formation of deoxyfructoserotonin (Fig. 1) and deoxyfructolysine derivatives in the blood, in this second case poly-L-lysine being used as a model compound (Fig. 2).

This observation was soon followed by the discovery of deoxyfructohemoglobin by Flückiger and Winterhalter (4). In normal individuals this sugar derivative represents 4% of the total amount of hemoglobin, but this value can be increased up to 20% in diabetics. More recently detrimental effects of the Maillard

0097-6156/83/0215-0451$06.00/0

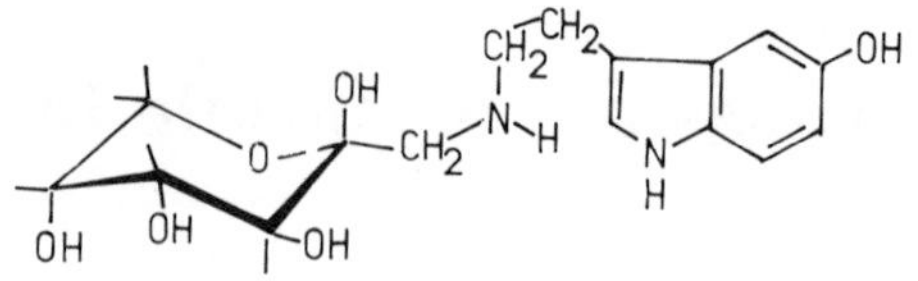

Figure 1. Deoxyfructoserotonin (DFS).

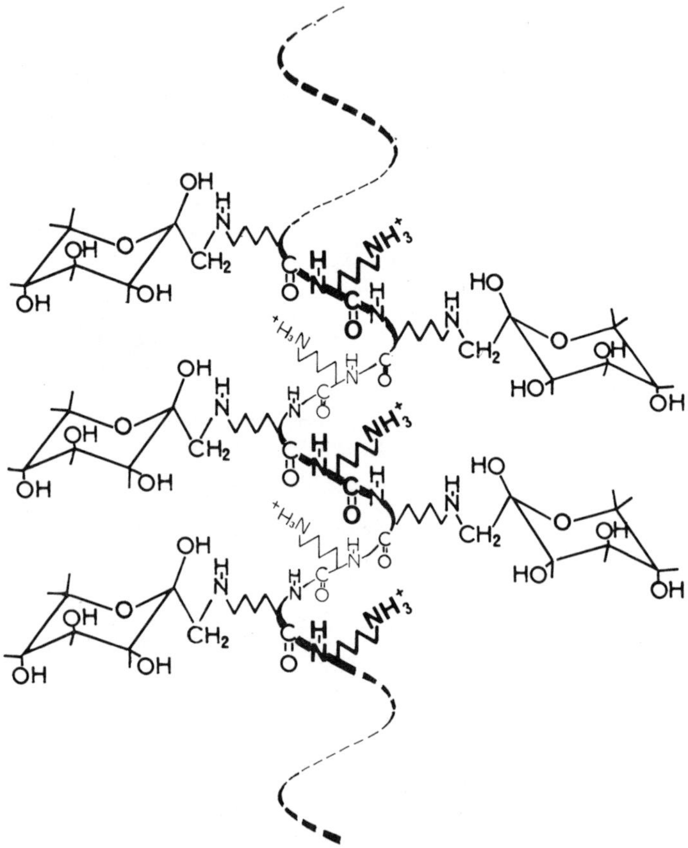

Figure 2. Helicoidal structure of Amadori-type sugar derivatives of poly-L-lysine.

reaction in vivo have also been observed in diabetes, such as cataract formation and thickening of the blood vessel walls (5).

Enzymic Cleavage of Maillard-type Compounds

The chemical synthesis of Maillard-type compounds proceeds readily in vivo; however, the chemical cleavage of these compounds requires drastic conditions that certainly do not exist in vivo (6). Nevertheless, a slow liberation of serotonin from deoxyfructoserotonin is observed in blood and brain, suggesting the existence of an enzyme system for the cleavage of Maillard-type sugar-amine derivatives (7).

It has been established that the enzymic cleavage of Maillard-type compounds depends on the concentration of cytochrome P-450 and of the coenzyme NADPH (8) (Fig. 3). The mechanism suggested for the enzymic cleavage of deoxyfructoserotonin is shown in Figure 4. Therefore, factors interfering with this enzyme system may prevent the detrimental effects of the Maillard reaction in diabetes.

Maillard Reaction Products of Therapeutic Interest

Some Maillard-type compounds formed in vitro or in vivo have well established therapeutic effects.

Effects of Maillard Compounds on the Growth of Micro-organisms. Maillard-type compounds affect the development of micro-organisms in various ways. Generally, the early Maillard reaction products promote such development, whereas late reaction products may have an inhibiting effect on the growth of micro-organisms. *N*-Glucosylglycine stimulates the growth (9) of *Lactobacilli* and only late premelanoidins inhibit growth. Nearly the same effects have been observed with *Aspergilli* (10).

Effects of Maillard Compounds on Platelet Aggregation Maillard reaction products of lysine-rich proteins prevent platelet aggregation, e.g., poly-L-lysine aggregates platelets, but upon Maillard reaction loses this ability (11).

Therapeutic Effects of Deoxyfructoserotonin Deoxyfructoserotonin was first synthesized in 1975 (12) in our laboratory from D-glucose and serotonin, coffee wax being an abundant source of this amine (13).

Deoxyfructoserotonin has a strong reducing power and ability for complexation. The sugar derivative shows the following biological properties:

- it displays "serotonin-like" activity, being in competition with serotonin for the same tryptaminergic receptors (14),
- the rate of its metabolism by monoamine oxidase (MAO) is very low (15),

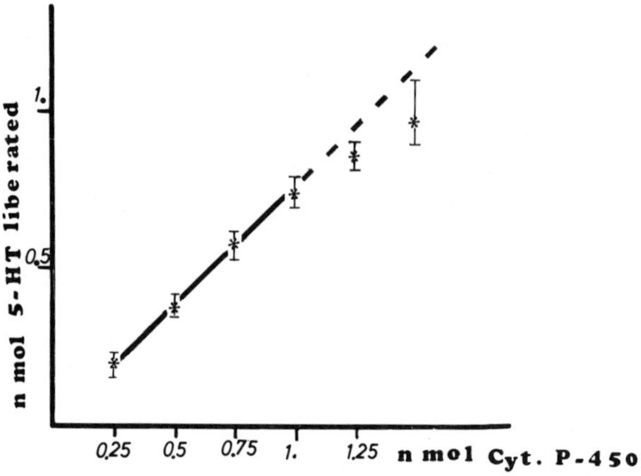

Figure 3. Enzymic cleavage of deoxyfructoserotonin (DFS) as a function of cytochrome P-450 concentration.

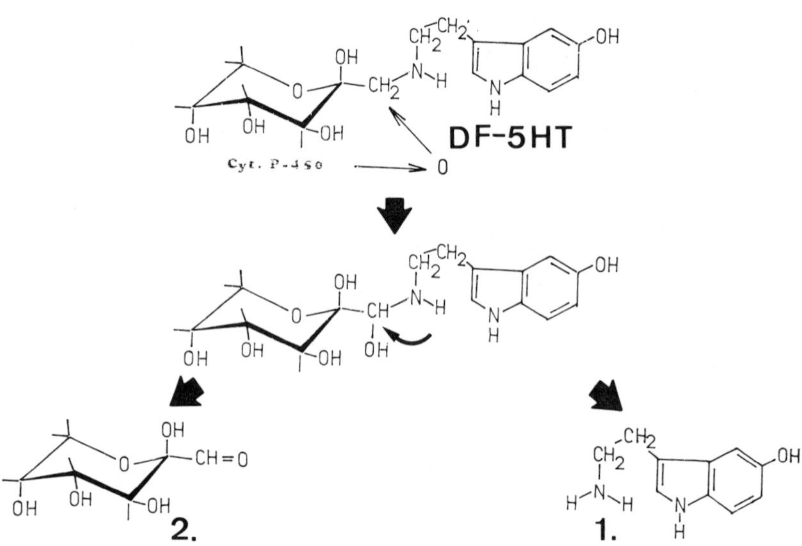

Figure 4. Mechanism of the enzymic cleavage of deoxyfructoserotonin.

- it has a low toxicity (LD_{50} = 1200 mg/kg oral adminstration, 400 mg/kg i.v.) in mice (16), and
- it inhibits the release of serotonin from platelets by neurotoxins (e.g., 5,6-dihydroxytryptamine) (17).

In leprosy, depletion of serotonin and low uptake of the bioamine by blood platelets has been observed. The sugar derivative counteracts these phenomena, reestablishing the serotonin level necessary for the nearly normal function of platelets and neurons (18).

During our investigation on the metabolism of deoxyfructoserotonin we have observed a strong inhibition of mushroom DOPA-oxidase (19) (Fig. 5).

It has been reported that Mycobacterium leprae contains a specific DOPA-oxidase and the the metabolites of DOPA are essential for the growth of M. leprae in vivo (20). In vitro we have observed that deoxyfructoserotonin inhibits the incorporation of [^{3}H]-DOPA in Mycobacterium leprae (21).

The competition of deoxyfructoserotonin with the enzyme-controlled utilization of DOPA may then suppress the viability of leprosy bacteria. In fact, 20 mg/kg body weight/day deoxyfructoserotonin incorporated in the diet of mice totally inhibited the multiplication of M. leprae on mouse foot pad, 4 months after incubation, from both sulfone-sensitive as well as sulfone-resistant cases (22). These results are shown in Table I.

The clinical effectiveness of deoxyfructoserotonin in leprosy has been tested in the Institute Marchoux in Bamako (Rep. of Mali). Seven patients suffering from leprosy were treated during 6 months with deoxyfructoserotonin at a daily rate of 5 - 20 mg/kg body weight. Considerable improvement was observed in a few weeks in cases classified "borderline" and in a few months in cases of lepromatous leprosy (23).

A young girl with typical "Facies leonin" is shown in Plate 1. After 5 months treatment with deoxyfructoserotonin, complete flattening of all nodules is observed (Plate 2).

Plate 3 shows a young man with skin patches on the nape of the neck. After 6 months treatment with deoxyfructoserotonin, the skin patches had disappeared (Plate 4).

The improvement was not only clinical, but also bacteriological, reducing the morphological index (percent of viable bacteria) to 0. During the treatment, no side-effects were observed either clinically or by biological tests (24). Continued treatment over a period of another 6 months confirmed the definite effectiveness of the drug against leprosy.

In addition to its antileprosy effect, deoxyfructoserotonin shows a considerable anti-stress activity, especially on ulcer formation by restraint (25).

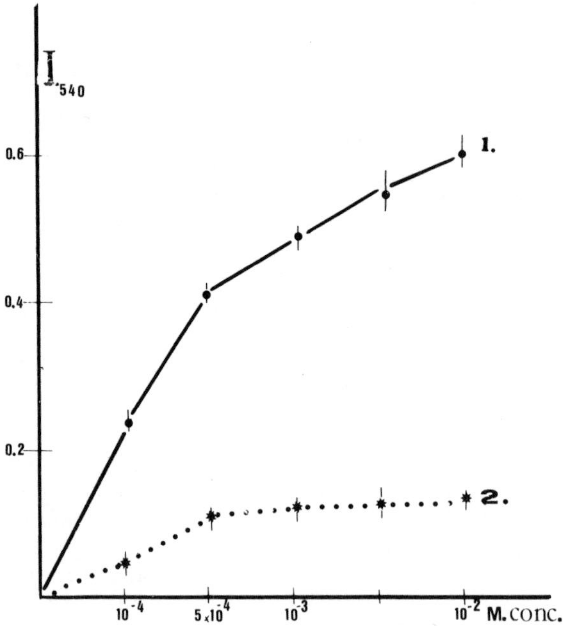

Figure 5. Inhibition of DOPA-oxidase by deoxyfructoserotonin. Key: 1, oxidation of DOPA by DOPA-oxidase; and 2, inhibition by DFS of the oxidation of DOPA by DOPA-oxidase.

Table I. Studies on the Antileprosy Activity of Deoxyfructo-serotonin (DFS) by the Mouse Foot Pad

Time after Inoculation (months)	M. Leprae Levels Found	
	Control	DFS-Treated
4	19.4 X[a]	5.3 X
6	22.8 X	Negative[b]
8	39.4 X	Negative
10	20.0 X	Negative

Note: Studies done at Central Leprosy Teaching and Research Institute, Chingleput, India. Eight groups of 6 mice each inoculated with biopsy material from a LL patient with BI over 2+ and MI 1%. Each treated animal received 0.5 mg DFS/day (20 mg/kg body weight) in the diet.

[a]X = times more than at the beginning.

[b]Negative means no acid-fast bacilli detected in the 20-22 microscopic fields in any of the smears.

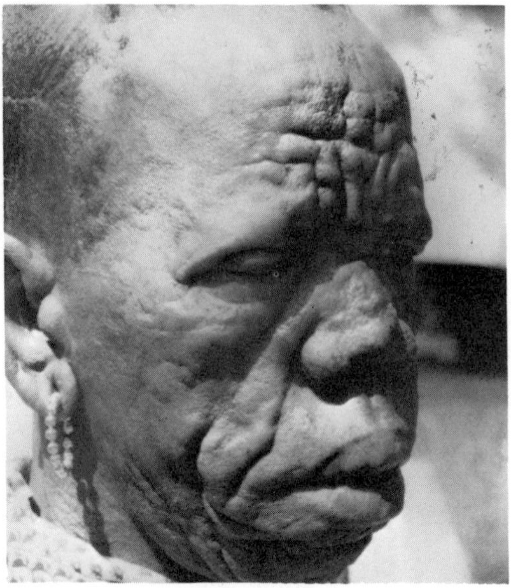

Plate 1. Patient 6 with "lion face".

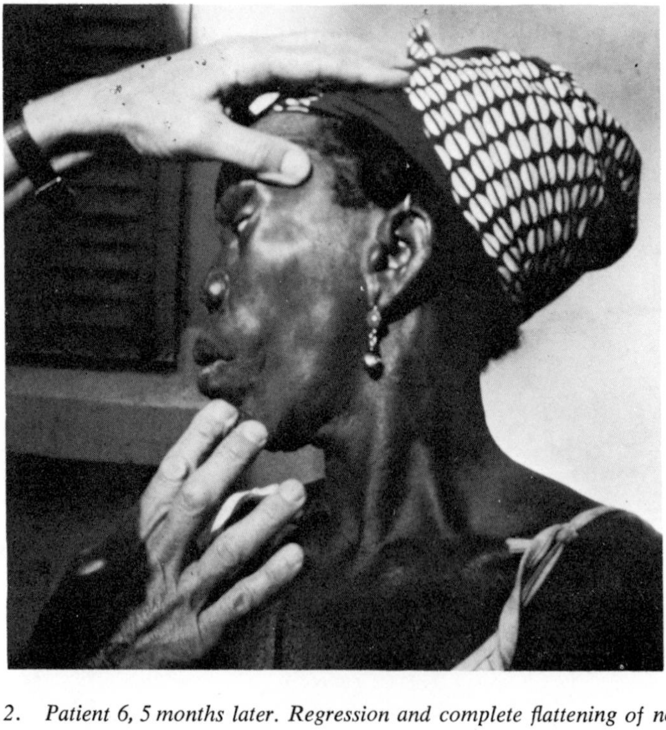

Plate 2. Patient 6, 5 months later. Regression and complete flattening of nodules.

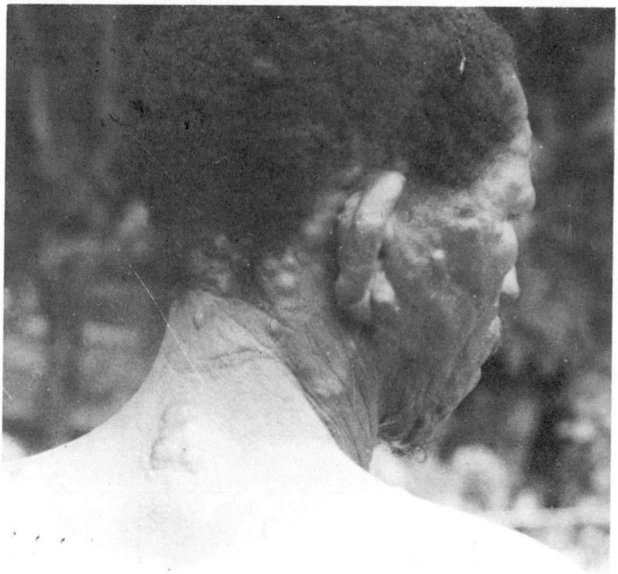

Plate 3. Patient 7 skin patches on nape of neck.

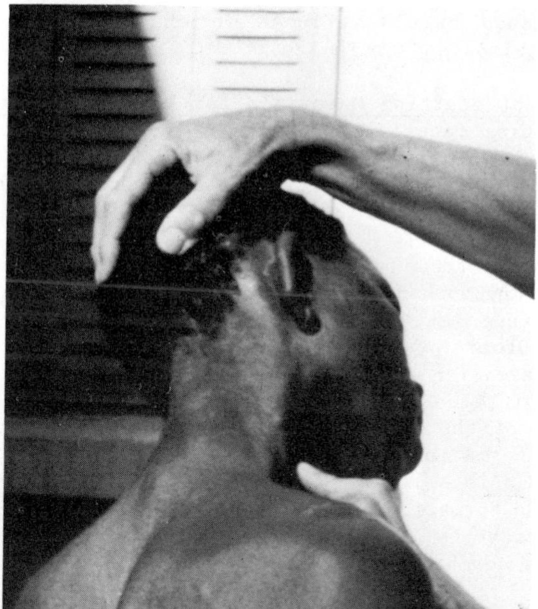

Plate 4. Patient 7, 6 months later. Patches on nape of neck had disappeared.

Relationship of Nutrition to Maillard Reactions in Vivo

Nutrients can modify Maillard reactions in vivo by many different ways.

Change of pH of Blood Plasma A slightly acid medium is usually needed for the Maillard reaction, but in some cases, e.g., with lysine-rich proteins, a slightly alkaline pH is also operating. Thus a small change in pH, due to the nature of food consumed or the pathologic metabolism of foodstuffs, can affect Maillard reactions in vivo.

The cytochrome P-450 enzyme system, responsible for the cleavage of Maillard compounds, is highly sensitive to pH change, maximum activity being at about pH 7.2. A small change of the pH in either direction decreases considerably the activity of the enzyme system.

Food as Source of Starting Materials for Maillard Reaction in Vivo Nutrients are, of course, the origin of many different reducing sugars and amines or proteins, which are suspected to undergo a Maillard reaction in vivo.

A high consumption of D-glucose or polysaccharides easily degradable into glucose can result in a temporarily high level of reducing sugar in the blood. Since the specificity of the reaction is pH dependent, the reducing sugar can subsequently undergo a Maillard reaction with one or more amines or proteins. It is also well established that tryptophan in the food is the source of serotonin in blood and brain (26).

Availability of Starting Materials for Maillard Reaction in Vivo Not all starting materials present in the body are available for Maillard reaction. For example, trypotophan in blood, formed by digestion and absorption of tryptophan-rich proteins, is easily bound by plasma albumin, making it unavailable for further transformation.

Unsaturated free fatty acids, having a higher affinity for albumin then tryptophan has, can help to liberate tryptophan (27). Unsaturated free fatty acids in the blood also, of course, come from food. Other nutritional factors, such as vitamin E can prevent autoxidation or enzymic oxidation, processes that inactivate the unsaturated free fatty acids (28).

In view of the antileprosy activity of deoxyfructoserotonin, the in vivo formation of this compound can also probably prevent leprosy. A possible route by which nutritional factors can favor such in vivo formation of deoxyfructoserotonin and in this way prevent leprosy is given in Scheme 1. The validity of this hypothesis is now being tested in India.

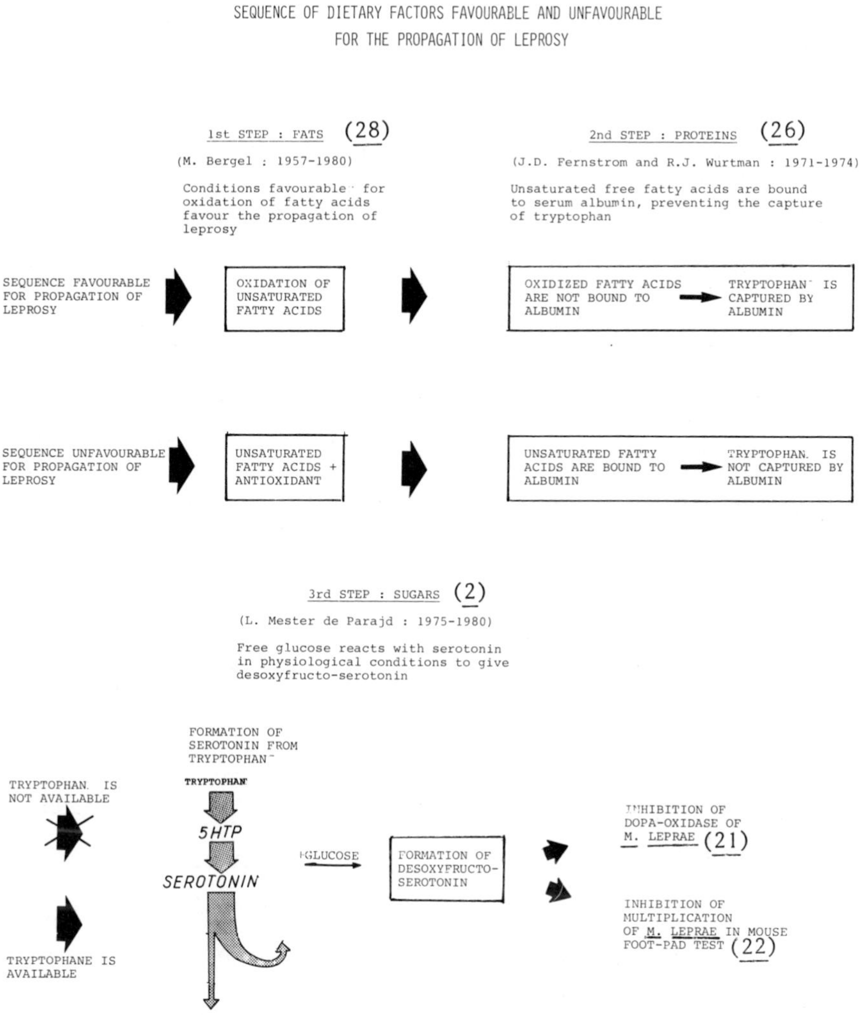
SEQUENCE OF DIETARY FACTORS FAVOURABLE AND UNFAVOURABLE FOR THE PROPAGATION OF LEPROSY
1st STEP : FATS (28)
(M. Bergel : 1957-1980)
Conditions favourable for oxidation of fatty acids favour the propagation of leprosy
2nd STEP : PROTEINS (26)
(J.D. Fernstrom and R.J. Wurtman : 1971-1974)
Unsaturated free fatty acids are bound to serum albumin, preventing the capture of tryptophan
SEQUENCE FAVOURABLE FOR PROPAGATION OF LEPROSY
OXIDATION OF UNSATURATED FATTY ACIDS
OXIDIZED FATTY ACIDS ARE NOT BOUND TO ALBUMIN
TRYPTOPHAN IS CAPTURED BY ALBUMIN
SEQUENCE UNFAVOURABLE FOR PROPAGATION OF LEPROSY
UNSATURATED FATTY ACIDS + ANTIOXIDANT
UNSATURATED FATTY ACIDS ARE BOUND TO ALBUMIN
TRYPTOPHAN IS NOT CAPTURED BY ALBUMIN
3rd STEP : SUGARS (2)
(L. Mester de Parajd : 1975-1980)
Free glucose reacts with serotonin in physiological conditions to give desoxyfructo-serotonin
FORMATION OF SEROTONIN FROM TRYPTOPHAN
TRYPTOPHAN IS NOT AVAILABLE
TRYPTOPHAN
5HTP
SEROTONIN
+GLUCOSE
FORMATION OF DESOXYFRUCTO-SEROTONIN
INHIBITION OF DOPA-OXIDASE OF M. LEPRAE (21)
INHIBITION OF MULTIPLICATION OF M. LEPRAE IN MOUSE FOOT-PAD TEST (22)
TRYPTOPHANE IS AVAILABLE

Scheme I

Conclusion

The formation in vivo of Maillard-type compounds depends mainly on a favorable concurrence of nutritional factors, but a pathologic status, as in diabetes, can also favor the reaction. Some of these compounds, such as deoxyfructoserotonin, are of therapeutic interest.

Acknowledgements

This work was supported by the Institut National de la Santé et de la Recherche Médicale (Contrat Nr. 123008) and by the Oeuvres Hospitalières Françaises de l'Ordre Souverin de Malte.

Literature Cited

1. Maillard, L.C. C.R. Soc. Biol. 1912, 72, 599.
2. Mester, L.; Szabados, L.; Mester, M.; Yadav, N. Prog. Food Nutr. Sci. 1981, 5, 295.
3. Mester, L.; Kraska, B.; Crisba, J.; Mester, M. "Sugar-amine Interactions in the Blood Clotting System and Their Effects on Haemostasis", Proc. 5th Intern. Congr. Thrombosis Haemostasis, Paris, 1975.
4. Flückiger, R.; Winterhalter, K.H. FEBS Lett. 1976, 71, 356.
5. Cerami, A.; Stevens, J.V.; Monnier, V.M. Metabolism 1979, 28, 431.
6. Gotschalk, A. Biochem. J. 1952, 52, 455.
7. Szabados, L.; Mester, M.; Mester, L.; Bhargava, K.P.; Parvez, S.; Parvez, H. Biochem. Pharm. 1982, 31, 2121.
8. Mester, L.; Szabados, L.; Mester, M. Thromb. Haem. 1981, 46, 393.
9. Rogers, D.; King, T.E.; Cheldelin, V.H. Proc. Soc. Exp. Biol. Med. 1953, 82, 140.
10. Jemmali, M. Thesis, Univ. Paris, Series A, Nr. 942, 1965.
11. Mester, L.; Kraska, B.; Crisba, J.; Mester, L. Haemostasis 1976, 5, 115.
12. Mester, L.; Mester, M. J. Carbohyd. Nucleos. Nucleot. 1975, 2, 141.
13. Hirsbrunner, P.; Brambilla, E. 1976, (Nestlé S.A.), French Patent Nr. 76 05453.
14. Mester, L.; Mester, M.; Rendu, F.; Labrid, C.; Dureng, G. Thromb. Res. 1978, 12, 783.
15. Bhargava, K.P.; Gujrati, V.R.; Gashee Ali; Mester, M.; Mester, L. Thromb. Res. 1978, 12, 791.
16. Rapport Pharmacologique, C.E.R.M.; 63201-Riom, France, 1977.
17. Launay, J.M.; Mester, L. Thromb. Res. 1978, 13, 805.
18. Szabados, L.; Launay, J.M.; Mester, M.; Cottenot, F.; Pennec, J.; Mester, L. Intern. J. Lepr. 1981, 49, 42.
19. Szabados, L.; Mester, L., unpublished results.
20. Prabhakaran, K. Lepr. Rev. 1973, 44, 112.

21. Jayaraman, P.; Mahadevan, P.R.; Mester, M.; Mester, L. Biochem. Pharm. 1980, 29, 2526.
22. Mester de Parajd, L.; Balakrishnan, S. Acta Lepr. 1981, 83, 21.
23. Mester de Parajd, L.; Bakakrishnan, S.; Saint-André, P.; Mester de Parajd, M. Ann. Microb. Inst. Pasteur 1982, 133 B, 149.
24. Mester de Parajd, L.; Balakrishnan, S.; Saint-André, P.; Mester de Parajd, M. Ann. Microb. Inst. Pasteur 1982, in press.
25. Mester, L.; Mester, M.; Bhargava, K.P.; Labrid, C.; Dureng, G.; Cheucle, M. Thromb. Res. 1979, 15, 245.
26. Fernstrom, J.D.; Wurtman, R.J. Science 1971, 173, 149.
27. Lipset, D.; Madras, B.K.; Wurtman, R.J.; Munro, H.N. Life Science 1973, 12, (Part II), 57.
28. Bergel, M. Lepr. Rev. 1959, 39, 153.

RECEIVED October 5, 1982

TOXICOLOGICAL ASPECTS

Nutritional and Toxicological Effects of Maillard Browned Protein Ingestion in the Rat

STEPHEN J. PINTAURO[1], TUNG-CHING LEE, and CLINTON O. CHICHESTER

University of Rhode Island, Department of Food Science and Technology, Nutrition, and Dietetics, Kingston, RI 02881

The Maillard, or nonenzymatic browning, reaction between reducing sugars and proteins is known to cause serious deterioration of the nutritional quality of foods during processing and storage (1-7). Recently, considerable attention has focused on the physiological effects of the ingestion of Maillard browned compounds beyond those that can be attributed to nutritionally related causes (7,8,9). In addition, the food industry often employs the reaction to produce desirable aromas, colors, and flavors. Thus, if there is indeed a food safety risk associated with Maillard browning, priorities in the food industry may have to be redirected to minimize and control the reaction, rather than promote it.

The purpose of this study was to separate the nutritionally related effects of the long-term feeding of Maillard browned protein, from the toxicological effects. This was done by eliminating, wherever possible, variables that might lead to nutritional problems secondary to the effects of feeding the Maillard proteins. The diets were not only of equal protein quantity, but also of equal and high protein quality.

PROCEDURE

Maillard Browned Proteins Three types of browned proteins were used in the long-term feeding study; egg albumin, hydrolyzed egg albumin, and a commercial, casein-based, instant breakfast product. The hydrolyzed egg albumin was chosen to examine the effects of feeding severely browned hydrolyzed proteins, such as are commonly used in a variety of processed foods to produce characteristic flavors and colors. The instant breakfast product was included in the feeding study as a representative, commercial food product which may undergo some degree of browning due to the processing or storage.

The hydrolyzed egg albumin was prepared by dissolving 500 g of egg albumin (ICN Nutritional Biochemicals, Cleveland, Ohio) in

[1] Current address: University of Vermont, Department of Human Nutrition & Foods, Burlington, VT 05405

0097-6156/83/0215-0467$06.00/0

3 l of distilled water. The solution was adjusted to pH 1.5 with 1N HCl and 1 g of pepsin (1:10,000, Sigma Chemical Co., St. Louis, MO) was added to the solution. The mixture was allowed to incubate at 37°C with mixing for 24 h. Following incubation, the solution was adjusted to pH 7.0 with 1N NaOH and freeze-dried.

The egg albumin and hydrolyzed egg albumin samples were prepared for browning by mixing 3 parts of the protein with 2 parts D-(+)-glucose (ICN Nutritional Biochemicals, Cleveland, Ohio) and adjusting the moisture content to 15%. The instant breakfast product (Carnation Company, Waverly, Iowa) was prepared for browning by simply adjusting the moisture content to 9%. All samples were browned by storage at 37°C in a sealed glass chamber. The humidity was maintained at 68% by placing a small beaker of 40% sulfuric acid in the chamber. After storage for 40 days, the samples were freeze-dried and milled for incorporation into the diets. Control samples for all proteins were prepared identically to the browned samples, but immediately freeze-dried rather than stored as described above.

Determination of Protein Quality The Protein Efficiency Ratio (PER) method was used for determining the protein quality of the treatment diets (10).

It was the purpose of this study to eliminate, as much as possible, nutritional factors from the treatment diets. It was necessary, therefore, to design the diets so as to contain sufficient levels of brown protein, yet at the same time keep the protein quality relatively high. Protein Efficiency Ratios were determined for the three types of protein at intervals from 0 to 40 days of browning under the conditions described earlier. PER's were then run on various combinations of the 40-day browned proteins and non-browned protein. It was determined that a diet could be formulated that would contain no less than 3% 40-day browned protein and still result in a PER of no lower than 2.0. The PER of the control diets were adjusted to match the protein quality of the experimental (browned) diets by substituting appropriate levels of gelatin as a protein source. The resulting protein compositions for the long-term feeding experiment and their respective Protein Efficiency Ratios are given in Table I.

Diets All diets were designed to contain exactly 10% protein, as measured by the Kjeldahl nitrogen determination method (10). The composition of the remainder of the diet is described in Table II.

Animals Male, Sprague-Dawley rats (COBS-CD strain) were purchased at five weeks of age from Charles River Breeding Laboratories, Wilmington, MA. The animals were maintained for two weeks on Rat Chow (Purina Lab Chow, Ralston Purina Company, St. Louis, MO). Following the two-week pretreatment, the animals were weighed and distributed into treatment groups as described in Table I.

Animals in the control groups were pair-fed to their corresponding browned diet treatment animals to insure equal protein intake in both groups. Water was supplied ad libitum and weight gain was monitored throughout the study.

Five animals from each group were sacrificed at the end of six

TABLE I

Composition and Protein Quality of Proteins Used in 18-Month Feeding Study

Group	Number of Rats	Protein Composition of Diets	PER
IB-C	20	7% 0-Day Instant Breakfast[a] 3% Gelatin	2.01 ± .07
IB-B	20	10% 40-Day Browned Instant Breakfast	1.94 ± .11
EA-C	20	7% 0-Day Egg Albumin[b] 3% Gelatin	2.34 ± .20
EA-B	20	3% 40-Day Browned Egg Albumin 7% 0-Day Egg Albumin	2.48 ± .23
HEA-C	20	3% 0-Day Hydrolyzed Egg Albumin 3% Egg Albumin 4% Gelatin	2.10 ± .39
HEA-B	20	3% 40-Day Browned, Hydrolyzed Egg Albumin 7% Egg Albumin	2.28 ± .29
			(Mean ± S. D.)

[a]Carnation Company, Waverly, Iowa

[b]ICN Nutritional Biochemicals, Cleveland, Ohio

TABLE II

Composition of Diets for 18-Month Feeding Study

Protein[a]	10%
Corn Oil	10%
Cellulose	5%
Vitamin Mix[b]	1.2%
Salt Mix[c]	4%
Dextrose[d]	17.45%
Dextrin	17.45%
Sucrose	17.45%
Corn Starch	17.45%

[a]Refer to Table I for composition of protein portion of diet.

[b]Vitamin diet fortification mixture, ICN Nutritional Biochemicals, Cleveland, Ohio.

[c]Mineral mixture, ICN Nutritional Biochemicals, Cleveland, Ohio.

[d]Dextrose content of diet was corrected for dextrose added in preparing browned proteins.

and twelve months of feeding. The remainder were sacrificed at the end of 18 months. All animals were fasted for 18 hours prior to being sacrificed.

Serum Preparation Whole blood was collected after decapitation of the animal and allowed to clot at room temperature for 30 min. The serum was then separated from the clot by centrifugation at 5000 rpm at 4°C. Aliquots of the serum were immediately taken and refrigerated for the determination of glutamic-pyruvic transaminase and lactic dehydrogenase. An additional 0.5 ml of serum from each animal was treated with a few milligrams of sodium fluoride and refrigerated for later glucose determinations. The rest of the serum was frozen for the remainder of the clinical chemistry determinations.

Samples of whole blood were obtained by heart puncture using and EDTA-treated syringe. The whole blood was immediately assayed for hematocrit (percent packed cell volume) and hemoglobin by the method of Drabkin and Austin (11).

Tissue Preparation Following decapitation and exsanguination, the liver, kidneys, heart, lungs, spleen, testes, stomach, and proximal third of the small intestine were removed, freed of fat, blotted, and weighed. In addition, the cecum was removed, cleaned and rinsed in saline, blotted, and weighed.

Small samples of the liver, kidney, spleen, lung, heart, small intestine, and brain were fixed in 10% buffered formalin for histopathological examination. The remainder of the tissue was kept in 0.25M sucrose and frozen in liquid nitrogen.

Clinical Biochemical Determinations of the Serum Serum lactic dehydrogenase (LDH) and glutamic-pyruvic transaminase (GPT) activities were measured on fresh, refrigerated serum within 48 h of sacrificing the animal. Lactic dehydrogenase was measured according to the method of Amador, Dorfman, and Wacker (12). Serum GPT activity was measured according to the method of Wroblewski and LaDue (13).

Serum glutamic-oxalacetic transaminase (GOT) activity was measured according to the method of Karmen (14), serum albumin by the bromocresol green method of Doumas and Biggs (15), alkaline phosphatase by the method of Young, Pestaner, and Gibberman (16), urea nitrogen by the method of Crocker (17), total iron and total iron binding capacity by the method of Persijn, Van Der Stik, and Riethorst (18), triglycerides by the method of Bercolo and David (19), and cholesterol by the method of Wybenga, et al. (20). Glucose was determined in serum on a Model 23A, YSI Glucose Analyzer.

Determination of Small Intestine Disaccharidases The proximal third of the small intestine from rats fed treatment diets for 18 months was rinsed in ice-cold 0.25M sucrose, cut open, and the mucosa scraped off with a glass slide. The mucosa were then homogenized in cold 0.25M sucrose and centrifuged at 5000 rpm for 10 min. The method of Dahlqvist (21) was used for determination of maltase and sucrase activity.

Determination of Liver Glutamic-Oxalacetic Transaminase A sample of liver (approximately one g) was homogenized in ice cold 0.25M sucrose and centrifuged at 5000 rpm for 10 min to remove nuclei and cell debris. The method of Reitman and Frankel (22) was used for determination of glutamic-oxalacetic transaminase activity.

Protein Determinations The biuret assay (23) was employed for determining protein in tissue samples.

Statistical Analysis All data were analyzed for significance between treatment groups using the Student's *t* test.

RESULTS

Weight Gain No significant difference in weight gain was detected between browned and control instant breakfast product, egg albumin, and hydrolyzed egg albumin fed animals, through 18 months.

Relative Organ Weights The results for the relative organ weights of rats fed the three browned proteins or their controls are listed in Table III. The browned instant breakfast product resulted in a significantly enlarged liver after six months, and a significantly smaller liver after 18 months. After 12 months of feeding the browned instant breakfast product, a significant enlargement of the lung and heart was noted.

The browned egg albumin diets resulted in a significant enlargement of all tissues after 12 months of feeding. The relative weight of the cecum was nearly double that of the control group. It should be noted, however, that with the possible exception of the cecum, all of these organ weights fall within the normal ranges for a rat (24,25).

The results for the relative organ weights of rats fed the hydrolyzed egg albumin show few significant changes. Again, the cecum is enlarged after six and eighteen months of feeding, and there is a slight enlargement of the kidney in the browned protein fed rats after six months.

Although we observed an increase in the relative weight of the cecum of rats fed browned proteins, the increase is much less dramatic than previously reported (7,9,26). This appears to be due to the fact that we cleaned and rinsed the cecum before weighing. Earlier investigators weighed the cecum plus its contents. Nevertheless, the enlargement of the cecum is still the most definitive and consistent organ weight change associated with the feeding of Maillard browned proteins.

Hemoglobin and Hematocrit The values for hemoglobin and hematocrit for all treatment diets through 18 months of feeding are listed in Table IV. The rats fed browned, hydrolyzed egg albumin demonstrated significantly depressed hemoglobin levels at six months, and significantly elevated hematocrit values at 12 months. The browned-instant breakfast-fed rats showed depressed hemoglobin levels after 12 months of feeding. Again, however, all values for both hemoglobin and hematocrit are within the normal ranges for rats (27,28). It is likely, therefore, that any apparent difference between brown and control groups has little clinical significance.

TABLE III

Ratios of Internal Organ Weight to Body Weight for Rats Fed Browned or Control Instant Breakfast Product, Egg Albumin, and Hydrolyzed Egg Albumin Diets

		% Body Wt.					
	Body Wt.(g)	Liver	Kidney	Cecum	Spleen	Lung	Heart
6 Months							
IB-C	463 ± 21.7	2.84 ± .22	0.57 ± .03	0.44 ± .09	0.13 ± .01	0.40 ± .02	0.28 ± .02
IB-B	439 ± 22.6	3.13 ± .11[b]	0.61 ± .03	0.46 ± .10	0.13 ± .01	0.40 ± .04	0.30 ± .01
EA-C	454 ± 40.4	2.23 ± .12	0.57 ± .03	0.33 ± .09	0.13 ± .01	0.30 ± .01	0.29 ± .04
EA-B	446 ± 22.8	2.09 ± .08[a]	0.56 ± .05	0.40 ± .03	0.13 ± .01	0.31 ± .01	0.27 ± .03
HEA-C	516 ± 16.0	2.33 ± .32	0.53 ± .03	0.32 ± .02	0.13 ± .02	0.35 ± .02	0.26 ± .03
HEA-B	438 ± 65.9	2.37 ± .08	0.58 ± .05[a]	0.46 ± .07[c]	0.14 ± .02	0.37 ± .01	0.28 ± .03
12 Months							
IB-C	467 ± 23.6	2.17 ± .18	0.63 ± .08	0.56 ± .07	0.18 ± .03	0.48 ± .04	0.36 ± .03
IB-B	472 ± 22.1	2.27 ± .03	0.66 ± .04	0.59 ± .08	0.20 ± .02	0.56 ± .06[b]	0.41 ± .01[c]
EA-C	476 ± 30.3	2.05 ± .14	0.62 ± .03	0.35 ± .06	0.13 ± .03	0.34 ± .06	0.31 ± .02
EA-B	423 ± 77.4	2.29 ± .10[b]	0.73 ± .10[a]	0.64 ± .18[b]	0.19 ± .02[c]	0.58 ± .10[b]	0.44 ± .05[c]
HEA-C	435 ± 77.5	2.48 ± .19	0.73 ± .13	0.55 ± .18	0.15 ± .05	0.54 ± .07	0.36 ± .10
HEA-B	502 ± 20.9	2.36 ± .18	0.61 ± .05	0.51 ± .07	0.11 ± .02	0.44 ± .07	0.31 ± .02
18 Months							
IB-C	558 ± 49.9	2.09 ± .15	0.54 ± .06	0.27 ± .07	0.12 ± .02	0.37 ± .04	0.26 ± .05
IB-B	647 ± 28.4[c]	1.89 ± .21[b]	0.48 ± .05[b]	0.28 ± .07	0.12 ± .04	0.37 ± .06	0.22 ± .01[b]
EA-C	628 ± 49.8	2.03 ± .12	0.52 ± .04	0.24 ± .06	0.11 ± .02	0.43 ± .09	0.23 ± .02
EA-B	553 ± 129.7	2.14 ± .16	0.58 ± .11	0.34 ± .12[a]	0.14 ± .02[a]	0.45 ± .10	0.26 ± .06
HEA-C	687 ± 134.2	2.08 ± .32	0.51 ± .13	0.27 ± .09	0.12 ± .02	0.42 ± .09	0.23 ± .06
HEA-B	647 ± 87.3	2.31 ± .22	0.54 ± .08	0.34 ± .05[a]	0.14 ± .02	0.41 ± .06	0.24 ± .04

Mean ± S.D.

Significantly different from control: a, at $p<0.05$; b, at $p<0.025$; c, at $p<0.01$

TABLE IV

Hemoglobin and Hematocrit Values for Rats Fed Treatment Diets for 6, 12, and 18 Months

Treatment Group	6 Months		12 Months		18 Months	
	Hb (g%)	HCT (%)	Hb (g%)	HCT (%)	Hb (g%)	HCT
IB-C	14.3 ± 1.06	38 ± 1.4	15.2 ± 1.17	39 ± 1.7	13.2 ± 0.96	34 ± 2.8
IB-B	14.1 ± 1.32	38 ± 3.3	13.4 ± 0.55[a]	37 ± 2.7	14.3 ± 1.86	36 ± 3.0
EA-C	16.4 ± 0.43	43 ± 2.1	14.8 ± 1.21	36 ± 2.7	14.5 ± 1.62	36 ± 3.6
EA-B	15.3 ± 1.12	42 ± 3.3	14.2 ± 0.39	38 ± 1.6	13.7 ± 0.63	36 ± 1.6
HEA-C	14.6 ± 0.83	37 ± 2.0	14.6 ± 1.06	32 ± 2.8	13.9 ± 1.5	37 ± 2.2
HEA-B	12.9 ± 0.85[a]	35 ± 1.8	14.3 ± 0.26	37 ± 1.9[a]	13.5 ± 0.65	37 ± 2.0

(Mean ± S.D.)

[a]Significantly different from control at $p>0.01$

Clinical Biochemistry The results of the clinical biochemical analysis of the serum from rats fed treatment diets for up to 18 months are listed in Table V. Rats fed browned instant breakfast product for 12 months exhibited significantly elevated blood urea nitrogen levels over control rats. After 18 months of feeding, serum globulin and albumin were clearly lower than in controls.

The results for rats fed browned or control egg albumin are also listed in Table V. The only significant change evident was a depressed serum glutamic-oxalacetic transaminase activity in the browned fed rats after six months.

The only significant changes observed in rats fed browned hydrolyzed egg albumin were a slightly elevated serum glutamic-pyruvic transaminase activity after 6 months, and a slightly elevated serum globulin level after 12 months.

All values for all serum assays in this section fall within the normal ranges for rats, reported in the literature (25,27). No pattern in the clinical biochemistry profile associated with the feeding of Maillard browned proteins is apparent. The few assays that did show significant differences between browned and control groups are still within the normal ranges and, therefore, probably have no clinical significance.

Serum, Cholesterol and Triglyceride Serum triglyceride and cholesterol levels were measured in rats fed Maillard browned proteins or control diets for 18 months. The results are given in Table VI. The feeding of Maillard browned egg albumin resulted in a significantly lower level of both cholesterol and triglycerides over animals fed the control egg albumin diet. Similar results of serum cholesterol values were observed by Gomyo (29). He suggested that the effect may be due to an interference in enterohepatic circulation and/or an inhibition of the absorption of dietary cholesterol. The latter hypothesis is likely the more accurate explanation, as we found no other evidence of an interference in enterohepatic circulation, such as changes in serum bilirubin levels.

Serum Total Iron and Total Iron Binding Capacity The results for the determination of serum total iron and total iron-binding capacity of rats fed treatment diets for 18 months are also listed in Table VI. A significant increase in serum total iron was detected in rats fed the Maillard browned egg albumin over their control group. Increased serum total iron with normal total iron binding capacity is associated with hemolytic anemia, hemochromatosis, hemosiderosis, and hepatitis (30). On the basis of other clinical and histopathological data, however, none of these causes are likely.

Intestinal Disaccharidase Activity The results of the determination of small intestine mucosal maltase and sucrase activity for all animals fed treatment diets for 18 months are given in Table VII. No difference was detected in the activities of these enzymes from the animals fed the instant breakfast product or the hydrolyzed egg albumin. The animals fed the browned egg albumin,

TABLE V

Clinical Serum Analysis of Rats Fed Browned or Control Proteins

6 Months	Glucose (mg %)	Globulin (g %)	Albumin (g %)	B.U.N. (mg %)	SGOT[a]	SGPT[a]	AP[b]
IB-C	ND	ND	ND	6.0 ± 6.6	47 ± 11.2	65 ± 3.00	1.2 ± .05
IB-B	104 ± 11.7	2.5 ± 0.30	2.0 ± 1.1	6.0 ± 1.6	53 ± 9.9	74 ± 10.7	1.3 ± .30
EA-C	80 ± 12.1	2.7 ± 0.34	3.2 ± 0.49	7.1 ± 0.72	54 ± 5.5	74 ± 14.6	0.99 ± .26
EA-B	67 ± 12.8	2.5 ± 0.33	3.2 ± 0.23	5.4 ± 1.7	41 ± 8.4[d]	53 ± 23.4	1.0 ± .41
HEA-C	58 ± 16.3	2.6 ± 0.53	3.7 ± 0.28	5.2 ± 0.68	53 ± 9.4	44 ± 7.9	1.3 ± .18
HEA-B	63 ± 6.7	2.4 ± 0.25	3.4 ± 0.19	5.7 ± 0.91	53 ± 3.8	66 ± 14.3[e]	1.2 ± .07
12 Months							
IB-C	73 ± 12.7	3.5 ± 0.15	2.4 ± 0.10	8.8 ± 3.1	92 ± 17.2	29 ± 5.2	1.1 ± .22
IB-B	82 ± 9.8	3.1 ± 0.37	2.3 ± 0.34	12.3 ± 1.8[c]	100 ± 21.5	34 ± 9.1	1.2 ± .31
EA-C	78 ± 7.8	3.9 ± 0.54	2.5 ± 0.18	8.9 ± 3.2	103 ± 28.2	33 ± 4.7	1.5 ± .38
EA-B	67 ± 17.4	3.8 ± 0.74	2.4 ± 0.36	12.4 ± 1.2	94 ± 43.1	26 ± 7.0	1.8 ± .52
HEA-C	75 ± 11.2	3.4 ± 0.39	2.4 ± 0.23	11.9 ± 0.76	100 ± 16.0	36 ± 9.0	1.6 ± .60
HEA-B	79 ± 14.6	4.0 ± 0.38[c]	2.3 ± 0.24	12.1 ± 1.58	104 ± 16.2	34 ± 6.4	1.3 ± .18
18 Months							
IB-C	81 ± 9.6	2.1 ± 0.30	3.8 ± 0.40	8.8 ± 3.0	179 ± 27.3	40 ± 8.8	0.87 ± .40
IB-B	76 ± 17.9	1.5 ± 0.46[e]	3.2 ± 0.52[d]	8.6 ± 2.3	175 ± 46.8	40 ± 15.6	0.82 ± .13
EA-C	78 ± 10.6	2.4 ± 0.25	4.0 ± 0.25	11.5 ± 2.3	225 ± 58.4	44 ± 4.7	0.52 ± .29
EA-B	87 ± 9.3	2.4 ± 0.34	4.0 ± 2.6	10.7 ± 1.6	231 ± 38.4	47 ± 15	0.96 ± .65
HEA-C	73 ± 17.8	2.1 ± 0.33	4.0 ± 0.33	10.7 ± 1.9	243 ± 91.6	62 ± 33.4	0.84 ± .33
HEA-B	73 ± 14.1	2.1 ± 0.25	3.9 ± 0.24	10.9 ± 1.5	248 ± 49.9	45 ± 15.5	1.15 ± .61
	(Mean ± S.D.)						

ND - Not Determined

[a] Serum glutamic-oxalacetic transaminase and glutamic-pyruvic transaminase (units/ml)

[b] Serum alkaline phosphatase (units/ml)

Significantly different from control: c, at $p<0.05$; d, at $p<0.025$; e, at $p<0.01$

TABLE VI

Serum Cholesterol, Triglyceride, Total Iron, and Total Iron-Binding Capacity In Rats Fed Treatment Diets for 18 Months

Treatment Group	Cholesterol (mg %)	Triglyceride (mg %)	Total Iron (mg %)	TIBC (mg %)
IB-C	144 ± 30.2	145 ± 61.4	141 ± 40.8	535 ± 58.6
IB-B	132 ± 28.6	138 ± 68.6	160 ± 34.4	552 ± 59.3
EA-C	175 ± 19.4	163 ± 54.8	139 ± 22.7	486 ± 44.2
EA-B	150 ± 28.7[a]	108 ± 20.9[b]	191 ± 48.2[b]	497 ± 48.6
HEA-C	160 ± 22.2	182 ± 48.5	139 ± 33.9	541 ± 78.6
HEA-B	168 ± 31.1	150 ± 30.8	135 ± 18.9	534 ± 62.6

(Mean ± S.D.)

Significantly different from control: a, at $p<0.05$; b, at $p<0.025$

TABLE VII

Tissue Enzyme Activities From Rats Fed Maillard Browned or Control Proteins for 18 Months

Treatment Group	Liver GOT (Units/mg Protein)	Small Intestine Mucosa Disaccharidase Activities[a]	
		Maltase	Sucrase
IB-C	965 ± 153.1	0.15 ± .020	0.16 ± .032
IB-B	1010 ± 146.4	0.15 ± .017	0.16 ± .030
EA-C	1196 ± 163.2	0.15 ± .005	0.15 ± .007
EA-B	1118 ± 90.0	0.13 ± .006[b]	0.13 ± .022
HEA-C	902 ± 208.6	0.12 ± .023	0.12 ± .023
HEA-B	921 ± 94.8	0.12 ± .029	0.12 ± .022
	(Mean ± S.D.)		

[a] μmoles substrate hydrolyzed/min/mg protein

[b] Significantly different from control at $p<0.01$

however, exhibited significantly lower maltase activity over the non-browned control group. A similar decrease in mucosal sucrase activity was noted, but this difference was not significant.

The observed decrease in disaccharidase activity is in agreement with earlier published reports (5,6,7,29). The mechanism of this inhibition is unclear. Lee, et al. (5) found that Amadori compounds were competitive inhibitors of maltase activity, while melanoidin compounds inhibited maltase in a noncompetitive manner. Gomyo (29) also found that the melanoidin compounds were noncompetitive inhibitors of sucrase, maltase, and lactase. He also observed these melanoidin compounds to adsorb irreversibly to the mucosa of the small intestine.

Again, the clinical significance of this effect is probably minimal. Although the maltase activity is decreased in rats fed the Maillard browned egg albumin, the activity is still above that found for rats fed either browned or control hydrolyzed egg albumin. In addition, no other clinical consequences of the lowered disaccharidase activity were detected in these animals, such as effects on weight gain or serum glucose levels.

Liver Glutamic-Oxalacetic Transaminase Activity The results of the determination of liver glutamic-oxalacetic transaminase activity are listed in Table VII. No significant difference was detected between browned and control groups for any of the treatment diets. This supports the serum data, which also showed no change in this enzyme between browned and control groups. Both the serum and liver data indicate the absence of any severe liver damage.

Urine Specific Gravity No significant change in the specific gravity of the urine was noted between browned and control animals after 18 months. Again, the results were within the range normally found in rats (28).

Histopathology

a) Six Months

Histopathological examination of tissues after six months of feeding revealed no significant lesions in any of the treatment groups. Some slight intracytoplasmic, basophilic pigment was detected in the liver hepatocytes but this was interpreted as being endoplasmic reticulum. In addition, both the control and browned-instant-breakfast-fed rats exhibited mild fatty changes in the liver, but still within normal ranges.

b) Twelve Months

After twelve months of feeding, two of the five rats fed browned egg albumin exhibited mild fatty metamorphosis. This was not detected in the control egg albumin group. Again, this fatty change was within a normal range. No significant lesions were detected in the instant breakfast or hydrolyzed egg albumin groups.

c) Eighteen Months

The feeding of browned or control diets for 18 months resulted in the appearance of a number of mild lesions. The incidence of these lesions, however, were equally distributed between the browned-protein-fed animals and their non-browned controls.

Essentially no liver fatty change was detected in animals fed the instant breakfast product. Animals fed the browned or control egg albumin showed mild fatty metamorphosis and animals fed the hydrolyzed egg albumin (browned and control) exhibited mild to severe fatty metamorphosis. Most of the kidneys from all groups showed some degree of proteinaceous casts in the lumen of the tubules. The spleens of all animals contained some megakaryocytes and significant amounts of what appeared to be ceroid pigment. When the tissues were stained with Perl's Prussian Blue stain, all spleens exhibited a large number of hemosiderin-laden macrophages. None of the other tissues contained any significant amount of hemosiderin.

In summary, no significant lesions due to the feeding of Maillard browned proteins could be detected through 18 months of feeding. There was a slightly higher incidence of mild fatty change detected in animals fed browned egg albumin for 12 months, but this was still within a normally expected range and its pathological significance is not supported by clinical biochemical changes. All of the lesions detected in the tissues of rats fed for 18 months (fatty change in the liver, proteinaceous casts, and hemosiderin in the spleen) were equally distributed between brown and control animals. In addition, all of these lesions have been well documented as being age-related in the rat (28,31).

DISCUSSION

The results of the long-term feeding study lead to the conclusion that Maillard browned proteins are not toxic to rats. In reaching this conclusion, consideration is given not only to the data generated by the present study, but also to a reexamination of earlier data that suggested a toxic effect.

Sgarbieri, et al. (9) found that when they supplemented Maillard browned egg albumin with those amino acids destroyed, the nutritional value could be restored to only about 84% of the non-browned control diet. Rogers and Harper (32), however, reported that free amino acids could not support a rate of growth equal to intact proteins. These researchers found that it required 22% free amino acids (each at 1.5 times the requirement for the essential amino acids, plus a mixture of nonessential amino acids) plus 12% casein to equal the growth rate of rats receiving a 20% casein diet. This represents an additional 14% of the diet as amino acid nitrogen required in the amino acid-supplemented group to match the growth rate of the intact-casein-fed animals

An interesting comparison can also be made between the study of Lee, et al. (26) and Kimiagar, et al. (7). Lee, et al. (26) fed a diet containing 71% browned apricot powder and 13% non-browned casein. The control diet contained 10% casein and 74% non-browned apricot powder. While a slight growth depression was observed in the browned-fed group, weight gain for both the control and browned groups was very good. After three months of feeding, the browned-apricot-fed animals exhibited a slight increase in serum GPT, decreased serum glucose, and urine specific gravity, and no significant histopathological changes. Kimiagar, et al. (7) fed a diet containing 10% browned egg albumin and a control of equal protein quality (PER = 1.10) containing 5% egg albumin plus 5% nonessential amino acids. This study resulted in very poor weight gain for both groups, although the browned egg albumin group was still significantly lower than the control. In addition, serum glucose levels were elevated in the browned group, and the specific gravity of the urine was increased. Thus, although both studies found significant changes between browned-protein-fed animals and the non-browned controls, the direction of these changes was opposite in the two studies.

Finally, all the observed clinical and histopathological findings of the study by Kimiagar, et al. (7) can be attributed to diet and nutrition. A study by Schwartz, et al. (33) found that restriction of food intake to rats resulted in depressed weight gain, elevated serum glucose, urea nitrogen, glutamic-pyruvic transaminase, and alkaline phosphatase, increased ratios of internal organs to body weight, increase in liver fatty metamorphosis, and increase in the deposition of hemosiderin in the liver and spleen. All of these changes are identical to the results reported by Kimiagar, et al. (7) for rats fed Maillard browned protein. However, it should be noted that 10% of 10-day-browned egg albumin was used in the test diet by Kimiagar, et al. (7), whereas in the present study 3% of 40-day-browned egg albumin and other type of browned proteins were used in the test diets. Based on the chemical kinetics of Maillard browning reaction, a three-stage development mechanism was proposed by Hodge (34): 1) Initial stage (sugar-amine condensation, Amadori rearrangement), 2) intermediate stage (sugar dehydration, sugar fragmentation, amino acid degradation) and 3) final stage (aldol condensation, aldehyde-amine polymerization, formation of heterocyclic nitrogen compounds). The qualitative and quantitative differences of the chemicals formed in different browned proteins in the test diets may cause the drastic different effects in long-term feeding. It may be useful in the future to isolate the principal products of various Maillard reactions and conduct acute and chronic feeding studies with those purified compounds.

In conclusion, it appears that all reported anthropometric, clinical biochemical, and histopathological changes resulting from the feeding of Maillard browned proteins in the present study can be attributed to nutritional and/or dietary factors.

Acknowledgment

This research was supported in part by research grant AER77-10200 from the National Science Foundation and the Rhode Island Agricultural Experiment Station.

Rhode Island Agricultural Experiment Station Contribution No. 2083.

Literature cited

1. Mauron, J.; Motter, E.; Engli, R.H. Arch. Biochem. Biophys. 1955, 59, 433.
2. Nesheim, M.C.; Carpenter, K.J. Br. J. Nutr. 1967, 21, 399.
3. Tanaka, M.: Lee, T-C; Chichester, C.O. Agric. Biol. Chem. 1975, 39, 863.
4. Amaya, J.; Lee, T-C.; Chichester, C.O. Nutr. Rep. Internat. 1976, 14, 229.
5. Lee, C.M.: Chichester, C.O.; Lee, T-C. J. Agric. Food Chem. 1980, 25, 775.
6. Tanaka, M.; Kimiagar, M.; Lee, T-C.; Chichester, C.O. "Protein Crosslinking-B. Nutritional and Medical Consequences."; M. Friedman, Ed.; Plenum Publishing Co.: New York., 1977.
7. Kimiagar, M.; Lee, T-C.; Chichester, C.O. J. Agric. Food Chem. 1980, 28, 150.
8. Rao, M.N.; Screenivas, H.; Swaminathan, M.; Carpenter, K.J.; Morgan, C.B. J. Sci. Food Agric. 1963, 14, 554.
9. Sgarbieri, V.C.; Amaya, J.; Tanaka, M.; Chichester, C.O. J. Nutr. 1973, 103, 1731.
10. "Official Methods of Analysis of the Association of Official Analytical Chemists.", 12th Ed.; W. Horwitz, Ed.; The Association of Official Analytical Chemists: Washington, D.C. 1975.
11. Drabkin, D.L.; Austin, J.H. J. Biol. Chem. 1935, 112, 51.
12. Amador, E.; Dorfman, L.E.; Wacker, W.E.C. Clin. Chem. 1963, 9, 391.
13. Wroblewski, F.; LaDue, J.S. Proc. Soc. Exp. Biol. Med. 1956, 91, 569.
14. Karmen, A. J. Clin. Invest. 1955, 34, 131.
15. Doumas, B.T.; Biggs, H.G. "Standard Methods of Clinical Chemistry," Vol 7; G.R. Cooper, Ed; Academic Press: New York, 1972.
16. Young, D.S.; Pestaner, L.C.; Gibberman, V. Clin. Chem. 1975, 21, 10.
17. Crocker, C.L. Am. J. Med. Technol. 1967, 33, 361.
18. Persijn, J.P.; Van Der Stik, W.; Riethorst, A. Clin. Chim. Acta. 1971, 35, 91.
19. Bercolo, G; David, H. Clin. Chem. 1973, 19, 476.
20. Wybenga, D.R.; Pileggi, V.J.; Dirstine, P.H.; Digiorgio, J. Clin. Chem. 1970, 16, 1980.
21. Dahlqvist, A. Anal. Biochem., 1964, 7, 18.
22. Reitman, S.; Frankel, S. Am. J. Clin. Path. 1957, 28, 56.
23. Gornall, A.G.; Bardavill, C.J.; David, M.M. J. Biol. Chem. 1949, 177, 757.
24. Caster, W.O.; Poncelet, J.; Simon, A.B.; Armstrong, W.D. Proc. Soc. Exp. Biol. Med. 1956, 91, 112.
25. Waynforth, H.B. "Experimental and Surgical Technique in the Rat"; Academic Press: London, 1980.
26. Lee, C.M.; Lee, T-C.; Chichester, C.O. Internat. Cong. Food Sci. Technol. 1974, 1, 587.

27. Mitruka, B.M.; Rawnsley, H.M. "Clinical Biochemical and Hematological Reference Values in Normal Experimental Animals"; Masson Publishing Co.: New York, 1977.
28. Baker, H.J.; Lindsey, J.R.; Weisbroth, S.H. "The Laboratory Rat, Volume I. Biology and Diseases"; Academic Press: New York, 1979.
29. Gomyo, T. Prog. Food Nutr. Sci. 1981, 5, 223.
30. Tietz, N.W. "Fundamentals of Clinical Chemistry" 2nd ed.; W.B. Saunders Co.: Philadelphia, 1976.
31. Cotchin, E.; Roe, F.J.C. "Pathology of Laboratory Rats and Mice"; Blackwell Scientific Publications: Oxford, 1967.
32. Rogers, Q.R.; Harper, A.E. J. Nutr., 1965, 87, 267-273.
33. Schwartz, E.; Turnaben, J.A.; Boxill, G.C. Toxicol. Appl. Pharmacol. 1973, 25, 515.
34. Hodge, J. E. 1953. J. Agr. Food Chem. 1, 928-943.

RECEIVED January 20, 1983

Mutagens in Cooked Foods: Possible Consequences of the Maillard Reaction

WILLIAM BARNES, NEIL E. SPINGARN,[1] CLAIRE GARVIE-GOULD, LORETTO L. VUOLO, Y. Y. WANG,[2] and JOHN H. WEISBURGER
American Health Foundation, Naylor Dana Institute for Disease Prevention, Valhalla, NY 10595

Fried or broiled meat contains powerful mutagens not present in cooked meat. Most of the mutagenic activity can be partitioned into a basic fraction and the structure of several of these mutagens is now known. All the mutagens are N-heterocyclic primary amines and are carboline or imidazoquinoline derivatives. Another common feature of these compounds is their requirement for metabolic activation and their ability to act as extremely potent frameshift mutagens in bacteria. A number of these compounds can induce genetic damage in mammalian cells in vitro. From the limited data available at present, some of these food components may be weak carcinogens. The mode of formation of such mutagens may involve Maillard reactions, because the crude product from model browning systems exhibits the same mutagenic characteristics as those in cooked meat. Experiments on the inhibition of the formation of mutagens in fried meat have shown that the addition of soy protein or antioxidants inhibit the formation of such mutagens. This area is important since it can be postulated that the genotoxic carcinogens for important human diseases such as colon and breast cancer may be the result of ingestion of such compounds.

[1] Current address: West Coast Technical Services, 17605 Fabrica Way, Suite D, Cerritos, CA 90701
[2] Current address: SRI International, 333 Ravenswood Avenue, Menlo Park, CA 94025

0097-6156/83/0215-0485$06.00/0

The majority, or as much as 70-90%, of human cancers have been associated with environmental causes (1,2) . Our environment is complex. Cancer causes are often misunderstood and misconstrued as consisting primarily of ubiquitous chemicals due to modern technology and industrial development. It is true that a number of food additives, pesticides, insecticides and industrial chemicals introduced commercially in the last 40 years have exhibited carcinogenic properties in animal models (3) . However, most of the main human cancers in the Western world do not stem from such chemical contaminants. It is, therefore, important to identify the actual causes of cancer in developing an effective basis for cancer prevention.

Table I. Causes of Human Cancer*

Type	% of total
I. Occupational cancers - Varied organs	1 - 5
II. Cryptogenic cancers - Lymphomas, leukemias, sarcomas (cervix?), (virus?)	10 - 15
III. Life-style	
A. Tobacco-related	
Lung, pancreas, bladder, kidney	21
B. Diet-related	
1. Nitrate-nitrite, low vitamin C, mycotoxin, stomach, liver	5
2. High fat, low fiber, broiled or fried foods	
Large bowel, pancreas, breast, prostate	45
C. Multi-factorial	
1. Tobacco and alcohol	
Oral cavity, esophagus	5
2. Tobacco-asbestos, tobacco-mining, tobacco-uranium-radium	
Lung, respiratory tract	5
IV. Iatrogenic-radiation, drugs	1

* Calculated from the 1978 incidence figures published by the American Cancer Society.

Data from Ref. 4.

The major causes of human cancer are shown in Table 1, together with an estimate of their relative significance (4). It should be noted that occupational cancers due to specific chemicals account for no more than 5% of the total; life-style factors are associated with most of the remainder. A numerically large group of these life-style cancers are those of the large bowel, pancreas, breast and prostate. There is considerable epidemiological and experimental evidence that diets high in fat and low in fiber are risk factors for these diseases. These elements appear to operate mainly through non-genotoxic, promoting, or inhibiting mechanisms. It is our working hypothesis that broiled and fried foods may be key contributory factors and provide sources of the genotoxic carcinogens associated with the nutritionally related cancers (5).

Age-adjusted mortality rates for cancer at different sites in U.S. males are shown in Figure 1. It is evident from these data that cancer of the lung has steadily increased in males since 1935 corresponding to an approximate 20-year lag period after cigarette smoking first became popular. In marked contrast to this, the incidence of cancers of the colon and rectum, breast and prostate have increased only at a low rate since 1940. We logically conclude that industrial pollution, food additives, synthetic chemical products, etc., the levels of which have increased dramatically in our environment, are not associated with the development of these three cancer types. On the other hand, the mode of cooking has remained similar over this period, and thus is consonant with the cancer incidence data.

Another element which underpins our working hypothesis has been the discovery of very potent bacterial mutagens in pyrolysates of amino acids and the charred surfaces of fish and meat (6,7,8) . Commoner and we (9,10) noted that mutagens were present in the basic fraction of ground beef fried at the lower temperatures typical of home-cooking. Since that time, the structure of a number of these mutagenic compounds has been determined (11,12,13). Figure 2 shows that two of the compounds, 3-amino-1,4-dimethyl-5*H*-pyrido[4,3*b*]indole and 3-amino-1-methyl-5*H*-pyrido[4,3*b*]indole, (Trp-P-1 and Trp-P-2) are structurally similar to the potent carcinogens 2-acetylaminofluorene (2AAF) and 2-aminofluorene (2AF), while 2-Amino-3-methylimidazo [4,5-*f*]quinoline (IQ) bears a structural and steric resemblance to the potent colon carcinogen,3,2'-dimethyl -4-aminobiphenyl.

Formation of Mutagens in Cooked Meat

The first intimation that mutagens could be formed from natural food substances came from the laboratory of Sugimura, where it was found that mutagenic activity was found in smoke condensates or in DMSO extracts of the charred surface of fish and meat. This activity could not be accounted for by the amounts of BaP and PAH known to be present. Extracts of pyrolysates of various proteins and amino acids were also mutagenic (14).

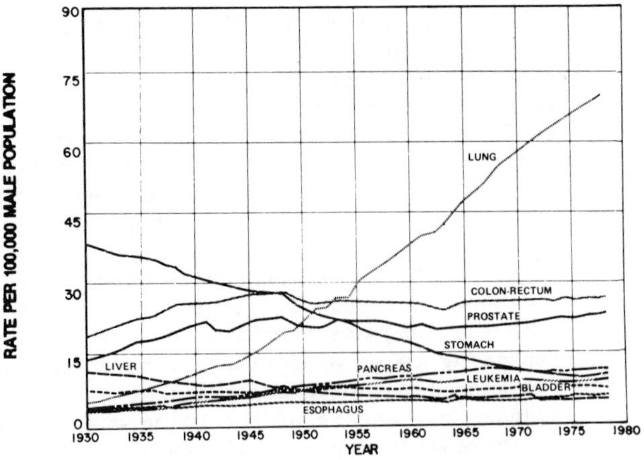

Figure 1. Age-adjusted cancer death rates for selected sites, males, United States, 1930 to 1978. Standardized on the age distribution of the 1970 U.S. census population. Sources of data: National Vital Statistics Division and U.S. Bureau of the Census. (Reproduced with permission from Ref. 70. Copyright 1982, American Cancer Society.)

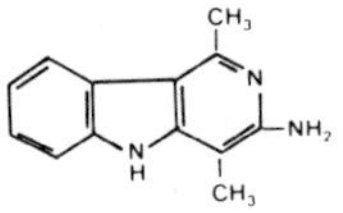

I
3-Amino-1,4-dimethyl-5H-pyrido[4,3-b]indole

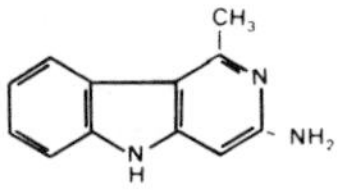

II
3-Amino-1-methyl-5H-pyrido-[4,3-b]indole

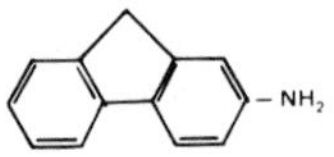

III
2-Aminofluorene

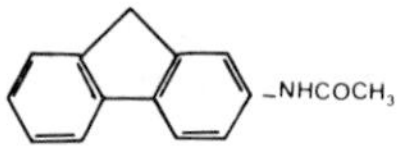

IV
2-Acetylaminofluorene

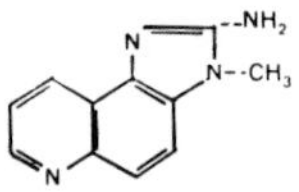

V
2-Amino-3-methylimidazo-[4,5-f]quinoline

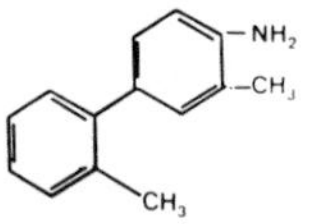

VI
3,2'-Dimethyl-4-amino-biphenyl

Figure 2. Structures of some known carcinogens and food mutagens.

Commoner et al (9) then reported that mutagenic activity was also found in basic fractions of a commercial beef extract and fried hamburgers. Formation of the mutagens was dependent upon cooking time and that they were chromatographically distinguishable from both BaP and amino acid pyrolysates. Perhaps the most important result of this work was the demonstration that normal cooking temperatures were sufficient for mutagen formation and the suggestion that browning reactions might be involved (9,15).

A number of papers subsequently investigated the relationship between mutagen formation, cooking time, and temperature in ground beef (10,16,17,18). It was found that mutagen formation is a complex function of cooking time and temperature. Representative results from Spingarn and Weisburger (10) are shown in Figure 3, where mutagen formation and temperature at the surface of the patty are shown as a function of time. A lag period of approximately five min precedes the appearance of mutagenic activity. This is probably due to the temperature plateau at 100°C as the water content of the meat is reduced. Pariza (18) also showed that meat can be cooked to the "medium well-done" stage without the formation of mutagens if cooking temperatures are kept low. Mode of cooking is also important. Frying is much more effective in mutagen formation than broiling (9,19). Microwave cooking does not result in any extractable mutagenic activity (19). Volatilized mutagens are found in smoke condensates of cooking beef (6,7,20,21), but there is disagreement about how much of the total mutagenic activity is volatilized, and uncertainty as to whether these mutagens differ from those remaining in the meat.

Mutagenic activity is also present in broiled herring, mackerel and sardine (11,12,22). In part, these mutagens were later identified as Trp-P-1 and Trp-P-2 (12) and IQ and 2-amino-3,4-dimethylimidazo [4,5-*f*]quinoline (Me-IQ)(13). In addition, Bjeldanes et al. (23) have also reported mutagen formation in pan-fried fish, such as rock cod, sole, halibut, trout, salmon and red snapper. However, the mutagenic activity is low compared to that in beef.

A limited amount of information is available on red meats other than ground beef (24) and is summarized in Table II. The relationship between mutagen formation and cooking time and temperature extends to these foods as well. Baking and broiling, as well as frying, are effective in producing mutagenic activity.

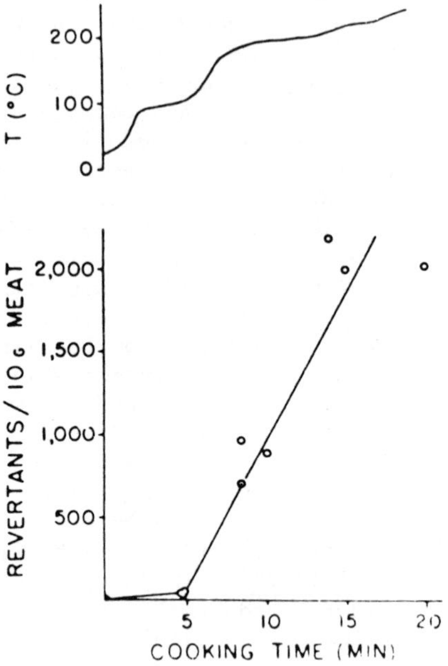

Figure 3. Mutagenic activity on TA98 with S9 mix of the basic organic extract of fried beef patties. (10).

The ordinate shows nonspontaneous his⁺ revertants/initial 10 g of meat. Each point is calculated from the slope of the linear portion of a dose–response curve. Temperature was recorded at the surface of the patty. Average beef patty weight was 92 ± 3 g.

Table II Mutagen Formation in Meat and Fowl with various cooking Procedures

Meat	Mode of Cooking	TA1538 revertants/ 100gE *
Beefsteak	Broiled 9 min/side at 300°C	1,500
	Broiled 11 min/side at 300°C	10,000
Ground beef	Fried, 300°C	30,000
Ham	Fried at 200°C	5,400
	Fried at 245°C	22,000
Pork chop	Fried at 200°C	4,600
	Fried at 245°C	12,000
	Fried at 280°C	23,000
Sausage	Fried at 200°C	9,000
	Fried at 245°C	18,000
Chicken (white meat)	Deep fried, 12 min. at 101°C	1,000
	Baked, 50 min. at 190°C	320
	Broiled, 17 min/side at 274°C	16,000
Lamb chop	Oven-broiled, 4 min/side at 225°C	390
	Pan-broiled, 5 min/side at 210°C	14,000

* revertant numbers are standardized by extrapolating dose-response data to a volume of extract equivalent to 100 g starting weight. This amount is referred to as 100 gram-equivalents (100gE).

Data adapted from Reference 24.

The chemical structure of a number of the mutagens found in cooked foods is now known. Two very potent mutagens in beef are IQ and Me-IQx (25,26,27). Yamaguchi et al. (28) have reported the presence of Trp-P-1 in broiled beef and 2-amino-α-carboline (AαC) and 2-amino-3-methyl-α-carboline have been found in grilled beef by Matsumoto, Yoshida and Tomita (29).Although not mutagenic themselves, harman and norharman enhance the activity of Trp-P-1 and other mutagenic carbolines, and also are present in cooked beef (30). Table III shows the information available on mutagens of this kind in beef.

A number of these compounds have been identified in other foods as well. Trp-P-1 and Trp-P-2 are found in broiled sardines, chicken and horse mackerel (30). 2-Amino-dipyrido[1,2-a:3',2'-d]imidazole (Glu-P-2) was identified in broiled squid. (22). AαC and Me-AαC are found in grilled chicken and grilled mushroom (29). IQ and Me-IQ also occur in broiled sardines (13).

Genotoxicity of N-Heterocycles from Proteinaceous Foods

Genotoxic compounds are those which are capable of causing genetic damage by interaction with DNA. Examples of genotoxic endpoints are chromosome aberrations (CA), sister chromatid ex-

Table III. Mutagens Found in Beef

COMPOUND	STRUCTURE	MUTAGENICITY TA 98 REVERT/ G	COOKING METHOD	CONTENT G/KG
2-AMINO-3-METHYL-IMIDAZO{4,5-F} QUINOLINE (IQ)	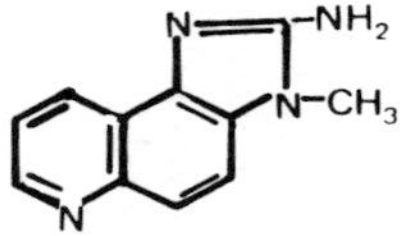	433,000	FRIED WELL-	0.9
2-AMINO-3,8-DI-METHYL IMIDAZO {4,5F} QUINOLINE (MeIQ$_X$)	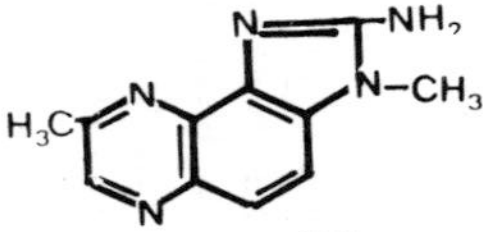	661,000	-	-
3-AMINO-1,4-DI METHYL 5H-PYRIDO {4,3-B}INDOLE (Trp-P-1)	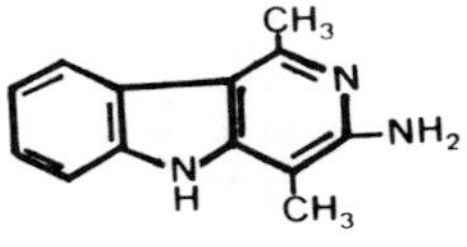	39,000	BROILED WELL-DONE	53
2-AMINO-α-CARBOLINE (A$_\alpha$C)	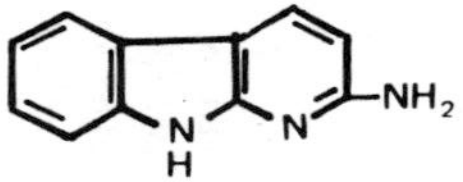	300	GRILLED OVER HIGH GAS FLAME (PYROLYSED?)	650.8
2-AMINO-3-METHYL-α-CARBOLINE(MeA C)	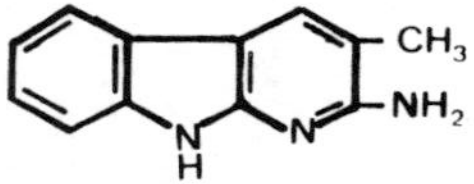	200	"	63.5
β-CARBOLINE (NORHARMAN)	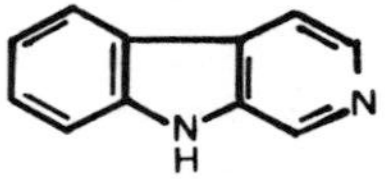	0	CHARRED PARTS	-
1-METHYL-β-CARBOLINE (HARMAN)	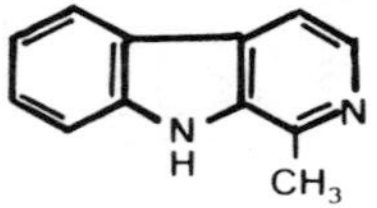	0	"	-

changes (SCE), mutagenicity and DNA repair. There is a plethora of information on the bacterial mutagenicity of these compounds and their literature is wellknown and accessible (30,31,32). Therefore, other genetic end-points only will be discussed here.

Chromosome aberrations have been reported in Syrian hamster embryo (SHE) cells after treatment with a DMSO extract of tryptophan pyrolysate. At a dose of 30 μg/ml, 69 exchanges and 29 chromosome or chromatid breaks were observed among 200 metaphases. Gaps and minutes were noted as well (33). Sasaki, *et al.* (34) also observed chromosome aberrations in human and Chinese hamster cell lines after exposure to Trp-P-1 and Trp-P-2.

Tryptophan pyrolysates have also been found to induce sister chromatid exchanges in cultured mammalian cell lines. Tohda *et al.* (35) tested Trp-P-1, Trp-P-2, 2-Amino-6-methyldipyrido[1,2-α:3',2'-*d*]imidazole (Glu-P-1), and AαC in a permanent cell line of human lymphoblastoid cells. All four compounds were active at a concentration of 10^{-5}M, but inclusion of S9 mix was necessary. Similar results are reported by Sasaki, *et al.* (34) in 2 human and Chinese hamster cell lines.

Hatch *et al.* (36) compared the activity of IQ and Trp-P-2 in assays for mutagenicity, CA, and SCE in repair-deficient and repair-proficient CHO cells. Trp-P-2 was positive for all end-points in repair-proficient cells at 1 μg/ml, and repair-deficient cells were even more sensitive. IQ, on the other hand, was negative for all endpoints in repair-proficient cells at concentrations in excess of 300 μg/ml. Repair-deficient cells did exhibit a positive response for mutagenicity and SCE in the range from 15-75 μg/ml of IQ, but an increased frequency of chromosome aberrations was not observed.

Although not necessarily a genotoxic event, in vitro cell transformation is included in this section. The basic fraction of a tryptophan pyrolysate and synthetic Trp-P-1 and Trp-P-2 were tested for transforming activity in Syrian golden hamster embryo cells (37). The percent of morphologically transformed colonies was 1.5, 1.3 and 0.54 for Trp-P-2, Trp-P-1 and 3-methylcholanthrene (3MC), respectively. The optimum concentrations for Trp-P-1 and Trp-P-2 were 0.5 μg/ml versus 1.0 μg/ml for 3MC. No morphologically transformed colonies were seen in control cultures. Cells transformed in this way also exhibited the ability to grow in soft agar and formed anaplastic fibrosarcomas when transplanted into the cheek pouches of four-week old male hamsters (38). Tsuda *et al.* (33) also demonstrated morphological transformation in primary cultures of Syrian hamster embryo cells by crude tryptophan pyrolysates. In this experiment, a concentration of 50 μg/ml of pyrolysate induced 1.8% transformed colonies. The control transformation frequency was 0.028%.

Carcinogenic Effects of N-Heterocycles from Proteinaceous Foods

Carcinogenicity data are presently rather limited because the compounds concerned have been so recently discovered. Nonetheless, some information is becoming available and more can be expected in the next several years, as several bioassay programs in the U.S.and Japan are completed. Matsukura et al. (39) have tested Trp-P-1 and Trp-P-2 in CDF1 mice (BALB/cAnNxDBA/2N) in a lifetime study. Initially, forty mice of each sex were fed 200 ppm of either Trp-P-1 or Trp-P-2 or a basal CE-2 diet for 621 days. There was a large incidence of hepatocarcinomas, and the effect is sex specific (Table IV). In male mice treated with Trp-P-2, there is also a slight, but statistically significant, excess of pulmonary adenocarcinomas (39).

The basic fraction of tryptophan pyrolysate was also tested by oral administration to Wistar rats over 2 years (40). The most advanced lesions obtained at the higher (but not the lower) dose were neoplastic nodules in the liver, an intermediate step in the process of hepatic carcinogenesis. The authors state that, historically, pre-neoplastic lesions in rats of this strain have never been seen in their laboratory without administration of carcinogens. Trp-P-2, as a pure compound, has also been tested in ACI rats in a lifetime (870 days) study (41). In this study, 10 male and 9 female animals survived more than 400 days on a diet containing 0.01% Trp-P-2. Neoplastic nodules were again seen in the livers of treated females, but none occurred in the males of the experimental group or any of the control animals.

A number of limited in vivo bioassays have been made for carcinogenicity. After subcutaneous injections of Trp-P-1 into Syrian golden hamsters or Fischer rats, sarcomas were observed in 3 of 8 surviving hamsters and 5 of 20 surviving rats. No tumors were observed in controls or in animals injected with Trp-P-2 (42). Skin painting experiments in female ICR mice were likewise negative for Trp-P-1 and Trp-P-2 (42). The induction of enzyme-altered foci in rat liver has also been studied in Sprague-Dawley males (42) and in Fisher 344 males (43). Trp-P-1, Glu-P-1 and Glu-P-2 were positive ($p<0.001$) while Trp-P-2 was negative.

Trp-P-1 and Trp-P-2 are certainly the best studied of the food mutagens. Both compounds are potent carcinogens in mice. However, this may reflect the sensitivity of mice to liver tumor induction as much as the carcinogenic potency of Trp-P-1 and Trp-P-2. This feeling is borne out by the generally weak response of two rat strains in feeding studies. Indications of precancerous changes were found in the liver, more so in females, but actual tumors were not. It seems justified to conclude that Trp-P-1 and Trp-P-2 are not strong carcinogens. However, because of the small experimental sizes, it is not possible to give a definitive answer as to actual potency. It would be advantageous

Table IV. Incidence of Hepatic Tumors in Mice Fed on Diets with Trp-P-1 or Trp-P-2 (200 ppm) for up to 621 Days

			Number of mice with hepatic tumors				
			Hepatocellulor tumor				
Treat-ment	Sex	N*	Adenoma	Carcinoma**	Heman-gioma	Total	P***
None	M	25	0	0	1	1 (4)+	
	F	24	0	0	0	0	
Trp-P-1	M	24	1	4	0	5(21)	<.179
	F	26	2	14	0	16(62)	<.001
Trp-P-2	M	25	1	3	0	4(16)	<.348
	F	24	0	22 (2)++	0	22(92)	<.001

* Number of mice surviving on day 402, when the first hepatic tumor was found.

** Mice with both hepatocellular adenoma and hepatocellular carcinoma are included under hepatocellular carcinoma.

*** Statistical significance of the difference in incidence of hepatic tumors between control and Trp-P-1 or Trp-P-2 groups by x^2 test.

\+ Numbers in parentheses are percentages.

++ Number in parentheses is number of mice with pulmonary metastases of hepatocellular carcinomas.

to have more information on carcinogenicity in other species. It is possible, for example, that differences observed in the response of mice and rats to Trp-P-1 may stem from the relatively low N-hydroxylating activity of rats (44,45), which seems to be necessary for the activation of heterocyclic amines. The possibility should also be borne in mind that these two compounds may be incomplete carcinogens; that is, strong initiators with little promoting activity. If this were true, other dietary components might be required to enhance the activity. Future experiments should test this hypothesis.

Modifying Factors in Mutagen Formation and Activity

A number of factors may influence the amount of mutagenic activity observed in extracts of cooked foods. These factors may alter the amount of mutagen which is formed during cooking or which would be available in vivo to bind to cellular DNA. However, care should be exercised to assure that modifying factors are not artefactual by way of changing the response of the in vitro test system in a manner unlikely to be duplicated in vivo.

Spingarn et al. (46) showed that the amount of fat present in ground beef was an important variable in mutagen formation. As beef tallow is added into lean ground meat and fried, the amount of mutagenic activity in the basic extracts peaks and thereafter declines (Figure 4). Figure 5 demonstrates that additives such as soy protein, butylated hydroxyanisole (BHA) and chlorogenic acid also decrease the amount of mutagenic activity observed (47). This decrease is probably not attributable to an inhibition of S9 activation or of mutagenesis by antioxidant because BHA and chlorogenic acid are partitioned into different fractions during the extraction procedure. It has also been reported that fatty acids, especially oleic acid, inhibit mutagenicity when added to samples of basic beef extract for Ames testing (48). Other substances known to decrease the in vitro mutagenicity of pyrolysis products are hemin, chlorophyllin, chlorophyll and aqueous extracts from vegetables such as cabbage, radish, turnip and ginger (49,50-53).

Several new facets of the chemical and physical behavior of mutagens isolated from food and pyrolysates have been noted recently. Trp-P-1, Trp-P-2, and Glu-P-1 are rapidly deaminated upon incubation with nitrite at acid pH (54). At pH 1.6, in 50 µM nitrite, the half lifetime of Trp-P-1 and Trp-P-2 is approximately 100 min, but less than 5 min for Glu-P-1. AαC is also deaminated in 1 mM sodium nitrite at pH less than four with the difference that longer incubation, for 1.5 h, leads to the formation of a directly mutagenic nitroso derivative (55). These reaction conditions approach those in the stomach (pH 1-2, 0-10 µM nitrite), but careful kinetic studies in vivo will be required

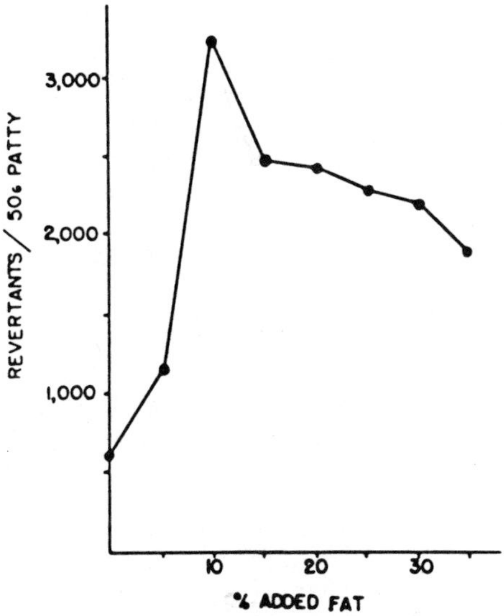

Figure 4. Mutagenicity of fried beef patties to TA98 with added S9 as a function of percentage fat in the beef (Ref. 46).

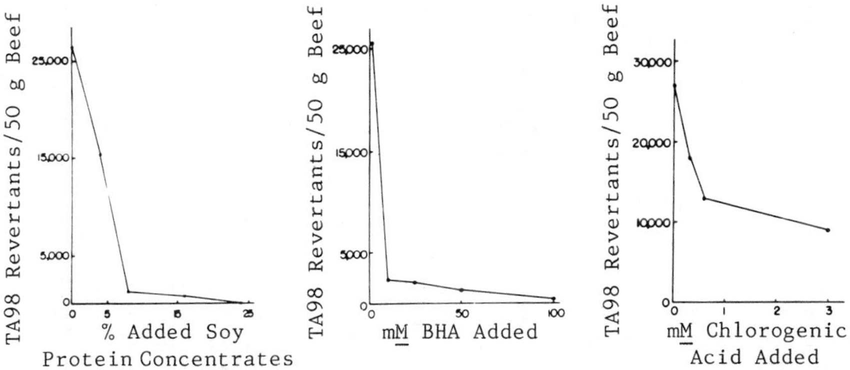

Figure 5. Reduction in observed mutagenic activity of ground beef basic extracts from patties cooked with several additives. Extracts were tested with TA98 and S9 mix (Ref. 47).

to assess the relevance of these observations. It is interesting that the imidazoquinolines are not sensitive to deamination although they are rapidly oxidized by hypochlorite (56).

Harman and norharman are β-carbolines that are found in charred parts of fish and beef (30). Although not mutagenic themselves, they have been found to increase the mutagenic activity of benzo(a)pyrene, and the azo dye, Yellow OB, and have, therefore, been termed co-mutagens (57). They seem to be particularly effective with aromatic amines such as aniline and o-toluidine, or Trp-P-1 and Trp-P-2. Inasmuch as the mutagens now being found in cooked foods are all N-heterocycles, it follows that the mutagenicity of a crude extract may not be the simple sum of all the mutagens present in it.

Genotoxic Effects of Maillard Reaction Products

Model browning systems can result in the production of substances which are mutagenic in the Ames test with S9. When arabinose, 2-deoxyglucose, galactose, glucose, rhamnose or xylose are refluxed with aqueous ammonium hydroxide, substances which revert Ames strain TA98 are produced. In another model browning experiment the amino acids, rather than the sugars, were varied (58). Equimolar amounts (0.01M) of fructose or glucose and amino acid wat a pH value of either 7 or 10, were autoclaved for 1 h at 121°C and tested for mutagenicity in the Ames strain TA100 without S9 (Fig. 6). These Maillard products also exhibited clastogenic activity in CHO cells. Shibamoto, Nishimura and Mihara also found that browning reaction products of maltol and ammonia were positive in TA98 with S9 (59). Aeschbacher *et al.* (60) found no bacterial mutagenicity in an arginine-glucose browning system.

The genetic activity of pure synthetic Maillard products has also been studied. 2-Methylpyrazine, 2-ethylpyrazine, 2,5-dimethylpyrazine and 2,6-dimethylpyrazine all induced a significant increase in chromosome aberrations in Chinese hamster ovary cells (61). Furfural, furfuryl alcohol, 5-methylfurfural, 2-methylfuran, 2,5-dimethylfuran, and 2-furyl methyl ketone also caused chromosome aberrations in CHO cells (62). The significance of these results requires investigation, since all of these compounds, when tested simultaneously in the Ames assay, were negative. This agrees with the findings of Aeschbacher *et al.* (60) on 2,5-dimethylpyrazine and 5-hydroxymethylfurfural. It appears that fried food mutagens may be products of advanced Maillard reactions, and that these simple heterocycles, if they are involved at all in mutagen formation, participate only as intermediates.

Caramel and caramelized sugars have also been investigated. Caramelized sucrose, glucose, mannose, arabinose, maltose and fructose induce high frequencies of chromosome aberrations (63),

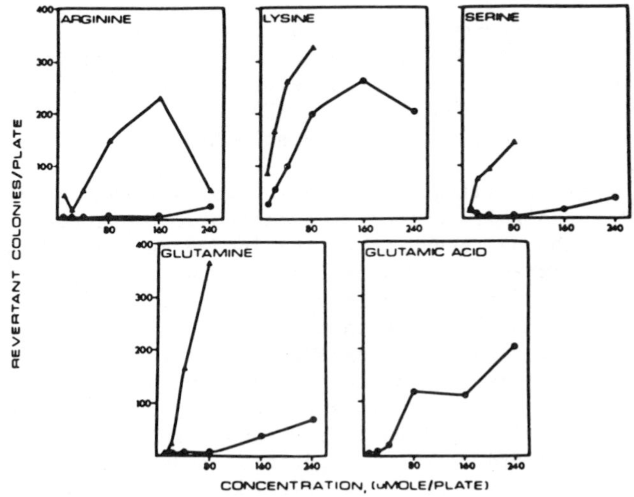

Figure 6. Mutagenic activity of fructose–amino acid mixtures in Ames strain TA100. (Reproduced with permission from Ref. 58. Copyright 1981, Institute of Food Technologists.)

but here also the positive result is difficult to interpret in the absence of structural chemical data since caramelized sugars are negative in Ames assays (60,64).

As would be expected from the results of studies on model systems, it has been found that foods with a high starch or sugar content may form genotoxic substances, but at a much lower level than meats or fish. Spingarn et al. (65) showed that several common foods, in addition to beef, contained mutagens active for TA98 in the presence of S9 (Table V) Pariza et al. (66) found mutagenic activity in basic fractions of chicken broth, beef broth, rice cereal, bread crust, crackers, corn flakes, toast and cookies.

Implications

It is most interesting to find not only that mutagens are present in many common cooked foods, but that in most cases they are some of the most potent mutagens so far identified. It is important to realize that, for a number of reasons, there is no way to translate this finding into an estimate of human risk at this time.

The severest limitation at this time is a lack of reliable data about the significance of the genotoxic properties of those compounds which have been identified in foodstuffs. With the exception of Trp-P-1 and Trp-P-2, hardly any work has been done on the activity of these bacterial mutagens in other in vitro systems. It is now accepted that a battery of such short-term assays is needed when screening for potential genotoxins and the availability of this information would do much to place the bacterial mutagenicity data in a clearer light (67). The same cautionary note also applies in any consideration of the clastogenicity of Maillard reaction products. Indeed, chromosomal aberration is a very complex genetic end-point and may be induced by phenomena other than those which signify possible carcinogenic risk. Thus, interpretation of such results awaits further testing for genotoxicity in other well validated assays. We also need to know more about the metabolism and pharmacokinetics of these compounds in vivo.

Long-term chronic bioassays for carcinogenicity will be the most relevant source of information for future risk assessment and such studies are now under way. However, the number of animals involved in these experiments needs to be considered in the evaluation. Data so far suggest that powerful bacterial mutagens of this type may not be strong carcinogens, and future experiments should be designed with this possibility in mind. For example, it is extremely important, when considering the data on carcinogenicity of tryptophan pyrolysates in rats, to determine the minimum increase in tumor incidence detectable by the experiments. Classifying results as "weakly positive" or "negative" is

Table V. Mutagen Formation in a Variety of Foods[a]

Food	Sample	Cooking procedure	Cooking time(min)	Revertants/ sample	Revertants/ m^2
White bread	1 slice	Broiling	6	205	18,600
Pumpernickel bread	1 slice	Broiling	12	945	96,500
Biscuit	1 each	Baking	20	735	N.C.[b]
Pancake	1 each	Frying	4	2,500	153,000
Potato	1 sm. sl.	Frying	30	200	329,000
Beef	1 patty	Frying	14	21,700	3,830,000

a Mutagen levels obtained after cooking food just beyond normal range of edibility. Basic fraction was tested on TA98 with S-9 as described in the text.

b No calculation due to the difficulty in measuring surface area of biscuits

Reprinted from (65)

useful only when the sensitivity of the experiment which generates them is defined. In relation to colon cancer, it is especially important to remember that this disease is related to lifestyle; other dietary components such as dietary fat level may be important modulators of carcinogenicity for food carcinogens, but these are not often taken into account in standard carcinogen bioassays. In the case of the mutagens in fried food, we have postulated that they may be the carcinogens associated with cancer of the colon, breast and prostate (4,68). In these diseases, dietary fat levels, through specific promotional mechanisms, exert powerful controlling influences on the overall cancer induction process.

Thus, the mechanism of colon cancer induction is in need of detailed research. Although genotoxic agents are likely to be important, promotion is also intimately involved. In fact, a great deal of effort at our Institute has been directed at precisely this question (69). Unfortunately, studies on promotional aspects of those nutritionally linked cancers are more difficult to do because of a somewhat cumbersome technology and the complexity of the biological phenomenon. For one thing, there is a need for improved accepted short-term assays for promoters, such as bile acids or certain hormones corresponding to the test batteries which are available for mutagens.

Conclusions

Much of the information we have on food mutagens derives from only one or two tests. This is highly suggestive, but insufficient for making judgments about relevance to human health. Some of these mutagens may turn out to be false-positives (i.e. mutagenic, but not carcinogenic). On the other hand, several of these compounds are carcinogenic in animal tests, although they are weaker than would have been predicted by a naive extrapolation from mutagenic potency. Underemphasis as well as overemphasis of risk should, therefore, be avoided.

Future work is needed, not only on a broader range of genetic end-points, but on in vivo metabolism and pharmacokinetics. The role of other factors modifying mutagenicity and/or carcinogenicity, such as nitrite, hypochlorite, antioxidants, inhibitors of mutagenicity, fiber, the gut microflora, carcinogens, and promoters needs elucidation.

Literature Cited

1. Wynder, E.L.; Gori, G.B. J. Nat. Cancer Inst. 1977, 58, 825-32.
2. Doll, R., Peto, R.; J. Nat. Cancer Inst. 1981, 66, 1191-1308.
3. IARC Monographs, (series), "On the Evaluation of the Carcinogenic Risk of Chemicals to Humans"; International Agency for Research on Cancer, Lyon, France, 1972.
4. Weisburger, J.; Reddy, B.S.; Hill, P.; Cohen, L.A.; Wynder, E.L. Bull. N.Y. Acad Med. 1980, 56, 673-96.
5. Weisburger, J.; "Pathophysiology of Carcinogenesis in Digestive Organs"; Univ. of Tokyo Press: Tokyo Univ. and Park Press: Baltimore, 1977; p. 1-17.
6. Nagao, M.; Honda, M.; Seino, Y.; Yahagi, T.; Sugimura, T. Cancer Lett. 1977, 2, 221-26.
7. Nagao, M.; Honda, M.; Seino, Y.; Yahagi, T.; Kawachi, T.; Sugimura, T. Cancer Lett. 1977, 2, 335-40.
8. Matsumoto, T.; Yoshida, D.; Mizusaki, S.; Okamoto, H. Mutat. Res. 1978, 56, 281-88.
9. Commoner, B.; Vithayathil, A.J.; Dolara, P.; Nair, S.; Madyastha, P.; Cuca, G.C. Science 1978, 201, 913-16.
10. Spingarn, N.E.; Weisburger,J.H. Cancer Lett. 1979, 7, 259-64.
11. Kasai, H.; Nishimura, S.; Nagao, M.; Takahashi, Y.; Sugimura,T. Cancer Lett. 1979, 7, 343-48.
12. Yamaizumi, Z.; Shiomi, T.; Kasai, H.; Nishimura, S.; Takahashi, Y.; Nagao, M.; Sugimura, T. Cancer Lett. 1980, 9, 75-83.
13. Kasai, H.; Yamaizumi, Z.; Wakabayashi, K.; Nagao, M.; Sugimura, T.; Yokoyama, S.; Miyazawa, T.; Spingarn, N.E.; Weisburger, J.H.; Nishimura, S. Proc. Jap. Acad. 1980, 56B, 278-83.
14. Nagao, M.; Yahagi, T.; Kawachi, T.; Seino, Y.; Honda, M.; Matsukura, N.; Sugimura, T.; Wakabayashi, K.; Tsuji, K.; Kosuge, T. "Progress in Genetic Toxicology"; Elsevier/North Holland, 1977; p. 259-64.
15. Spingarn, N.E.; Garvie, C.T. J. Agric. Food Chem. 1979, 27, 1319-21.
16. Weisburger,J.H.; Spingarn, N.E. "Mutagens as a Function of Mode of Cooking of Meats"; Japan Sci. Soc. Press: Tokyo/ Baltimore, 1979; p. 177-84.
17. Dolara, P.; Commoner, B.; Vithayathil, A.; Cuca, G.; Tuley, E.; Madyastha, P.; Nair, S.; Kriebel, D. Mutat. Res. 1979, 60, 231-37.
18. Pariza, M.W.; Ashoor, S.H.; Chu, F.S.; Lund, D.B. Cancer Lett. 1979, 7, 63-69.
19. Nader, C.J.; Spencer, L.K.; Weller, R.J. Cancer Lett. 1981, 13, 147-51.
20. Rappaport, S.M.; McCartney, M.C.; Wei, E.T. Cancer Lett. 1979, 8, 139-45.

21. Felton, J.S.; Healy, S.; Stuermer, D.; Berry, C.; Timourian, H.; Hatch, F.T. Mutat. Res. 1981, 88, 33-44.
22. Yamaguchi, K.; Shudo, K.; Okamoto, T.; Sugimura, T.; Kosuge, T. Gann 1980, 71, 743-44.
23. Bjeldanes, L.F.; Morris, M.M.; Felton, J.S.; Healy, S.; Stuermer, D.; Berry, P.; Timourian, H.; Hatch, F.T. Lawrence Livermore National Laboratory, UCRL Preprint 86279, 1981.
24. Bjeldanes, L.F.; Morris, M.M.; Felton, J.S.; Healy, S.; Stuermer, D.; Berry, P.; Timourian, H.; Hatch, F.T. Lawrence Livermore National Laboratory, UCRL Preprint 86278, 1981.
25. Spingarn, N.E.; Kasai, H.; Vuolo, L.L.; Nishimura, S.; Yamaizumi, Z.; Sugimura, T.; Matsushima, T.; Weisburger, J.H. Cancer Lett. 1980, 9, 177-83.
26. Kasai, H.; Nishimura, S.; Wakabayashi, K.; Nagao, M.; Sugimura, T. Proc. Jap. Acad. 1980, 56B, 382-84.
27. Kasai, H.; Yamaizumi, Z.; Shiomi, T.; Yokoyama, S.; Miyazawa, T.; Wakabayashi, K.; Nagao, M.; Sugimura, T.; Nishimura, S. Chem. Lett. 1981, p. 485-88.
28. Yamaguchi, K.; Shudo, K.; Okamoto, T.; Sugimura, T.; Kosuge,T. Gann 1980, 71, 745, 745-46.
29. Matsumoto, T.; Yoshida, D.; Tomita, H. Cancer Lett. 1981, 12, 105-110.
30. Sugimura, T.; Kawachi, M.; Nagao, M.; Yamada, M.; Takayama, S.; Matsukawa, N.; Wakabayashi, K. "Biochemical and Medical Aspects of Tryptophan Metabolism", Elsevier/North Holland, 1980, p. 297-310.
31. Sugimura, T; Nagao, M. CRC Crit. Rev. Toxicol. 1979, 6, 189.
32. Sugimura, T.; Nagao, M. "Carcinogenic and Mutagenic N-substituted Aryl Compounds" National Cancer Institute Monograph 58, NIH Publ. No. 81-2379, 1980, p. 27-33.
33. Tsuda, H.; Kato, K.; Matsumoto, T.; Yoshida, D.; Mizusaki, S. Mutat. Res. 1978, 49, 145-48.
34. Sasaki, M.; Sugimura, K.; Yoshida, M.; Kawachi, T. Proc. Jap. Acad. 1980, 56B, 332-37.
35. Tohda, H.; Oikawa, A.; Kawachi, T.; Sugimura, T. Mutat. Res. 1980, 77, 65-69.
36. Hatch, F.T.; Felton, J.S.; Thompson, L.H.; Carrano, A.V.; Healy, S.K.; Salazar, E.P.; Minkler, J.L. "13th Annual Meeting, U.S. Environmental Mutagen Soc. (abstract)"; Boston, Mass., 1982, p. 131.
37. Takayama, S.; Katoh, Y.; Tanaka, M.; Nagao, M.; Wakabayashi, K.; Sugimura, T. Proc. Jap. Acad. 1977, 53B, 126-29.
38. Takayama, S., Hirakawa, T.; Sugimura, T. Proc. Jap. Acad. 1978, 54B, 418-22.
39. Matsukura, N.; Kawachi, T.; Morino, K.; Ohgaki, H.; Sugimura, T. Science 1981, 213, 346-47.

40. Matsukura, N.; Kawachi, T.; Wakabayashi, K.; Ohgaki, H.; Morino, K.; Sugimura, T.; Nukaya, H.; Kosuge, T. Cancer Lett. 1981, 13, 181-86.
41. Hosaka, S.; Matsushima, T.; Hirono, I.; Sugimura, T. Cancer Lett. 1981, 13, 23-28.
42. Ishikawa, T.; Takayama, S.; Kitagawa, T.; Kawachi, T.; Kinebuchi, M.; Matsukura, N.; Uchida, E.; Sugimura, T. "Naturally Occurring Carcinogens-Mutagens and Modulators of Carcinogenesis"; Japan Sci. Soc. Press: Tokyo/Baltimore, 1979; p. 159-167.
43. Tamano, S.; Tsuda, H.; Tatematsu, M.; Hasegawa, R.; Imaida, K.; Ito, N. Gann 1981, 72, 747-53.
44. Matsushima, T.; Yahagi, T.; Takamoto, Y.; Nagao, M.; Sugimura, T. "Proceedings of the Fourth International Symposium on Microsomes and Drug Oxidations", 1979.
45. Thorgeirsson, S.; Weisburger, E.K.; King, C.M.; Scribner, J.D. (eds.) "Carcinogenic and Mutagenic N-substituted Aryl compounds". National Cancer Institute Monograph 58, NIH Publ. No. 81-2379, 1980.
46. Spingarn,N.E.; Garvie-Gould, C.; Vuolo, L.L.; Weisburger,J.H. Cancer Lett. 1981, 12, 93-97.
47. Wang, Y.Y.; Vuolo, L.L.; Spingarn, N.E.; Weisburger, J.H. Cancer Lett., In Press.
48. Hayatsu, H.; Inoue, K.; Ohta, H.; Namba, T.; Togawa, K.; Hayatsu, T.; Makita, M.; Wataya, Y. Mutat. Res. 1981, 91, 437-42.
49. Morita, K.; Hara, M.; Kada, T. Agric. Biol. Chem. 1978, 42, 1235-38.
50. Kada, T.; Morita, K.; Inoue, T. Mutat. Res. 1978, 53, 351-53.
51. Arimoto, S.; Ohara, Y.; Namba, T.; Negishi, T.; Hayatsu, H. Biochem. Biophys. Res. Commun. 1980, 92, 662-68.
52. Arimoto, S.; Negishi, T.; Hayatsu, H. Cancer Lett. 1980, 11, 29-33.
53. Lai, C.; Butler, M.A.; Matney, T.S. Mutat. Res. 1980, 77, 245-50.
54. Tsuda, M.; Takahashi, Y.; Nagao, M.; Hirayama, T.; Sugimura, T. Mutat. Res. 1980, 78, 331-39.
55. Tsuda, M.; Nagao, M.; Hirayama, T.; Sugimura, T. Mutat. Res. 1981, 83, 61-68.
56. Sugimura, T. Symposium on Environmental Aspects of Cancer: The Role of Macro and Micro Components of Foods (abstract) American Health Foundation: New York, 1982.
57. Sugimura, T.; Nagao, M. "Chemical Mutagens, Vol. 6."; Plenum Publ. Corp.: New York, 1980; p. 41-60.
58. Powrie, W.D.; Wu, C.H.; Rosin, M.P.; Stich, H.F. J. Food Sci. 1981, 46, 1433-45.
59. Shibamoto, T.; Nishimura, O.; Mihara, S. J. Agric. Food Chem. 1981, 29, 641-46.

60. Aeschbacher, H.U.; Chappius, C.H.; Manganel, M.; Aeschbacher, R. "Progress in Food and Nutrition Science. Maillard Reactions in Food"; Pergamon Press: New York, 1981; p. 279-93.
61. Stich, H.F.; Stich, W.; Rosin, M.P. Food Cosmet. Toxicol. 1980, 18, 581-84.
62. Stich, H.F.; Rosin, M.P.; Wu, C.H.; Powrie, W.D. Cancer Lett. 1981, 13, 89-95.
63. Stich, H.F.; Stich, W.; Rosin, M.P.; Powrie, W.D. Mutat. Res. 1981, 91, 129-136.
64. Stich, H.F.; Rosin, M.P.; San, R.H.C.; Wu, C.H.; Powrie, W.D. "Banbury Report 7: Gastrointestinal Cancer: Endogenous Factors", Cold Spring Harbor Laboratory, 1981; p. 247-63.
65. Spingarn, N.E.; Slocum, L.A.; Weisburger, J.H. Cancer Lett. 1980, 9, 7-12.
66. Pariza, M.W.; Ashoor, S.H.; Chu, F.S. Food Cosmet. Toxicol. 1979, 17, 429-30.
67. Weisburger, J.H.; Williams, G. Science 1981, 214, 401-07.
68. Weisburger, J.H.; Horn, C. Bull. N.Y. Acad. Med. 1982, 58, 296-312.
69. Reddy, B.S.; Cohen, L.; McCoy, G.D.; Hill, P.; Weisburger, J.H.; Wynder, E.L. Adv. Cancer Res. 1980, 32, 237-345.
70. "1982 Cancer Facts and Figures," The American Cancer Society, New York, 1982.

RECEIVED October 5, 1982

Creatinine and Maillard Reaction Products as Precursors of Mutagenic Compounds Formed in Fried Beef

M. JÄGERSTAD, A. LASER REUTERSWÄRD, R. ÖSTE, and A. DAHLQVIST—University of Lund, Department of Applied Nutrition, Chemical Center, P.O. Box 740, S-220 07 Lund, Sweden

S. GRIVAS—Swedish Meat Research Institute, P.O. Box 504, S-244 00 Kavlinge, Sweden

K. OLSSON and T. NYHAMMAR—Swedish University of Agricultural Sciences, Department of Chemistry and Molecular Biology, S-750 07 Uppsala, Sweden

Evidence of formation of mutagens from creatinine and Maillard reaction products was obtained from meat experiments and model reaction systems. Lean beef containing varying amounts of glucose after frying showed mutagenic activity with Ames test, which increased with the glucose content. Varying amounts of creatinine were formed from creatine during frying and the mutagenic activity was found to be related to the creatinine levels. In model experiments solutions of creatinine, glucose and glycine or alanine in diethylene glycol: H_2O were refluxed for 4 h. The resulting mutagenic activity was around 12 - 20 x 10^3 revertants (TA98 + S9) per ml solution depending on which amino acid was used.

In recent years, several reports have demonstrated mutagenic activity by use of Ames test in meat crust after frying (for ref. see 1). A high proportion of the mutagenic activity found in fried meat and also in broiled fish has been identified by Kasai et al. (2,3) and Spingarn et al. (4) as due to imidazo-quinolines (IQ and MeIQ) and an imidazo-quinoxaline (MeIQx). The structure of these so-called IQ-compounds are shown in Figure 1. IQ and MeIQx have been demonstrated to occur in fried beef (2-4). These compounds are among the most potent mutagens so far identified for TA98 and require S9 activation. One µg corresponds to 433 x 10^3 and 100 x 10^3 revertants, respectively (2,3).

Hitherto no explanation has been given concerning the mechanism for the formation of these compounds. The Maillard reaction has, however, been proposed to be involved in the formation of mutagenic activity in the meat crust during frying or broiling (5-7). In the present investigation a tentative route is formulated to explain the formation of the IQ-compounds starting from creatinine and Maillard reaction products originating from glucose and certain amino acids. Increased mutagenic activity in meat

0097-6156/83/0215-0507$06.00/0

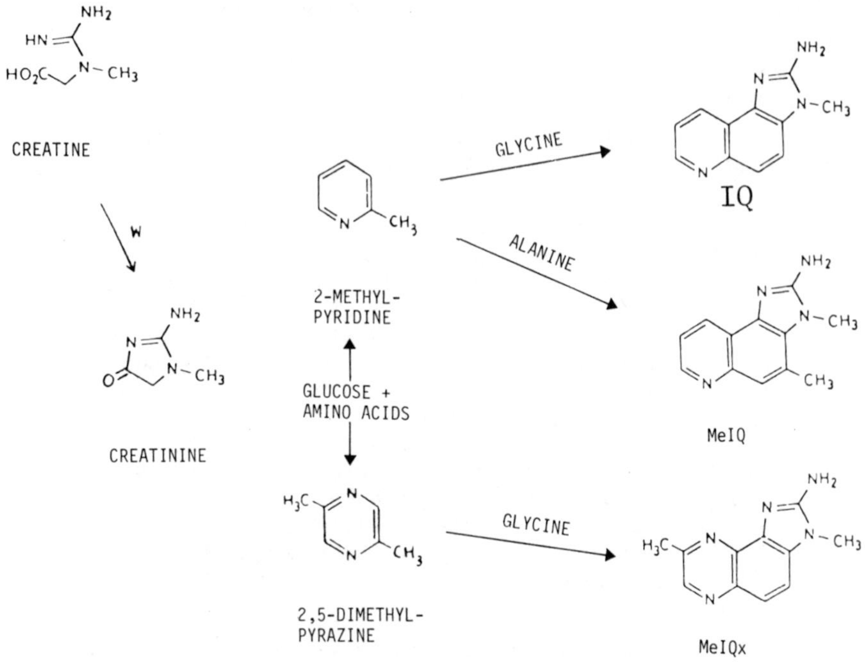

Figure 1. Suggested route for the formation of the imidazo-quinoline compounds.

experiments by addition of creatine or glucose seemed to support the route. Moreover, very high mutagenic activities were observed in model reaction systems containing the postulated precursors. From one of these systems, IQ was isolated in surprisingly high yield and found to be identical with a synthetic sample. So far, we have not been able to repeat that experiment, however. The observed mutagenic activity may therefore be due to species other than the IQ-compounds. Work is in progress to identify the mutagens.

Formation of the IQ-compounds

The imidazole part of the IQ-compounds suggests creatinine as a common precursor. The remaining parts of the IQ-compounds could arise from Maillard reaction products, e.g., 2-methylpyridine or 2,5-dimethylpyrazine. These two compounds could be formed through Strecker degradation. In Maillard reactions, this is induced by α-dicarbonyl compounds derived from carbohydrates, which are thereby converted to pyrroles, pyridines, pyrazines, etc. (8).

Thus, IQ may arise from creatinine, 2-methylpyridine and formaldehyde or a related Schiff base, formed from glycine through Strecker degradation. The initial step may be a Mannich reaction or an aldol condensation. By analogy MeIQ may arise from creatinine, alanine and 2-methylpyridine, and MeIQx from creatinine, glycine and 2,5-dimethylpyrazine according to the scheme in Figure 1.

Meat experiments

Effect of frying temperature and time on development of mutagenic activity and brown color (Maillard reaction products). In the frying experiments, lean muscle from beef, received directly from a slaughter house, was ground and formed to flat 50-g patties. The muscle contained 2.0% fat and was fried without adding fat on a thermostatic double-sided Teflon-coated plate. For each parameter 4-6 patties were fried, one at a time. A crust was peeled off, thickness (mm) was estimated and the crust and crumb were separately thoroughly minced to a fine powder. Water content was estimated, and some samples were lyophilized prior to analyses. The surface temperature of the fryer was 120, 180 or 250°C (varying -10°C) and the frying time varied between 1.5 min and 9 min. Batches (20.0 g) of each crust and crumb were suspended in water, which was acidified to pH 2. The proteins were precipitated with sodium sulfate and filtered off. The filtrate was made basic with sodium hydroxide and the mutagens were extracted with methylene chloride from the basic fraction of the water suspension. The organic phases were evaporated to dryness and dissolved in 0.7 ml DMSO (dimethyl sulfoxide). The mutagenic activity of the meat samples was assayed in triplicates at two different concentrations according to Ames (9) by using the Salmonella strain TA98. S9 mix was prepared according to Ames (9) and contained 5 per cent rat liver homogenate from rats treated with PCB. Preincubation of samples and S9 mix (0.5 ml) was performed for 20 min in a water bath at 37°C.

Table I summarizes the mutagenic activity in the meat experiment expressed in revertants per plate. Raw beef as well as the meat crumb contained no significant levels of mutagenic activity, that is below 20 revertants per plate, when corrected for spontaneous revertants. The mutagenic activity was found in the meat crust and required S9 activation. None of the samples assayed without S9 activation exceeded the values of spontaneous revertants and were therefore not included in the table.

The mutagenic activity of the meat crust increased with frying temperature being without significance when frying was performed at 120°C irrespective of frying time. Frying at 180°C for 3 min resulted in a mutagenic activity slightly higher than in samples of 120°C. A longer frying time at 180°C did not increase the mutagenic activity, which probably could be because analyses were done without separating crust and crumb; the patty was very hard fried. At 250°C the mutagenic activity in meat crust increased markedly with time.

The degree of Maillard reaction was estimated as amount of brown color produced. Samples were extracted in buffer of pH 8.0, proteins were precipitated by TCA and absorbance was measured at 375 nm spectrophotometrically according to a method by Laser Reutersward and Johansson (unpublished).

As shown in Figure 2 the mutagenic activity (expressed as revertants per g dry matter of crust) as well as the browning color increased in parallel to increasing heating conditions. All samples mentioned above could be considered as extremely fried as estimated from the very high water losses which mainly occurred because the patties were very thin when frying began. A surface temperature of 180°C on a Teflonfryer could otherwise be considered as realistic, as well as the color that was developed by frying at 180°C for 3 min.

Effect of glucose

Glucose and glucose-6-phosphate are the main reducing sugars occurring in meat. During certain conditions an animal can develop strong stress before slaughter, resulting in extremely low contents of glycogen in the living muscle and thus extremely low contents of reducing sugars in the meat (10). Analyses according to Laser Reuterswärd (10) showed that the contents of glucose and glucose-6-P were 0.01% and 0.00% and 0.15% and 0.24% in low-glucose beef and normal beef respectively after frying. The low-glucose beef did not develop the typical brown crust and thus showed in very low contents of extractable brown color (Table II). These patties also lacked the typical meat aroma which usually develops from Maillard reactions.

Table I. Mutagenic activity of meat crust, expressed in revertants per plate, assayed with TA98 in the presence of S9 (spontaneous revertants substracted). Meat samples (20.0 g[2]) were extracted and the residue dissolved in 0.7 ml DMSO. Triplicates of samples of 25 and/or 50 µl per plate were tested.

Frying conditions °C - min	Water loss in % of initial weight.	Thickness of crust (mm)	Mutagenic activity TA98+S9 25 µl	50 µl
120-3	40	0.50	0	-
			0	-
			0	-
120-6	51	0.55	6	-
			7	-
			4	-
120-9	54	0.67	0	-
			0	-
			4	-
180-3	55	0.55	37	-
			36	-
			45	-
180-6	66	4.29[1)]	17	-
			16	-
			24	-
250-1.5	48	0.57	38	135
			43	166
			49	161
250-3	60	0.81	235	447
			223	538
250-6	75	4.10[1)]	367	628
			335	664

1) crust and crumb could not be separated

2) lyophilized weight, except for 120-3, 6, 9, which were wet-weight.

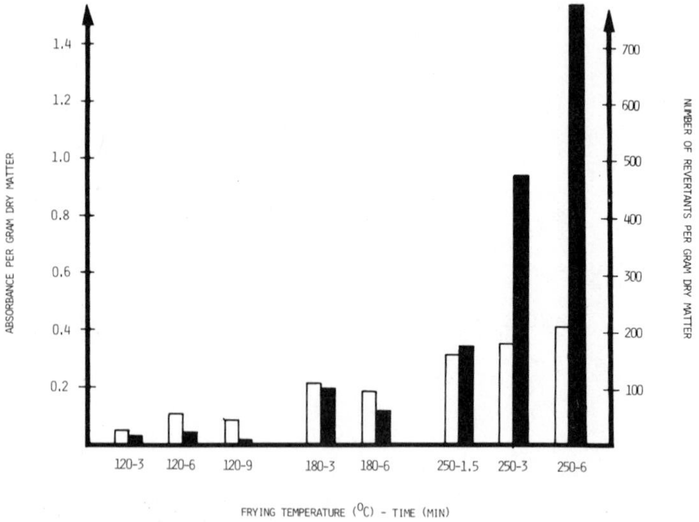

Figure 2. Mutagenic activity (filled bars) expressed in revertants (TA98 + S9) per gram dry meat crust and browning reaction (unfilled bars) estimated spectrophotometrically (375 nm) as a function of increasing frying time and temperature.

The meat crust from the low-glucose meat showed very low mutagenic activity compared to normal beef when frying conditions were the same (Figure 3). When one ml of a 5 per cent D-glucose solution was spread over the upper surface of the low-glucose patty just before frying, the brown color increased (Table II), as well as the mutagenic activity (Figure 3). These results indicate that low contents of glucose can be a limiting factor for Maillard reaction and that the Maillard reaction in some way participatesin the formation of mutagenic activity during frying of meat.

Effect of creatinine. Raw beef containscreatine which by heating is converted to creatinine (11, 12). Chemical analyses, to determine creatine and creatinine (13), were performed on raw low-glucose beef (A) and raw normal beef (B). The contents of creatine and creatinine were for (A) 0.45%, 0.00% and for (B) 0.42% and 0.00% respectively. During frying of different beef samples, varying amounts of creatinine were produced from creatine as seen in Table II. On some beef patties a 2 per cent solution of creatine was spread over the upper surface just before frying. As seen in Table II the mutagenic activity, expressed as revertants per gram dry meat crust, increased with increasing levels of formed creatinine. It is noteworthy that creatine is present only in foods of animal origin (12). Mutagenic tests of a variety of heated foods have shown that meat produces mutagens at levels an order of a magnitude higher than plant food (6), a fact that could be due to the lack of creatine in plants.

Table II. Effect of creatinine concentrations on the mutagenic activity in crusts (dry matter) of fried beef. Creatine concentrations within parenthesis. Brown color as a measure of Maillard reaction products is given as absorbance at 375 nm per g dry matter

	Creatinine %	Number of revertants (TA98+S9) per g	Absorbance at 375 nm per g
Low-glucose beef (A)	0.42 (1.46)	11 ± 12	0.03
A+D-glucose	0.43 (1.56)	41 ± 17	0.12
A+D-glucose+creatine	0.58 (1.47)	88 ± 23	0.14
Normal beef (B)	1.05 (1.19)	99 ± 69	0.32
B+creatine	1.32 (1.18)	164 ± 32	0.38

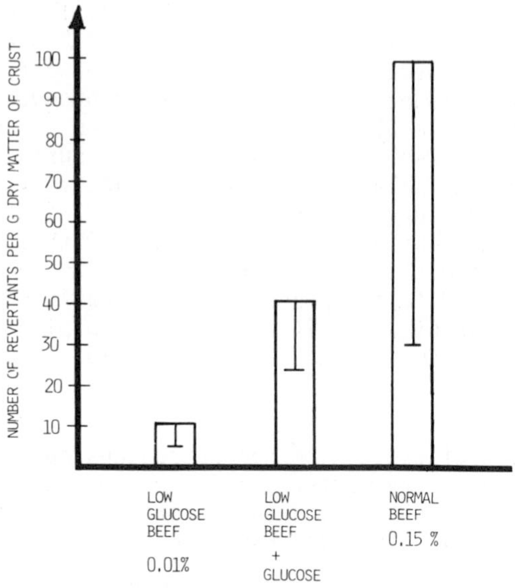

Figure 3. Effect of glucose concentration on the mutagenic activity in fried beef (180 °C for 3 min).

Model reaction systems

Reflux boiling of creatinine, glucose and amino acid. In order to test the suggested hypothesis for the formation of the IQ-compounds, mixtures of creatinine (1.25 g), D-glucose (0.8 g), glycine (1.07 g) or alanine (1.07 g) were dissolved in 10 ml of water doubly distilled in glass and 50 ml of diethylene glycol were added. The pH was stable at 7.2 and the boiling temperature was around 130°C. Mixtures were refluxed and samples were taken out after 1, 2 and 4 h. Assays for mutagenic activity were performed as described above for the meat samples. Both TA98 and TA100 were used without and with S9 activation. Triplicates of samples directly from the mixtures were assayed. As seen in Table III and Figure 4, glycine gave higher mutagenic activity compared to alanine, about 20×10^3 revertants per ml for TA98 in the presence of S9 mix versus 12×10^3 revertants per ml for alanine. Using glycine instead of alanine always resulted in higher mutagenic activity but different batches using the same levels of reactants varied greatly in their mutagenic activity. The highest level reached for glycine after reflux for 4 h was about 100×10^3 rev/ml. (TA98+S9).

Table III. Mutagenic activity expressed as revertants per plate (TA98+S9) of model reaction systems (creatinine, glucose, amino acid in diethylene glycol-water 6:1, v/v). Samples in triplicates were directly withdrawn from the mixtures after 1, 2 and 4 h of reflux. 50 µl were used per plate. Comparison of mutagenic activity after corrections for spontaneous revertants between model mixtures containing alanine or glycine.

	Number of revertants per plate (TA98+S9)		
reflux time (h)	1	2	4
model mixture containing alanine	153	404	640
	151	334	408
	104	313	450
model mixture containing glycine	777	903	970
	461	712	1146
	732	810	980

No mutagenic activity was obtained without S9 activation or when the reactants were heated separately. Boiling all reactants in water, without diethylene glycol, was found to produce mutagenic activity. The addition of diethylene glycol made it possible to raise the temperature of the solution to 130-140°C and thus enhance the reaction rate significantly. A low water content and a high temperature is known to enhance the Maillard reaction. Boiling extract from meat demands several hours for obtaining mutagenic activity while frying of beef in some 10-20 min gives high mutagenic activity (14).

As seen in Figure 4 the mutagenic activity increased with boiling time and was considerably higher for TA98 compared with TA100. The powerful mutagenic activity, the higher response for TA98 and the need of S9 activation agree with the conditions typical for the IQ-compounds (2, 3).

Effect of added Maillard reaction products. The suggested route for formation of the IQ-compounds included participation of 2,5-dimethylpyrazine or 2-methylpyridine (see Figure 1). Therefore 2,5-dimethylpyrazine (0.5 ml, Alpharetta Aromatics Inc, redistilled before use) or 2-methylpyridine (0.5 ml) was added to refluxing model systems containing creatinine, D-glucose and an amino acid in diethylene glycol: water 6:1 (v/v). As seen in Table IV the addition of either compound enhanced the mutagenic activity with an average of around 50 per cent.

Table IV. Effect of Maillard reaction products (2-methylpyridine or 2,5-dimethylpyrazine) on the mutagenic activity of different model reaction mixtures (creatinine, D-glucose, amino acid in diethylene glycol-water 6:1, v/v) after 4 h of refluxing, 10-µl samples being directly withdrawn from the reaction mixtures and assayed with TA98 after S9 activation.

	Number of revertants per plate (TA98+S9)
Model mixture containing glycine	843
	922
	692
Model mixture containing glycine plus 2-methylpyridine	1113
	1267
	944
Model mixture containing alanine	336
	365
	498
Model mixture containing alanine plus 2,5-dimethylpyrazine	690
	585
	480

Chemical identification of IQ-compounds in the model system

After refluxing creatinine, glycine and D-glucose in diethylene glycol: water 6:1 (v/v) for 20 h, IQ was isolated from the reaction mixture in 2.6 per cent yield (7.6 mg calculated on glucose) and found to be identical (EI-MS, H-NMR, HPLC) with a synthetic sample given by T. Sugimura and also with IQ synthe-

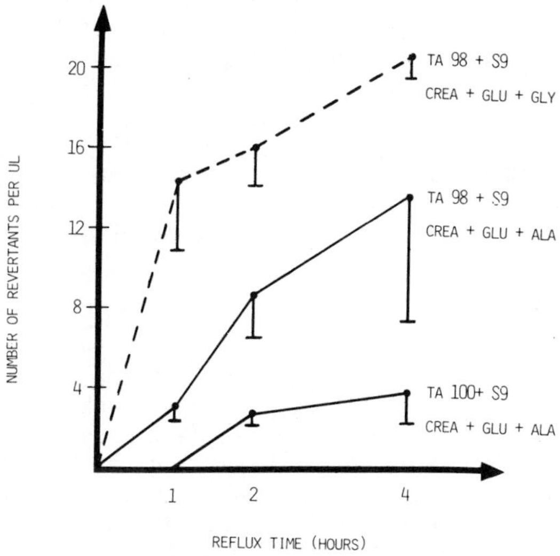

Figure 4. Effect of reflux time, amino acid (glycine or alanine), and Salmonella *strain on the mutagenic activity formed in model reaction systems of creatinine (1.25 g),* D*-glucose (0.8 g), and amino acid (1.07 g) in diethylene glycol–water (6:1 v/v).*

sized in our laboratory (to be published). Several attempts to repeat this experiment were without success, however.

Conclusion

The meat experiments as well as the model system thus supported that the Maillard reaction is of importance for the formation of mutagenic compounds. The mutagenic activity obtained in meat crust was found to be influenced by the glucose and creatinine levels of the beef. The model system showed that besides special reactants producing Maillard reaction products, creatinine also was essential for the formation of mutagens. All reactants in the model system are present in beef in adequate amounts to explain the mutagenic activity produced in the meat crust during frying. The suggested route for formation of the IQ-compounds needs further substantiation. The difficulties in repeating the successful experiments demonstrating IQ in the model system suggest that other mutagens could be produced as well.

Further research is necessary to identify these mutagens formed in the model system and to demonstrate their occurrence in the meat crust of fried or broiled beef. Also important is to show if these mutagens hav tumor-inducing effects in animal studies.

Acknowledgement

This work was supported by grants from the Swedish Board for Technical Development which are gratefully acknowledged. We thank Ms Anne Marie Lindberg, Swedish Meat Research Institute, Mrs Holmfridur Thorgeirsdottir, Department of Chemistry and Molecular Biology, Uppsala, and Mrs Anna Westesson, Department of Nutrition, Lund, for their skilled technical assistance. Professor T. Sugimura is greatly owed thanks for kindly supplying synthetic IQ and MeIQ and Professor O. Theander for his kind interest in the present work. Salmonella strains of TA98 and TA100 were kindly supplied by Dr B. Ames, University of California, and purified (redistilled) 2.5-dimethylpyrazine was a gift from Ms Susan Fors, Swedish Food Institute.

Literature Cited

1. Sugimura, T.; Nagao, M. CRC Crit. Rev. Toxicol. 1979, 6, 189-209.
2. Kasai, H.; Yamaizumi, Z.; Wakabayashi, K.; Nagao, M.; Sugimura, T.; Yokoyama, S.; Miyazawa, T.; Nishimura, S. Chem. Lett. 1980, 1391-1394.
3. Kasai, H.; Yamaizumi, Z.; Shiomi, T.; Yokoyama, S.; Miyazawa, T.; Wakabayashi, K.; Nagao, M.; Sugimura, T.; Nishimura, S. Chem. Lett. 1981, 485-488.

4. Spingarn, N.E.; Kasai, H.; Vuolo, L.L.; Nishimura, S.; Yamaizumi, Z.; Sugimura, T.; Matsushima, T.; Weisburger, J.H. Cancer Lett. 1980, 9, 177-183.
5. Weisburger, J.H.; Spingarn, N.E. "Naturally occurring carcinogens-mutagens and modulators of carcinogenesis". Japan Sci. Soc. Press, Tokyo and Press: Baltimore, 1979, p. 177-184.
6. Spingarn, N.E.; Slocum, L.A.; Weisburger, J.H. Cancer Lett. 1980, 9, 7-12.
7. Spingarn, N.E.; Garvie, C.T. J. Agric. Food Chem. 1979, 27, 1319.
8. Mauron, J. Prog. Food Nutr. Sci. 1981, 5, 5-35.
9. Ames, B.N.; McCann, J.; Yamasaki, E. Mutat. Res. 1975, 31, 347-364.
10. Laser Reuterswärd, A.; Johansson, G.; Kullmar, B.; Rudérus, H. Proc. 27th European Meat Research Workers, Vienna, 1981, A:43.
11. Hughes, R.B. J. Sci. Food Agric. 1960, 11, 700-705.
12. Sulser, H. "Die Extraktstoffe des Fleisches". Wissenschaftliche Verlagsgesellschaft MBH, Stuttgart, 1978, pp. 1-169.
13. Bernt, E.; Bergmeyer, H.U.; Möllering, H. "Methods of enzymatic analyses"; Bergmeyer, H.U., Ed.; Verlag Chemie GmbH, Weinheim/Bergstr. 1974, Vol. IV, p 1772.
14. Dolara, P,: Commoner, B.; Vihayathil, A.; Cuca, G.; Tuley, E.; Mayastha, P.; Nair, S.; Kriebel, D. Mutat. Res. 1979, 60, 231.

RECEIVED December 7, 1982

28

Mutagens Produced by Heating Foods

M. NAGAO, S. SATO, and T. SUGIMURA

National Cancer Center Research Institute, Biochemistry Division, 1-1, Tsukiji 5-chome, Chuo-ku, Tokyo, 104 Japan

Grilling fish or meat on a naked gas flame or an electric hot plate produced mutagens. Mutagenic activity of such products to *S. typhimurium* TA98 in the presence of S9 mix increased with an increase of cooking time. 2-Amino-3-methylimidazo[4,5-*f*]quinoline (IQ) and 2-amino-3,4-dimethylimidazo[4,5-*f*]quinoline (MeIQ) were identified as mutagens in sun-dried sardine grilled over a bare flame. From fried beef, 2-amino-3,8-dimethylimidazo[4,5-*f*]quinoxaline (MeIQx) was isolated as a mutagen. IQ and MeIQ induced diphtheria toxin-resistant mutation in Chinese hamster lung cells. The presence of other mutagenic heterocyclic amines, which had been isolated from pyrolysates of amino acids and protein, was confirmed in cooked food. Some of them were proved to be carcinogenic. Roasting coffee beans resulted in the formation of genotoxic substance(s) towards *E. coli*, *S. typhimurium*, and Chinese hamster lung cells. Methylglyoxal was identified as one of the major mutagens.

Humans have used fire for cooking since about 10^5 years ago. The presence of polycyclic aromatic hydrocarbons in broiled foods such as beefsteak and broiled fish has been reported (1,2). The presence of genotoxic substances in broiled foods other than typical carcinogenic hydrocarbons was detected after the development of Ames' *Salmonella*/mammalian-microsome test (3). Charred parts of broiled fish and meat showed mutagenicity towards *Salmonella typhimurium* TA98, a frameshift mutant of histidine auxotroph, in the presence of S9 mix, a mixture of rat liver

0097-6156/83/0215-0521$06.00/0

postmitochondrial supernatant fraction and NADPH (4,5). Products of pyrolysis of proteins and amino acids also showed strong mutagenicity towards TA98 with S9 mix (6,7). Since mutagens in pyrolyzed amino acids and broiled fish or meat have several common characteristics (specific induction of revertants of TA98, requirement of S9 to exhibit mutagenicity, and basic nature), we first tried to isolate and determine the structures of mutagens from pyrolysates of amino acids which showed strong mutagenic activity—tryptophan, glutamic acid, lysine, and ornithine (8-11). Then the structures of mutagens in broiled fish and meat were also determined (12-15). All the mutagenic compounds isolated from the various pyrolysis products were heterocyclic amines with one exception. Carcinogenic activities of some of these mutagens were proved with long-term in vivo animal experiments.

Heating carbohydrates also resulted in the formation of mutagens but the mutagenic properties were different from those produced by heating protein foods, proteins, or amino acids. Mutagens from carbohydrates were direct-acting towards *S. typhimurium* TA100, a base-pair change mutant of histidine auxotroph. Mutagen(s) present in coffee fell in this type (16,17). Freshly brewed and instant, regular, and caffeine-free coffees were all mutagenic. Mutagenic activity of coffee was produced by roasting green coffee beans. Recently, associations between human pancreatic and ovarian cancer, and coffee intake were reported (18,19,20), although contradictory reports are also available (21,22).

Heating or cooking foods makes them easily digestible, and increases their nutritional value. However, heating foods can yield various mutagens. Isolation of mutagens from pyrolysis products, and examination of their carcinogenic effects on laboratory animals, may be useful in finding a clue to human cancer prevention.

Mutagenicity of Broiled Fish and Meat

Broiling or grilling fish and meat produces mutagens. Dimethyl sulfoxide or methanol extracts of smoke condensates and charred parts of various fish obtained by grilling were mutagenic to *S. typhimurium* TA98 in the presence of S9 mix (4,5). Charred parts of beefsteak and the hamburger were also mutagenic (4,23). The mutagenic activities of methanol extracts of sun-dried sardine and hamburger increased with cooking time, as shown in Figure 1. A naked gas flame and an electric hot plate were used for cooking the fish and the hamburger, respectively. Broiling sun-dried sardine and hamburger for 5-7 min and 15-20 min, respectively, yielded the best taste. Methanol extracts of sun-dried sardine were 10^3 times as mutagenic as those of hamburger (Figure 1). The difference may not be due to the difference in the quality of protein but to the water content (24) and the cooking method.

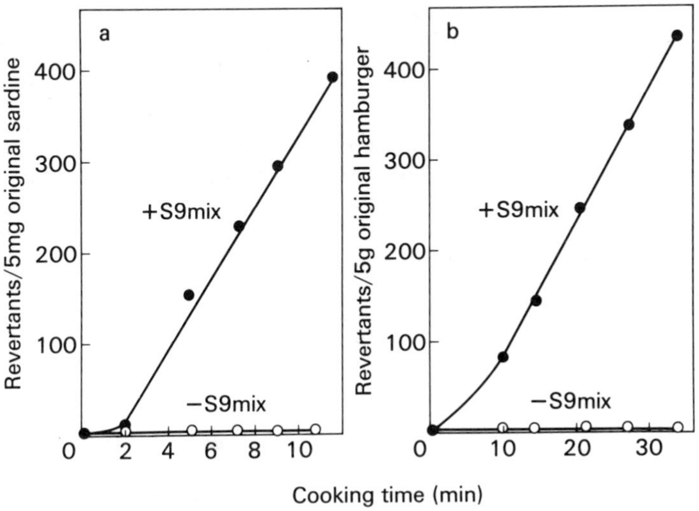

Figure 1. Mutagenicities of sun-dried sardine broiled over naked gas flame (a) and hamburger cooked in electric hamburger cooker (b). Methanol extracts of cooked materials were tested for their mutagenicities with S. typhimurium *TA98 with or without S9 mix.*

Mutagens Isolated from Pyrolysates of Amino Acids and Protein, and from Cooked Foods

During an early stage of our studies of mutagens in cooked foods, we found that mutagens which mainly induced frameshift mutation in the presence of S9 mix were produced by pyrolyzing proteins or amino acids (Table I). Pyrolysis of tryptophan, glutamic acid, lysine, and ornithine produced mutagenic activity which was stronger than that found for other amino acids. From a basic fraction of tryptophan pyrolysate, mutagenic compounds were purified and two crystalline substances were obtained. Their structures were determined, by X-ray crystallography, to be 3-amino-1,4-dimethyl-5*H*-pyrido[4,3-*b*]indole (Trp-P-1) and 3-amino-1-methyl-5*H*-pyrido[4,3-*b*]indole (Trp-P-2) (8). These structures were confirmed by chemical synthesis (8). Similarly, the following compounds were isolated: 2-amino-6-methyldipyrido-[1,2-*a*:3',2'-*d*]imidazole (Glu-P-1) and 2-aminodipyrido[1,2-*a*:3',2'-*d*]imidazole (Glu-P-2), both from glutamic acid pyrolysate; 3,4-cyclopentenopyrido[3,2-*a*]carbazole (Lys-P-1), from lysine pyrolysate (10); 4-amino-6-methyl-1*H*-2,5,10,10b-tetraazafluoranthene (Orn-P-1), from ornithine pyrolysate (11); and 2-amino-5-phenylpyridine (Phe-P-1), from phenylalanine pyrolysate (8). All these are new compounds.

Table I. Mutagenic activity of condensate of smoke formed by pyrolysis of biomacromolecules and of amino acids

Substrate pyrolyzed	Revertants/mg smoke condensate			
	TA98		TA100	
	+S9 mix	-S9 mix	+S9 mix	-S9 mix
Lysozyme	8,311	0	2,319	0
Histone	5,012	0	1,311	0
DNA	278	0	170	0
RNA	83	0	0	0
Starch	0	0	70	338
Vegetable oil	0	0	85	0
Tryptophan[a]	22,420	0	2,500	0
Glutamic acid[a]	13,800	0	3,400	0
Lysine[a]	5,250	0	608	0
Ornithine[a]	8,290	0	560	0
Cysteine[a]	324	0	965	0
Valine[a]	0.9	0	0	0

Adapted from Nagao, et al. (6) and Kosuge, et al. (60)

a. Tar obtained by dry distillation was subjected to test.

Yoshida et al (25) isolated 2-amino-9*H*-pyrido[2,3-*b*]indole (AαC) and 2-amino-3-methyl-9*H*-pyrido[2,3-*b*]indole (MeAαC) from pyrolysate of soybean globulin.

From grilled sun-dried sardines, two mutagenic fractions were obtained with high performance liquid chromatography using μBondapak C_{18}. Their structures were determined, by 270 MHz ^{1}H-NMR and low- and high-resolution mass spectra, to be 2-amino-3-methylimidazo[4,5-*f*]quinoline (IQ) and 2-amino-3,4-dimethylimidazo[4,5-*f*]quinoline (MeIQ) (12,13,14). The structures of these new compounds were confirmed by chemical synthesis.

From fried beef, another new compound, 2-amino-3,8-dimethylimidazo[4,5-*f*]quinoxaline (MeIQx), was isolated, and its structure was confirmed by chemical synthesis (15). The structures of all the compounds isolated from pyrolysates of amino acids and a protein, and from cooked fish and meat, are shown in Table II.

Quantitative analyses of these newly found mutagenic compounds in cooked foods are important for estimating the hazards for humans. Although methods of analysis have not been standardized yet, partial purification of methanol extracts of cooked foods with acid-base partition, silica-gel or Sephadex LH-20 column chromatography and high performance liquid chromatography are recommended. For the identification of compounds, gas chromatography/mass spectrometry (GC/MS) with multiple ion detection (MID) and high performance liquid chromatography monitored with fluorometry are useful. One gram of broiled sun-dried sardine was found to contain 13.3 ng and 13.1 ng of Trp-P-1 and Trp-P-2, respectively (26). Trp-P-1 was also detected in beef grilled over a naked flame (27). A deuterium dilution technique using a [^{2}H]CH_3 derivative as the standard compound made analyses more precise. The amounts of IQ in sun-dried sardine and fried beef were 158 ng/g and 0.59 ng/g, respectively (28). Quantitative analyses of mutagens in cooked foods are summarized in Table III.

Genotoxic Activity of Pyrolyzed Products

Trp-P-1, Trp-P-2, Glu-P-1, IQ, MeIQ and MeIQx were very strongly mutagenic to *S. typhimurium* TA98 (a frameshift mutant of histidine auxotroph which has plasmids pKM101) in the presence of S9 mix. MeIQ and MeIQx were also strongly mutagenic to *S. typhimurium* TA100 (a base-pair change mutant of histidine auxotroph which has plasmids pKM101) with S9 mix, but Trp-P-1, Trp-P-2, and Glu-P-1 were not as strongly mutagenic to TA100. Specific mutagenic activities of the compounds isolated from pyrolysis products including those of typical carcinogens, are shown in Table IV. It is interesting that *S. typhimurium* TA1535, which has the same genetic properties as TA100 except that it lacks plasmid, did not respond to any of these pyrolysis products. These pyrolysis products were mutagenic towards *Escherichia coli* WP2 *uvrA* (*trp*$^-$, a base-pair change mutant), and their mutagenicities were enhanced by introducing plasmids pKM101 to *E. coli* WP2 *uvrA* (29).

Table II. Mutagens isolated from pyrolysates

Chemical name	Abbreviation	Structure	Source of isolation
3-Amino-1,4-dimethyl-5*H*-pyrido[4,3-*b*]indole	Trp-P-1	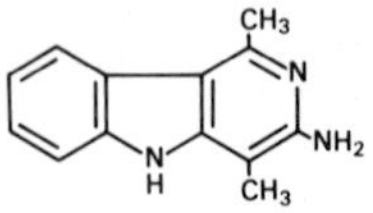	Tryptophan pyrolysate (8)
3-Amino-1-methyl-5*H*-pyrido[4,3-*b*]indole	Trp-P-2	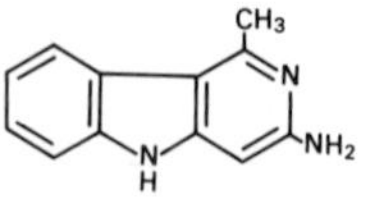	Tryptophan pyrolysate (8)
2-Amino-6-methyldipyrido-[1,2-*a*:3',2'-*d*]imidazole	Glu-P-1	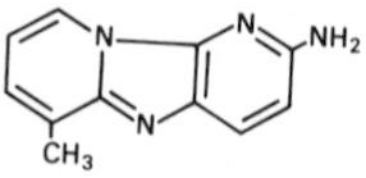	Glutamic acid pyrolysate (9)
2-Aminodipyrido[1,2-*a*:3',2'-*d*]imidazole	Glu-P-2	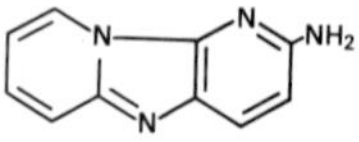	Glutamic acid pyrolysate (9)
3,4-Cyclopentenopyrido-[3,2-*a*]carbazole	Lys-P-1	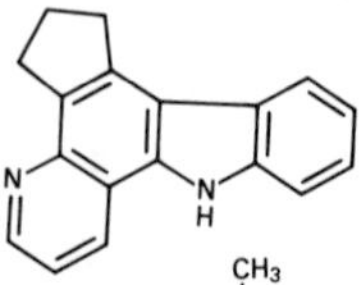	Lysine pyrolysate (10)
4-Amino-6-methyl-1*H*-2,5,10,10b-tetraaza-fluoranthene	Orn-P-1	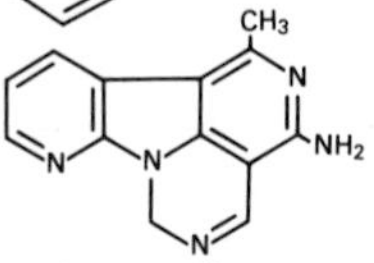	Ornithine pyrolysate (11)
2-Amino-5-phenyl-pyridine	Phe-P-1	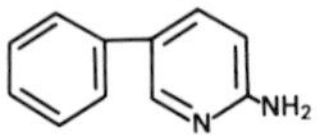	Phenylalanine pyrolysate (8)
2-Amino-9*H*-pyrido-[2,3-*b*]indole	AαC	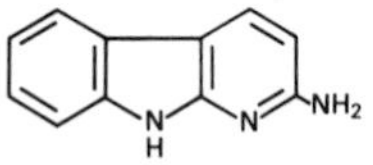	Soybean globulin pyrolysate (25)
2-Amino-3-methyl-9*H*-pyrido[2,3-*b*]indole	MeAαC	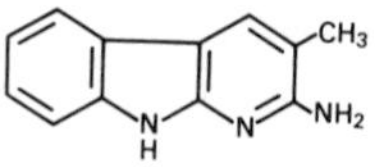	Soybean globulin pyrolysate (25)
2-Amino-3-methylimidazo-[4,5-*f*]quinoline	IQ	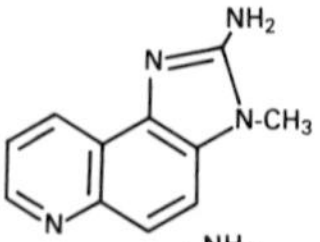	Broiled sardine (12,13)
2-Amino-3,4-dimethyl-imidazo[4,5-*f*]quinoline	MeIQ	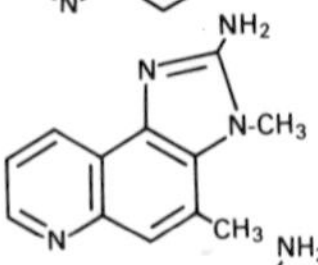	Broiled sardine (12,13)
2-Amino-3,8-dimethyl-imidazo[4,5-*f*]quinoxaline	MeIQx	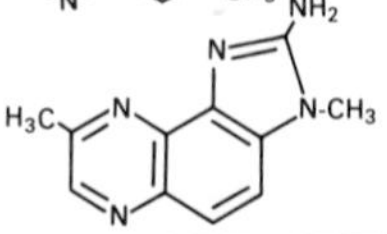	Fried beef (15)

Table III. Example of quantitative analyses for mutagens in cooked foods

Food material	Method of cooking	Methods of identification and quantification	Mutagens in cooked material (ng/g)	
Sun-dried sardine	Naked flame	GC/MS with MID	Trp-P-1 Trp-P-2	13.3 (26) 13.1
Sun-dried sardine	Naked flame	GC/MS with MID by addition of CD_3-substituted MeIQ	IQ MeIQ	158 (28) 72
Beef	Electric hot plate	GC/MS with MID	IQ MeIQx	0.59 (28) 2.4
Beef	Naked flame	Identified by fluorescence and MF, quantified by GC/MS	Trp-P-1	53[a] (27)
Chicken	Naked flame	Fluorescence	AαC MeAαC	180 (61) 15
Sun-dried cuttlefish	Naked flame	Identified by UV, quantified by MF	Glu-P-2	280[b] (62)

a. ng/g of raw beef
b. ng/g of sun-dried material

Table IV. Mutagenic activities of mutagens isolated from pyrolysis products and typical carcinogens

	Mutagenic activity (revertants/μg)	
	TA98	TA100
Pyrolysis product		
MeIQ	661,000	30,000
IQ	433,000	7,000
MeIQx	145,000	14,000
Trp-P-2	104,200	1,800
Orn-P-1[a]	56,800	-
Glu-P-1	49,000	3,200
Trp-P-1	39,000	1,700
Glu-p-2	1,900	1,200
AαC	300	20
MeAαC	200	120
Lys-P-1	86	99
Phe-P-1	41	23
Typical carcinogen		
AF-2	6,500	42,000
AFB_1	6,000	28,000
4NQO	970	9,900
B[a]P	320	660
DEN	0.02	0.15
DMN	0.00	0.23

Mutagenicity was tested with optimal amount of S9.
a. S9 at the rate of 150 μl/plate was used.
AF-2, 2-(2-furyl)-3-(5-nitro-2-furyl)acrylamide; AFB_1, aflatoxin B_1; 4NQO, 4-nitroquinoline 1-oxide; B[a]P, benzo[a]-pyrene; DEN, diethylnitrosamine; DMN, dimethylnitrosamine.

Pyrolysis products were also found to induce prophage λ in lysogenic *E. coli* K12, strain GY5027 (30), by Inductest III, which was developed by Moreau et al. (31) (Figure 2). The genotoxic effects of pyrolysis products on cultured mammalian cells were also investigated. Trp-P-1, Trp-P-2, IQ and MeIQ were found to induce diphtheria toxin-resistant mutants of Chinese hamster lung cells in the presence of S9 mix. The mutation frequencies were 33, 160, 40 and 38 per 10^6 survivors per μg of Trp-P-1, Trp-P-2, IQ and MeIQ, respectively (32). These mutagenic activities were comparable to those of benzo[a]pyrene and methyl methanesulfonate.

Trp-P-1, Trp-P-2, Glu-P-1, and AαC induced sister chromatid exchanges in human lymphoid cells, NL-3 (33). Trp-P-1, Trp-P-2, Glu-P-1, Glu-P-2, AαC, and MeAαC induced chromosome aberrations in Chinese hamster fibroblasts (34).

Carcinogenic activities of pyrolysis products were studied in vitro and in vivo. Trp-P-1, Trp-P-2 (35), and Glu-P-1 (36) induced malignantly transformed foci in the embryonic cells of cryopreserved Syrian golden hamsters.

The in vivo carcinogenic activities of Trp-P-1, Trp-P-2, Glu-P-1, Glu-P-2, AαC, and MeAαC were detected in CDF_1 mice by feeding. Mice which were fed a diet containing 0.02% Trp-P-1 or Trp-P-2 for an entire life span developed specifically hepatocellular carcinomas; females were more susceptible than males (37) (Table V). Female rats but not males fed a diet containing 0.01% Trp-P-2 developed hemangioendothelial sarcoma and neoplastic nodules of the liver (38).

Table V. Numbers of CDF_1 mice with hepatic tumors induced by Trp-P-1 and Trp-P2[a]

Treatment	Sex	No. effective animals	Hepatocellular carcinoma	Total hepatic tumor[b] (%)
Trp-P-1	Male	24	4	5 (21)
(0.02%)	Female	26	14	16 (62)
Trp-P-2	Male	25	3	4 (16)
(0.02%)	Female	24	22 (2[c])	22 (92)
Control	Male	25	0	1 (4)
	Female	24	0	0 (0)

a. Adapted from Matsukura et al (37) (Copyright 1981 by the American Association for the Advancement of Science)
b. Total hepatic tumor contains hepatocellular carcinomas, hepatocellular adenoma, and hemangioma.
c. Two mice had metastasis to the lung.

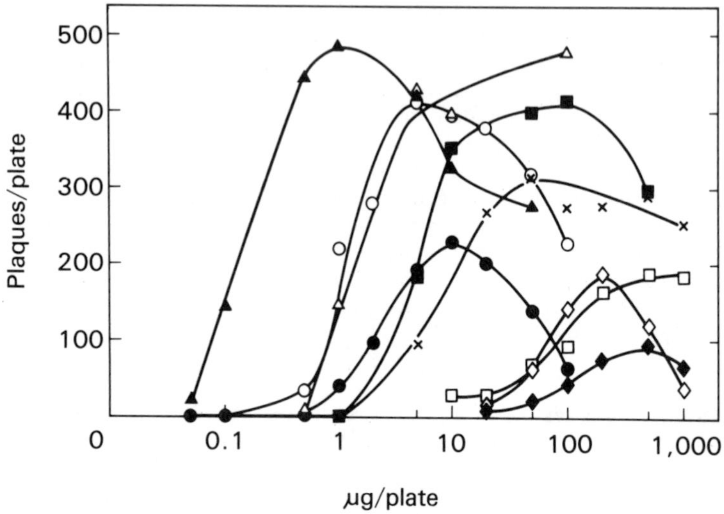

*Figure 2. Phage-inducing activities of mutagens isolated from pyrolysates (*E. coli *GY5027 was used). Key (S9 at 10 µL/plate): ▲, MeIQ; △, IQ; ●, Trp-P-1; ○, Trp-P-2; and ■, Glu-P-1. Key (S9 at 30 µL/plate): □, Glu-P-2. Key (S9 at 50 µL/plate): ♦, MeAαC; ◇, AαC; and ×, Lys-P-1.*

Mice fed a diet containing 0.05% of Glu-P-1 or Glu-P-2, or 0.08% of MeAαC, mainly developed hepatic tumors and hemangiosarcomas in the brown fat tissue of the interscapula (39). The incidence of hepatic tumors produced by these three compounds was higher in females than in males, but the incidence of hemangiosarcoma in males and females was similar. AαC also induced hepatic tumors and hemagiosarcomas by feeding.

Mutagens in Coffee

Freshly brewed and instant, regular, and caffeine-free coffees were mutagenic to *S. typhimurium* TA100 without S9 mix (16,17). The addition of S9 mix abolished or reduced mutagenicity. *E. coli* WP2 *uvrA*/pKM101 also responded to these types of coffee, but those strains which had no plasmids, such as *S. typhimurium* TA1535 and *E. coli* WP2 *uvrA*, did not. All coffees also induced prophage λ in *E. coli* K12, strain GY5027 (Table VI). One cup of coffee contained mutagens which induced 4-10 $\times 10^4$ revertants of *S. typhimurium* TA100 by the Ames method (3) with some modifications (40). One cup of coffee is equivalent to 800 μg of *N*-methyl-*N'*-nitro-*N*-nitrosoguanidine in prophage λ-inducing activity. These genotoxic activities were shown to be produced by roasting coffee beans; coffee prepared from green coffee beans did not show any mutagenic activity or prophage λ-inducing activity (41).

Recently, some of the mutagenic compounds in coffee have been identified. They are dicarbonyl compounds, methylglyoxal, diacetyl and glyoxal (42). One gram of instant coffee contains 150 μg of methylglyoxal and 100 μg of diacetyl and glyoxal. Specific mutagenic activity of methylglyoxal was highest among these, and one mg of methylglyoxal induced 1 $\times 10^5$ revertants of TA100. Mutagenic activities of diacetyl and glyoxal are much

Table VI. Genotoxicity of coffee in bacteria

Amount (mg/plate)	Mutagenicity (revertants/plate) *S. typhimurium* TA100	*S. typhimurium* TA1535	*E. coli* WP-2 *uvrA*/pKM101	*E. coli* WP-2 *uvrA*	Phage-inducing activity (plaques/plate)
Instant coffee					
0	89	59	58	17	22
2	225	40	97	23	
5	361	51	106	18	
10	621	60	127	19	46
20					146
50					576

Table VII. Effects of sulfite on the mutagenicity and phage-inducing activity of coffee

Coffee (mg/plate)	Revertants/10^8 survivors[a] None	Sodium sulfite[b]	Coffee (mg/plate)	Plaques/plate[c] None	Sodium sulfite[b]
Instant "regular"			Instant "caffeine-free"		
2.5	43	0	2	8	23
5.0	120	0	10	93	37
7.5	238	0	20	305	52
10.0	365	0			

a. *S. typhimurium* TA100 was used. The number of spontaneous revertants (163) was subtracted.
b. 300 μg of sulfite were added.
c. *E. coli* GY5027 was used. The number of spontaneous plaques (62) was subtracted.

less, being 360 and 9,000 revertants per mg, respectively. This is in accordance with a paper by other investigators (43). A simple calculation suggested that these dicarbonyl compounds can account for about half of the mutagenicity of coffee.

Both regular and caffeine-free instant coffees, without S9 mix, also induced diphtheria toxin-resistant mutants of Chinese hamster lung cells (44). It is urgent that carcinogenicity of methylglyoxal be tested with animals.

Mutagenic activity and phage-inducing activity of coffee were suppressed by sulfite or bisulfite (45). Both activities can be suppressed by adding about 300 ppm of sulfite or bisulfite to coffee served at home (Table VII). Sulfite and bisulfite have already gained wide acceptance as food preservatives and anti-oxidants. However, ascorbic acid, which is also a reducing agent, enhanced the mutagenicity of instant coffee several fold.

Epidemiological studies showed positive association of coffee drinking and cancers of pancreas and ovary (18,19,20), although reports describing negative association are also available (21,22). In an experimental animal study, no carcinogenic activity of coffee was reported (46,47). However, this data should be reconsidered in terms of the genotoxic activities of coffee revealed in bacteria and cultured mammalian cells.

Discussion

The Salmonella mutagenicity test is a very powerful method for detecting genotoxic substances in crude materials, including cooked foods. We have already identified 10 new compounds from pyrolysates of amino acids and cooked foods. Among those, all 6 compounds which we tested were proved to be carcinogenic. However, there was no quantitative correlation between the carcinogenic potency of these compounds and the mutagenic activity to Salmonella. Information on the carcinogenic potency of IQ, MeIQ, and MeIQx, which were isolated from cooked foods, is especially important for estimating the hazards that cooked foods hold for humans, and animal experiments with these three compounds are now under way in our laboratory.

Most mutagens are carcinogenic in long-term animal tests (48,49,50), but we observed exceptional cases in which mutagens were not carcinogenic. For instance, quercetin, which is widely distributed in vegetables, is moderately mutagenic towards bacteria (51-54) and cultured mammalian cells (55,56). Nevertheless, intensive studies in Japan failed to demonstrate carcinogenicity (57,58,59).

There have been many epidemiological studies on coffee intake and human cancer. Since we have indentified one of major genotoxic substance in coffee, thorough experimental studies are urgently required to evaluate the hazards of coffee.

Acknowledgements

This work was supported in part by Grants-in-Aid for Cancer Research from the Ministry of Education, Science, and Culture, and the Ministry of Health and Welfare of Japan, and by a grant from the Nissan Science Foundation.

Literature Cited

1. Lijinsky, W.; Shubik, P. Science 1974, 145, 53-54.
2. Masuda, Y.; Mori, K.; Kuratsune, M. Gann 1966, 57, 133-142.
3. Ames, B.N.; McCann, J.; Yamasaki, E. Mutat. Res. 1975, 31, 347-364.
4. Nagao, M.; Honda, M.; Seino, Y.; Yahagi, T.; Sugimura, T. Cancer Lett. 1977, 2, 221-226.
5. Sugimura, T.; Nagao, M.; Kawachi, T.; Honda, M.; Yahagi, T.; Seino, Y.; Sato, S.; Matsukura, N.; Matsushima, T.; Shirai, A.; Sawamura, M.; Matsumoto, H. "Origins of Human Cancer"; Hiatt, H.H., Watson, J.D., Winsten, J.A., Eds.; Cold Spring Harbor Laboratories: Cold Spring Harbor, 1977; p. 1561.
6. Nagao, M.; Honda, M.; Seino, Y.; Yahagi, T.; Sugimura, T. Cancer Lett. 1977, 2, 335-340.
7. Matsumoto, T.; Yoshida, D.; Mizusaki, S.; Okamoto, H. Mutat. Res. 1977, 48, 279-286.
8. Sugimura, T.; Kawachi, T.; Nagao, M.; Yahagi, T.; Seino, Y.; Okamoto, T.; Shudo, K.; Kosuge, T.; Tsuji, K.; Wakabayashi, K.; Iitaka, Y.; Itai, A. Proc. Japan Acad. 1977, 53, 58-61.

9. Yamamoto, T.; Tsuji, K.; Kosuge, T.; Okamoto, T.; Shudo, K.; Takeda, K.; Iitaka, Y.; Yamaguchi, K.; Seino, Y.; Yahagi, T.; Nagao, M.; Sugimura, T. Proc. Japan Acad. 1977, 54B, 248-250.
10. Wakabayashi, K.; Tsuji, K.; Kosuge, T.; Takeda, K.; Yamaguchi, K.; Shudo, K.; Iitaka, Y.; Okamoto, T.; Yahagi, T.; Nagao, M.; Sugimura, T. Proc. Japan Acad. 1978, 54, 569-571.
11. Yokota, M.; Narita, K.; Kosuge, T.; Wakabayashi, K.; Nagao, M.; Sugimura, T.; Yamaguchi, K.; Shudo, K.; Iitaka, Y.; Okamoto, T. Chem. Pharm. Bull. 1981, 29, 1473-1475.
12. Kasai, H.; Yamaizumi, Z.; Wakabayashi, K.; Nagao, M.; Sugimura, T.; Yokoyama, S.; Miyazawa, T.; Spingarn, N.E.; Weisburger, J.H.; Nishimura, S. Proc. Japan Acad. 1980, 56B, 278-283.
13. Kasai, H.; Nishimura, S.; Wakabayashi, K.; Nagao, M.; Sugimura, T. Proc. Japan Acad. 1980, 56B, 382-384.
14. Kasai, H.; Yamaizumi, Z.; Wakabayashi, K.; Nagao, M.; Sugimura, T.; Yokoyama, S.; Miyazawa, T.; Nishimura, S. Chem. Lett. 1980, 1391-1394.
15. Kasai, H.; Yamaizumi, Z.; Shiomi, T.; Yokoyama, S.; Miyazawa, T.; Wakabayashi, K.; Nagao, M.; Sugimura, T.; Nishimura, S. Chem. Lett. 1981, 485-488.
16. Nagao, M.; Takahashi, Y.; Yamanaka, H.; Sugimura, T. Mutat. Res. 1979, 68, 101-106.
17. Aeschbacher, H.U.; Würzner, H.P. Toxicol. Lett. 1980, 5, 139-145.
18. MacMahon, B.; Yen, S.J.; Trichopoulos, D.; Warren, K.; Nardi, G. N. Engl. J. Med. 1981, 630-633.
19. Nomura, A.; Stemmermann, G.N.; Heilbrun, L.K. Lancet 1980, 2, 415.
20. Trichopoulos, D.; Papapostolou, M.; Polychronopoulow, A. Int. J. Cancer 1981, 28, 691-693.
21. Feinstein, A.R.; Horwitz, R.I.; Spitzer, W.D.; Battista, R.N. J. Am. Med. Assoc. 1981, 246, 957-961.
22. Jick, H.; Dian, B.J. Lancet 1981, 1, 92.
23. Commoner, B.; Vithayathil, A.J.; Dolara, P.; Nair, S.; Madyastha, P.; Cuca, G.C. Science 1978, 201, 913-916.
24. Ueta, M.; Kaneda, T.; Mazaki, M.; Taue, S.; Takahashi, S. "Naturally Occurring Carcinogens-Mutagens and Modulators of Carcinogenesis": Miller, E.C., Miller, J.A., Hirono, I., Sugimura, T., Takayama, S. Eds.; Japan Scientific Societies Press: Tokyo/University Park Press: Baltimore, 1979; p. 169.
25. Yoshida, D.; Matsumoto, T.; Yoshimura, R.; Matsuzaki, T. Biochem. Biophys. Res. Commun. 1978, 83, 915-920.
26. Yamaizumi, Z.; Shiomi, T.; Kasai, H.; Nishimura, S.; Takahashi, Y.; Nagao, M.; Sugimura, T. Cancer Lett. 1980, 9, 75-83.
27. Yamaguchi, K.; Shudo, K.; Okamoto, T.; Sugimura, T.; Kosuge, T. Gann, 1980, 71, 745-746.

28. Sugimura, T.; Nagao, M.; Wakabayashi, K. "Environmental Carcinogens Selected Methods of Analysis"; Egan, H., Fishbein, L., Castegnaro, M., O'Neill, I.K., Bartsh, H. Eds.; International Agency for Research on Cancer: Lyon, 1981; Vol. 4, p. 251.
29. Matsushima, T. personal communication.
30. Kosugi, A.; Suwa, Y.; Nagao, M.; Wakabayashi, K.; Sugimura, T. Proc. 10th Environ Mutagen Soc. Japan Mtg. 1981, p. 62.
31. Moreau, P.; Bailone, A.; Devoret, R. Proc. Natl Acad. Sci. USA 1976, 73, 3700-3704.
32. Nakayasu, M. personal communication.
33. Tohda, H.; Oikawa, A.; Kawachi, T.; Sugimura, T. Mutat. Res. 1980, 77, 65-69.
34. Ishidate, M.Jr.; Sofuni, T.; Yoshikawa, K. Gann Monogr. Cancer Res. 1981, 27, 95-108.
35. Takayama, S.; Katoh, Y.; Tanaka, M.; Nagao, M.; Wakabayashi, K.; Sugimura, T. Proc. Japan Acad. 1977, 53B, 126-129.
36. Takayama, S.; Hirakawa, T.; Tanaka, M.; Kawachi, T.; Sugimura, T. Toxicol. Lett. 1979, 4, 281-284.
37. Matsukura, N.; Kawachi, T.; Morino, K.; Ohgaki, H.; Sugimura, T.; Takayama, S. Science 1981, 213, 346-347.
38. Hosaka, S.; Matsushima, T.; Hirono, I.; Sugimura, T. Cancer Lett. 1981, 13, 23-28.
39. Sugimura, T. "Genetic Toxicology: An Agricultural Perspective"; Hollaender, A., Ed.; Plenum: New York, 1982, in press.
40. Sugimura, T.; Nagao, M. "Chemical Mutagens". Vol. 6; de Serres, F.J., Hollaender, A., Eds.; Plenum, New York, 1980; p. 41.
41. Kosugi, A.; Nagao, M.; Suwa, Y.; Wakabayashi, K.; Sugimura, T. Mutat. Res. 1982, in press.
42. Kasai, H.; Kumeno, K.; Yamaizumi, Z.; Nishimura, S.; Nagao, M.; Fujita, Y.; Sugimura, T.; Nukaya, H.; Kosuge, T. Gann 1982, 73, 681-683.
43. Bjeldanes, L.F.; Chew, H. Mutat. Res. 1979, 67, 367-371.
44. Nakayasu, M. personal communication.
45. Suwa, Y.; Nagao, M.; Kosugi, A.; Sugimura, T. Mutat. Res. 1982, in press.
46. Würzner, H.P.; Lindström, E.; Vuataz, L.; Luginbühl, H. Fd Cosmet. Toxicol. 1977, 15, 289-296.
47. Zeitlin, B.R. Lancet, 1972, 1, 1066.
48. McCann, J.; Ames, B.N. Proc. Natl Acad. Sci. USA 1976, 73, 950-954.
49. Purchase, I.F.H.; Longstaff, E.; Ashby, J.; Styles, J.A.; Anderson, D.; Lefevre, P.A.; Westwood, F.R. Nature 1976, 264, 624-627.
50. Sugimura, T.; Sato, S.; Nagao, M.; Yahagi, T.; Matsushima, T.; Seino, Y.; Takeuchi, M.; Kawachi, T. "Fundamentals in Cancer Prevention"; Magee, P.N., Takayama, S., Sugimura, T., Matsushima, T., Eds.; Japan Scientific Societies Press: Tokyo/University Park Press: Baltimore, 1976; p.191.

51. Bjeldanes, L.F.; Chang, G.W. Science 1977, 197, 577-578.
52. Brown, J.B. Mutat. Res. 1980, 75, 243-277.
53. MacGregor, J.; Jurd, L. Mutat. Res. 1978, 54, 297-309.
54. Nagao, M.; Morita, N.; Yahagi, T.; Shimizu, M., Kuroyanagi, M.; Fukuoka, M.; Yoshihira, K., Natori, S.; Fujino, T.; Sugimura, T. Environ. Mutagen. 1981, 3, 401-419.
55. Amacher, D.E.; Paillet, S.C.; Turner, G.N.; Ray, V.A.; Salsburg, D.S. Mutat. Res. 1980, 72, 447-474.
56. Meltz, M.L.; MacGregor, J.T. Mutat. Res. 1981, 88, 317-324.
57. Hirono, I.; Ueno, I., Hosaka, S.; Takanashi, H.; Matsushima, T.; Sugimura, T.; Natori, S. Cancer Lett. 1981, 13, 15-21.
58. Saito, D.; Shirai, A.; Matsushima, T.; Sugimura, T.; Hirono, I. Teratogen. Carcinogen. Mutagen. 1980, 1, 213-221.
59. Morino, K.; Matsukura, N.; Kawachi, T.; Ohgaki, H.; Sugimura, T.; Hirono, I. Carcinogenesis 1982, 3, 93-97.
60. Kosuge, T.; Tsuji, K.; Wakabayashi, K.; Okamoto, T.; Shudo, K.; Iitaka, Y.; Itai, A.; Sugimura, T.; Kawachi, T.; Nagao, M.; Yahagi, T.; Seino, Y. Chem. Pharm. Bull. 1978, 26, 611-619.
61. Matsumoto, T.; Yoshida, D.; Tomita, H. Cancer Lett. 1981, 12, 105-110.
62. Yamaguchi, K.; Shudo, K.; Okamoto, T.; Sugimura, T.; Kosuge, T. Gann 1980, 71, 743-744.

RECEIVED November 17, 1982

29

Formation of Mutagens by the Maillard Reaction

HIROHISA OMURA, NAZMA JAHAN, KAZUKI SHINOHARA, and HIROKI MURAKAMI

Kyushu University, Food Chemistry Laboratory, Department of Food Science and Technology, Fukuoka, 812, Japan

Several weak mutagens were produced by Maillard reactions of 20 different amino acids with sugars at 100 °C. Mutagenicity was studied by Ames test with Salmonella typhimurium TA 100 and TA 98. Mutagenic activity varied with the nature of amino acids or sugars, the tester strains and the presence or absence of S9 mixture. The mode of mutagenic action was categorized into 5 types. Mutagenicity varied with pH values at the reaction and increased with prolongation of heating time. Rec-assay and test on pupal oocytes of silkworms also gave positive mutagenicity. Several principal mutagens were identified as 5-hydroxymethylfurfural, ε-(2-formyl-5-(hydroxymethyl)pyrrol-1-yl)norleucine and 2-methylthiazolidine. Products of triose reductone with amino acids or nucleic acid-related compounds also showed mutagenicity.

It is now well known that mutagenic substances are produced by pyrolysis of food or foodstuffs which contain proteins, amino acids and sugars (1). This fact attracted public attention in view of the potential mutagenicity and carcinogenicity of foods. The original pyrolysis was carried out at unusually high temperatures, over 200 °C, although these temperatures are involved in cooking processes particularly such as roasting or frying. Recently, the contribution of food industries to the supply of processed foods has been extensively increasing. Therefore, mutagen formation under milder condition, at least below 100 °C, needed investigation because of the possible relation to daily life.

Many of the reactions between foodstuffs take place under mild conditions in processing, storage, and cooking of foods. The Maillard reaction appears to be the one to occur most commonly, accompanied not only by browning but also by change of aroma, loss of nutritional value, and development of antioxidative ac-

0097-6156/83/0215-0537$07.75/0

tivity. In this reaction, a number of intermediates such as reductones and osones are produced. Since these compounds are quite reactive and are expected to have some physiological effects on living organisms, we have investigated their biochemical properties and confirmed several activities, including nucleic acid-breaking ability and mutagenicity of triose reductone, its condensation products with amino acids, and ascorbic acid (2, 3, 4). We then examined the formation of mutagens by Maillard reactions.

As a typical model reaction, mixtures of 1 M glucose and equimolar amounts of various amino acids were heated at 100 °C for 10 hr under reflux and proper aliquots of the browned solution were subjected to mutagenic tests (5). By Ames test on Salmonella typhimurium TA 100 and TA 98 with or without S9 mixture, dose-dependent mutagenicity was detected in several such reaction mixtures, although it was much lower than that of well known chemical mutagens or pyrolysates of foods (Table I). In addition, the mutagenicity varied with the amino acids used, which were divided tentatively into 5 categories, by calling the result positive when the number of revertants formed in the test was over twice that in the control (6). No mutagenic activity was detected with the browned solution of glucose and Trp on both strains TA 100 and TA 98, with or without S9 mixture, while Trp alone was the most effective among amino acids in producing mutagenic products by pyrolysis. Reaction mixtures of glucose and $(Cys)_2$, Tyr, Asp, Asn, or Glu were also included in this group (group E). With products from glucose and other amino acids, mutagenic activity was observed on TA 100 without S9 mixture. However, for most of them, the activity was decreased by S9 mixture. The reaction mixture of glucose and Lys did not show any mutagenicity on TA 98 either with or without S9 mixture. Similarly, no mutagenic activity on TA 98 was produced with Leu, Ser, Thr, Met, and Gln (group A). However, such activity on TA 98 was observed with S9 mixture for reaction mixtures of glucose and Arg, Gly, Ala, Val, and Ile (group B). On the other hand, glucose and Cys products showed mutagenicity on both strains, TA 100 and TA 98, without S9 mixture and the activity was enhanced by it (group C). Reaction between glucose and Phe gave a product with activity on TA 100 without S9 mixture and even more with it in lower doses of the sample, but no activity was shown on TA 98 with and without S9 mixture, unlike the case of Cys (group D).

Instead of glucose, other sugars were examined for the formation of mutagens by the Maillard reaction using a typical amino acid from each of the 5 groups : Lys, Arg, Cys, Phe, and Trp. Several sugars such as fructose, galactose, and xylose showed ability to form mutagens with all amino acids except Trp, though their activities varied depending on the sugars as well as the amino acids. The result with Lys and sugars is shown in Fig. 1.

Mutagen formation was then studied at varying pH values. Browned samples were prepared by heating glucose and Lys, Arg,

Table I. MUTAGENICITY OF BROWNED SOLUTION FROM GLUCOSE AND AMINO ACID ON SALMONELLA

Strain	TA 100								TA 98							
S9 Mix.	-				+				-				+			
Dose, ml/plate	0.1	0.2	0.3	0.4	0.1	0.2	0.3	0.4	0.1	0.2	0.3	0.4	0.1	0.2	0.3	0.4
Glysine	275	451	230	187	188	130	144	186	35	47	20	13	246	240	146	-
Alanine	530	566	123	119	134	133	135	134	30	23	18	13	38	36	38	42
Valine	203	277	218	109	165	138	148	168	26	0	6	17	51	87	33	24
Leucine	192	269	336	400	150	167	185	194	33	29	17	28	44	52	28	21
Isoleucine	199	319	246	157	240	224	329	332	48	30	19	0	68	36	28	0
Serine	195	257	262	267	141	161	138	121	15	14	13	6	39	37	38	38
Threonine	223	315	104	48	164	103	99	165	41	15	9	7	20	0	0	0
Aspartic acid	153	173	140	106	91	86	123	144	34	41	30	20	43	56	50	47
Asparagine	139	146	162	242	160	164	178	194	27	27	39	32	36	39	30	31
Glutamic acid	119	103	89	68	165	153	131	108	47	37	0	0	32	25	31	36
Glutamine	208	309	455	353	183	138	145	147	51	60	39	15	41	49	38	33
Lysine	645	739	152	30	252	297	119	106	15	16	20	12	52	32	34	39
Arginine	683	754	676	559	218	235	156	145	17	34	0	0	85	97	82	66
Phenylalanine	200	253	281	313	372	365	343	260	25	21	11	7	51	36	28	24
Tryptophan	129	139	118	101	141	153	155	153	24	16	7	0	17	9	0	0
Tyrosine	134	178	186	246	150	142	164	186	33	19	17	13	15	19	18	16
Cysteine	369	213	30	30	763	1669	769	396	77	76	48	25	128	159	146	88
Cystine	166	198	134	130	138	131	131	120	27	24	22	36	51	82	53	50
Methionine	196	262	340	416	165	179	187	253	42	20	11	5	46	26	39	49
Proline	163	183	314	294	129	153	156	214	27	30	35	20	39	37	36	34
Control	129				136				28				33			
4-NQO* 0.1 µg	1800								130							
DEN** 70 µmol	198				352				38				70			

* ; 4-Nitroquinoline 1-oxide, ** ; Diethylnitrosamine.

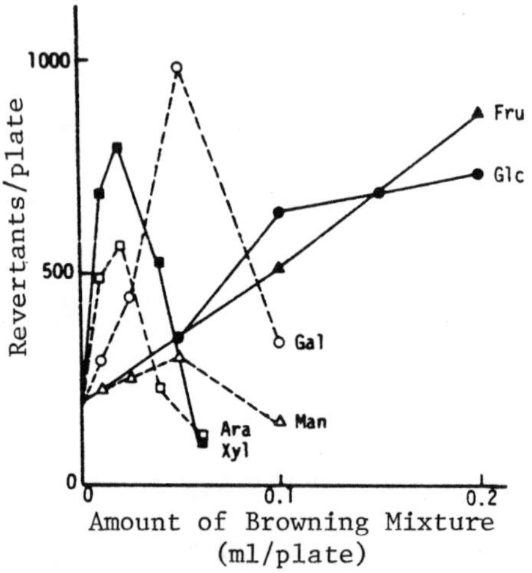

Figure 1. Mutagenicity of the browned solutions from varying sugars and Lys. Key: ▲, fructose; ●, glucose; ○, galactose; △, mannose; □, arabinose; and ■, xylose.

Cys or Phe at 100 °C for 10 hr at pH 2, 7, and 12. As an example, mutagen formation from glucose and Phe is shown in Fig. 2. A browning sample made at pH 7 showed the highest mutagenicity on TA 100, while on TA 98 strong activity was produced at pH 2, i.e. in strongly acidic medium. On the other hand, mutagenic activity on TA 100 was the highest in the reaction mixture of glucose and Lys at pH 2, whereas those were detected in the case of glucose and Arg or Cys at pH 7. Furthermore, it was observed that mutagens were formed proportionally to heating time and coloration of the reaction mixture of glucose and Lys or Phe. However, activity was sharply produced after preliminary heating for a few hours of the sample of glucose with Arg or Cys (Fig. 3). Then, we have confirmed that the activities of the mutagens formed between glucose and amino acids were decreased by the action of several reducing substances; ascorbic acid and triose reductone exhibited the most significant effect, while Cys, dithiothreitol, or penicillamine had lower ones (Fig. 4). On the contrary, the browned mixture of glucose with Lys or Phe depressed the activities of other mutagens such as nitrofuran and N-ethyl-N'-nitro-N-nitrosoguanidine (ENNG), although the sample itself contains some mutagenic activity (Fig. 5).

By Rec-assay with Bacillus subtilis H 17 Rec^+ and M 45 Rec^-, positive activity was also shown by reaction mixtures of glucose and amino acids, while the activity varied depending on amino acids, as shown in Table II. This suggests that the browned solutions are able to injure bacterial DNA. On the other hand, we have often demostrated that nucleic acids are degraded in vitro by the action of several carcinogens, mutagens, and virus inducers as well as antitumor substances by studies with viscosimetry, sucrose density gradient centrifugation, and electrophoresis (7). Many reductones brought about strand breaks of DNA, especially in the presence of Cu^{2+} (2, 3, 4). Therefore, DNA-breaking ability of the browned solutions was investigated by means of agarose-gel electrophoresis, and such activity of the samples was observed (Fig. 6). Thus, it can be assumed that mutagens formed by the Maillard reaction may have DNA-breaking activity.

Attempts were then made to isolate and identify the principal mutagens in the browned samples. Since properties of the mutagens vary with amino acids, it was supposed that those formed by the Maillard reaction may differ. Therefore, mutagens were sought in the typical reaction mixtures of glucose and Phe, Lys, and Cys. After dialysis of the browned solution against distilled water at 5 °C overnight, outer and inner fractions were separated and evaporated to dryness in vacuo. Only the outer portion of the dialysate showed both mutagenicity on TA 100 and DNA-breaking activity (Fig.7). This observation indicates that the mutagens are of low molecular weight, probably intermediates in the Maillard reaction between glucose and amino acids.

The outer portion of the browned solution of glucose and Phe was evaporated to dryness and the residue extracted with absolute

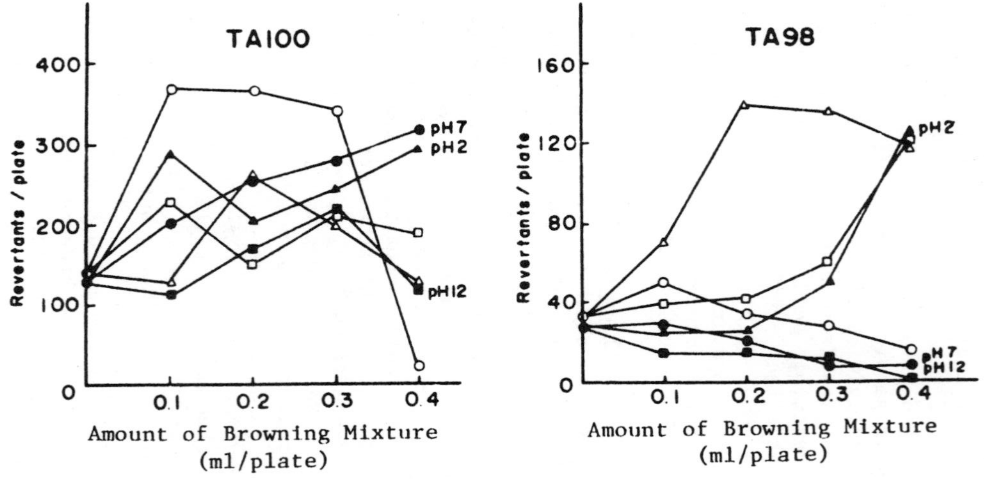

Figure 2. Effect of pH value on the mutagenic action of the browning mixture of Glc and Phe on TA 100 and TA 98. Key: ○, with S9 mix; ●, without S9 mix.

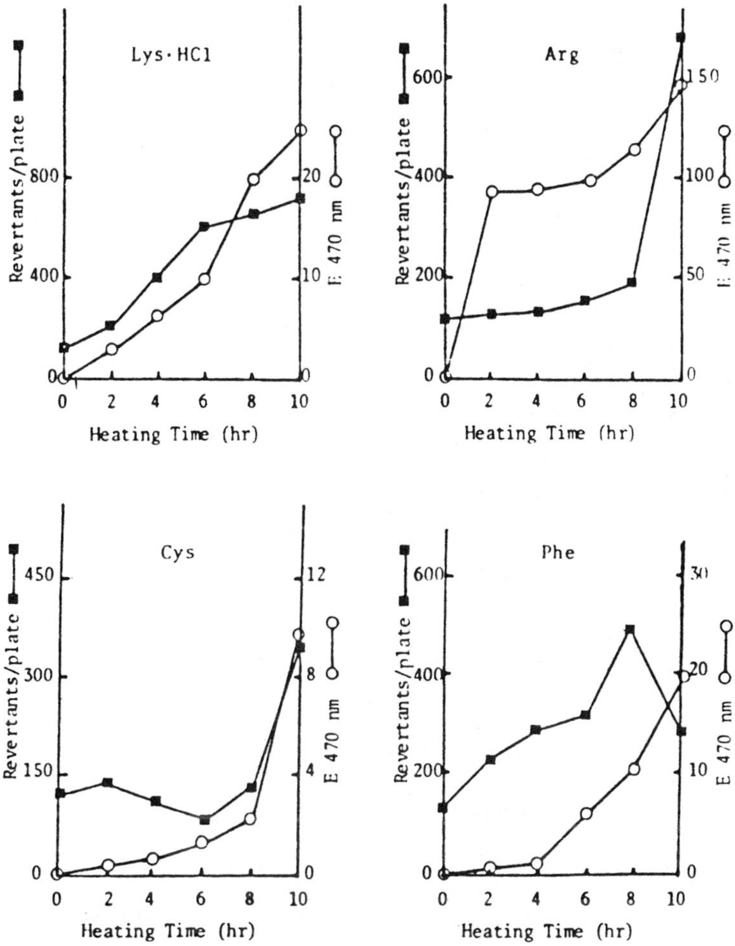

Figure 3. Browning intensity and mutagenicity of the glucose–amino acid mixtures on TA 100 as a function of time.

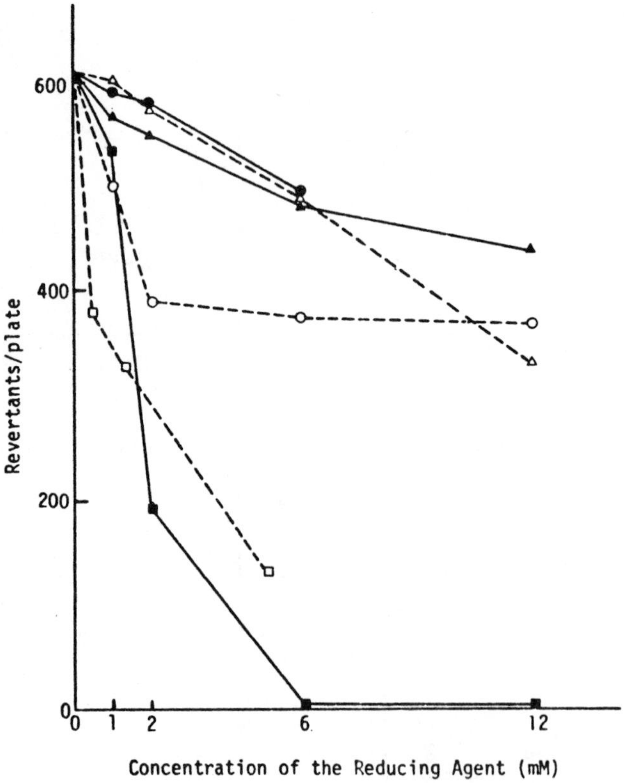

Figure 4. Effect of reducing agents on the mutagenic action of the browning mixture of Glc and Lys. Key: ●*, cysteine;* ○*, N-acetylcysteine;* ▲*, penicillamine;* △*, dithiothreitol;* ■*, triose reductone; and* □*, ascorbic acid.*

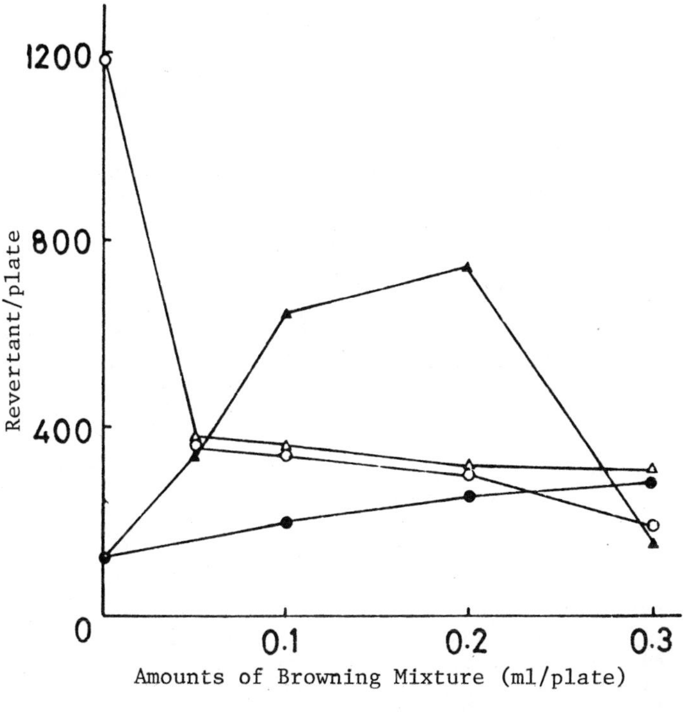

Figure 5. Effect of browning solutions on mutagenic action of N-*ethyl*-N′-*nitro*-N-*nitrosoguanidine (ENNG) on TA 100. Key:* ●, *browned sample from Glc and Phe;* ○, *ENNG and browned sample from Glc and Phe;* ▲, *browned sample from Glc and Lys; and* △, *ENNG and browned sample from Glc and Lys.*

Table II. REC-ASSAY OF BROWNED SUBSTANCE FROM GLUCOSE AND AMINO ACIDS

Amino acid + (Glucose)	Dose (mg)	Inhibition zone (mm) Rec+ (H-17)	Rec− (M-45)	Differ-ence	Amino acid + (Glucose)	Dose (mg)	Inhibition zone (mm) Rec+ (H-17)	Rec− (M-45)	Differ-ence
Glycine	980	3	8	5	Lysine	760	5	12	7
	780	3	6	3		525	4	9	5
	330	0	0	0		106	0	0	0
Alanine	415	1	4	3	Arginine	354	3	10	7
	260	0	0	0		266	2	7	5
Valine	170	1	2	1		222	1	5	4
	136	0	0	0		58	0	0	0
Leucine	904	1	4	3	Phenylalanine	530	2	5	3
	565	0	0	0		260	0	0	0
Isoleucine	608	5	2	3	Tyrosine	980	0	0	0
	318	0	0	0	Tryptophan	280	2	5	3
Serine	350	1	2	1		68	0	0	0
	245	0	0	0	Cysteine	251	4	10	6
Threonine	352	10	15	5		188	3	7	4
	155	0	0	0		46	0	0	0
Aspartic acid	148	2	5	3	Cystine	375	0	0	0
	74	1	2	1	Methionine	115	0	0	0
	37	0	0	0	Proline	920	10	13	3
Asparagine	715	6	10	4		744	0	0	0
	532	0	0	0	Histidine	239	3	10	7
Glutamic acid	359	8	15	7		205	1	4	3
	179	8	13	5		170	1	2	1
	90	2	4	2		60	0	0	0
	45	0	0	0	4-Nitro-quinoline-1-oxide	40	3	14	11
Glutamine	480	2	3	1		20	4	6	2
	110	0	0	0		10	0	4	4
						4	0	2	2

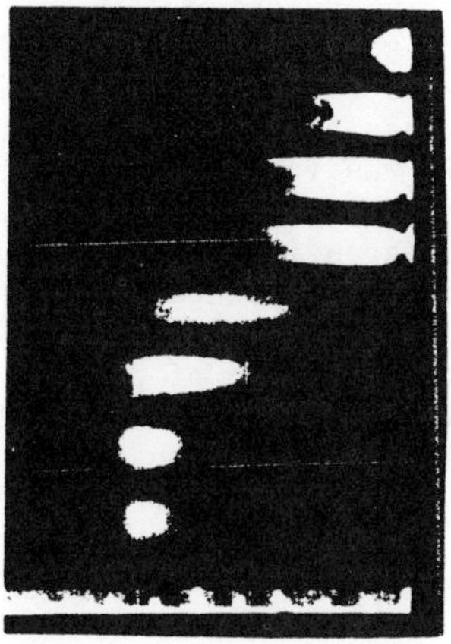

Figure 6. Gel electrophoresis to determine DNA-breaking effect of dialyzed outer solution from the browning mixture of Gly and Phe after heating at 100 °C for 10 h with Cu^{2+}.

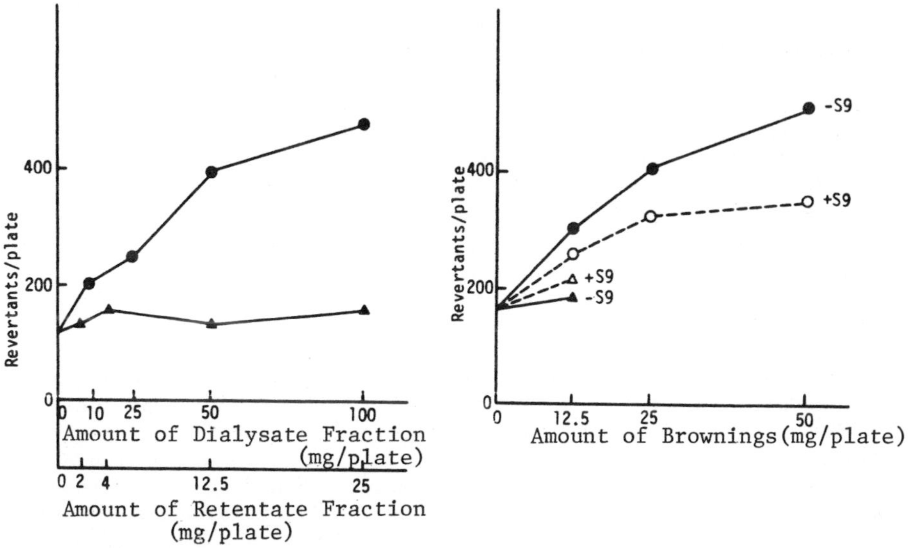

Figure 7. Dose–response curve of mutagenic effect of dialyzed fractions from browning mixture of Glc and Lys after heating at 100 °C for 10 h. (Assay with Salmonella typhimurium *TA 100 with or without S9 mixture.) Key: ●, dialysate; ○, fraction; ▲, retentate; and △, fraction.*

ethanol. From the soluble portion, the mutagenic substance was purified by successive chromatographic separations on a DEAE-cellulose column, a Dowex-50 column (Fig. 8), and silica gel in thin layer. On the basis of positive reaction with 2,4-dinitrophenylhydrazine, R_f value (Fig. 9), and UV, NMR (Fig. 10) and mass (Fig. 11) spectrometries, the mutagen was identified as 5-hydroxymethylfurfural (HMF) by comparing results with those obtained with authentic HMF. In addition, it was found that authentic HMF has mutagenic and DNA-breaking activity and that its mutagenic activity coincided with that of the mutagen isolated (Fig. 12). Thus, it can be concluded that the principal mutagen formed by the Maillard reaction between glucose and Phe is HMF.

The mutagen in the browned mixture of glucose and Lys was also examined. From the ethanol-soluble portion of the dialysate fraction, the mutagen was isolated by successive chromatographic separations on a DEAE-cellulose column, a CM-cellulose column (Fig. 13), and thin-layer silica. From R_f value, positive reactions with 2,4-dinitrophenylhydrazine and ninhydrin reagents (Fig. 14), NMR (Fig. 15), and UV spectrum having a shoulder at 265 nm and a peak at 297 nm in water, it was suggested to be related to pyrrole-2-carbaldehyde as reported by Nakayama, et al. (8). They isolated the intermediate formed by the Maillard reaction between glucose and Lys and proved it to be ε-(2-formyl-5-(hydroxymethyl)pyrrol-1-yl)norleucine. By comparing the above properties and the mutagenicity of Nakayama's standard sample with those of ours, it was confirmed that the samples were the same. Furthermore, thin-layer chromatography of the sample gave another positive spot with 2,4-dinitrophenylhydrazine; this was found to be mutagenically active and to agree in properties with HMF. Thus, it was at least suggested that the main mutagens formed by the Maillard reaction between glucose and Lys are a pyrrole compound, probably ε-(2-formyl-5-hydroxymethyl)pyrrol-1-yl)norleucine and HMF.

Finally, isolation of the mutagen in the browned solution of glucose and Cys was carried out. From pyrolyzed Cys or $(Cys)_2$, Fujimaki et al. (9) identified 7-8 volatile compounds including 2-methylthiazolidine. Mihara and Shibamoto (10) separated the browning mixture of glucose and cysteamine into 11 fractions having some mutagenicity at certain concentrations, 7 volatile fractions from the methylene chloride extract by high-pressure liquid chromatography and 4 from the residual aqueous solution by ion-exchange chromatography. Among them, thiazolidine and 2-methylthiazolidine were found in the former and 2-(1,2,3,4,5-pentahydroxy-n-pentyl)thiazolidine in the latter. Since it was supposed that a similar reaction may occur when Cys is employed instead of cysteamine, mutagens were surveyed as follows. The dialyzed outer part of the dark brown reaction mixture from glucose and Cys was distilled in vacuo to about half its volume, and the distillate was shaken with twice its volume of methylene chloride. The methlene chloride layer was evaporated at room temperature in vacuo to remove the solvent and the residual aqueous phase was dried over

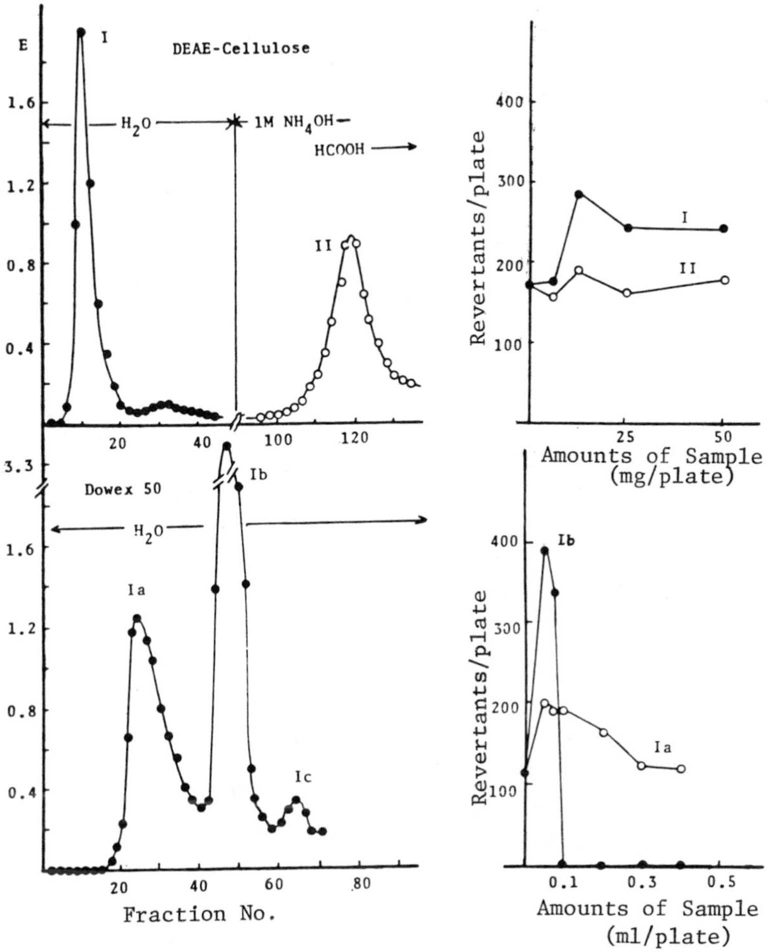

Figure 8. Chromatograms on DEAE-cellulose column and Dowex 50 column of browned mixture from Glc and Phe and mutagenicity of fractions eluted.

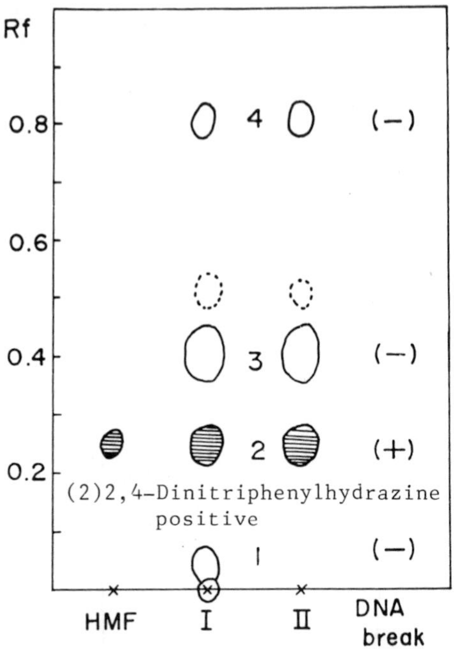

Figure 9. TLC of samples from Glc and Phe. Solvent: chloroform–methanol (50:1).

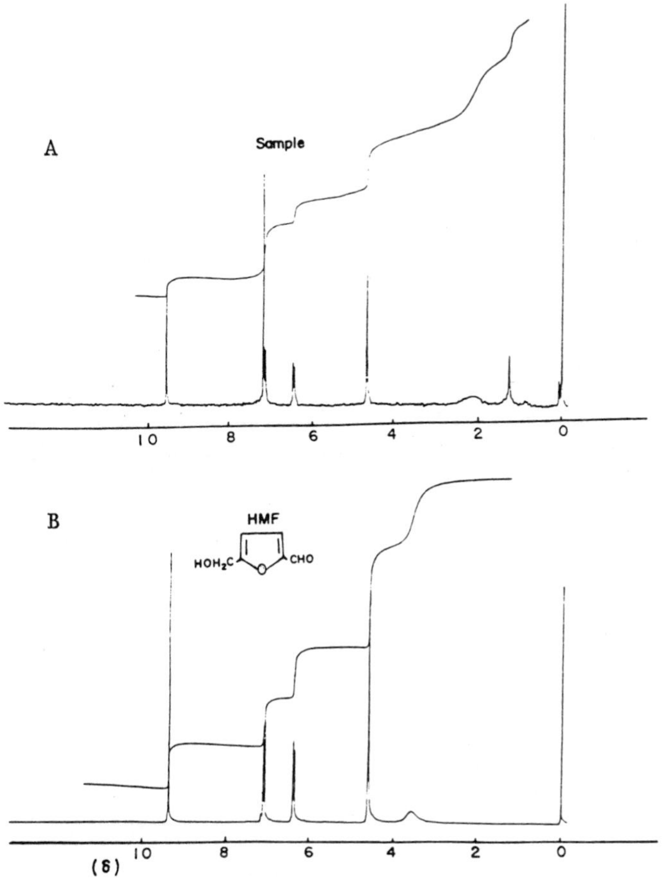

Figure 10. Comparison of NMR spectra of (A) the browned sample from Glc and Phe and (B) 5-hydroxymethylfurfural (HMF).

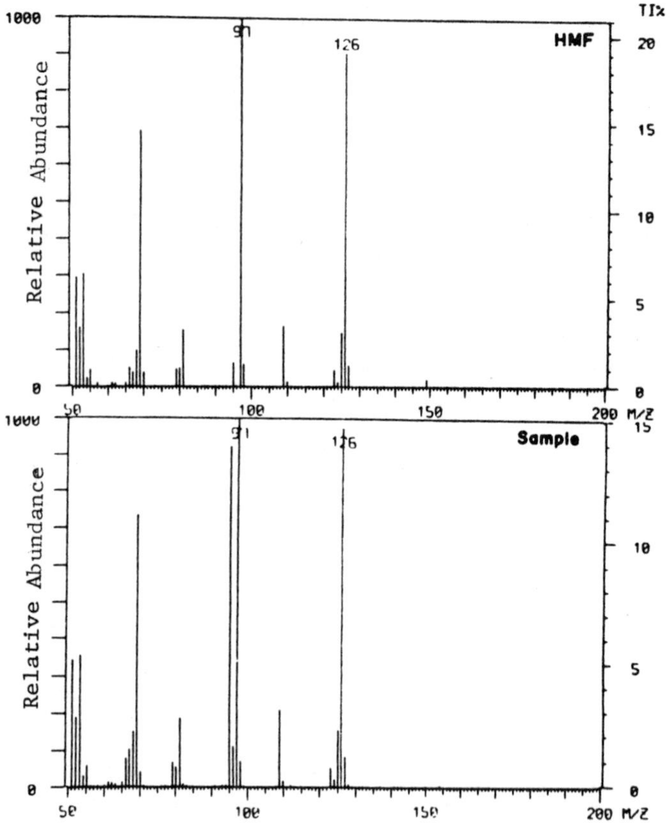

Figure 11. Comparison of mass spectra of HMF (top) and the browned sample from Glc and Phe (bottom).

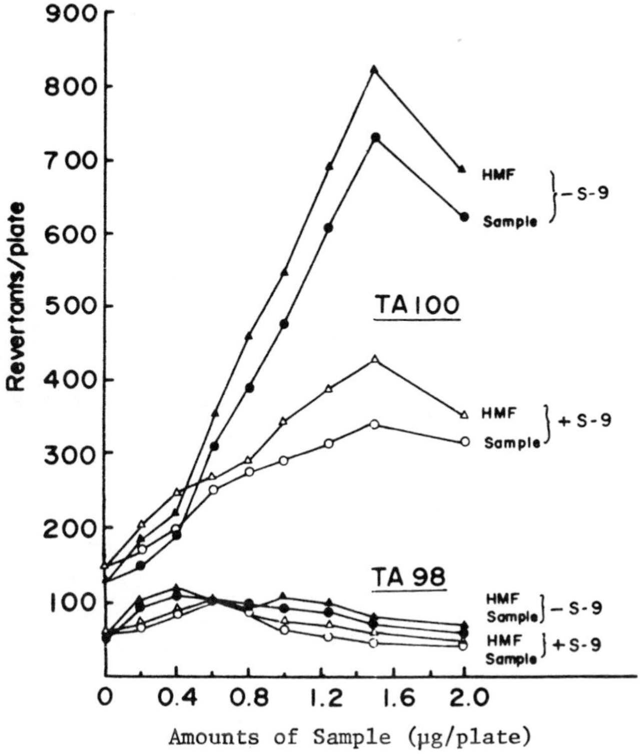

Figure 12. Comparative mutagenicities of the browned sample from Glc and Phe, and HMF. Key: ▲, *HMF without S9;* △, *HMF with S9;* ●, *sample without S9; and* ○, *sample with S9.*

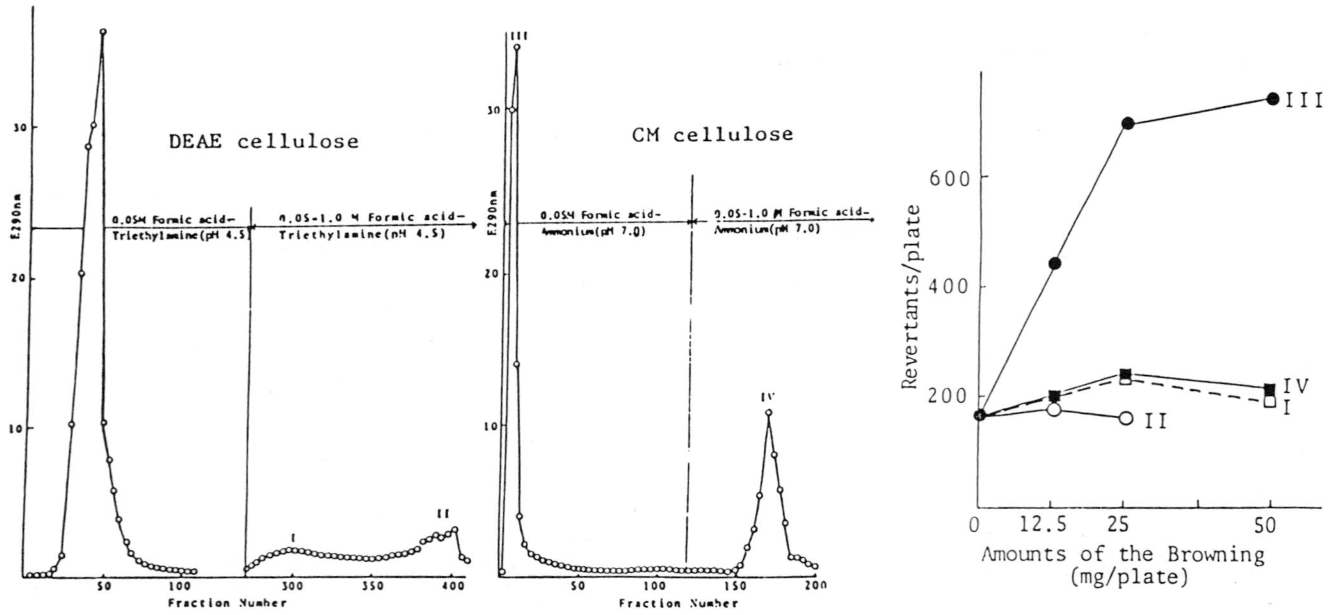

Figure 13. Chromatograms on DEAE-cellulose column and CM-cellulose column of the browned sample from Glc and Lys, and mutagenicity of fractions.

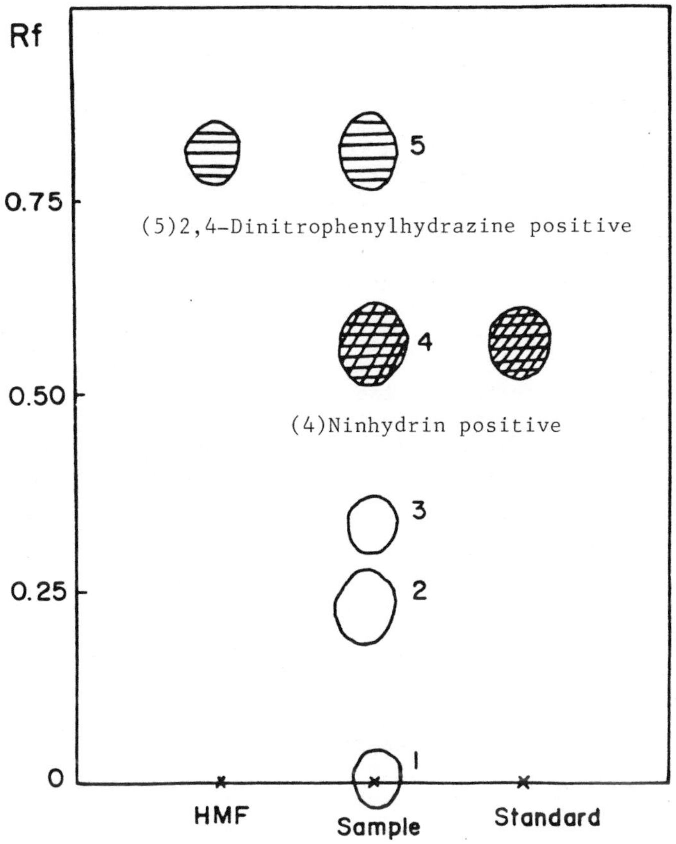

Figure 14. TLC of mutagenic fraction (peak III in figure 13) of the browned sample from Glc and Lys. Solvent: n-*BuOH–acetic acid–water (4:1:1).*

anhydrous $MgSO_4$. The resultant oily material was then dissolved in ether and subjected to gas chromatography. Retention time of the major peak was the same with that of authentic 2-methylthiazolidine (Fig. 16). Furthermore, GC/MS data for the main peak coincided with that of the authentic compound (Fig. 17). In addition, they both showed mutagenicity on TA 100 and TA 98 in the presence or absence of S9 mixture, just as did the browned mixture from glucose and Cys (Fig. 18). Thus, at least one of the volatile mutagenic substances formed by the Maillard reaction between glucose and Cys is 2-methylthiazolidine.

To test mutagenicity on the animal level, we used silkworm oocytes (11). Browned solutions were injected into the body cavity of the female at mid-pupal stage. Mutagenicity was detected as pink or red eggs among dark colored wild-type ones. Mutagenicity of the browned reaction mixtures between glucose and amino acids was observed on pupal oocytes of the silkworm, but at lower rates than those produced by a typical mutagen, mitomycin C, as shown in Table III.

Triose reductone contributes more readily to the browning reaction with amino acids than sugars do (12). This reaction is a special one in the Maillard reaction and is referred as the "Reductone-Amino Reaction". Amino reductones or enaminol compounds are first produced as the intermediate during the browning by reaction of the OH group in triose reductone with the NH_2 group of amino acids. We have isolated such amino reductones from reaction mixtures of triose reductone and Gly, Ala, Leu, Met, Phe, Trp and Ile, and confirmed that some of them are mutagenic on TA 100 (Fig. 19). On the other hand, browning reactions of triose reductone with several nucleic acid-related compounds were also observed (13). Two types of reductive intermediates, linear and tricyclic forms, were isolated (Fig. 20). The reaction of triose reductone with guanine produced a brown tricyclic compound, 1,N^2-(2-hydroxypropenylidene)guanine (cyclic TR-Gua) and a labile yellow intermediate, N^3-(3-oxo-2-hydroxypropenyl)guanine (TR-Gua), whereas such reaction with guanosine, 2'(3')- or 5'-guanylic acid gave N^2-(3-oxo-2-hydroxypropenyl)guanosine (TR-Guo), N^2-(3-oxo-2-hydroxypropenyl)-2'(3')-guanylic acid (TR-2'(3')-GMP) or N^2-(3-oxo-2-hydroxyprpenyl)-5'-guanylic acid (TR-5'-GMP), respectively (14). These intermediates are a kind of amino reductones too and showed evident mutagenicity on TA 100 without S9 mixture (Fig. 21), but not on TA 98 on *Bacillus subtilis*, H 17 Rec^+ and M 45 Rec^-, used for Rec-assay (15).

Thus, by the Maillard reaction in different browning systems of sugars and amino compounds, some mutagenic substances were formed, although their activities are quite weak compared with those formed by pyrolysis of amino acids. They were confirmed as intermediates and some of them were identified as furan, pyrrole, or thiazolidine derivatives formed from glucose and amino acids

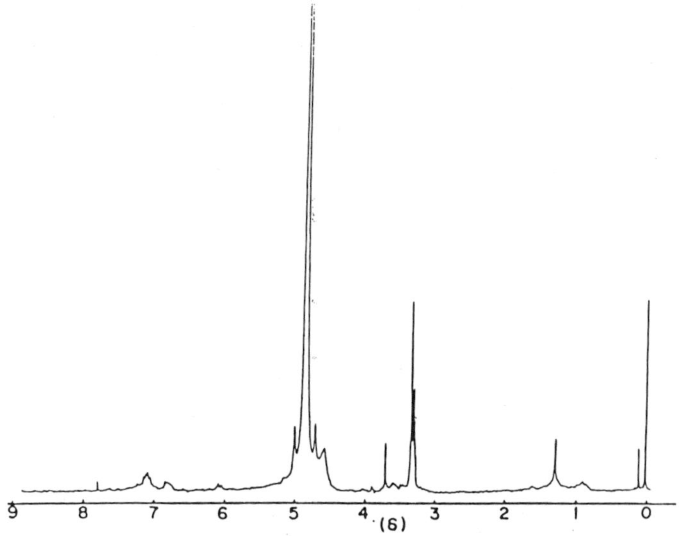

Figure 15. NMR spectrum of browned sample from Glc and Lys.

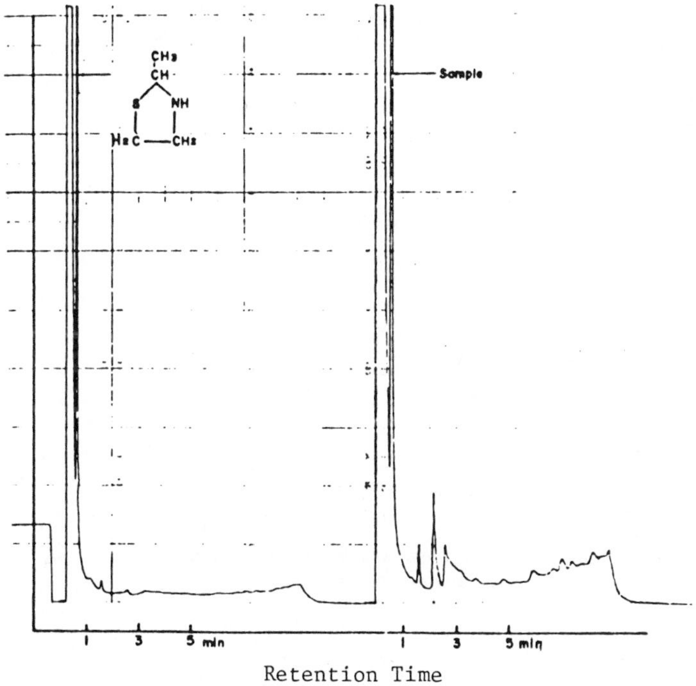

Figure 16. Comparison of gas chromatograms of 2-methylthiazolidine and volatile sample from Glc and Cys.

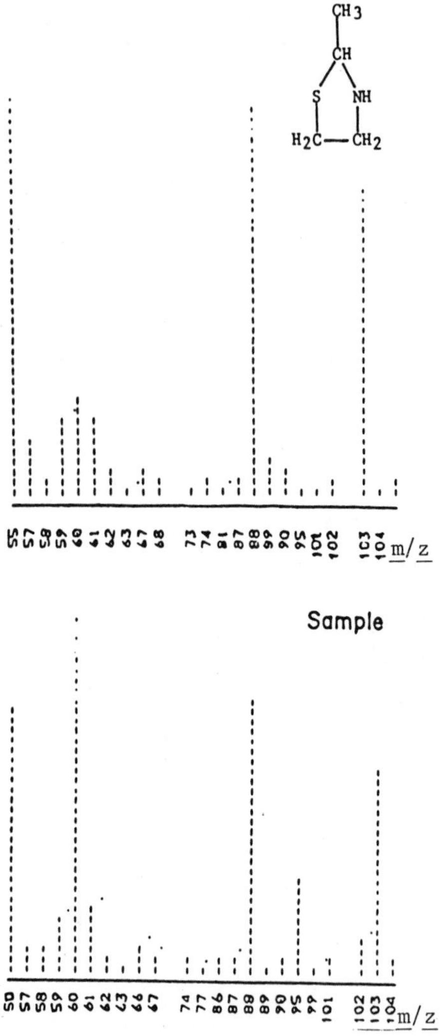

Figure 17. Comparison of mass spectra of 2-methylthiazolidine and volatile sample from Glc and Cys.

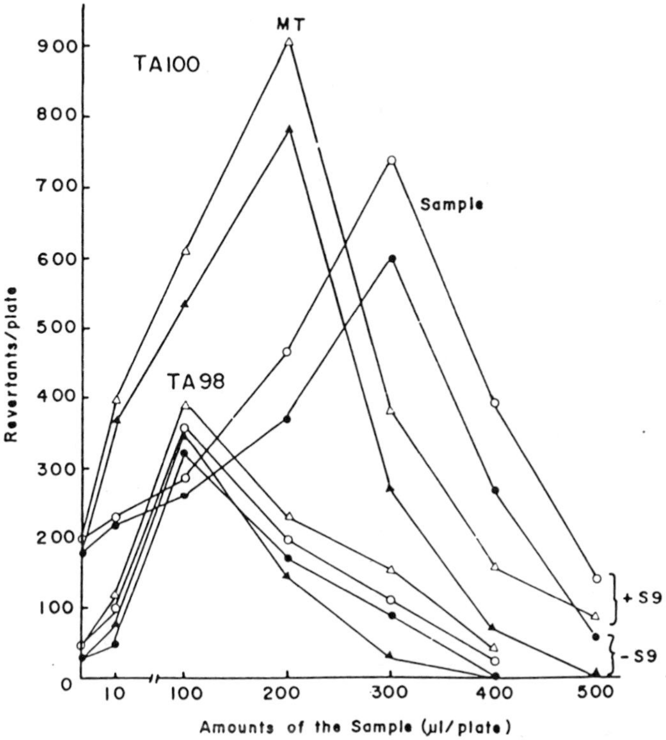

Figure 18. Comparison of mutagenicity of 2-methylthiazolidine and volatile sample from Glc and Cys. Key: ○, sample with S9; ●, sample without S9; △, 2-methylthiazolidine with S9; and ▲, 2-methylthiazolidine without S9.

Table III. MUTAGENICITY OF BROWNED SOLUTIONS FROM GLUCOSE AND AMINO ACIDS ON EGGS OF SILKWORM

Amino acid	No. of moths	No. of eggs observed	No. of mutants detected	Mut. frequency ($x10^{-4}$)
Glycine	26	10,889	4	3.4 (0 - 7.7)
Alanine	17	6,010	3	4.7 (0 -10.0)
Valine	16	7,015	2	3.2 (0 - 8.1)
Isoleucine	24	8,728	7	10.7 (6.5-14.9)
Serine	22	6,872	5	6.9 (0 -14.1)
Threonine	12	5,167	7	14.8 (0 -36.8)
Asparagine	24	9,944	7	6.8 (0.9-12.7)
Glutamic acid	22	7,748	3	2.8 (0 - 7.7)
Glutamine	18	8,140	7	11.7 (4.3-19.2)
Lysine	22	8,982	10	10.8 (4.6-16.9)
Arginine	26	9,229	6	6.1 (0.3-11.9)
Phenylalanine	28	11,361	10	9.0 (3.8-14.3)
Tyrosine	24	9,544	6	6.8 (1.5-12.1)
Tryptophan	20	8,188	3	3.6 (0 - 7.7)
Cysteine	31	14,415	11	8.1 (3.5-12.7)
Cystine	22	9,624	9	10.0 (1.8-18.2)
Methionine	31	11,089	6	6.8 (1.0-12.6)
Histidine	28	11,173	9	7.8 (2.6-13.0)
H_2O	43	17,352	2	1.2 (0 - 2.8)
MNNG*, 0.01 µg	13	5,889	3	7.2 (0 -16.2)
Mitomycin C, 6 µg	10	3,146	617	2511.3 (1,695-3,327)

* N-Methyl-N'-nitro-N-nitrosoguanidine. An aliquot of 10 µl of sample was injected into a pupa.

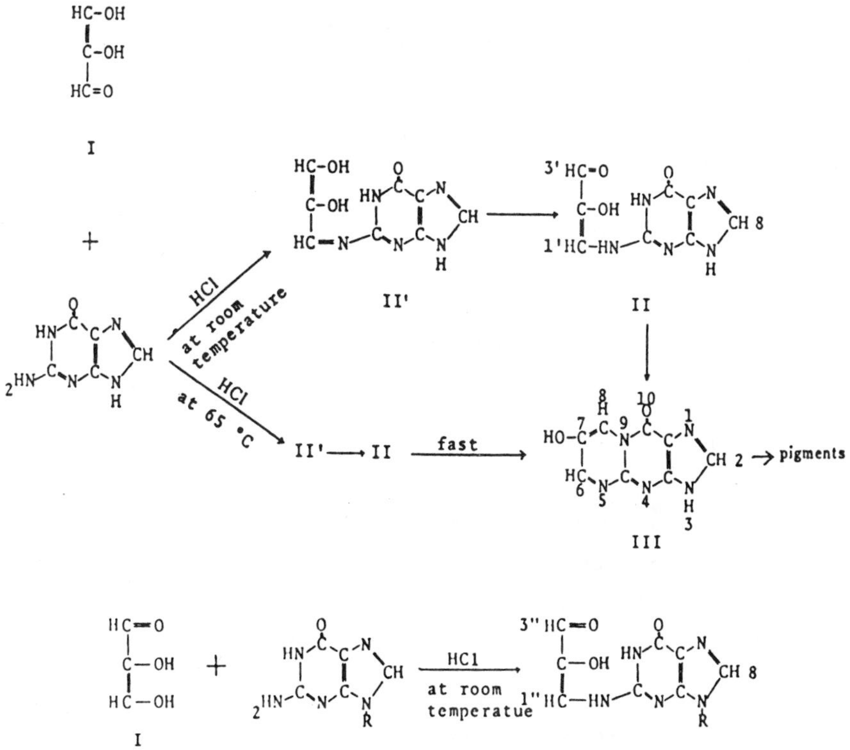

(IV) R = β-D-ribofuranosyl
(V) R = 2′(3′)-ribotide
(VI) R = 5′-ribotide

Figure 19. Reaction between triose reductone and nucleic acid-related bases.

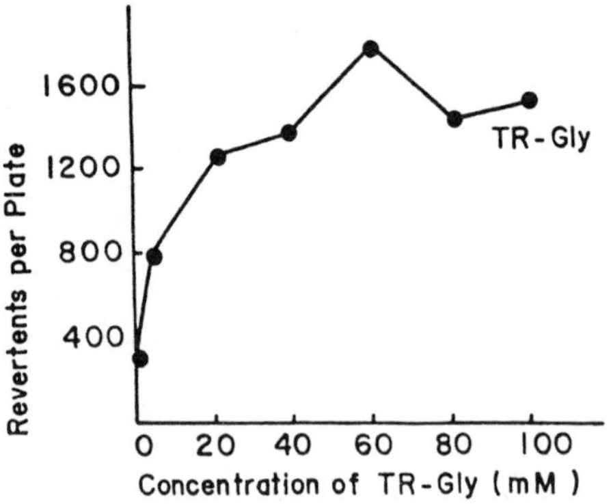

Figure 20. Mutagenicity of the product from triose reductone and Gly.

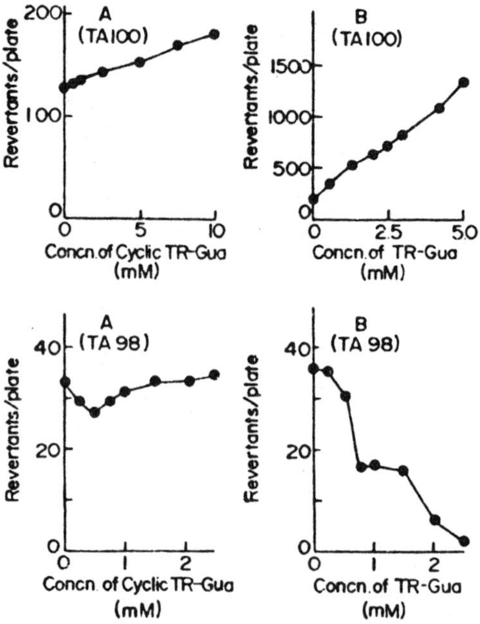

Figure 21. Mutagenic effects of cyclic TR-Gua (A) or TR-Gua (B) in the presence of Cu^{2+} on TA 100 or TA 98.

and as combined amino reductones formed from triose reductone and amino acids or other bases. In addition, it was found that the mutagens have DNA-breaking activity in vitro. Furthermore, it is expected that other mutagens will be studied in several Maillard reaction mixtures.

Acknowledgements

The studies were supported by a Grant-in-Aid for Scientific Research from the Japanease Ministry of Education, Science and Culture. The authors thank Dr. E. Kuwano for giving much advice in identification of the mutagens and are grateful to Professor H. Kato of Tokyo University for the gift of standard sample of ε-(2-formyl-5-(hydroxymethyl)pyrrol-1-yl)-norleucine.

Literature Cited

1. Sugimura, T.; Kawachi, T.; Nagao. M.; Yahagi, T.; Seino, Y.; Okamoto, T.; Shudo, K.; Kosuge, T.; Tsuji, K.; Wakabayashi, K.; Iitaka, A.; Itai, A. Proc. Japan. Acad. 1977, 53, 58.
2. Shinohara, K.; Fukumoto, Y.; Tseng, Y-K.; Inoue, Y.; Omura, H. J. Agr. Chem. Soc. Japan 1974, 48, 499.
3. Omura, H.; Iiyama, S.; Narazaki, Y.; Shinohara, K.; Murakami, H. J. Nutr. Sci. Vitaminol. 1975, 21, 237.
4. Omura, H.; Tomita, Y.; Fujiki, H.; Shinohara, K.; Murakami, H. J. Nutr. Sci. Vitaminol. 1978, 24, 263.
5. Shinohara, K.; Wu, R-T.; Jahan, N.; Tanaka, M.; Morinaga, N.; Murakami, H.; Omura, H. Agric. Biol. Chem. 1980, 44, 671.
6. Seiler, J.P.; Mattern, I.E.; Gree, M.H.L.; Anderson, D. Mutation Res. 1980, 74, 71.
7. Yamafuji, K. "Food, Cancer and Cytodifferentiation" (40th Anniversary of Prof. K. Yamafuji's research), Shukosha Co. Ltd., Fukuoka, 1970.
8. Nakayama, T.; Hayase, F.; Kato, H. Agric. Biol. Chem. 1980, 44, 1201.
9. Fujimaki, M.; Kato, H.; Kurata, T. Agric. Biol. Chem. 1969, 33, 1144.
10. Mihara, S.; Shibamoto, T. J. Agric. Food Chem. 1980, 28, 62.
11. Yamashita, N., Shinohara, K.; Jahan, N.; Torikai, Y.; Doira, H.; Omura, H. J. Japan. Soc. Food Nutr. 1981, 34, 367.
12. Shinohara, K.; Tseng, Y-K.; Inoue, Y.; Sato, M.; Omura, H. Sci. Bull. Fac. Agr., Kyushu Univ. 1974, 28, 139.
13. Lee, J-H.; Shinohara, K.; Murakami, H.; Omura, H. J. Agr. Chem. Soc. Japan 1978, 52, 11.
14. Lee, J-H.; Shinohara, K.; Murakami, H.; Omura, H. Agric. Biol. Chem. 1979, 43, 279.
15. Shinohara, K.; Lee, J-H.; Tanaka, M.; Murakami, H.; Omura, H. Agric. Biol. Chem. 1980, 44, 1737.

RECEIVED November 17, 1982

INDEX

INDEX

B

F

H

I

K

L

M

R

S

T

U

V

W

X

Z

Jacket Design by Kathleen Schaner
Editing and production by Susan Futchs Robinson

Elements typeset by Service Composition Co., Baltimore, MD
Printed and bound by Maple Press Co., York, PA